Solving
Problems
in
Analytical Chemistry

Solving Problems in Analytical Chemistry

Stephen Brewer

Eastern Michigan University, Ypsilanti

John Wiley & Sons,
New York · Chichester · Brisbane · Toronto

Library of Congress Cataloging in Publication Data:

Brewer, Stephen, 1941-
   Solving problems in analytical chemistry.

   Includes index.
   1. Chemistry, Analytic—Problems, exercises, etc.
I. Title.
QD75.9.B73   543'.0076   79-17164
ISBN 0-471-04098-3

Printed in the United States of America

10 9 8 7 6 5 4

To
R. L. M.

Preface

A knowledge of analytical chemistry is important to students of many different disciplines. Thus in a typical class there will not only be students majoring in chemistry and chemical engineering but also those who are studying soil science, poultry science, dietetics, medical technology, and biology. Often, the biggest difficulty that these students, chemistry majors included, encounter is that of doing the necessary calculations in laboratory and classroom work.

The purpose of this book is to give students a readable, detailed, and useful guide to the calculations required in a beginning analytical chemistry course. Some knowledge of algebra and graphing is assumed, but no knowledge of calculus is required. Moreover, no calculus is introduced, although many topics in algebra and graphing are reviewed.

Generally, the best way to learn how to solve problems in any discipline is to work through detailed examples and then to try to solve similar problems, whose solutions are available in instances where difficulty is encountered. Thus I divided each section of almost every chapter into three parts: (1) a presentation with some descriptive material of typical calculations, which are worked out in detail; (2) a number of problems for the student; and (3) detailed solutions to those problems.

The student using the book should first buy a hand calculator, preferably one featuring scientific notation, logarithms, and a y^x key. Next he or she should get plenty of paper and sharp pencils. Each one of the calculations in the first part of a given section should be worked out with pencil, paper, and calculator. (Marking the text with a yellow pen almost never does any good.) Then the student should do each of the problems at the end of the section. An earnest attempt of at least five minutes' duration should be made on any given problem. If *no* progress is made after this attempt, the student should turn to the detailed solutions that follow the problems and learn enough at least to get a start. Once the student has worked all of the problems, he or she should try to work them once again, to determine whether the knowledge just acquired has been retained. Finally, two or more people can usually secure a quiet classroom with a blackboard and quiz each other on the problems. The process requires time, it is true, but these detailed examples and solutions should make learning how to calculate a less fearsome task.

This book may be assigned as a supplementary text by an instructor, or it may, in combination with a detailed laboratory manual, be used as the lecture text for a course. I have followed the latter course of action, perhaps because calculations are very heavily stressed in my own lectures. There are many correct ways to teach analytical chemistry, and therefore each instructor in the discipline should make his or her own choice.

In this book, I cover the calculations most frequently encountered in most beginning analytical chemistry courses. There are six chapters in the book: Chapter 1 on basic material, covering basic graphing, statistics, and stoichiometry; Chapter 2 on acid-base chemistry, in which solutions by approximation methods are stressed, and in which polyprotic acids with closely spaced dissociation constants are given their just due; Chapter 3 on solubility control, with approximation methods and a stress on complexation and acid dissolution; Chapter 4 on spectrophotometry, which includes, after basic analytical material, calculations of physical constants; Chapter 5 on potentiometry, ranging from simple cells through titrations, to physical constants and activities to ion-selective electrodes; and Chapter 6 on separations, beginning with separatory funnel extractions, going through multistage (Craig) separations, to liquid-liquid chromatography, through gas chromatography to sorption chromatography.

It would be best if the chapters were taken in the order just listed. However, it is possible to study each of the first five chapters independently. The sixth chapter, that on separations, *does* require knowledge of topics covered in Chapters 1, 2, and 4.

Acknowledgments

I acknowledge first the generous contribution of Professor John Walters, of the Department of Chemistry at the University of Wisconsin-Madison. Professor Walters was my mentor during my graduate school days at Madison, and thus made a great contribution to the shaping of this book, although neither of us had such a project in mind at that time. In later years, Professor Walters wrote a set of highly detailed notes, which he then distributed to his classes in analytical chemistry. He very kindly sent me sets of these massive notes and urged me to use his material freely, which I was pleased to do.

In 1974, Professor Walters and I decided, together with Gary Carlson of Wiley, that the notes should be used as the basis for a textbook. The textbook was to emphasize problem solving for analytical chemistry, and was to contain vast numbers of solved problems.

Time went by quickly. In 1976, Professor Walters, as busy as a chemist could possibly be, generously offered me, for use in this book, any of the material contained in his notes and problem sets. The material represents a labor of several years on his part and represents an intense dedication to making complicated matters accessible to the beginning student in analytical chemistry. Indeed, the material is a symbol of his dedication to students.

Professor Walters' influence is found particularly in the section on stoichiometry and in the chapters on acids, solubility, spectrophotometry, and potentiometry. I have taken much that appears in these parts of this book directly from his notes. In the chapters on acids and solubility, his approach using approximations and α- and β-functions is closely followed. (Professor Walters has told me that Professor H. L. Laitinen, then at the University of Illinois, introduced him to the α- and β-functions) I should like, then, gratefully to acknowledge my debt to Professor Walters, without whose efforts, dedication, and generosity this book would not have appeared.

Professors J. Schrag and J. Wright, also of the Department of Chemistry at Madison, have made considerable contributions both to Professor Walters' notes and to this book. Professor D. Coleman, of Wayne State University, read the typescript and offered valuable suggestions.

Assistance in computer programming was given by Dr. A. Scheeline, then of the University of Wisconsin, who helped me write a program executed at the Madison Academic Computer Center. Also, B. Finzel, of Eastern Michigan's Chemistry Department, and R. Frownfelter, of University Computing at Eastern Michigan, made a plotting program produce the desired graphs with Eastern Michigan's instructional computer system.

Several patient and able typists helped me to prepare the typescript. They are Ms. P. Quinn, Ms. J. Wright, Ms. L. Mundt, Ms. J. Haynie, Ms. L. Myers, and Ms. C. Bashawaty.

Gary Carlson, Chemistry Editor at Wiley, gave me much help and coordinated the work of three other analytical chemists who reviewed the typescript. Nina R. West, production supervisor, gave her best efforts to the sometimes trying task of getting the book from manuscript to printed pages.

Finally, there were my student helpers—about 200 of them both at Madison and at Ypsilanti. From two of them, Vicki McGuffin and Kena Helkaa, I borrowed quires of data. Others read preliminary editions of the book, and eagerly sought and reported errors. Patrick Harrington helped proofread the galleys. (With this company of diligent proofreaders, I hope that errors are few but would be grateful to anyone who discovers and informs me of errors.)

My colleagues in the Chemistry Department at Eastern Michigan and the Eastern Michigan University administration gave me a sabbatical leave for the writing. The chemistry faculty at Madison very kindly furnished me working facilities during that leave and invited me to teach a summer session and spring term, doing everything to make me feel welcome.

S. B.

Contents

Chapter One
Basic Material

Chapter Two
Acid-Base Chemistry

Chapter Three

The Control of Solubility

Chapter Four

Absorption Spectro-photometry

Chapter Five

Potentiometry

Chapter Six

Separations Not Involving Precipitation

Solving
Problems
in
Analytical Chemistry

chapter 1
Basic Material

$\equiv$ 1.1 USEFUL STATISTICS $\equiv$

In 1954, a charming little volume entitled *How to Lie with Statistics* appeared.* The author's purpose was to educate the literate public in all the statistical ruses that could be used to influence them to buy a particular product or vote for a particular office seeker. The members of the author's high school debating team eagerly sought the book out. The purpose of *our* discussion of statistical matters is not chiefly to protect students from purveyors of snake oil, but to aid them in interpreting and evaluating their own and others' data. A rudimentary knowledge of statistics, intelligently applied, can give fairly satisfactory answers to such questions as:

¤ **1.** How many figures in an answer are significant (have meaning)?

¤ **2.** What do the terms *accuracy* and *precision* mean, and how do they differ?

¤ **3.** What are some measures of accuracy and precision?

¤ **4.** What sorts of errors plague results?

¤ **5.** How may one mathematically describe distributions of random errors?

¤ **6.** Can it be determined whether two averages are significantly different from each other?

¤ **7.** Can a result that looks bad, or deviant, be cast out?

* Darrell Huff, *How to Lie with Statistics*, Norton, New York, 1954.

☐ **8.** How can one draw the best straight line through points that ought to (but do not quite) line up?

1.2 Significant Figures

Let us address the first question. Modern-day computing techniques and pocket-sized calculators make it likely that workers will retain a great many more significant figures than they really should.

EXAMPLE:

Multiply

$$(1.236) \cdot (4.859)$$

on a pocket calculator.

$$(1.236) \cdot (4.859) = 6.005724$$

The multiplier and multiplicand each contain four digits. The product, displayed on the dial of a calculator, contains 7. If numbers are simply being multiplied together, this is all right, but what happens when each number is part of a measurement? Is the area of a square 1.236 in. by 4.859 in. exactly 6.005724 in.^{2}? Can one start with four-figure numbers, and finish with a seven-figure number, all of whose digits are significant? Of course not! The following rules will help the student avoid at least the sillier mistakes in manipulation.

RULE 1 / Retain *no* digits beyond the first uncertain one.

EXAMPLE:

On a balance, you might read a mass as 21.4238 grams. This number means, in practice, 21.4238 ± 0.0002 grams because ± 0.0002 grams is the uncertainty stated by the manufacturer. The last figure, 8, is uncertain.

You should not guess one more figure. In a number such as 21.42387 grams, guessed on the same balance, the 7 would be meaningless.

RULE 2 / Rounding off. The *residue* is the digit to be dropped. If the residue > 5, increase the digit to the left of the residue by 1.

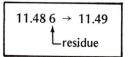

If the residue < 5, do not change the digit to the left of the residue.

```
┌─────────────────────────┐
│   11.39 2  →  11.39      │
│        ↑                 │
│        └residue          │
└─────────────────────────┘
```

If the residue = 5, and the digit to the left of the residue is even, do not change the digit to the left of the residue. If

```
┌─────────────────────────┐
│   41.28 5  →  41.28      │
│        ↑residue          │
└─────────────────────────┘
```

the digit to left of residue is odd, increase the digit to the left of the residue by 1.

```
┌─────────────────────────┐
│   37.13 5  →  37.14      │
│         ↑                │
│         └residue         │
└─────────────────────────┘
```

RULE 3 / In addition or subtraction, one may safely retain only as many digits to the right of the decimal as there are in addend or subtrahend with the smallest number of digits to the right of the decimal point.

```
┌────────────────────────────────────┐
│  EXAMPLE:                           │
│  ────────                           │
│              48.7                   │
│               2.941                 │
│              33.13                  │
│              ─────                  │
│              84.771  →  84.8        │
└────────────────────────────────────┘
```

RULE 4 / In multiplication and division, the relative uncertainty in the product or quotient must be the same as the relative uncertainty in the factor with the lowest number of significant figures. In simple terms, this means roughly that the number of significant figures in the answer may not exceed the number of significant figures in the factor with the lowest number of significant figures.

```
┌────────────────────────────────────────────────────┐
│  EXAMPLES:                                          │
│  ─────────                                          │
│        (1.0923)(2.07) = 2.261061 → 2.26             │
│  three significant figures    also three significant figures │
│        (1.0923)(207) = 226.1061 → 226               │
└────────────────────────────────────────────────────┘
```

```
┌────────────────────────────────────────────────────┐
│  CAUTION:                                           │
│  ────────                                           │
│                                                     │
│    What is the average of these four replicates done on one │
│  chloride sample?                                   │
└────────────────────────────────────────────────────┘
```

$$\overline{\% \text{ Cl}}$$

14.68
14.66
14.70
14.68

$$\text{average} = \frac{14.68 + 14.66 + 14.70 + 14.68}{4}$$

Is this "4" only *one* significant figure?

ANSWER:

The 4 is like 4.00000000 ad infinitum. There is *no* uncertainty in the number of samples. Who, for example, ever heard of 3.9 or 4.1 samples?

$$\text{average} = 14.68\% \text{ Cl}$$

RULE 5 / Sometimes it is not altogether clear which zeros in a number are significant and which are not. Exponential notation makes the task of assigning significance easier.

Example: In 0.000250, the leftmost four zeros are not significant. The rightmost zero *is* significant. This is more easily seen with exponential notation

$$0.000250 = 2.50 \times 10^{-4}$$

The zero to the right of the five is there for a reason—namely, because some other number is not. The number is 0.000250, *not* 0.000251 nor 0.000252.

In a number written like this

420

it is not clear whether

$$4.2 \times 10^2 \quad \text{or} \quad 4.20 \times 10^2$$

is the correct representation. A decimal point after the last zero would indicate that the last zero is significant. For example,

$$220. = 2.20 \times 10^2$$

In a number like

340.0

the zero after the decimal point is significant.

$$340.0 = 3.400 \times 10^2$$

RULE 6 / There should be as many significant figures in the mantissa of the logarithm of a number as there are in the number. The characteristic of the logarithm does not count in enumerating significant figures.

In logarithms, the numbers to the left of the decimal are called the characteristic, and those to the right the mantissa.

Example:

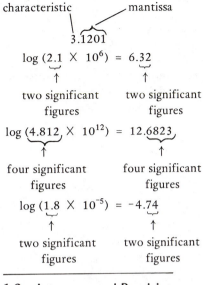

characteristic / mantissa

$$3.1201$$

$$\log (2.1 \times 10^6) = 6.32$$

two significant   two significant
figures   figures

$$\log (4.812 \times 10^{12}) = 12.6823$$

four significant   four significant
figures   figures

$$\log (1.8 \times 10^{-5}) = {-4.74}$$

two significant   two significant
figures   figures

1.3 Accuracy and Precision

In answer to the second question, that about the meaning of accuracy and precision, it should first be noted that a good many people use the terms interchangeably—and thus incorrectly. They are different.

> *Accuracy* is the closeness of a result, or average of a set of results, to the true value.
>
> *Precision* is the closeness of results in a set to each other.

In 1936, A. A. Benedetti-Pichler, for the benefit of his fellow analytical chemists, graphically illustrated accuracy and precision[1], as shown in Figure 1.1.

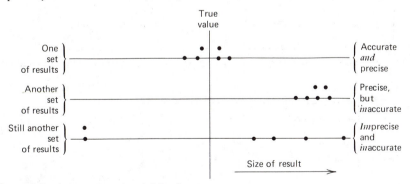

Figure 1.1 *Accuracy and precision*.*

* *A. A. Benedetti-Pichler,* Ind. and Eng. Chem., Anal Ed. 8, *373 (1936).*

Thus results may be precise without being accurate, or both precise and accurate, but it is most unlikely that results could at once be *im*precise and accurate.

The third question deals with simple measurements of accuracy and precision. Some definitions are in order.

Suppose that n measurements of a quantity, for example, %Cl or % Pb, are made. Their values are

$$x_1, x_2, x_3, \ldots, x_n$$

$$\text{mean} = \text{average} = \bar{x} = \frac{x_1 + x_2 + x_3 + \ldots + x_n}{n}$$

or

$$\bar{x} = \frac{\sum\limits_{i=1}^{n} x_i}{n}$$

or, for short

$$\bar{x} = \frac{\sum x_i}{n} \tag{1.1}$$

The accuracy of any one result, x_i, may be expressed in terms of error. When μ is taken as the true value, the absolute error may be expressed as

$$\text{absolute error} = x_i - \mu \tag{1.2}$$

The relative error, a quotient, is also a useful quantity.

$$\text{relative error} = \frac{x_i - \mu}{\mu} \quad \text{as a fraction} \tag{1.3}$$

$$\text{relative error, percent} = \frac{x_i - \mu}{\mu}\,(100) \quad \text{in percent}$$

$$\text{relative error, ppt} = \frac{x_i - \mu}{\mu}\,(1000) \quad \text{in parts per thousand}$$

The absolute and relative errors of the mean of a set of results are given in analogous fashion.

$$\text{absolute error} = \bar{x} - \mu$$

$$\text{relative error} = \frac{\bar{x} - \mu}{\mu}$$

$$\text{relative error, percent} = \frac{\bar{x} - \mu}{\mu}\,(100)$$

$$\text{relative error, ppt} = \frac{\overline{x} - \mu}{\mu} (1000)$$

There are a number of measures of the precision of a set of results. Two of these measures are based on common sense; the other measure is more subtle, and, at times, more powerful.

The first measure of precision is the *range*, which gives an idea of the spread of results from highest to lowest. It is defined as the numerical difference of the highest and the lowest results.

$$\text{range} = w = x_{\text{highest}} - x_{\text{lowest}} \tag{1.4}$$

The *relative range* is defined as the ratio of the range to the mean.

$$\text{relative range} = \frac{w}{\overline{x}} \tag{1.5}$$

$$\text{relative range, percent} = \frac{w}{\overline{x}} (100)$$

$$\text{relative range, ppt} = \frac{w}{\overline{x}} (1000)$$

The *deviation* of a given result, x_i, from the mean, $\overline{x}$, is easily defined.

$$\text{deviation} = d_i = x_i - \overline{x}$$

We can sum the absolute values of these deviations, dividing by the number of values, n, and get a quantity called the average deviation, $\overline{d}$.

$$\text{average deviation} = \overline{d} = \frac{\Sigma |d_i|}{n} \tag{1.6}$$

(It is instructive to note that one must sum the absolute values of the deviations, rather than their signed values. The sum of the signed (+ and −) values of deviations is invariably zero.)

The relative average deviation is easily expressed.

$$\text{relative average deviation} = \frac{\overline{d}}{\overline{x}} \tag{1.7}$$

$$\text{relative average deviation, percent} = \frac{\overline{d}}{\overline{x}} (100)$$

$$\text{relative average deviation, ppt} = \frac{\overline{d}}{\overline{x}} (1000)$$

Even though the average deviation is a quantity of almost obvious usefulness, there is another measure of precision, the standard deviation, which must be understood. Under the right conditions it can be part of a powerful means of data analysis. These conditions are just those of a Gaussian distribution of results and will be discussed in detail later.

The standard deviation, for a set consisting of a great many results, when a true value, μ, is known, is given by the following formula:

$$\text{standard deviation} = \sigma = \left[\frac{\Sigma\,(x_i - \mu)^2}{n - 1}\right]^{\frac{1}{2}} \tag{1.8}$$

$$n = \text{number of results, as before}$$

When a true value, μ, is unknown, as is often the case in the laboratory, and when the number of results is small, the standard deviation is given the symbol s, and its formula is

$$\text{standard deviation} = s = \left[\frac{\Sigma\,(x_i - \bar{x})^2}{n - 1}\right]^{\frac{1}{2}}$$

or

$$d_i = x_i - \bar{x}$$

$$\text{standard deviation} = s = \left(\frac{\Sigma\,d_i^2}{n - 1}\right)^{\frac{1}{2}} \tag{1.9}$$

The standard deviation is not so easy a quantity to calculate as the average deviation, but with today's pocket calculators, the objection to the arithmetic drudgery involved in its computation is easily overcome. Moreover, there are forms of the standard deviation expression that make the computation of s even quicker. Equation 1.10 is one such form.

$$s = \left(\frac{\Sigma\,d_i^2}{n - 1}\right)^{\frac{1}{2}} = \left[\frac{\Sigma\,x_i^2 - \dfrac{(\Sigma\,x_i)^2}{n}}{n - 1}\right]^{\frac{1}{2}} \tag{1.10}$$

The relative standard deviation is given by

$$\text{relative standard deviation} = \frac{s}{\bar{x}}$$

$$\text{relative standard deviation, percent} = \frac{s}{\bar{x}}\,(100) \tag{1.11}$$

$$\text{relative standard deviation, ppt} = \frac{s}{\overline{x}}(1000)$$

The usefulness of standard deviation will be discussed at length later; an example of its calculation will be given here.

EXAMPLE:

At some time in the middle 1950s, a number of students at Eastern Michigan University, then Michigan State Normal College, analyzed a soluble chloride sample and found the following values of percent cloride:

% Cl	% Cl
34.34	34.43
34.31	34.36
34.33	34.58
34.39	34.24

Calculate the standard deviation of this set of results.

SOLUTION:

Use Eq. 1.10.

x_i	x_i^2	
34.34	1179.2356	
34.31	1177.1761	We really
34.33	1178.5489	must carry
34.39	1182.6721	all these
34.43	1185.4249	figures,
34.36	1180.6096	rounding off
34.58	1195.7764	only at the
34.24	1172.3776	end of the
		problem

$$\Sigma x_i = 274.98 \qquad \Sigma x_i^2 = 9451.8212$$

$$(\Sigma x_i)^2 = 75614.0004$$

$$s^2 = \frac{\Sigma x_i^2 - \frac{(\Sigma x_i)^2}{n}}{n-1} = \frac{9451.8212 - \frac{75,614.0004}{8}}{8-1}$$

$$= \frac{0.0712}{7}$$

$$s = 0.10\% \text{ Cl}$$

It will be noted that a great many significant figures were used in the example. All were needed. Only two were kept in the final answer, because %Cl is only measured to two decimal places. The standard deviation may be kept in mind during the discussion of errors in results, particularly in the discussion of random error.

1.4 Errors

Errors are usually divided into two sorts: determinate and indeterminate (random) errors.

Determinate errors are those errors for which a cause may be established.

Indeterminate, or random, errors are those errors for which no cause can readily be found. They are usually small, and may be both positive and negative.

The difference between the two sorts of errors may be exemplified by a scene in a market, where cabbages are being weighed. Suppose that the cabbages always appear, on the scale, to weigh more than they actually do. These errors may be traced to a faulty scale or a dishonest grocer, who may be hauled before a magistrate. These errors are plainly determinate. Indeterminate errors in the weight of the cabbage may, on the other hand, be caused by insects crawling on and off the leaves or by trains rumbling over poorly maintained tracks past the store. The former events are certainly random, and, given the state of U.S. railways, so are the latter.

1.5 Mathematical Description of Random Error Distributions

Random errors, or indeterminate errors, cannot readily be traced. Unlike determinate errors, which are always unidirectional—either always positive or always negative, but never both—random errors may be either positive or negative within a set of results. In fact, in a given set of results, we can often expect as many positive random errors as negative random errors. Random errors may, in some cases be rationally (mathematically) described, but in other cases, they defy description. Let us take examples of both describable and utterly chaotic error distributions. Let us further show a simple method by which it may be determined whether indeterminate errors may be treated by a well-known function or not.

The first example to be taken is one of results gathered by students over a period of many years for a certain sample of potassium hydrogen phthalate (KHPh). All used the same method. Their results are displayed in Table 1.1.

One can look at this mass (or mess) of data, and marvel, but until the data are arranged in some rational order, there is little else one can do. The easiest ordering is by magnitude. The data, thus ordered, may be grouped into cells. About 13 to 20 cells (a cell being a limited range of % KHPh) should be sufficient to group any set of data. Each cell here will be made 0.10% KHPh wide. A histogram, or bar graph, may be plotted, with frequency, or the number of values within a cell on the vertical axis, and % KHPh on the horizontal axis. A *frequency polygon* may be drawn by connecting the

Table 1.1
KHPh RESULTS

% KHPh	% KHPh	% KHPh
26.18	26.36	25.54
26.02	27.18	25.94
25.53	26.18	26.23
25.86	25.29	25.29
25.65	25.50	25.61
26.00	25.47	25.90
25.80	26.13	26.21
25.67	26.02	27.09
26.08	26.09	25.81
25.83	26.49	25.95
25.66	25.47	25.81

midpoints of the bars with straight lines. Figure 1.2 shows the frequency polygon, which is jagged, and does not begin to approximate a smooth curve of any sort. Perhaps more samples would have yielded a better polygon, but without the results from such analyses, it can only be concluded that the errors are not smoothly distributed.

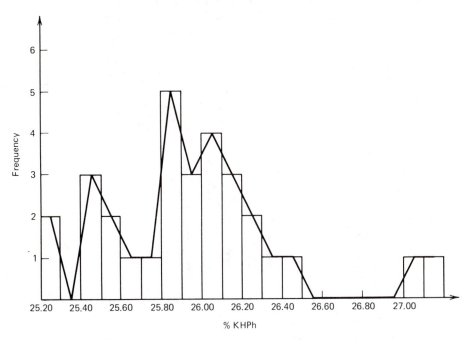

Figure 1.2 *Frequency polygon of data from Table 1.1.*

The next set of results is for another KHPh sample analyzed by another method. These unordered results are set down in Table 1.2.

Table 1.2 OTHER KHPh RESULTS		
% KHPh	% KHPh (cont.)	% KHPh (cont.)
27.30	27.81	28.40
27.71	27.42	27.64
27.67	27.61	28.01
27.84	27.80	28.22
27.60	27.59	28.78
27.96	27.77	27.51
27.52	27.81	27.74
27.74	27.72	27.80
27.43	27.67	
27.44	27.97	
27.88	27.79	

The frequency polygon (Figure 1.3) looks promising. The shape of the frequency polygon strongly suggests a familiar curve shape—the bell curve, as it is commonly known, or the Gaussian distribution.

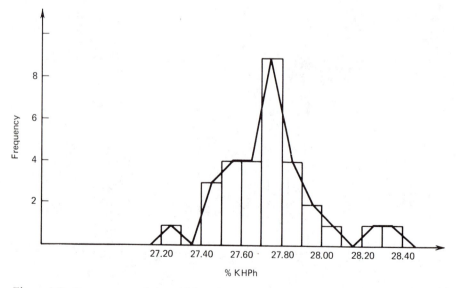

Figure 1.3 *Frequency polygon of data from Table 1.2.*

A Gaussian distribution of experimental results, if not a cause for rejoicing, is at least a cause for relief. Two logical questions are (1) what does a Gaussian curve look like and (2) what is so special about a Gaussian curve in comparison with a non-Gaussian curve?

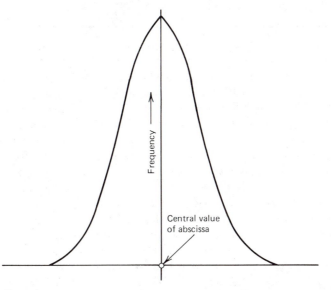

Figure 1.4 *Gaussian curve.*

In answer to the first question, a Gaussian curve looks like the one in Figure 1.4. The ordinate is frequency, and the abscissa may be plotted as: (1) the value of the result, (2) the deviation from the mean, or (3) the number of standard deviations from the mean.

Abscissa Plotted as:	Central Value of Abscissa
Value of the result	$\bar{x}$
Deviation from the mean	0
Number of standard deviations from the mean	0

No matter which value is represented in the abscissa the curve has a regular, symmetrical shape. Equation 1.12 describes the curve.

$$y = \frac{1}{s\sqrt{2\pi}} \exp \frac{-(x - \bar{x})^2}{2s^2}$$

(1.12)

y = frequency (relative)

x = size of result

$\bar{x}$ = mean value of a set of results

s = standard deviation of a set of results

π = 3.14159

The abbreviation exp simply means "e to the power of"; for example, e^Z = exp $[Z]$.

The second question, "what is so special about a Gaussian curve?" may now be partially answered. Although the equation describing the curve is formidable, *it is still an equation,* written in terms of the frequency, the standard deviation, and the mean. This means that rational, mathematical description of sets of data such as the one presented in Table 1.2 and displayed in Figure 1.3 is possible. *One* equation describes *all* such sets of data whose distribution approaches the Gaussian ideal. The graph of Figure 1.2 is not even approximately Gaussian. An equation could be found to describe the polygon of Figure 1.2, but it would be unique for that polygon, having no general applicability. Gaussian behavior of the distribution of results is a most desirable quality.

The shape of a Gaussian curve may tell much about the precision of the results, just as the precision of the results dictates the shape of the Gaussian curve. A rule may be very simply stated: The greater the value of the standard deviation, the wider and less sharp the curve; the lower the value of the standard deviation, the narrower and sharper the curve. This qualitative rule is illustrated quantitatively in Figure 1.5. In Figure 1.5, the values of y versus x are plotted for two sets of chloride results, both with $\bar{x}$ = 42.50% Cl. The standard deviation of one set is 0.08% Cl; that of the other is 0.16% Cl. The smaller value of standard deviation yields a sharp, narrow curve. The larger value of standard deviation yields a fat, spread-out curve.

A number of observations can be made about Gaussian distributions of data.

    □ **1.** Small errors are much more likely to occur than large errors.

    □ **2.** The curve is symmetrical so that positive and negative errors are equally likely.

    □ **3.** If the postulate that the area under the curve is proportional to the total probability of finding a result within that area is accepted, a number of other conclusions may be drawn:

        68.26% of all results fall within ±1 s of the mean,
        95.46% of all results fall within ±2 s of the mean,
        99.74% of all results fall within ±3 s of the mean.

In this entire discussion, no attempt has been made to equate μ, the true value, with $\bar{x}$, the experimental mean. (The student in analytical chemistry seldom has any knowledge of μ, anyway.) There is always the dismal possibility that a graph of the distribution of results might look like Figure 1.6. Here, there is a nice, narrow distribution, quite Gaussian, and a gaping distance between μ and $\bar{x}$ which can only be determinate error.

How do we detect determinate error? Sometimes, a true value is known. We can also

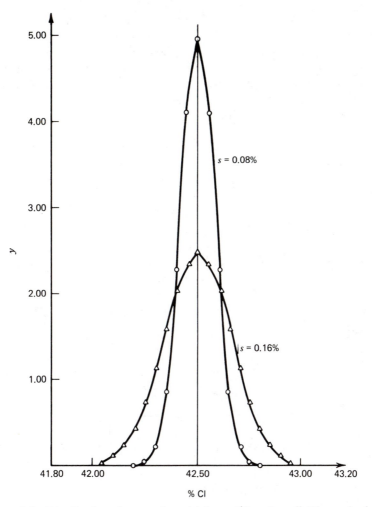

Figure 1.5 *Distributions for samples with large (△) and small (○) standard deviations.*

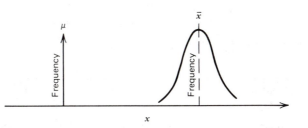

Figure 1.6 *Disagreement of μ and x̄.*

apply statistics, with caution, to a set of results and establish *confidence limits*. We should undertake this procedure—and believe the numbers generated by it—only when every source of determinate error has been firmly discounted. Why is this so?

The establishment of *confidence limits* is, in the first place, based on a Gaussian distribution, or presumed Gaussian distribution, of data. Moreover, common sense dictates that (1) poor precision offers precious little chance of good accuracy; and (2) good precision, although it does *not* generate good accuracy, at least allows one the hope of accuracy.

What is a confidence limit? Suppose that someone had run a number of determinations of chloride in an unknown and determined the confidence limits by a means that will be explained later.

$$\overline{x} = 30.56\% \text{ Cl}$$
$$95\% \text{ confidence limits} = 30.56\% \text{ Cl} \pm 0.12\% \text{ Cl}$$

Put very simply, there is a 95% chance that the true value of % Cl lies within ±0.12% Cl of the mean, or somewhere between 30.68% Cl and 30.44% Cl.

Now, how do we calculate a confidence limit?

$$\text{confidence limit} = \overline{x} \pm \frac{ts}{\sqrt{n}}$$

$$\overline{x} = \text{mean}$$
$$s = \text{standard deviation of results}$$

(1.13)

$$t = \text{a value found in Table 1.3}$$
$$n = \text{number of samples}$$

We must also know how to define a quantity known as "degrees of freedom," in order to find t from Table 1.3.

For one set of n results
degrees of freedom = $n - 1$

It would probably be instructive to work an example.

EXAMPLE:

A quantitative student grudgingly did a chloride determination. His results were

Table 1.1
TABLE OF t-VALUES FOR VARIOUS PROBABILITY LEVELS FOR USE IN CONFIDENCE LIMITS AND STUDENT'S t-TEST. PROBABILITY LEVEL →

Degrees of Freedom	80%	90%	95%	99%
1	3.078	6.314	12.706	63.657
2	1.886	2.920	4.303	9.925
3	1.638	2.353	3.182	5.841
4	1.533	2.132	2.776	4.604
5	1.476	2.015	2.571	4.032
6	1.440	1.943	2.447	3.707
7	1.415	1.895	2.365	3.499
8	1.397	1.860	2.306	3.355
9	1.383	1.833	2.262	3.250
10	1.370	1.812	2.228	3.169
11	1.363	1.796	2.201	3.106
12	1.356	1.782	2.179	3.055
13	1.350	1.771	2.160	3.012
14	1.345	1.761	2.145	2.977
15	1.341	1.753	2.131	2.947
16	1.337	1.746	2.120	2.921
17	1.333	1.740	2.110	2.898
18	1.330	1.734	2.101	2.878
19	1.328	1.729	2.093	2.861
20	1.325	1.725	2.086	2.845
∞	1.282	1.645	1.960	2.576

% CI	% CI
47.64	47.52
47.69	47.55

Calculate the confidence limits for the average of these determinations at the 80, 95, and 99% probability levels.

First, calculate $\bar{x}$ and s.

$$\bar{x} = \frac{47.64 + 47.69 + 47.52 + 47.55}{4} = 47.60$$

$$s^2 = \frac{\Sigma x_i^2 - \dfrac{(\Sigma x_i)^2}{n}}{n-1}$$

x_i	x_i^2
47.64	2269.5696
47.69	2274.3361
47.52	2258.1504
47.55	2261.0025

$\Sigma x_i = 190.40$

$(\Sigma x_i)^2 = 36{,}252.1600 \qquad \Sigma x_i^2 = 9063.0586$

$$s^2 = \frac{9063.0586 - \dfrac{36{,}252.1600}{4}}{3} \qquad = 0.0062$$

$$s = 0.08\% \text{ CI}$$

Next, calculate the confidence limits.
Remember: degrees of freedom $= n - 1 = 4 - 1 = 3$

$$80\% \text{ confidence limits} = \bar{x} \pm \frac{t_{80}s}{\sqrt{n}} = 47.60 \pm \frac{(1.638)(0.08)}{2}$$

$$= 47.60 \pm 0.06\% \text{ CI}$$

$$95\% \text{ confidence limits} = \bar{x} \pm \frac{t_{95}s}{\sqrt{n}} = 47.60 \pm \frac{(3.182)(0.08)}{2}$$

$$= 47.60 \pm 0.13\% \text{ CI}$$

$$99\% \text{ confidence limits} = \bar{x} \pm \frac{t_{99}s}{\sqrt{n}} = 47.60 \pm \frac{(5.841)(0.08)}{2}$$

$$= 47.60 \pm 0.23\% \text{ CI}$$

There is a trend in the confidence limits: As probability gets higher, confidence limits get wider.

80% confidence limits = 47.60 ± 0.06% CI
or
        47.54 to 47.66% CI

95% confidence limits = 47.60 ± 0.13% CI
or

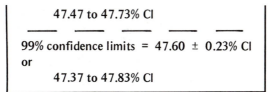

47.47 to 47.73% Cl

99% confidence limits = 47.60 ± 0.23% Cl
or

47.37 to 47.83% Cl

The trend is even more noticeable when graphed, as shown in Figure 1.7. The graph brings home these points: (1) There is an 80% chance that the true value of % Cl is within a short distance of the experimental mean; (2) there is a 95% chance that the true value of % Cl is within a greater distance of the average; and (3) there is a 99% chance that the true value of % Cl is within an even greater distance of the experimental average.

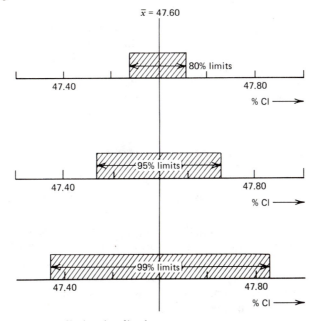

Figure 1.7 *Confidence limits visualized.*

All of the foregoing arguments do no violence to common sense. Another proposition, equally sound, is that the larger the standard deviation of a set of results, the wider the confidence limits must be.

1.6 Comparison of Two Means

Often, in analytical chemistry, the problem faced has not only to do with one mean, but also with a comparison of two means. The questions asked are whether two means are significantly different or whether they are not significantly different. Student's* *t*-

---

* *The nom de plume of its originator.*

Test, a calculation performed with the help of Table 1.3, can sometimes provide an answer. For comparison of two means, Equation 1.14 gives Student's t.

(1.14)

$$t_{exp} = \frac{(\bar{x} - \bar{y})}{s} \sqrt{\frac{n \cdot m}{(n + m)}}$$

t_{exp} = the quantity calculated from experimental data

$\bar{x}$ = the value of one mean

$\bar{y}$ = the value of the other mean

m = the number of replicates taken to give one mean

n = the number of replicates taken to give the other mean

s = the standard deviation of the method, often the pooled standard deviation of the two sets of data

The value of t_{exp} is computed and compared to the critical t-value found in Table 1.3. If $t_{exp} > t$, then there is a likelihood, expressed by the probability level (99, 95, and 90%, etc.) that the two means are different. An example of the calculation is given.

EXAMPLE:

In a good many Southeastern states, illicit distillers have adopted the repellent custom of condensing the distillate in truck radiators. Because these radiators are pasted together with Pb-Sn solder, lead salts, in fair quantity, are dissolved in the condensed distillate. The retailed product can contain a lot of lead. An outbreak of lead poisoning in two southern towns, 30 miles apart, prompts an investigation. Samples of liquor from each town are taken and analyzed for lead.

	From Town A	From Town B
Average [Pb], ppm:	$\bar{x}$ = 60 ppm	$\bar{y}$ = 40 ppm
Number of samples:	m = 6	n = 5
Pooled standard deviation	2 ppm	

Is there a 95% chance that the two batches of moonshine liquor are different?

$$t_{exp} = \frac{\bar{x} - \bar{y}}{s} \sqrt{\frac{n \cdot m}{n + m}}$$

$$= \frac{60 - 40}{2} \sqrt{\frac{6 \cdot 5}{6 + 5}} = 16.5$$

degrees of freedom $= n + m - 2 = 6 + 5 - 2 = 9$
For 9 degrees of freedom the critical value of t at a 95% level is found in Table 1.3.

$$t = 2.262$$

$$t_{exp} > t$$

$$16.5 > 2.262$$

We can conclude that there is a 95% chance that the two batches of liquor are significantly different in their lead content. Such a conclusion *does not* let us say that the two batches of liquor came from different stills, for there are many factors that can affect the lead content of illicit spirits. It is only common sense: One must be cautious in drawing sweeping conclusions.

The term *pooled standard deviation* appeared in the previous box.

$$\text{pooled } s = \left[\frac{\Sigma (x_i - \bar{x})^2 + \Sigma (y_i - \bar{y})^2}{(m - 1) + (n - 1)} \right]^{\frac{1}{2}} \tag{1.15}$$

or

$$\text{pooled } s = \left[\frac{\Sigma x_i^2 - \frac{(\Sigma x_i)^2}{m} + \Sigma y_i^2 - \frac{(\Sigma y_i)^2}{n}}{(m - 1) + (n - 1)} \right]^{\frac{1}{2}} \tag{1.16}$$

$\bar{x}$ = the mean of one set of results

x_i = the ith result from this set

m = the number of results in this set

$\bar{y}$ = the average of the other set

y_i = the ith result from the other set

n = the number of results in the other set

EXAMPLE:

Two sets of chloride determinations are run on samples from two different bottles. The results are

Set 1 (% Cl)	Set 2 (% Cl)
x_1 = 50.50	y_1 = 49.86
x_2 = 50.71	y_2 = 49.85
x_3 = 50.11	y_3 = 49.86
x_4 = 50.48	
m = 4	n = 3

Is there a 95% chance that the samples are significantly different? A lot of informaiton is needed before t_{exp} can be calculated: $\bar{x}$, $\bar{y}$, and s. Use Eq. 1.16 to calculate s, the pooled standard deviation.

x_i	x_i^2	y_i	y_i^2
50.50	2550.2500	49.86	2486.0196
50.71	2571.5041	49.85	2985.0225
50.11	2511.0121	49.86	2486.0196
50.48	2548.2304		
Σx_i = 201.80	Σx_i^2 = 10180.9966	Σy_i = 149.57	Σy_i^2 = 7457.0617

$$\bar{x} = \frac{\Sigma x_i}{m} = 50.45 \qquad \bar{y} = \frac{\Sigma y_i}{n} = 49.86$$

$$(\Sigma x_i)^2 = 40723.2400 \qquad (\Sigma y_i)^2 = 22371.1849$$

$$\frac{(\Sigma x_i)^2}{m} = 10180.8100 \qquad \frac{(\Sigma y_i)^2}{n} = 7457.0616$$

$$s = \left\{\frac{[10180.9966 - 10180.8100]}{(4-1)} + \frac{[7457.0617 - 7457.0616]}{(3-1)}\right\}^{\frac{1}{2}}$$
$$= 0.19$$

$$t_{exp} = \frac{\bar{x} - \bar{y}}{s} \sqrt{\frac{n \cdot m}{(n+m)}}$$

$$= \frac{(50.45 - 49.86)}{0.19} \sqrt{\frac{(4)(3)}{(3+4)}} = 4.06$$

degrees of freedom = $n + m - 2 = 4 + 3 - 2 = 5$

For 5 degress of freedom, at 95% level, from Table 1.3,

$$t = 2.571$$

$$t_{exp} > t$$

$$4.06 > 2.571$$

Thus there is a 95% chance that the chloride contents of the two samples are significantly different.

1.7   Rejection of a Discordant Datum

A question that arises even more often in a quantitative analysis course than the comparison of two means is the question of whether or not to reject a seemingly discordant datum. Look at the following set of data, all for % Cl in the same sample:

% Cl$^-$
56.46
56.77
56.50
56.47

One datum is obviously discordant. Obviously? Yes, when the data are arranged in order of magnitude, it stands out.

% Cl
56.77
56.50
56.47
56.46

There are several methods by which one could decide whether or not to discard the 56.77% value:

◻ **1** The discordant value may be cast out at once if a determinate error is discovered. If, for example, dirt was seen to fall on the final weighing form of the sample, that would constitute a reason for rejection.

◻ **2** We could discard the discordant datum if we simply didn't like it. The validity of this approach is open to question. It is generally only taken by people who are (a) clairvoyant, (b) pigheaded, or (c) both of the above.

¤ 3 The Q-test may be applied to the set of data. We need a table of critical Q-values and a knowledge of how to calculate an experimental value for Q.

Table 1.4
CRITICAL Q-VALUES AT
90% PROBABILITY LEVEL[a]

Number of Replicates	Q_{crit}
3	0.94
4	0.76
5	0.64
6	0.56
7	0.51
8	0.47
9	0.44
10	0.41
∞	0.00

[a]Reprinted with permission from R. B. Dean and W. J. Dixon, *Anal. Chem.* 23., 636 (1951). Copyright © by the American Chemical Society.

Table lists critical values of a quantity called "Q_{crit}." If a value of Q_{exp} calculated from experimental data exceeds the value of Q_{crit} given in the table for that number of replicates, then there is a 90% chance that the questionable datum ought to be rejected. How does one calculate a value of Q_{exp} from experimental data?

$$Q_{exp} = \frac{\Delta_{dev}}{w}$$

(1.17)

w = range of results

Δ_{dev} = difference between the discordant value and its nearest neighbor

EXAMPLE:

The following data were taken for a chloride sample:

$$\frac{\% \text{ Cl}}{}$$

56.77
56.50
56.47
56.46

Should the 56.77% Cl value be thrown out? Use Eq. 1.17:

$$Q_{exp} = \frac{56.77 - 56.50}{56.77 - 56.46}$$

$$= 0.87$$

For four samples, $Q_{crit} = 0.76$.

$$Q_{exp} > Q_{crit}$$

$$0.87 > 0.76$$

The value of 56.77% Cl may be dropped with a 90% probability that it *ought* to be dropped.

1.8 Graphing

In a good many quantitative analysis courses in earlier days, the discussion of the Q-test would have concluded the discussion of statistics. Samples issued then bore names like "gravimetric chloride," "volumetric chloride," "volumetric sodium carbonate in an ash," "soluble arsenic" and "limestone." The instruments used by students in those early days were mostly the two-pan balance, the buret, and the transfer pipet. Such analyses and reports of analytical results did not depend upon the use of pH meters, flame photometers, and spectrophotometers. Reports of results from these last three classes of instruments require graphing. Graphing is an important part of data treatment in today's laboratories for two reasons: (1) Many results depend upon graphs, and (2) poor graphs yield poor results.

Graphing is a simple operation. Certain rules that ought to be observed follow.

1. Choose the proper graph paper. For this course, rectilinear coordinate paper will answer handsomely. (One should, however, remember that semilogarithmic, logarithmic, polar, and triangular coordinate papers exist, and have their uses.) Papers that are very coarsely ruled, for example, 5 or 10 squares to the inch, are of limited utility. Finely ruled (20 squares to the inch) paper which does not have lines that are slightly darker than the rest every 10 squares or so, is capable of inducing eyestrain, frustration, and occasional violent behavior in persons who attempt to use it. The best kinds of rectilinear coordinate paper for our purposes are ruled 20 squares to the inch or 10

squares to the centimeter, and have dark lines running the length and width of the paper every 5 or 10 spaces.

    2. Select the coordinate scales.

        (a) First set the axes. The independent variable should be plotted on the abscissa (x-axis), and the dependent variable on the ordinate (y-axis). The independent variable is the one set by the experimenter, such as the concentration of a standard solution or the milliliters of titrant added. The dependent variable is the one read on the dial of the pH meter or spectrophotometer.

        (b) The scale used should be of such a size that all the significant figures of the measurements made can be accommodated without guesswork and extensive interpolating. Also, the curve or line plotted should fill the entire paper, or as much of it as possible.

    3. Label the coordinate scales. The values represented by each main division line should be placed on the axes, as well as the names of the quantities being represented, for example, milliliters pH, and the like.

    4. Plot the data. Each point should be surrounded by a circle. The radius of the circles should correspond to the uncertainty of the data.

    5. Draw the appropriate smooth curve through the points, using a French curve, if you are expecting other than a straight line. The curve does not have to pass through every point but should be reasonably close to each one. The plotting of straight lines will be dealt with in more detail later.

Generally speaking, the best way to plot a straight line is to draw the best possible line through the points and to examine the graph for gross deviations from this best straight line. If there are no inexplicable deviations, the data may be further treated to give a most probably correct straight line by one of several methods.

The reason that a line should be drawn before treating the data to find the best

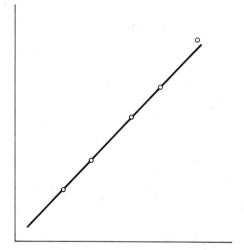

Figure 1.8 Showing one deviant point.

straight line is that, if you treat data like those shown in Figure 1.8, where four points fall into a straight line and the fifth is most likely the result of a determinate error, with a least-squares or other procedure, you are apt to get a line like that shown in Figure 1.9. This line is less accurate than the one that was drawn with only four points, because it gives undue weight to the plainly aberrant point.

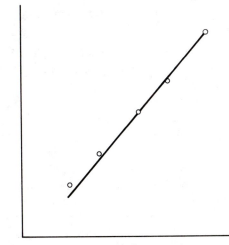

Figure 1.9 Data of Figure 1.8 mathematically treated.

On the other hand, a line like the one in Figure 1.10, where the points seem to fall on either side, probably indicates that a mathematical treatment (least squares is the method we shall use here) can give an even more probably correct straight line.

The derivation of the equations for slope and intercept of a line passing close to or through a set of points depends on certain assumptions. These assumptions are

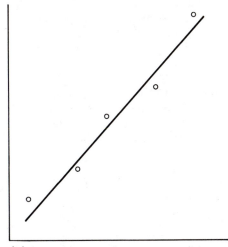

Figure 1.10 Scatttered data.

☐ **1** The uncertainties in the x-values are negligible compared to those expected for the y-values.

☐ **2** The errors in the y-values are random and are distributed in a Gaussian manner.

☐ **3** You can minimize error in y-values for slope and intercept. These proper values will insure that the squares of the deviations of the points from the line are at a minimum.

The derivation based on these assumptions is not complex if you are on friendly terms with calculus. Because not everyone who will be reading this is, we shall skip the derivation and state the results.

Straight lines are described by the relationship,

$$y = mx + b$$

where y is the dependent variable, x the independent variable, m the slope of the line, and b the y-intercept. A graphical illustration is given in Figure 1.11.

Suppose that we have made a series of n measurements, each consisting of a value x_i and a value y_i, where subscript i simply indicates the ordinal number of the measurement. If the plot of y versus x theoretically yields a straight line, then the most probable slope and the y-intercept for that straight line are given by

$$m = \frac{(n)\Sigma x_i y_i - \Sigma x_i \Sigma y_i}{(n)\Sigma x_i^2 - (\Sigma x_i)^2} \tag{1.18}$$

$$b = \frac{\Sigma x_i^2 \Sigma y_i - \Sigma x_i \Sigma x_i y_i}{(n)\Sigma x_i^2 - (\Sigma x_i)^2} \tag{1.19}$$

Obviously, a great deal of arithmetic is involved here. Let us plunge in, however, with an example. Let us suppose that we wish to apply a least-squares treatment to a sodium calibration plot.

Suppose that our data are as follows:

ppm Na	Intensity
1	20
2	40
3	60
4	80
5	100

Intensity for unknown solution = 50

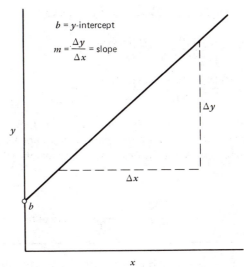

Figure 1.11 Linear relationships.

Plainly, sodium concentration is the independent (x-axis) variable and intensity the dependent (y-axis) variable. Then we assign some symbols to the data:

x_i (ppm Na)	y_i (intensity)
$x_1 = 1$	$y_1 = 20$
$x_2 = 2$	$y_2 = 40$
$x_3 = 3$	$y_3 = 60$
$x_4 = 4$	$y_4 = 80$
$x_5 = 5$	$y_5 = 100$

In least-squaring the data, the intensity (y-value) for the unknown is not included. It will be dealt with later.

Now, let us make a table for computation:

$x_i{}^2$	x_i	y_i	$x_i y_i$
$x_1{}^2 = 1$	$x_1 = 1$	$y_1 = 20$	$x_1 y_1 = 20$
$x_2{}^2 = 4$	$x_2 = 2$	$y_2 = 40$	$x_2 y_2 = 80$
$x_3{}^2 = 9$	$x_3 = 3$	$y_3 = 60$	$x_3 y_3 = 180$
$x_4{}^2 = 16$	$x_4 = 4$	$y_4 = 80$	$x_4 y_4 = 320$
$x_5{}^2 = 25$	$x_5 = 5$	$y_5 = 100$	$x_5 y_5 = 500$
$\Sigma x_i{}^2 = 55$	$\Sigma x_i = 15$	$\Sigma y_i = 300$	$\Sigma x_i y_i = 1100$
	$(\Sigma x_i)^2 = 225$		

Using Eqs. 1.18 and 1.19 for m and b, we see that

$$m = \frac{(5)(1100) - (15)(300)}{(5)\ (55)\ -\ 225} = 20$$

$$b = \frac{(55)(300) - (15)(1100)}{(5)\ (55)\ -\ 225} = 0$$

A line using these values of m and b may be plotted, or the un-known concentration x_u may be found with a knowledge of m, b, and the intensity for the unknown x_u. Remember that

$$y_u = mx_u + b$$

so that

$$x_u = \frac{y_u - b}{m} = \frac{50 - 0}{20} = 2.5 \text{ ppm}$$

It will be agreed that the arithmetic is a great pain. These days, however, program-mable calculators and computers make it possible to do least-squares analyses of data with a minimum of agony.

≡ 1.9 PROBLEMS ≡

1. The correct value of % Cl in a sample is 34.31%. A quantitative analysis student turns in 34.58% Cl as his answer.

   (a) What is the absolute error in his answer?
      Ans: *0.27% Cl*

   (b) What is the relative error in his answer?
      (1) As a decimal number.
         Ans: *0.0079*

      (2) In percent.
         Ans: *0.79%*

      (3) In parts per thousand.
         Ans: *7.9 ppt*

2. Another chlorine sample has 59.11% Cl. "Scattergun" Smith, a quantitative analysis student, runs four replicates of the sample. These are his results:

% Cl
58.90
58.43
59.15
59.33

(a) What is the mean of Smith's results?
   Ans: *58.95% Cl*
(b) What is the absolute error of his mean?
   Ans: *-0.16% Cl*
(c) What is the relative error in his mean, in ppt?
   Ans: *-2.7 ppt*

3. A chemist tests a sample of road salt for chlorine. He runs seven replicates gravimetrically. His results are

% Cl	% Cl
56.66	56.56
56.66	56.63
56.68	56.59
56.59	

(a) Calculate the mean of these results.
   Ans: *56.62% Cl*
(b) Calculate the range and relative range of the results, the latter in parts per thousand.
   Ans: *0.12% Cl; 2.1 ppt*
(c) Calculate the average deviation and relative average deviation of these results, the latter in parts per thousand.
   Ans: *0.04% Cl; 0.67 ppt*
(d) Calculate the standard deviation and relative standard deviation of these results, the latter in parts per thousand.
   Ans: *0.04% Cl; 0.80 ppt*

4. A sample of potassium acid phthalate (KHPh) is run. Are these data distributed in a Gaussian manner?

% KHPh	% KHPh	% KHPh
24.42	24.53	24.03
24.77	24.54	24.99
26.18	24.65	24.42
23.28	24.94	24.33
24.41	23.95	23.52
24.91	24.75	24.43
24.69	24.73	24.04
24.48	24.81	24.89
24.40	24.64	24.78
24.16		
24.14	23.80	
24.07	24.07	

Ans: *No*

5. A set of replicates of a chloride is run. Here are the results:

% Cl	% Cl	% Cl
52.03	50.84	52.19
52.87	51.99	52.33
52.35	52.17	52.20
52.24	52.05	52.36
52.25	52.29	52.23
52.37	51.93	53.12
52.76	52.52	52.30
51.80	52.32	
52.35	52.25	
52.23	51.44	

Are they distributed in a Gaussian manner?
Ans: *More closely than the data of Problem 4.*

6. Suppose that a certain analytical procedure for silica in rocks gives results that fit a Gaussian curve very nicely. The assertion is made that if the average of a set of replicates is 52.43% SiO_2, with a standard deviation of 0.06% SiO_2, 19 out of 20 replicate determinations of this same sample ought to fall between 52.31 and 52.55% SiO_2. Explain the basis for making the assertion.
Ans: *See detailed solutions.*

7. Suppose that two different methods are used to determine chloride in the same sample. One is gravimetric, the other volumetric. The results are

Gravimetric % Cl	Volumetric % Cl
53.06	52.91
53.08	53.12
53.10	53.16
53.09	53.18
53.07	53.08
53.10	52.95

What are the 80, 95, and 99% confidence limits of the gravimetric and volumetric methods?
Ans: *53.08 ± 0.01% Cl; 53.08 ± 0.02% Cl; 53.08 ± 0.03% Cl for grav. 53.07 ± 0.07% Cl; 53.07 ± 0.12% Cl; 53.07 ± 0.18% Cl for vol.*

8. Suppose that the gravimetric method is chosen for the task of analysis of chlorine. On the same sample, one analyst runs two sets of replicates, with the following results.

First Set	Second Set
49.40	49.40
49.42	49.42
49.46	49.44
49.48	49.48
	49.46
	49.44

What are the 80, 95, and 99% confidence limits of both sets of data? Is there a difference? Why?

Ans: *49.44 ± 0.03% Cl; 49.44 ± 0.06% Cl; 49.44 ± 0.12% Cl for the first set.*

49.44 ± 0.02% Cl; 49.44 ± 0.03% Cl; 49.44 ± 0.05% Cl for the second set.

9. Some old, unlabeled chloride samples were found in an unlabeled drawer in a chemistry laboratory at a midwestern teachers' college. The instructor suggested to an eager student that he run volumetric analyses of the chloride bottles to see if the two were different. The student agreed to do so. Here are his results.

% Cl, Bottle A	% Cl, Bottle B
51.52	51.15
51.41	51.05
51.43	51.15
	51.08

Are these results significantly different at the 95% probability level?

Ans: *Yes*

10. Two iron ore samples are taken and analyzed for % Fe. Three replicates of the first sample are run, giving an average of 54.19% Fe. Four replicates of the second sample are run, giving an average of 53.88% Fe. The pooled standard deviation of the results is 0.08% Fe. Are the two samples significantly different at the 95% probability level?

Ans: *Yes*

11. Chlorides, chlorides, chlorides! A weary student does an analysis of a chloride sample. Just to be sure, she does four replicates. These are her results:

% Cl
52.68
53.17
52.73
52.67

Could she justifiably reject the discordant datum?
Ans: *Yes*

12. Another student does an analysis of iron in an ore. These are his results:

% Fe
50.44
50.42
50.38
50.43

Can he justifiably reject the discordant datum?
Ans: *No*

13. A student analyzes a water sample for sodium. Here are the data obtained for calibration:

ppm Na of Standard	Intensity Reading (arbitrary units)
1.0	24
2.0	42
3.0	59
4.0	76
5.0	96

(a) Using a least-squares procedure, calculate the slope and intercept of the line.
Ans: m = *17.8*; b = *6.0*

(b) If the unknown sodium solution gives an intensity of 80 units, what is the concentration of sodium in the unknown?
Ans: *4.2 ppm*

14. Another student analyzes another water sample for sodium. Here are the calibration data:

ppm Na of Standard	Intensity Reading (arbitrary units)
1.0	17
2.0	42
3.0	60
4.0	83
5.0	99

(a) Using a least-squares procedure, calculate the slope and intercept of the line.
Ans: m = *20.5*; b = *-1.3*

(b) Plot the line, using this slope and intercept.

Ans: *See detailed solutions.*

(c) If the student got an intensity reading of 41 units for his unknown sample, what is its sodium concentration?

Ans: *2.1 ppm*

15. Using a device called a spectrophotometer, we can sometimes determine very small concentrations of metals in water. A plot of absorbance, A, versus concentration yields a straight line. The quantity read from the instrument dial, however, is the percent transmission (% T) and is related to absorbance by the formula

$$A = 2 - \log (\% \text{ T})$$

A student prepares a series of standard solutions and takes the following data:

Fe Concentration, ppm	% T
0.5	74.7
1.0	57.4
2.0	34.8
3.0	21.5
4.0	13.3

(a) By a least-squares procedure, determine the slope and intercept of a plot of A versus the concentration.

Ans: m = *0.214*; b = *0.025*

(b) If the unknown iron solution gives a reading of 32.2% T, what is the concentration of iron in the solution?

Ans: *2.2 ppm*

≡ 1.10 DETAILED SOLUTIONS TO PROBLEMS ≡

1. (a) μ = true value

absolute error = (experimental result) − μ = (34.58−34.31)% Cl = 0.27% Cl (1.2)

(b) relative error = $\dfrac{(\text{absolute error})}{\mu}$ (1.3)

(1) As a decimal:

relative error = $\dfrac{0.27}{34.31}$ = 0.0079

(2) In percent:

relative error, % = $\dfrac{0.27}{34.31}$ (100%) = 0.79%

(3) In parts per thousand:

$$\text{relative error, ppt} = \frac{0.27}{34.31}(1000) = 7.9 \text{ ppt}$$

2. (a) $\text{mean} = \bar{x} = \dfrac{x_1 + x_2 + x_3 + x_4 + \cdots}{n}$

$$= \frac{58.90 + 58.43 + 59.15 + 59.33}{4}$$

$$= 58.95\% \text{ Cl}$$

(b) $\text{absolute error} = \bar{x} - \mu = 58.95{-}59.11$

$$= -0.16\% \text{ Cl}$$

(c) $\text{relative error, ppt} = \dfrac{(\text{absolute error})}{\mu}(1000)$

$$= \frac{(-0.16)}{(59.11)}(1000) = -2.7 \text{ ppt}$$

3. (a) $\bar{x} = \dfrac{x_1 + x_2 + x_3 + x_4 + x_5 + x_6 + x_7}{7}$

or

$$\bar{x} = \frac{\sum\limits_{i=1}^{7} x_i}{7} \tag{1.1}$$

or just (for simplicity)

$$\bar{x} = \frac{\Sigma x_i}{7}$$

56.66
56.66
56.68
56.59 $\qquad \bar{x} = \dfrac{396.37}{7}$
56.56
56.63
56.59 $\qquad \bar{x} = 56.62\% \text{ Cl}$

$\Sigma x_i = 396.37$

(b) $\text{range} = w = x_{\text{highest}} - x_{\text{lowest}} = 56.68 - 56.56 = 0.12\% \text{ Cl}$ $\qquad$ (1.4)

$$\text{relative range, ppt} = \frac{w}{\bar{x}}(1000) = \frac{0.12}{56.62}(1000) = 2.1 \text{ ppt}$$

(c) $\text{average deviation} = \bar{d} = \Sigma\dfrac{|d_i|}{n}$ $\qquad$ (1.6)

$$|d_i| = |x_i - \bar{x}|$$

$$|d_1| = |56.66 - 56.62| = 0.04$$
$$|d_2| = |56.66 - 56.62| = 0.04$$
$$|d_3| = |56.68 - 56.62| = 0.06$$
$$|d_4| = |56.59 - 56.62| = 0.03$$
$$|d_5| = |56.56 - 56.62| = 0.06$$
$$|d_6| = |56.63 - 56.62| = 0.01$$
$$|d_7| = |56.59 - 56.62| = 0.03$$

$$\Sigma |d_i| = 0.27$$

$$\bar{d} = \frac{0.27}{7} = 0.038\% \text{ Cl} \approx 0.04\% \text{ Cl}$$

relative average deviation, ppt $= \dfrac{\bar{d}}{\bar{x}}(1000) = \dfrac{0.038}{56.62}(1000) = 0.67$ ppt

(d) One formula for standard deviation is

$$s = \sqrt{\frac{\Sigma d_i^2}{n-1}} \qquad\qquad (1.8)$$

Because we already have all the d_i values, let us use this formula.

Arithmetic:

$|d_i|$	d_i^2
$|d_1| = 0.04$	$d_1^2 = 16 \times 10^{-4}$
$|d_2| = 0.04$	$d_2^2 = 16 \times 10^{-4}$
$|d_3| = 0.06$	$d_3^2 = 36 \times 10^{-4}$
$|d_4| = 0.03$	$d_4^2 = 9 \times 10^{-4}$
$|d_5| = 0.06$	$d_5^2 = 36 \times 10^{-4}$
$|d_6| = 0.01$	$d_6^2 = 1 \times 10^{-4}$
$|d_7| = 0.03$	$d_7^2 = 9 \times 10^{-4}$

$$\Sigma d_i^2 = 123 \times 10^{-4}$$

$$s = \sqrt{\frac{123 \times 10^{-4}}{7-1}} = \sqrt{20.5 \times 10^{-4}} = \sqrt{4.53 \times 10^{-2}}\% \text{ Cl}$$

or

$$= 0.0453\% \text{ Cl}$$

In view of the fact that the % Cl values for our sample are only reported to *two* places, we round off s.

$$s = 0.04\% \text{ Cl}$$

relative standard deviation, ppt $= \dfrac{s}{\bar{x}}(1000) = \dfrac{0.0453}{56.62}(1000) = 0.80$ ppt

4. A three-step procedure is called for here:
   (a) Arrange the data in the order of magnitude.
   (b) Group the data into cells.
   (c) Graph the histogram and construct a frequency polygon from it.
   (a)

% KHPh	%KHPh	% KHPh
26.18	24.54	23.95
	24.53	
24.99		23.80
24.94	24.48	
24.91	24.43	23.52
	24.42, 24.42	
24.89	24.41	23.28
24.81		
	24.40	
24.78	24.33	
24.77		
24.75	24.16	
24.73	24.14	
24.69	24.07, 24.07	
24.65	24.04	
24.64	24.03	

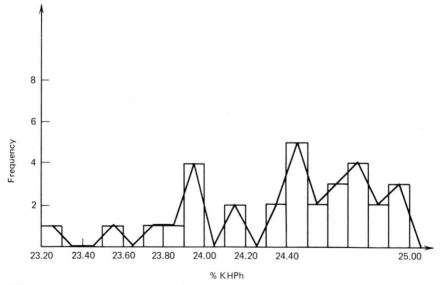

Figure 1.12 Frequency polygon from Problem 4.

(b) Let us take a cell width of 0.10% KHPh.

Cell Boundaries	Center of Cell	Frequency
23.20		
	23.25	1
23.30		
	23.35	0
23.40		
	23.45	0
23.50		
	23.55	1
23.60		
	23.65	0
23.70		
	23.75	1
23.80		
	23.85	1
23.90		
	23.95	4
24.00		
	24.05	0
24.10		
	24.15	2
24.20		
	24.25	0
24.30		
	24.35	1
24.40		
	24.45	5
24.50		
	24.55	2
24.60		
	24.65	3
24.70		
	24.75	4
24.80		
	24.85	2
24.90		
	24.95	3
25.00		
	25.05	1
25.10		

(c) The resultant plot, shown in Figure 1.12, does not look as if it even approximates a Gaussian distribution. Perhaps more replicates would help, but we couldn't draw many conclusions from this plot as it stands.

5. Again, a three-step procedure is called for here:
   (a) Arrange the data in the order of magnitude.
   (b) Group the data into cells.
   (c) Graph the histogram and construct a frequency polygon from it.
   (a)

%Cl	%Cl	%Cl	%Cl
52.12	52.35, 52.35	52.24	51.99
52.87	52.33	52.23, 52.23	51.93
52.76	52.32	52.20	51.80
52.52	52.30	52.19, 52.19	51.44
52.37	52.29	52.05	51.84
52.36	52.25, 52.25	52.03	

(b)

Cell Boundaries	Center of Cell	Frequency
51.60		
	51.65	0
51.70		
	51.75	1
51.80		
	51.85	0
51.90		
	51.95	2
52.00		
	52.05	2
52.10		
	52.15	3
52.20		
	52.25	7
52.30		
	52.35	8
52.40		
	52.45	0
52.50		
	52.55	2
52.60		
	52.65	0
52.70		
	52.75	1
52.80		
	52.85	1
52.90		
	52.95	0
53.00		
	53.05	0
53.10		
	53.15	1
53.20		
	53.25	0
53.30		

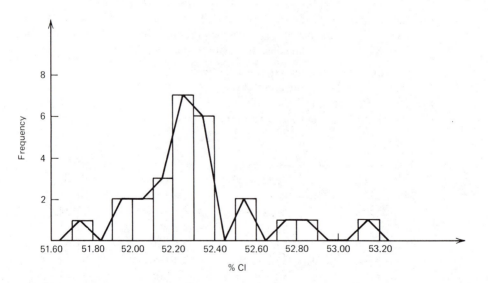

Figure 1.13 *Frequency polygon from Problem 5.*

(c) The resultant plot, shown in Figure 1.13, looks as if it may approximate a Gaussian distribution. There is more hope here than in Problem 4. Again, more replicates would be useful.

6. Ninety-five percent of all replicates fall within $\pm 2s$ of the mean when the data follow a Gaussian distribution.

$$(19/20)(100\%) = 95\%$$

The limits $52.43\% \pm 2s$ are

$$52.43\% \pm 0.12$$

or from

$$52.31 \text{ to } 52.55\%$$

7. We need to calculate:
   (a) Means.
   (b) Standard deviations for both sets of determinations.
   (c) Then, we can calculate confidence limits from Eq. 1.13.

$$\text{confidence limits} = \bar{x} \pm \frac{ts}{\sqrt{n}} \tag{1.13}$$

   (a) Means
   (b) Standard deviation

Gravimetric

x_i	x_i^2
53.06	2815.3636
53.08	2817.4864
53.10	2819.6100
53.09	2818.5481
53.07	2816.4249
53.10	2819.6100
$\Sigma x_i = 318.50$	$\Sigma x_i^2 = 16,907.0430$

$$(\Sigma x_i)^2 = 101442.2500$$

$$\text{mean} = \bar{x} = \frac{\Sigma x_i}{n} = \frac{318.50}{6} = 53.08\% \text{ Cl}$$

$$s_x = \left[\frac{\Sigma x_i^2 - \frac{(\Sigma x_i)^2}{n}}{n-1} \right]^{1/2} \tag{1.10}$$

$$= \left(\frac{16{,}907.0430 - \dfrac{101{,}442.2500}{6}}{6 - 1} \right)^{\frac{1}{2}} \quad = 0.02\% \text{ Cl}$$

Volumetric

y_i	y_i^2
52.91	2799.4681
53.12	2821.7344
53.16	2825.9856
53.18	2828.1124
53.08	2817.4864
52.95	2803.7025

$\Sigma y_i = 318.40 \qquad\qquad \Sigma y_i^2 = 16{,}896.4894$

$(\Sigma y_i)^2 = 101{,}378.5600$

$$\text{mean} = \overline{y} = \frac{\Sigma y_i}{n} = \frac{318.40}{6} = 53.07\% \text{ Cl}$$

$$s_y = \left[\frac{\Sigma y_i^2 - \dfrac{(\Sigma y_i)^2}{n}}{n - 1} \right]^{\frac{1}{2}} \tag{1.10}$$

$$= \left(\frac{16{,}896.4894 - 101{,}378.5600/6}{6 - 1} \right)^{\frac{1}{2}} = 0.11\% \text{ Cl}$$

Note that $s_y > s_x$. This means that the volumetric method gives a larger standard deviation (lower precision) than the gravimetric method, at least in this case.

(c)

(1) Confidence limits for the gravimetric data:

at 80% level, for six samples (5 degrees of freedom), $t = 1.476$

$$\text{confidence limits} = 53.08 \pm \frac{(1.476)(0.02)}{\sqrt{6}} = 53.08 \pm 0.01\% \text{ Cl}$$

at 95% level, for six samples (5 degrees of freedom), $t = 2.571$

$$\text{confidence limits} = 53.08 \pm \frac{(2.571)(0.02)}{\sqrt{6}} = 53.08 \pm 0.02\% \text{ Cl}$$

at 99% level, for six samples (5 degrees of freedom), $t = 4.032$

$$\text{confidence limits} = 53.08 \pm \frac{(4.032)(0.02)}{\sqrt{6}} = 53.08 \pm 0.03\% \text{ Cl}$$

(2) Confidence limits for the volumetric data:

at 80% level, for six samples (5 degrees of freedom), $t = 1.476$

$$\text{confidence limits} = 53.07 \pm \frac{(1.476)(0.11)}{\sqrt{6}} = 53.07 \pm 0.07\% \text{ Cl}$$

at 95% level, for six samples (5 degrees of freedom), $t = 2.571$

$$\text{confidence limits} = 53.07 \pm \frac{(2.571)(0.11)}{\sqrt{6}} = 53.07 \pm 0.12\% \text{ Cl}$$

at 99% level, for six samples (5 degrees of freedom), $t = 4.032$

$$\text{confidence limits} = 53.07 \pm \frac{(4.032)(0.11)}{\sqrt{6}} = 53.07 \pm 0.18\% \text{ Cl}$$

Note how much wider the confidence limits are for the volumetric than for the gravimetric results. The reason is clear: The standard deviation of the volumetric results is higher than that of the gravimetric results, and equal numbers of replicates are run by both methods.

8. Again, we calculate
   (a) Means
   (b) Standard deviations for both sets of determinations.
   (c) Then we can calculate confidence limits from Eq. 1.13.

$$\text{confidence limits} = \bar{x} \pm \frac{ts}{\sqrt{n}} \qquad (1.13)$$

   (a) Means
   (b) Standard deviations

First Set

x_i	x_i^2
49.40	2440.3600
49.42	2442.3364
49.46	2446.2916
49.48	2448.2704

$\Sigma x_i = 197.76 \qquad\qquad \Sigma x_i^2 = 9777.2584$

$(\Sigma x_i)^2 = 39{,}109.0176$

$$\text{mean} = \bar{x} = \frac{\Sigma x_i}{n}$$

$$\bar{x} = \frac{197.76}{4} = 49.44\% \text{ Cl}$$

$$s_x = \left[\frac{\Sigma x_i^2 - \dfrac{(\Sigma x_i)^2}{n}}{n-1} \right]^{\frac{1}{2}} \qquad (1.10)$$

$$= \left[\frac{9777.2584 - \dfrac{39{,}109.0176}{4}}{4-1} \right]^{\frac{1}{2}} = 0.04\% \text{ Cl}$$

Second Set

y_i	$y_i{}^2$
49.40	2440.3600
49.42	2442.3364
49.44	2444.3136
49.48	2448.2704
49.46	2446.2916
49.44	2444.3136

$$\Sigma y_i = 296.64 \qquad \Sigma y_i{}^2 = 14{,}665.8856$$

$$(\Sigma y_i)^2 = 87{,}995.2896$$

$$\text{mean} = \bar{y} = \frac{\Sigma y_i}{n} = \frac{296.64}{6} = 49.44\% \ \text{Cl}$$

$$s_y = \left[\frac{\Sigma y_i{}^2 - \dfrac{(\Sigma y_i)^2}{n}}{n-1} \right]^{\frac{1}{2}}$$

$$= \left(\frac{14{,}665.8856 - \dfrac{87{,}995.2896}{6}}{6-1} \right)^{\frac{1}{2}} = 0.03\% \ \text{Cl}$$

(c)

(1) Confidence limits for the first set:

at 80% level, for four samples (3 degrees of freedom), $t = 1.638$

$$\text{confidence limits} = 49.44 \pm \frac{(1.638)(0.04)}{\sqrt{4}} = 49.44 \pm 0.03\% \ \text{Cl}$$

at 95% level, for four samples (3 degrees of freedom), $t = 3.182$

$$\text{confidence limits} = 49.44 \pm \frac{(3.182)(0.04)}{\sqrt{4}} = 49.44 \pm 0.06\% \ \text{Cl}$$

at 99% level, for four samples (3 degrees of freedom), $t = 5.841$

$$\text{confidence limits} = 49.44 \pm \frac{(5.841)(0.04)}{\sqrt{4}} = 49.44 \pm 0.12\% \ \text{Cl}$$

(2) confidence limits for the second set:

at 80% level, for six samples (5 degrees of freedom), $t = 1.476$

$$\text{confidence limits} = 49.44 \pm \frac{(1.476)(0.03)}{\sqrt{6}} = 49.44 \pm 0.02\% \ \text{Cl}$$

at 95% level, for six samples (5 degrees of freedom), $t = 2.571$

$$\text{confidence limits} = 49.44 \pm \frac{(2.571)(0.03)}{\sqrt{6}} = 49.44 \pm 0.03\% \ \text{Cl}$$

at 99% level, for six samples (5 degrees of freedom), $t = 4.032$

$$\text{confidence limits} = 49.44 \pm \frac{(4.032)(0.03)}{\sqrt{6}} = 49.44 \pm 0.05\% \text{ Cl}$$

Note that the confidence intervals are wider for the first set (four replicates) than for the second set (six). This is because the first set contains *fewer* replicates, and because $s_x > s_y$.

9. The *t*-test is probably the most useful to apply to this problem:
   (a) Calculate the means for each set.
   (b) Calculate a pooled standard deviation for the two sets.
   (c) Calculate *t* from Eq. 1.14.

$$t_{\text{exp}} = \frac{\overline{x} - \overline{y}}{s} \sqrt{\frac{nm}{n + m}} \tag{1.14}$$

and compare it to t_{crit} for this number of degrees of freedom at this probability level.

(a)
Bottle A

x_i	x_i^2
52.52	2654.3104
51.41	2642.9881
51.43	2645.0449
$\Sigma x_i = 154.36$	$\Sigma x_i^2 = 7942.3434$
$(\Sigma x_i)^2 = 23{,}827.0096$	

$$\overline{x} = \frac{\Sigma x_i}{m} = \frac{154.36}{3}$$

$$= 51.45\% \text{ Cl}$$

Bottle B

y_i	y_i^2
51.15	2616.3225
51.05	2006.1025
51.15	2616.3225
51.08	2609.1664
$\Sigma y_i = 204.43$	$\Sigma y_i^2 = 10{,}447.9139$
$(\Sigma y_i)^2 = 41{,}791.6249$	

$$\overline{y} = \frac{\Sigma y_i}{n} = \frac{204.43}{4}$$

$$= 51.11\% \text{ Cl}$$

(b) Calculate a pooled s from Eq. 1.16.

$$s = \left[\frac{\Sigma x_i^2 - \frac{(\Sigma x_i)^2}{m} + \Sigma y_i^2 - \frac{(\Sigma y_i)^2}{n}}{(m-1) \quad + \quad (n-1)} \right]^{\frac{1}{2}} \quad (1.16)$$

$$= \frac{7942.3434 - \frac{23,827.0096}{3} + 10,447.9139 - \frac{41,791.6249}{4}}{(3-1) \quad + \quad (4-1)}$$

$$= 0.05\% \text{ Cl}$$

(c) Calculate t_{exp}.

$$t_{exp} = \frac{\overline{x} - \overline{y}}{s} \sqrt{\frac{nm}{n+m}} \quad (1.14)$$

$$= \frac{51.45 - 51.11}{0.05} \sqrt{\frac{(3) \cdot (4)}{(3+4)}}$$

$$= 8.903$$

degrees of freedom $= n + m - 2 = 3 + 4 - 2 = 5$

For 5 degrees of freedom at the 95% probability level,

$$t_{crit} = 2.571$$

$$t_{exp} > t_{crit}$$

$$8.903 > 2.571$$

So there is a 95% chance that the two means are different.

10. The t-test is called for here, too. We are, however, lucky because some kind soul has figured out the standard deviation for us.

$$t_{exp} = \frac{\overline{x} - \overline{y}}{s} \sqrt{\frac{nm}{n+m}} \quad (1.14)$$

Let $\overline{x}$ = 54.19%
$\overline{y}$ = 53.88%
s = 0.08%
n = 4
m = 3

$$t_{exp} = \frac{54.19 - 53.88}{0.08} \sqrt{\frac{(4) \cdot (3)}{4+3}} = 5.074$$

degrees of freedom $= n + m - 2 = 3 + 4 - 2 = 5$

For 5 degrees of freedom at the 95% probability level,

$$t_{crit} = 2.571$$

$$t_{exp} > t_{crit}$$

$$5.074 > 2.571$$

46 / Basic Material

So there is a 95% chance that the two means are different.

11. This is a case for the Q-test. Arrange the data:

$$53.17$$
$$52.73$$
$$52.68$$
$$52.67$$

$$Q_{exp} = \frac{\Delta \text{ dev}}{w} = \frac{53.17 - 52.73}{53.17 - 52.67} = 0.88 \quad (1.17)$$

For 4 samples, $Q_{crit} = 0.76$ at the 90% level

$$Q_{exp} > Q_{crit}$$

$$0.88 > 0.76$$

The discordant value may safely be discarded.

12. One more Q-test follows. First, arrange the data:

$$50.44$$
$$50.43$$
$$50.42$$
$$50.38$$

$$Q_{exp} = \frac{\Delta \text{ dev}}{w} = \frac{50.42 - 50.38}{50.44 - 50.38} = 0.67 \quad (1.17)$$

For four samples $Q_{crit} = 0.76$ at the 90% level.

$$Q_{exp} < Q_{crit}$$

$$0.67 < 0.76$$

The discordant value had probably better be retained.

13. (a) There is much arithmetic to be done. The independent (x) variable is sodium concentration, which is set by the experimenter. The dependent (y) variable is intensity reading.

$n = 5$

x_i^2	x_i	y_i	$x_i y_i$
$x_1^2 = 1$	$x_1 = 1$	$y_1 = 24$	$x_1 y_1 = 24$
$x_2^2 = 4$	$x_2 = 2$	$y_2 = 42$	$x_2 y_2 = 84$
$x_3^2 = 9$	$x_3 = 3$	$y_3 = 59$	$x_3 y_3 = 177$
$x_4^2 = 16$	$x_4 = 4$	$y_4 = 76$	$x_4 y_4 = 304$
$x_5^2 = 25$	$x_5 = 5$	$y_5 = 96$	$x_5 y_5 = 480$
$\Sigma x_i^2 = 55$	$\Sigma x_i = 15$	$\Sigma y_i = 297$	$\Sigma x_i y_i = 1069$
	$(\Sigma x_i)^2 = 225$		

$$\text{slope} = m = \frac{n\Sigma x_i y_i - \Sigma x_i \Sigma y_i}{n\Sigma x_i^2 - (\Sigma x_i)^2} = \frac{(5)(1069) - (15)(297)}{(5)(55) - 225} = 17.8 \qquad (1.18)$$

$$y\text{-intercept} = b = \frac{\Sigma x_i^2 \Sigma y_i - \Sigma x_i \Sigma x_i y_i}{n\Sigma x_i^2 - (\Sigma x_i)^2} = \frac{(55)(297) - (15)(1069)}{(5)(55) - 225} \qquad (1.19)$$

$$= 6.0$$

(b) $y = mx + b$ is the equation

Let x = concentration of Na
y = intensity
m = slope
b = intercept

$$x = \frac{y - b}{m} = \frac{80 - 6.0}{17.8} = 4.2 \text{ ppm of Na}$$

14. (a) There is, again, much arithmetic to be done. The independent and dependent variables will be the same as in Problem 13.

x_i^2	x_i	y_i	$x_i y_i$	$n = 5$
$x_1^2 = 1$	$x_1 = 1$	$y_1 = 17$	$x_1 y_1 = 17$	
$x_2^2 = 4$	$x_2 = 2$	$y_2 = 42$	$x_2 y_2 = 84$	
$x_3^2 = 9$	$x_3 = 3$	$y_3 = 60$	$x_3 y_3 = 180$	
$x_4^2 = 16$	$x_4 = 4$	$y_4 = 83$	$x_4 y_4 = 332$	
$x_5^2 = 25$	$x_5 = 5$	$y_5 = 99$	$x_5 y_5 = 495$	
$\Sigma x_i^2 = 55$	$\Sigma x_i = 15$	$\Sigma y_i = 301$	$\Sigma x_i y_i = 1108$	
	$(\Sigma x_i)^2 = 225$			

$$\text{slope} = m = \frac{n\Sigma x_i y_i - \Sigma x_i \Sigma y_i}{n\Sigma x_i^2 - (\Sigma x_i)^2} = \frac{(5)(1108) - (15)(301)}{(5)(55) - 225} \qquad (1.18)$$

$$= 20.5$$

$$y\text{-intercept} = b = \frac{\Sigma x_i^2 \Sigma y_i - \Sigma x_i \Sigma x_i y_i}{n\Sigma x_i^2 - (\Sigma x_i)^2} = \frac{(55)(301) - (15)(1108)}{(5)(55) - 225} \qquad (1.19)$$

$$= -1.3$$

(b) Calculate two values of y for two values of x, using the m and b just found. These two points will determine the straight line. If

$$x = 1.0$$

$$y = mx + b = (20.5)(1.0) - 1.3 = 19.2$$

If

$$x = 5.0$$

$$y = mx + b = (20.5)(5.0) - 1.3 = 101.2$$

Plot the line (see Figure 1.14). Also indicate the positions of the experimental points.

(c) $y = mx + b$ is still the equation

Let x = concentration of Na

$\quad y$ = intensity

$\quad m$ = slope

$\quad b$ = intercept

$$x = \frac{y - b}{m} = \frac{41 - (-1.3)}{20.5} = 2.1 \text{ ppm}$$

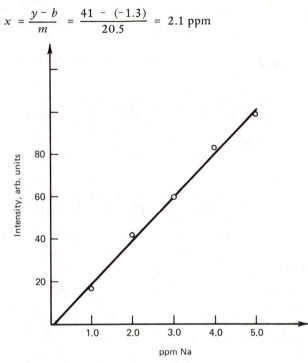

Figure 1.14 Least-squares plot of data from Problem 14. O = exp't'l point.

15. This problem is not too hard, but before the data may be least-squared, the % T values must be converted into A-values by use of the equation given.

(a)

ppm Fe Concentration	% T	A	A = 2 - log % T
0.5	74.7	0.127	
1.0	57.4	0.241	
2.0	34.8	0.458	
3.0	21.5	0.668	
4.0	13.3	0.876	

Let the Fe concentrations be the x-values and the A-values be the y-values.

x_i^2	x_i (concentrations)	y_i (A-values)	$x_i y_i$
0.25	0.5	0.127	0.0635
1.00	1.0	0.241	0.2410
4.00	2.0	0.458	0.9160
9.00	3.0	0.668	2.0040
16.00	4.0	0.876	3.5040
$\Sigma x_i^2 = 30.25$	$\Sigma x_i = 10.5$	$\Sigma y_i = 2.37$	$\Sigma x_i y_i = 6.7285$
	$(\Sigma x_i)^2 = 110.25$	$n = 5$	

(Here, we have carried a lot of significant figures. These can be rounded off later).

$$\text{slope} = m = \frac{n\Sigma x_i y_i - \Sigma x_i \Sigma y_i}{n\Sigma x_i^2 - (\Sigma x_i)^2} = \frac{(5)(6.7285) - (10.5)(2.37)}{(5)(30.25) - (110.25)} \qquad (1.18)$$

$$= 0.214$$

$$y\text{-intercept} = b = \frac{\Sigma x_i^2 \Sigma y_i - \Sigma x_i \Sigma x_i y_i}{n\Sigma x_i^2 - (\Sigma x_i)^2} \qquad (1.19)$$

$$= \frac{(30.25)(2.37) - (10.5)(6.7285)}{(5)(30.25) - (110.25)} = 0.025$$

(Actually, according to theory, $b = 0$. Little problems can arise and give b a nonzero value.)

(b) For the unknown, calculate the absorbance

$$y_u = A_u = 2 - \log \% \ T_u$$

$$= A_u = 2 - \log (32.2) = 0.492$$

Let x = the concentration of unknown in ppm Fe

$$y_u = mx_u + b$$

$$x_u = \frac{y_u - b}{m} = \frac{0.492 - 0.025}{0.214}$$

$$= 2.2 \ \text{ppm Fe}$$

≡ 1.11 STOICHIOMETRY ≡

Stoichiometry is the study of weight (or, more properly, mass) relationships in chemistry. Anyone who does any chemical analysis, or who might interpret the results of analyses, needs a very clear understanding of stoichiometry because the calculation of amounts of substances produced or consumed in chemical reactions is a basic act of quantitative chemistry. If stoichiometry is not understood from the very beginning, further study in chemistry is a waste of time. It would be easy to say to the beginning student in analytical chemistry: "We won't cover much stoichiometry in this course; you've already learned it in your freshman course or in high school." This statement would not apply to most students. Thus stoichiometry has to be considered at the very beginning of an analytical course. This section, which deals with stoichiometry, is organized into three parts:

    ¤   1.   A study of relatively simple gravimetric problems.

    ¤   2.   A study of indirect analyses.

    ¤   3.   A study of systems involving solutions and coupled chemical reactions.

1.12   Simple Gravimetric Calculations

It is a good idea to begin most studies with definitions, to cut down confusion before it arises. The student ought to memorize the following definition and keep it in mind in dealing with any stoichiometric problem.

DEFINITION:

    A mole is the *amount* of a *substance* of [any] specified chemical formula units (atoms, molecules, ions, electrons, quanta, or other entities) containing the *same number of formula units* as there are in [exactly] 12 grams of the pure nuclide ^{12}C. Because there are 6.023×10^{23} atoms of carbon in 12 grams of ^{12}C,
One mole is the amount of material that contains 6.023×10^{23} particles or entities.

The basic, simplest, stoichiometric unit is the mole. Other units are used for convenience, for example, equivalents.

The Laws of Definite and Multiple Proportions, developed early in the history of chemistry, coupled with the concept of the mole, allow us to state that atoms, molecules, ions, or electrons react with each other in the *ratio of simple whole numbers*. Thus moles of atoms, molecules, ions, or electrons must also react with each other in the *ratio of simple whole numbers*. The reaction that produces the silver chloride precipitate furnishes a good example:

$$AgNO_3 + NaCl \rightleftharpoons AgCl\downarrow + NaNO_3$$

The written reaction says:

| 1 mole of $AgNO_3$ | will react with 1 mole of NaCl | to produce 1 mole of AgCl precipitate | and 1 mole of $NaNO_3$ |

The stoichiometry of the reaction is

$$1 : 1 \theta 1 : 1$$

The **symbol** θ helps to distinguish between the mathematical = and the chemical $\rightleftharpoons$. The stoichiometry of the reactions is expressed by writing down the numbers of moles of substances consumed and produced by the reaction.

If it were possible to count directly the reactant and product molecules (or atoms, or ions, etc.) involved, calculations more involved than the one just above would *not* be needed. No counting apparatus that is adequate to this task exists, however, and none is apt to be devised soon. Simple measurements of mass and volume must be relied upon, and more elaborate stoichiometric calculations done. Molecular, atomic, or ionic weights of reacting substances are used to express the stoichiometry of reactions in terms of grams of substances reacting.

$$AgNO_3 + NaCl \rightleftharpoons AgCl\downarrow + NaNO_3$$

$$\left. \begin{array}{cccccc} 1 & : & 1 & \theta & 1 & : & 1 \\ 169.87 & : & 58.44 & \theta & 143.32 & : & 84.99 \end{array} \right\} \begin{array}{l} \text{in moles} \\ \text{in grams} \end{array}$$

and

$$228.31 = 228.31 \text{ grams}$$

Mass is conserved.

A problem involving this reaction is easy to solve.

PROBLEM:

How many grams of AgCl can be formed from 2.00 grams of NaCl and an excess of $AgNO_3$?

There are at many ways of working the problem. It is up to the student to choose the way that suits him or her best.

The first way is probably the simplest. We ask the question: If 58.44 grams of NaCl produce 143.32 grams of AgCl (these numbers are just the molecular weights of NaCl and AgCl),* how many grams of AgCl will be produced by 2.00 grams of NaCl? Simple algebra and arithmetic are applied to the solution of the problem.

Let x = the number of grams of AgCl that will be produced by 2.00 grams of NaCl

$$\frac{x \text{ grams of AgCl}}{2.00 \text{ grams of NaCl}} = \frac{143.32 \text{ grams of AgCl}}{58.44 \text{ grams of NaCl}}$$

---

* Tables of atomic and molecular weights may be found inside the book covers.

$$x \text{ grams of AgCl} = \cancel{(2.00 \text{ grams of NaCl})} \frac{(143.32 \text{ grams of AgCl})}{\cancel{(58.44 \text{ grams of NaCl})}}$$

$$= 4.90 \text{ grams of AgCl}$$

This problem is like the familiar one posed to students of algebra: "If two apples cost 25 cents, how much will six apples cost?" The solution is pretty easy.

Let y = the cost of six apples

$$\frac{y \text{ cents}}{6 \text{ apples}} = \frac{25 \text{ cents}}{2 \text{ apples}}$$

$$y \text{ cents} = (25 \text{ cents}) \frac{(6 \cancel{\text{apples}})}{(2 \cancel{\text{apples}})} = 75 \text{ cents}$$

Simple proportion, then, can be applied to chemical problems as well as to the cost of algebra-book apples.

The second way of working the problem involves the mole rather more directly. One starts out in a very simple-minded way (the best kind of way to start an endeavor!).

$$AgNO_3 + NaCl \rightleftharpoons AgCl\downarrow + NaNO_3$$

$$1 \quad : \quad 1 \quad \ominus \quad 1 \quad : \quad 1$$

or

$$2AgNO_3 + 2NaCl \rightleftharpoons 2AgCl\downarrow + 2NaNO_3$$

$$2 \quad : \quad 2 \quad \ominus \quad 2 \quad : \quad 2$$

or

$$3AgNO_3 + 3NaCl \rightleftharpoons 3AgCl\downarrow + 3NaNO_3$$

$$3 \quad : \quad 3 \quad \ominus \quad 3 \quad : \quad 3$$

A table is made from the information above, and a simple equation is derived from the table.

(Number of moles of AgCl)	(Number of moles of NaCl)
1	1
2	2
3	3
(number of moles of AgCl) =	(number of moles of NaCl)

The equation is put to use, with the fact that for any compound A:

$$(\text{Number of moles of } A) = \frac{(\text{number of grams of } A)}{(\text{molecular weight of } A)}$$

The rest of the solution follows:

Let x = the number of grams of AgCl formed.

(number of moles of AgCl) $= \dfrac{x}{143.32}$

(number of moles of NaCl) $= \dfrac{2.00}{58.44}$

(number of moles of AgCl) $=$ (number of moles of NaCl)

$$\dfrac{x}{143.32} = \dfrac{2.00}{58.44}$$

$$x = \dfrac{(2.00)(143.32)}{(58.44)} = 4.90 \text{ grams}$$

It is probable that the reader has noticed that the factor 143.32/58.44, numerically equal to 2.45 or so, appears in both the first and second methods of solution quite prominently. This number could be called a *gravimetric factor* because one can find out how many grams of silver chloride can be produced by a given amount of sodium chloride simply by multiplying the weight of NaCl by the gravimetric factor. The gravimetric factor was extremely useful in the last century; extensive tables may be found in analytical texts of the time. The author even owns a sort of gravimetric slide rule, made many years ago. Today, however, with accurate and accessible molecular and atomic weights, and cheap electronic calculators, there is not much use in printing a table of gravimetric factors.

It is instructive to examine another sort of simple problem, one in which the stoichiometry is not, as in the silver chloride example, one-to-one. In the following problem, moreover, no chemical reaction is given, and, for the purposes of solving the problem, none is needed.

PROBLEM:

A 0.2905-gram sample containing only Pb_3O_4 and inert matter gives a precipitate of $PbSO_4$ weighing 0.3819 grams. What is the purity of the sample?

The same two methods used before will be applied to this problem. The sample, we know, contains lead oxide and inert material. The purity of the sample is the percent or fraction if its weight that is lead oxide. The question must then be answered: Out of 0.2905 grams of sample, how much is Pb_3O_4 if the amount of $PbSO_4$ precipitate formed from the Pb_3O_4 is 0.3819 gram? In other words, how much Pb_3O_4 is in the sample?

Of course, it is not stated, and is hence not known, what chemical reaction produced the $PbSO_4$ from the Pb_3O_4. This is not necessary because a reaction containing the essence of the quantitative problem, called a stoichiometric reaction, can be written from the molecular formulas of the two compounds involved.

$$Pb_3O_4 \rightarrow 3PbSO_4$$

$$
\begin{array}{ccc}
1 & \ominus & 3 \\
685.57 & \ominus & 909.75
\end{array}
\left.\begin{array}{c} \\ \\ \end{array}\right\}
\begin{array}{l}
\text{in moles} \\
\text{in grams}
\end{array}
$$

weight of   weight of
1 mole of   3 moles of
Pb_3O_4 $PbSO_4$

The first method of solving the problem is the one involving simple proportion. We ask the question: If 685.57 grams of Pb_3O_4 produce 909.75 grams of $PbSO_4$, how many grams of Pb_3O_4 produced 0.3819 grams of $PbSO_4$?

Let x = number of grams of Pb_3O_4 that will produce 0.3819 grams of $PbSO_4$.

$$\frac{x \text{ grams of } Pb_3O_4}{0.3819 \text{ grams of } PbSO_4} = \frac{685.57 \text{ grams of } Pb_3O_4}{909.75 \text{ grams of } PbSO_4}$$

x grams of Pb_3O_4

$$= \frac{(0.3819 \text{ grams of } PbSO_4)(685.57 \text{ grams of } Pb_3O_4)}{909.75 \text{ grams of } PbSO_4}$$

$$= 0.2878 \text{ grams of } Pb_3O_4$$

$$\% \text{ purity} = \frac{(\text{number of grams of } Pb_3O_4)}{(\text{number of grams of sample})} \times 100\%$$

$$= \frac{0.2878}{0.2905} \times 100\% = 99.07\%$$

The method that involves the mole concept very directly works almost as easily as it does for the example of silver chloride.

$$Pb_3O_4 \rightarrow 3PbSO_4$$

$$1 \ominus 3$$

$$2Pb_3O_4 \rightarrow 6PbSO_4$$

$$2 \ominus 6$$

or

$$3Pb_3O_4 \rightarrow 9PbSO_4$$

$$3 \ominus 9$$

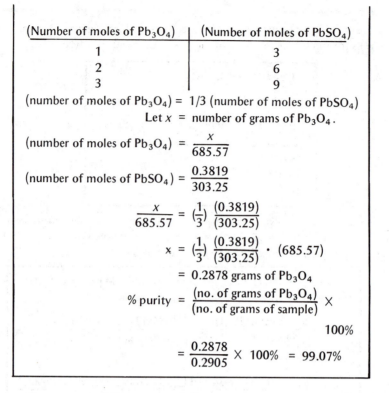

(Number of moles of Pb_3O_4)	(Number of moles of $PbSO_4$)
1	3
2	6
3	9

(number of moles of Pb_3O_4) = 1/3 (number of moles of $PbSO_4$)

Let x = number of grams of Pb_3O_4.

$$\text{(number of moles of } Pb_3O_4) = \frac{x}{685.57}$$

$$\text{(number of moles of } PbSO_4) = \frac{0.3819}{303.25}$$

$$\frac{x}{685.57} = \left(\frac{1}{3}\right)\frac{(0.3819)}{(303.25)}$$

$$x = \left(\frac{1}{3}\right)\frac{(0.3819)}{(303.25)} \cdot (685.57)$$

$$= 0.2878 \text{ grams of } Pb_3O_4$$

$$\% \text{ purity} = \frac{\text{(no. of grams of } Pb_3O_4)}{\text{(no. of grams of sample)}} \times 100\%$$

$$= \frac{0.2878}{0.2905} \times 100\% = 99.07\%$$

1.13   Indirect Analyses

Simple problems involve only one algebraic equation in one unknown in their solution. Slightly more complicated problems are encountered in indirect analyses. One measurement, in indirect analysis, can serve to define two quantities. The following problem will serve as an example.

PROBLEM:

A sample is composed only of $CaCO_3$ and CaO. It is found to contain 50% Ca by weight. What are the % CaO and % $CaCO_3$?

It might be asked first why we bother to tackle some slightly complicated mathematics when two measurements could probably be made and the mathematics thereby simplified. The answer to this question is threefold: (1) It is easier to calculate than to do laboratory work; (2) it is quicker to calculate than to do laboratory work; and (3) it is cheaper to calculate than to do laboratory work. How, then, are the calculations made?

The first thing to understand is that this is a problem with two unknowns: % CaO and % $CaCO_3$. Suppose that the total sample weight were 1 gram.

$$\boxed{\begin{aligned} \text{Let } x &= \text{grams of CaO} \\ \text{Let } y &= \text{grams of CaCO}_3 \\ x + y &= 1.00 \end{aligned}}$$

This *one* equation has *two* unknowns. Another equation is needed before a solution can be found. To get this other needed equation, remember that the sample is composed only of CaO and CaCO$_3$. Of the 1.00 gram of the sample, 0.50 gram, or 50%, is calcium.

$$\boxed{\begin{aligned} (\text{grams of Ca from CaO}) + (\text{grams of Ca from CaCO}_3) = \\ 0.50 \text{ grams} \end{aligned}}$$

This equation still does not contain x and y. The reasoning below will show how to turn the terms (grams Ca from CaO) and (grams Ca from CaCO$_3$) into quantities incorporating x and y.

$$(\text{no. of moles of Ca from CaO}) = (\text{no. of moles of CaO})$$

$$\underbrace{\frac{(\text{grams of Ca from CaO})}{40.08}}_{\text{atomic weight of Ca}} = \underbrace{\frac{(\text{grams of CaO})}{56.08}}_{\substack{\text{molecular weight} \\ \text{of CaO}}}$$

Let x and y mean the same things as before.

$$\frac{(\text{grams of Ca from CaO})}{40.08} = \frac{x}{56.08}$$

$$(\text{grams of Ca from CaO}) = \frac{40.08}{56.08} x$$

$$(\text{no. of moles of Ca from CaCO}_3) = (\text{no. of moles of CaCO}_3)$$

$$\underbrace{\frac{(\text{grams of Ca from CaCO}_3)}{40.08}}_{\text{atomic weight of Ca}} = \underbrace{\frac{(\text{grams of CaCO}_3)}{100.09}}_{\substack{\text{molecular weight} \\ \text{of CaCO}_3}}$$

$$\frac{(\text{grams of Ca from CaCO}_3)}{40.08} = \frac{y}{100.09}$$

$$(\text{grams of Ca from CaCO}_3) = \frac{40.08}{100.09} y$$

$$\frac{40.08}{56.08} x + \frac{40.08}{100.09} y = 0.50$$

The last-written equation is the second of the two that are needed. The two are written and solved as follows

$$x + y = 1.00$$

$$\frac{40.08}{56.08} x + \frac{40.08}{100.09} y = 0.50$$

$$y = 1.00 - x$$

$$\frac{40.08}{56.08} x + \frac{40.08}{100.09} y = 0.50$$

$$(100.09)(\cancel{56.08}) \frac{(40.08)}{\cancel{(56.08)}} x + (\cancel{100.09})(56.08) \frac{(40.08)}{\cancel{(100.09)}} y$$

$$= (0.50)(100.09)(56.08)$$

$$4012\, x + 2248 - 2248\, x = 2806$$

$$1764\, x = 558$$

$$x = 0.32 \text{ gram*}$$

$$y = 1.00 - 0.32 = 0.68 \text{ gram}$$

$$\% \text{ CaO} = \frac{(\text{grams of CaO})}{(\text{grams of sample})} \times 100\%$$

$$= \frac{0.32}{1.00} \times 100\% = 32\%$$

$$\% \text{ CaCO}_3 = \frac{\text{grams of CaCO}_3}{\text{grams of sample}} \times 100\%$$

$$= \frac{0.68}{1.00} \times 100\% = 68\%$$

Thus, with one measurement, two quantities are readily calculated.

1.14 Solutions and Coupled Chemical Reactions

Frequently, the student and the practicing chemist are called upon to perform dilutions, titrations, and precipitations. These operations involve stoichiometric knowledge. Often, the measurement of volume is performed, as well as the measurement of mass. More than one chemical reaction may be involved. Thus measurements of mass and volume, together with chemical manipulations, combine to yield analytical answers. The simplest sort of solution stoichiometry, that of dilutions, will be considered first.

* Significant figures and rounding off are treated in the section on "Useful Statistics".

The units that will be used in most of our calculations will be three: the mole per liter (often abbreviated as M), the gram per liter, and the equivalent per liter (often abbreviated as N).

The law governing dilutions is straightforward, a sort of restatement of the law of the conservation of matter.

> The number of moles of solute you start with equals the number of moles of solute you end up with, or
>
> $$(\text{number of moles}_1) = (\text{number of moles}_2)$$

To find the number of moles of solute in a known volume of solution, of a known molarity in that solute, use this definition.

> $$(\text{number of moles}) = \left(\frac{\text{number of moles}}{\text{liter}}\right)(\text{number of liters})$$

To find the number of millimoles (a millimole, or mmole, is 1/1000 as big as a mole), or grams, the same kind of reasoning is applied.

> $$(\text{number of millimoles})$$
> $$= \frac{(\text{number of moles})}{(\text{liter})} \cdot (\text{number of milliliters})$$

Similarly,

> $$(\text{number of grams}) = \frac{(\text{number of grams})}{(\text{liter})} \cdot (\text{number of liters})$$

> In dilutions:
> $$(\text{number of moles}_1) = (\text{number of moles}_2)$$
> $$(\text{number of millimoles}_1) = (\text{number of millimoles}_2)$$
> $$\left(M_1 \frac{\text{moles}}{\text{liter}}\right)(v_1 \text{ milliliters}) = \left(M_2 \frac{\text{moles}}{\text{liter}}\right)(v_2 \text{ milliliters})$$
> or
> $$M_1 v_1 = M_2 v_2$$
> This is a formula worth:
> (a) Understanding.
> (b) Remembering.

An example of the application of this principle would be very helpful.

PROBLEM:

How would we make 250 milliliters of a solution 6.0 M (moles per liter) in HNO_3 from a solution of 15.0 M in HNO_3?

$$M_1v_1 = M_2v_2$$

M_1 = 15.0 moles per liter

v_1 = ?

M_2 = 6.0 moles per liter

v_2 = 250. milliliters

$$v_1 = \frac{M_2v_2}{M_1} = \frac{(6.0)(250.)}{15.0} = 1.0 \times 10^2 \text{ milliliters}$$

One hundred milliliters of 15.0 M HNO_3 are to be diluted to 250. milliliters with distilled H_2O. In practice, we would pour the HNO_3 *cautiously* into some distilled H_2O, then bring the volume up to 250. milliliters with more distilled H_2O.

Not all problems are so easy. Sometimes concentrations are expressed in grams per liter.

PROBLEM:

How would we make 500. milliliters of a solution 0.100 M in $AgNO_3$ from a solution 50.0 grams per liter in $AgNO_3$?

M_1 = ?

v_1 = ?

M_2 = 0.100 moles per liter

v_2 = 500. milliliters

Find M_1:

$$M_1 \left(\frac{\text{moles}}{\text{liter}} \right) = \frac{(50.0 \text{ grams per liter})}{(169.87 \text{ grams per mole})}$$

Molecular weight of $AgNO_3$

M_1 = 0.294 moles per liter

$$M_1v_1 = M_2v_2$$

$$v_1 = \frac{M_2v_2}{M_1} = \frac{(0.100)(500.)}{(0.294)} = 170. \text{ milliliters}$$

170. milliliters of 50.0 grams per liter of $AgNO_3$ are diluted to 500. milliliters with distilled H_2O.

Another problem is slightly more intricate; yet, it too, can be solved.

PROBLEM:

How would you make 500. milliliters of 2.0 M H_2SO_4 from reagent-grade H_2SO_4? (Molecular weight = 98.08 grams per mole; weight % H_2SO_4 = 94.0; density = 1.831 grams per milliliter)

$$M_1 = ?$$

$$v_1 = ?$$

$$M_2 = 2.0 \text{ moles per liter}$$

$$v_2 = 500. \text{ milliliters}$$

Find M_1:

$$\frac{\text{number of grams of } H_2SO_4}{\text{grams of solution}}$$

$$= 0.940$$

$$\frac{\text{number of grams of } H_2SO_4}{\text{milliliters of solution}}$$

$$= \frac{0.940 \text{ grams of } H_2SO_4}{\text{grams of solution}} \cdot \frac{1.831 \text{ gram of solution}}{\text{milliliters of solution}}$$

$$\frac{\text{number of grams of } H_2SO_4}{\text{milliliters of solution}}$$

$$= 1.72 \frac{\text{grams of } H_2SO_4}{\text{milliliters of solution}}$$

$$\frac{\text{number of grams of } H_2SO_4}{\text{liter of solution}}$$

$$= \frac{1.72 \text{ grams of } H_2SO_4}{\text{milliliters of solution}} \cdot \frac{1000 \text{ milliliters}}{1 \text{ liter}}$$

$$\frac{\text{number of grams of } H_2SO_4}{\text{liter of solution}}$$

$$= 1720 \frac{\text{grams of } H_2SO_4}{\text{liter of solution}}$$

$$M_1 = \frac{\text{number of moles of } H_2SO_4}{\text{liter of solution}}$$

$$M_1 = \frac{(1720 \text{ grams of } H_2SO_4)}{\text{liter of solution}} \cdot \frac{(1 \text{ mole of } H_2SO_4)}{(98.08 \text{ grams of } H_2SO_4)}$$

$$M_1 = 17.5 \frac{\text{moles}}{\text{liter}}$$

$$M_1 v_1 = M_2 v_2$$

$$v_1 = \frac{M_2 v_2}{M_1} = \frac{(2.0)(500.)}{17.5} = 57 \text{ milliliters}$$

Thus we dilute 57 milliliters of concentrated H_2SO_4 to 500. milliliters with distilled H_2O. To avoid splashing, we slowly pour the H_2SO_4 into some H_2O, then add H_2O until 500. milliliters is the total solution volume.

Dilutions of solutions whose concentrations are expressed in normality, or equivalents per liter, are frequently encountered in all sorts of laboratories. Definitions are given in Table 1.5.

These problems exemplify the definitions:

PROBLEM / How many equivalents of H_2SO_4 are there in 2.0000 moles of H_2SO_4?

Answer: (number of equivalents) = (number of moles)(number of protons donated or accepted) = (2.0000)(2) = 4.0000 equivalents.

PROBLEM / What is the normality, N, of a solution 1.8 M in H_2SO_4?

Answer: (normality, N) = (molarity, M)(no. protons donated or accepted) = (1.8)(2) = 3.6 N = 3.6 equivalents per liter.

PROBLEM / How many equivalents of $KMnO_4$ are there in 0.01000 moles of $KMnO_4$? The redox half-reaction is

$$MnO_4^- + 5e^- + 8H^+ \rightleftharpoons Mn^{2+} + 4H_2O$$

Answer: (number of equivalents) = (number of moles)(number of electrons in half-reaction) = (0.01000)(5) = 0.2000 equivalents.

PROBLEM / What is the normality, N, of a solution 0.01667M in $K_2Cr_2O_7$? The redox half-reaction is

$$Cr_2O_7^{2-} + 6e^- + 14H^+ \rightleftharpoons 2Cr^{3+} + 7H_2O$$

Answer: (normality, N) = (molarity, M)(number of electrons in half-reactions) = (0.01667)(6) = 0.1000N = 0.1000 equivalents per liter.

PROBLEM / How many equivalents of Ca^{2+} are there in 0.025 moles of $CaCl_2$?

Answer: (number of equivalents) = (number of moles)(charge on ion) = (0.025)(2) = 0.050 equivalents.

Table 1.5
DEFINITIONS OF EQUIVALENTS AND NORMALITIES

For Acids and Bases	For Species in Redox Reactions
(number of equivalents) = (number of moles) (number of protons donated or accepted)	(number of equivalents) = (number of moles) (number of electrons in half-reaction)
$$\text{normality, } N = \frac{\text{(number of equivalents)}}{\text{(liter)}} = \frac{\text{(number of moles)}}{\text{(liter)}} \text{(number of protons donated or accepted)}$$	$$\text{normality, } N = \frac{\text{(number of equivalents)}}{\text{liter}} = \frac{\text{(number of moles)}}{\text{(liter)}} \text{(number of electrons in half-reaction)}$$
or	or
normality, N = (molarity, M) (number protons donated or accepted)	normality, N = (molarity, M) (number of electrons in half-reaction)

For Any Ionic Species
(number of equivalents) = (number of moles) (charge on ion)
$$\text{normality, } N = \frac{\text{(number of equivalents)}}{\text{(liter)}} = \frac{\text{(number of moles)}}{\text{(liter)}} \text{(charge on ion)}$$
or
normality, N = (molarity, M) (charge on ion)

PROBLEM / What is the calcium normality of a solution 0.032*M* in $CaCl_2$?

Answer: (normality, *N*) = (molarity, *M*)(charge on ion) = (0.032)(2) = 0.064*N* = 0.064 equivalents per liter.

The law governing dilutions of solutions whose concentrations are given in equivalents per liter is straightforward.

The number of equivalents of solute you start with equals the number of equivalents of solute you end up with,

or

$$(\text{number of equivalents}_1) = (\text{number of equivalents}_2)$$

To find the number of equivalents of solute in a known volume of solution, use this definition:

$$(\text{number of equivalents}) = \left(\frac{\text{number of equivalents}}{\text{liter}}\right)(\text{number of liters})$$

To find the number of milliequivalents (a milliequivalent, or meq, is 1/1000 as big as an equivalent), the same kind of reasoning is applied.

$$(\text{number of meq}) = \left(\frac{\text{number of equivalents}}{\text{liter}}\right)(\text{number of milliliters})$$

In dilutions:

$$(\text{number of equivalents}_1) = (\text{number of equivalents}_2)$$

$$(\text{number of meq}_1) = (\text{number of meq}_2)$$

$$\left(N_1 \frac{\text{equivalents}}{\text{liter}}\right)(v_1 \text{ ml}) = \left(N_2 \frac{\text{equivalents}}{\text{liter}}\right)(v_2 \text{ ml})$$

or

$$N_1 v_1 = N_2 v_2$$

PROBLEM / How would we make 300. milliliters of a solution 4.0*N* in HCl from a solution 12.0*N* in HCl?

$$N_1 v_1 = N_2 v_2$$

N_1 = 12.0 equivalents per liter

v_1 = ?

N_2 = 4.0 equivalents per liter

v_2 = 300. milliliters

$$v_1 = \frac{N_2 v_2}{N_1} = \frac{(4.0)(300.)}{12.0} = 1.0 \times 10^2 \text{ milliliters}$$

Answer: One hundred milliliters of 12.0 *N* HCl are to be diluted to 300 milliliters with distilled H_2O. In practice, we would pour the HCl *cautiously* into some distilled H_2O, then bring the volume up to 300.0 milliliters with more distilled H_2O.

PROBLEM / How would we make 200. milliliters of a solution 2.0 N in H_2SO_4 from 4.0M H_2SO_4?

$$N_1 v_1 = N_2 v_2$$

N_1 = ?

v_1 = ?

N_2 = 2.0 equivalents per liter

v_2 = 200. milliliters

Find N_1

N_1 = (normality, N) = (molarity, M)(number of protons donated or accepted)

N_1 = (4.0)(2) = 8.0 equivalents per liter

$$v_1 = \frac{N_2 v_2}{N_1} = \frac{(2.0)(200.)}{(8.0)} = 50. \text{ milliliters}$$

Answer: Fifty milliliters of 4.0M H_2SO_4 are diluted to 200. ml with distilled H_2O. In practice, the 4.0M acid is *cautiously* poured into some H_2O, and the final dilution to 200. milliliters is then made.

Some problems involve both solutions and solids. The student who does gravimetric analyses can attest to this. In gravimetric analysis, the sample is most often a solid, which is weighed out and dissolved. The ion to be analyzed in the sample—typically chloride or sulfate—is then precipitated by the addition of a solution of silver nitrate or barium chloride.

PROBLEM:

How many milliliters of 0.100 M $AgNO_3$ are required to precipitate all of the chloride in 4.00 grams of NaCl?

$$AgNO_3 + NaCl \rightleftharpoons AgCl\downarrow + NaNO_3$$

or

$$Ag^+ + Cl^- \rightleftharpoons AgCl\downarrow$$

(Number of moles of NaCl)	(Number of moles of $AgNO_3$)
1	1
2	2
3	3

---

so that

$$(\text{number of moles of } AgNO_3) = (\text{number of moles of } NaCl)$$

$$(\text{number of moles of } NaCl) = \frac{4.00 \cancel{\text{grams of NaCl}}}{58.44 \frac{\cancel{\text{grams of NaCl}}}{\text{moles of NaCl}}}$$

Let $x = (\text{number of milliliters of } AgNO_3)$

$$(\text{number of moles of } AgNO_3) = (x \cancel{\text{ml}}) (0.100 \frac{\text{moles}}{\cancel{\text{liter}}}) (1 \frac{\cancel{\text{liter}}}{1000 \cancel{\text{ml}}})$$

$$x \left(\frac{0.100}{1000}\right) = \frac{4.00}{58.44}$$

$$x = \frac{4.00}{(58.44)} \cdot \frac{(1000)}{(0.100)} = 684 \text{ milliliters}$$

Other problems, involving the same processes of precipitation, require more than one-to-one stoichiometry to solve.

PROBLEM:

How many milliliters of 0.100 M $AgNO_3$ are required to precipitate all of the bromide in 2.00 grams of $BaBr_2$?

$$2AgNO_3 + BaBr_2 \rightleftharpoons 2AgBr\downarrow + Ba(NO_3)_2$$

or

$$2Ag^+ + 2Br^- \rightleftharpoons 2AgBr\downarrow$$

or

$$Ag^+ + Br^- \rightleftharpoons AgBr\downarrow$$

$$BaBr_2 \xrightarrow[H_2O]{} Ba^{2+} + 2Br^-$$

(Number of moles of $AgNO_3$)	(Number of moles of $BaBr_2$)
2	1
4	2
6	3

$$(\text{number of moles of } AgNO_3) = (2)(\text{number of moles of } BaBr_2)$$

$$(\text{number of moles of } BaBr_2) = \frac{2.00 \cancel{\text{grams of } BaBr_2}}{297.16 \frac{\cancel{\text{grams of } BaBr_2}}{\text{moles of } BaBr_2}}$$

Let $x = $ number of milliliters of $AgNO_3$

$$\left| (\text{no. of moles of AgNO}_3) = (x \text{ milliliters})(0.100 \frac{\text{moles}}{\text{liter}})(\frac{1 \text{ liter}}{1000 \text{ milliliters}}) \right|$$

$$(x)\left(\frac{0.100}{1000}\right) = (2)\left(\frac{2.00}{297.16}\right)$$

$$x = \frac{(2)(2.00)}{(297.16)} \cdot \frac{(1000)}{(0.100)} = 135 \text{ milliliters}$$

Most often when we work with precipitates, we must add an *excess* of the precipitating reagent to the sample, just to insure that every possible bit of the analyzed substance is in the precipitate. Of course, the exact composition of an unknown is just that—unknown. A typical example would be the kind of soluble chloride unknown used in gravimetric experiments in many teaching laboratories. Student analysts may be told that the unknown consists of a mixture of sodium chloride and potassium chloride. Their job is to precipitate all of the chloride. How? First, they reflect on the nature and stoichiometry of NaCl and KCl. Then they consider mixtures, namely a number of 5.0000-gram unknowns, each with a different composition:

	Total weight = 5.0000 grams		
Unknown No.	Grams of KCl	Grams of NaCl	% Cl
1	5.0000	0.0000	47.62
2	3.7500	1.2500	50.88
3	1.5000	3.5000	56.74
4	0.0000	5.0000	60.66

Pure NaCl contains the highest % Cl of any of the samples. Thus any student who wants to precipitate *all* of the chloride in an unknown mixed from NaCl and KCl will have to assume that the sample may be all NaCl. The following problem provides an example.

PROBLEM:

It is desired to precipitate all the chloride in a sample of mixed NaCl and KCl as AgCl. A 10% molar excess of AgNO$_3$ is to be added to all the samples to insure complete precipitation. The concentration of the AgNO$_3$ solution is 0.30 moles per liter, and the sample weights 0.70 grams. How much AgNO$_3$ must be added?

Note: Assume that the sample is all NaCl.

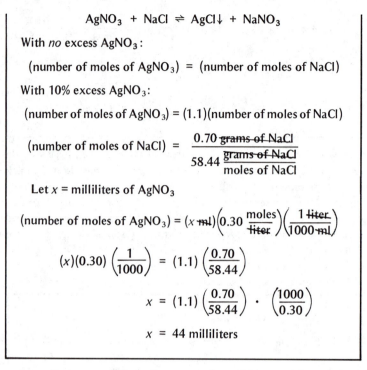

$$AgNO_3 + NaCl \rightleftharpoons AgCl\downarrow + NaNO_3$$

With *no* excess $AgNO_3$:

(number of moles of $AgNO_3$) = (number of moles of NaCl)

With 10% excess $AgNO_3$:

(number of moles of $AgNO_3$) = (1.1)(number of moles of NaCl)

$$(\text{number of moles of NaCl}) = \frac{0.70 \ \cancel{\text{grams of NaCl}}}{58.44 \ \dfrac{\cancel{\text{grams of NaCl}}}{\text{moles of NaCl}}}$$

Let x = milliliters of $AgNO_3$

$$(\text{number of moles of } AgNO_3) = (x \ \cancel{\text{ml}})\left(0.30 \ \frac{\text{moles}}{\cancel{\text{liter}}}\right)\left(\frac{1 \ \cancel{\text{liter}}}{1000 \ \cancel{\text{ml}}}\right)$$

$$(x)(0.30)\left(\frac{1}{1000}\right) = (1.1)\left(\frac{0.70}{58.44}\right)$$

$$x = (1.1)\left(\frac{0.70}{58.44}\right) \cdot \left(\frac{1000}{0.30}\right)$$

$$x = 44 \text{ milliliters}$$

Gravimetric analyses are not the only ones performed in most laboratories. Volumetric analyses, involving titrations, are even more frequently done. Titrations are useful not only in unknown determinations, but also in the *standardization* of reagents. *Standardization* is the process of determining the concentration of a solution that is to be used in analyzing an unknown sample. The mathematics involved in volumetric analysis is no more difficult than that encountered in gravimetry. As an example, let us consider the titration of unknown H_2SO_4 by the standard (known) NaOH solution.

PROBLEM:

What is the concentration of a solution of H_2SO_4, 25.00 milliliters of which required 36.08 milliliters of 0.1286 M NaOH to be titrated into it before an endpoint could be reached?

$$H_2SO_4 + 2NaOH \rightleftharpoons Na_2SO_4 + 2H_2O$$

(Number of moles of NaOH)	(Number of moles of H_2SO_4)
2	1
4	2
6	3

$$\text{(no. of moles of NaOH)} = (2)\,\text{(no. of moles of } H_2SO_4)$$

$$\text{(no. of moles of NaOH)} = (36.08 \text{ ml}) \left(0.1286\frac{\text{moles}}{\text{liter}}\right)\left(\frac{1\text{ liter}}{1000 \text{ ml}}\right)$$

Let x = the concentration of H_2SO_4.

$$\text{(no. of moles of } H_2SO_4) = \left(x\frac{\text{moles}}{\text{liter}}\right)(25.00 \text{ ml})\left(\frac{1\text{ liter}}{1000 \text{ ml}}\right)$$

$$(36.08)\left(\frac{0.1286}{1000}\right) = (2)(x)\left(\frac{25.00}{1000}\right)$$

$$x = \frac{(36.08)(0.1286)}{(2)(25.00)} = 0.09280 \frac{\text{moles}}{\text{liter}}$$

In titrations, equivalents and normalities may be used for working stoichiometric problems. Two terms are useful: (1) titrant, and (2) analyte. The titrant is the solution added from the buret, and the analyte is the solution in the titration vessel. At the endpoint of a titration,

$$\text{(number of equivalents of titrant)} = \text{(number of equivalents of analyte)}$$

With the use of equivalents, we do not need to use factors of 2, 3, or other integers as in the mole method. Let us restate the problem just worked: 36.08 ml of 0.1286 M NaOH were required to titrate 25.00 ml of H_2SO_4 to an endpoint. What is the concentration of the H_2SO_4 in equivalents per liter? In moles per liter?

$$\text{(number of equivalents of titrant)} = \text{(number of equivalents of analyte)}$$

$$\text{(number of meq of titrant)} = \text{(number of meq of analyte)}$$

$$\text{(number of meq of NaOH)} = \text{(number of meq of } H_2SO_4)$$

$$(N_{OH})\text{(number of milliliters of NaOH)} = (N_{H_2SO_4})\text{(number of milliliters of } H_2SO_4)$$

$$N_{H_2SO_4} = \frac{(N_{OH})\text{(number of milliliters of NaOH)}}{\text{(number of ml of } H_2SO_4)}$$

$$N_{OH} = ?$$

$$\text{(number of milliliters of NaOH)} = 36.08 \text{ milliliters}$$

$$\text{(number of milliliters of } H_2SO_4) = 25.00 \text{ milliliters}$$

$$\text{(normality, } N) = \text{(molarity } M)\text{(number of protons donated or accepted)}$$

$$N_{OH} = (0.1286)(1) = 0.1286 \text{ equivalents per liter}$$

$$N_{H_2SO_4} = \frac{(0.1286)(36.08)}{25.00} = 0.1856 \text{ equivalents per liter}$$

To find the molarity of H_2SO_4,

$$\text{(normality, } N) = \text{(molarity, } M)\text{(number of protons donated or accepted)}$$

$$(\text{molarity}, M) = \frac{(\text{normality}, N)}{(\text{number of protons donated or accepted})}$$

so that

$$M_{H_2SO_4} = \frac{N_{H_2SO_4}}{2} = \frac{0.1856}{2} = 0.09280 \text{ moles per liter}$$

Another problem involves the standardization of NaOH with a solid reagent, primary standard KHP.

PROBLEM:

A 0.8512-gram sample of KHP is titrated with 42.16 milliliters of NaOH of unknown molarity. What is the molarity of the NaOH solution?

KHP, sometimes written KHPh, is an abbreviation for potassium hydrogen phthalate (also known as potassium biphthalate), $KHC_8H_4O_4$, or

whose molecular weight is 204.23 grams per mole. It is *not* a compound of potassium, hydrogen, and phosphorus, of molecular weight 71 grams per mole.

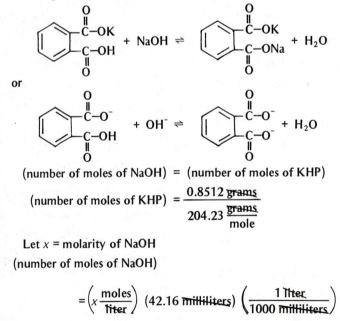

(number of moles of NaOH) = (number of moles of KHP)

$$(\text{number of moles of KHP}) = \frac{0.8512 \text{ grams}}{204.23 \frac{\text{grams}}{\text{mole}}}$$

Let x = molarity of NaOH

(number of moles of NaOH)

$$= \left(x \frac{\text{moles}}{\text{liter}}\right) (42.16 \text{ milliliters}) \left(\frac{1 \text{ liter}}{1000 \text{ milliliters}}\right)$$

$$(x) \left(\frac{42.16}{1000} \right) = \frac{(0.8512)}{(204.23)}$$

$$x = 0.09886 \, \frac{\text{moles}}{\text{liter}}$$

Not all analytical processes, as we glimpsed briefly in the Pb_3O_4-$PbSO_4$ gravimetric problem, involve just one reaction. Some involve a whole series of reactions, whose stoichiometry must be known before an analysis may be done. These reactions may be said to be coupled. An example may be given—the analysis of calcium in limestone by permanganate. The process of analysis follows.

A sample, containing $CaCO_3$, is dissolved in acid.

$$CaCO_3 + 2HCl \rightleftharpoons CaCl_2 + H_2O + CO_2\uparrow$$

$$(\text{or } CaCO_3 + 2H^+ \rightleftharpoons Ca^{2+} + H_2O + CO_2\uparrow).$$

The calcium is precipitated as CaC_2O_4.

$$CaCl_2 + Na_2C_2O_4 \rightleftharpoons CaC_2O_4\downarrow + 2NaCl$$

$$(\text{or } Ca^{2+} + C_2O_4^{2-} \rightleftharpoons CaC_2O_4\downarrow).$$

The calcium oxalate precipitate is filtered. It is then treated with acid.

$$CaC_2O_4 + H_2SO_4 \rightleftharpoons CaSO_4\downarrow + H_2C_2O_4$$

$$(\text{or } CaC_2O_4 + 2H^+ + SO_4^{2-} \rightleftharpoons CaSO_4\downarrow + H_2C_2O_4).$$

The resultant $H_2C_2O_4$ is titrated with a standard solution of potassium permanganate.

$$3H_2SO_4 + 2KMnO_4 + 5H_2C_2O_4 \rightleftharpoons 2MnSO_4 + K_2SO_4 + 10CO_2\uparrow + 8H_2O$$

$$(\text{or } 6H^+ + 2MnO_4^- + 5H_2C_2O_4 \rightleftharpoons 2Mn^{2+} + 10CO_2\uparrow + 8H_2O)$$

Somehow, the number of moles of calcium in the original sample is related to the number of moles of permanganate consumed in the final titration.

From this point, one can proceed in either of two ways. The first involves a step-by-step mental analysis, and the second involves writing a sort of overall reaction for the entire determination. Both give the same result, a relationship between the number of moles of calcium and the number of moles of permanganate. The first method is:

(number of moles of Ca) = (number of moles of $CaCO_3$)

(number of moles of $CaCO_3$) = (number of moles of $CaCl_2$)

(number of moles of $CaCl_2$) = (number of moles of CaC_2O_4)

(number of moles of CaC_2O_4) = (number of moles of $H_2C_2O_4$)

(number of moles of $H_2C_2O_4$)

$$= \left(\frac{5}{2}\right)(\text{number of moles of } KMnO_4)$$

Hence

$$(\text{number of moles of Ca}) = \left(\frac{5}{2}\right)(\text{number of moles of } KMnO_4)$$

The second method starts from the same assertion as the first and yields the same conclusion. In between, equations are multiplied and summed as in heat-of-reaction problems.

$$(\text{number of moles of Ca}) = (\text{number of moles } CaCO_3)$$

$$\begin{cases} 5CaCO_3 + 10HCl \rightarrow 5CaCl_2 + 5H_2O + 5CO_2 \\ 5CaCl_2 + 5Na_2C_2O_4 \\ \qquad \rightarrow 5CaC_2O_4 + 10NaCl \\ 5CaC_2O_4 + 5H_2SO_4 \\ \qquad \rightarrow 5CaSO_4 + 5H_2C_2O_4 \\ 3H_2SO_4 + 2KMnO_4 + 5H_2C_2O_4 \\ \qquad \rightarrow 2MnSO_4 + K_2SO_4 + 10CO_2 + 8H_2O \end{cases}$$

$$5CaCO_3 + 10HCl + 5Na_2C_2O_4 + 8H_2SO_4 + 2KMnO_4$$
$$\rightarrow 13H_2O + 15CO_2 + 10NaCl + 5CaSO_4 + 2MnSO_4$$
$$+ K_2SO_4$$

so:

$$(\text{number of moles of } CaCO_3)$$

$$= \left(\frac{5}{2}\right)(\text{number of moles of } KMnO_4)$$

$$(\text{number of moles of Ca})$$

$$= \left(\frac{5}{2}\right)(\text{number of moles of } KMnO_4)$$

Either method yields the same stoichiometric relationship. A problem will illustrate further the arithmetic of these coupled chemical reactions.

PROBLEM:

A sample weighs 0.6000 grams and contains mostly $CaCO_3$ and inert SiO_2. The $CaCO_3$ is dissolved into HCl, and the calcium precipitated as CaC_2O_4. The CaC_2O_4 is dissolved, and the $H_2C_2O_4$ formed is titrated with 0.03000M $KMnO_4$. The endpoint is reached when 35.00 milliliters of $KMnO_4$ solution have been added. What is the % Ca in the sample?

(number of moles of Ca)

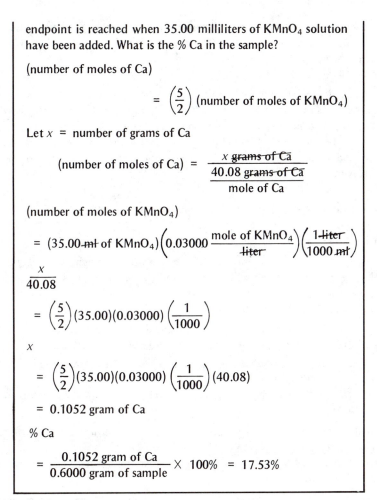

$$= \left(\frac{5}{2}\right) \text{(number of moles of } KMnO_4)$$

Let x = number of grams of Ca

$$\text{(number of moles of Ca)} = \frac{x \text{ grams of Ca}}{\dfrac{40.08 \text{ grams of Ca}}{\text{mole of Ca}}}$$

(number of moles of $KMnO_4$)

$$= (35.00 \text{ ml of } KMnO_4)\left(0.03000 \frac{\text{mole of } KMnO_4}{\text{liter}}\right)\left(\frac{1 \text{ liter}}{1000 \text{ ml}}\right)$$

$$\frac{x}{40.08}$$

$$= \left(\frac{5}{2}\right)(35.00)(0.03000)\left(\frac{1}{1000}\right)$$

$$x$$

$$= \left(\frac{5}{2}\right)(35.00)(0.03000)\left(\frac{1}{1000}\right)(40.08)$$

$$= 0.1052 \text{ gram of Ca}$$

% Ca

$$= \frac{0.1052 \text{ gram of Ca}}{0.6000 \text{ gram of sample}} \times 100\% = 17.53\%$$

This example of coupled chemical reactions and analyses is the end of the formal discussion of stoichiometry. Many problems have served as examples. It is only by faithfully working through these examples and many other problems that the student will achieve skill in stoichiometry.

≡ 1.15 PROBLEMS ≡

1. How many grams of $BaSO_4$ can be produced from 1.354 grams of $BaCl_2$ and an excess of sodium sulfate?
   Ans: *1.518 grams of $BaSO_4$*

2. A sample weighs 0.2500 grams and contains chloride. The chloride is precipitated as silver chloride. The precipitate weighs 0.7476 gram. What is the % Cl in the sample?
   Ans: *73.96% Cl*

3. A sample contains potassium and weighs 0.5742 gram. The potassium is precipitated as the perchlorate, $KClO_4$. The precipitate weighs 0.4240 gram. What is the % K in the sample?

   Ans: *20.83% K*

4. If one is desperate enough, one can determine magnesium by precipitating it as magnesium ammonium phosphate hexahydrate, $MgNH_4PO_4 \cdot 6H_2O$, which is then ignited to magnesium pyrophosphate, $Mg_2P_2O_7$. The magnesium pyrophosphate is then weighed. A sample weighs 0.6004 gram and contains magnesium. It is treated by the method described above and yields 0.4250 gram of $Mg_2P_2O_7$. What is the % *Mg* in the sample?

   Ans: *15.46% Mg*

5. How many grams of Fe_2O_3 can be prepared from 1.0000 gram of Fe_3O_4, assuming an unlimited oxygen supply?

   Ans: *1.0345 grams of Fe_2O_3*

6. A sample is composed only of KCl and NaCl. It weighs 0.1170 gram. The chlorides are precipitated with $AgNO_3$, and the AgCl weighs 0.2500 gram. What are the % KCl and the % NaCl in the sample?

   Ans: *40.43% NaCl; 59.57% KCl*

7. How could 250 milliliters of 1.0 *M* HCl be made from 12 *M* HCl?

   Ans: *Dilute 21 milliliters of 12 M HCl to 250 milliliters with water.*

8. How could 100 milliliters of 1.0 milligram per liter NaCl be made from 50.0 milligram per liter of NaCl?

   Ans: *Dilute 20 milliliters of 50.00 milligram per liter of NaCl to 100 milliliters with water.*

9. Calculate how to make 500 milliliters of a 0.100 *M* solution of $BaCl_2$ from a solution containing 42.0 grams of $BaCl_2$ per liter.

   Ans: *Dilute 247 milliliters of 42.0 gram per liter of $BaCl_2$ to 500 milliliters with water.*

10. How would 250.0 milliliters of a solution of 1000 milligrams per liter in sodium be made from solid NaCl and H_2O?

    Ans: *Dissolve 0.6355 gram of NaCl in water and dilute to 250 milliliters.*

11. Commercial phosphoric acid, H_3PO_4, is 85% by weight H_3PO_4 and has a density of 1.689 grams per milliliter. How would you make 1000 milliliters of 1.0 *M* H_3PO_4 from this reagent?

    Ans: *Dilute 68 milliliters of commercial H_3PO_4 to 1000 milliliters with water.*

12. How could 100 milliliters of 0.120*N* oxalic acid be made from 0.500*M* oxalic acid? Oxalic acid, $H_2C_2O_4$, is a diprotic acid.

    Ans: *Dilute 12.0 milliliters of 0.500M $H_2C_2O_4$ to 100 milliliters with distilled H_2O.*

13. How many milliliters of 0.10 *M* Na_2SO_4 are required to precipitate all the barium in a 0.2200-gram sample of $Ba(NO_3)_2$ as $BaSO_4$?

    Ans: *8.4 milliliters*

14. How many milliliters of a 60. grams per liter solution of $AgNO_3$ are required to precipitate all the iodine in a 0.4126-gram sample of KI as AgI?
    Ans: *7.1 milliliters*

15. How many milliliters of this same 60. grams per liter solution of $AgNO_3$ are required to precipitate *all* the chloride in a 0.4812-gram sample that consists of $CaCl_2$ and $MgCl_2$ mixed in an unknown proportion?
    Ans: *29 milliliters*

16. What is the molarity of an HCl solution, 25.00 milliliters of which require titration with 35.04 milliliters of $0.1123M$ NaOH before an endpoint is reached?
    Ans: *0.1574*M

17. What is the molarity of an NaOH solution if 0.7576 gram of KHP (potassium hydrogen phthalate) can be titrated to an endpoint by 42.05 milliliters of the NaOH solution?
    Ans: *0.08822*M

18. What is the molarity of an HCl solution if 40.36 milliliters of it can titrate 0.2345 gram of pure Na_2CO_3 completely, that is, to the second endpoint?
    Ans: *0.1096*M

19. Iron is often determined volumetrically. A sample of iron is dissolved and all the iron is converted to Fe^{2+}, which is then titrated with $K_2Cr_2O_7$ solution. What is the % Fe in a 0.5285-gram sample that requires 26.87 ml of $0.01524M$ $K_2Cr_2O_7$ solution before an endpoint is reached? The reaction is

$$Cr_2O_7{}^{2-} + 6Fe^{2+} + 14H^+ \rightleftharpoons 2Cr^{3+} + 6Fe^{3+} + 7H_2O$$

Ans: *25.96%*

20. Iron may be determined volumetrically with potassium permanganate. A sample of iron is dissolved and all the iron is converted to Fe^{2+}, which is then titrated with $KMnO_4$ solution. What is the % Fe in a 0.6282-gram sample that is titrated to an endpoint with 24.22 milliliters of $0.1012N$ $KMnO_4$ solution?
    The overall reaction is

$$MnO_4{}^- + 5Fe^{2+} + 8H^+ \rightleftharpoons Mn^{2+} + 5Fe^{3+} + 4H_2O$$

The two half-reactions are

$$MnO_4{}^- + 5e^- + 8H^+ \rightleftharpoons Mn^{2+} + 4H_2O$$
$$Fe^{3+} + e^- \rightleftharpoons Fe^{2+}$$

Ans: *21.79% Fe*

21. A protein was found to contain 0.58% tryptophan. The molecular weight of tryptophan is 204 grams per mole. Assuming that at least 1 mole of tryptophan must be present per mole of protein, calculate the minimum molecular weight of the protein.
    Ans: *3.5 $\times$ 10^4 grams per mole*

22. Copper is often determined by a method involving coupled chemical reactions. A copper-containing sample is dissolved and treated with an excess of iodide:

$$4I^- + 2Cu^{2+} \rightarrow 2CuI + I_2$$

The liberated iodine is titrated with standard thiosulfate solution:

$$I_2 + 2S_2O_3^{2-} \rightarrow S_4O_6^{2-} + 2I^-$$

What is the % Cu in a 0.7627-gram sample, which is dissolved, treated with excess iodide, and titrated with 40.74 milliliters of 0.1034M $Na_2S_2O_3$ to the correct endpoint?

Ans: **35.10%**

≡ 1.16 DETAILED SOLUTIONS TO PROBLEMS ≡

1. $BaCl_2 + Na_2SO_4 \rightleftharpoons BaSO_4\downarrow + 2NaCl$

    1   :   1   1   :   2 } in moles

208.25 : 142.04   233.40 : 116.88 } in grams

SOLUTION
METHOD 1

Let x = number of grams of $BaSO_4$ that can be produced by 1.354 grams of $BaCl_2$

$$\frac{x \text{ grams of } BaSO_4}{1.354 \text{ grams of } BaCl_2} = \frac{233.40 \text{ grams of } BaSO_4}{208.25 \text{ grams of } BaCl_2}$$

ARITHMETIC

$$x \text{ grams of } BaSO_4 = (1.354 \text{ grams of } BaCl_2) \frac{(233.40 \text{ grams of } BaSO_4)}{(208.25 \text{ grams of } BaCl_2)}$$

$$= 1.518 \text{ grams of } BaSO_4$$

SOLUTION
METHOD 2

$$BaCl_2 + Na_2SO_4 \rightleftharpoons BaSO_4\downarrow + 2NaCl$$

(Number of moles of $BaCl_2$)	(Number of moles of $BaSO_4$)
1	1
2	2
3	3

(number of moles of $BaSO_4$) = (number of moles of $BaCl_2$)

Let x = number of grams $BaSO_4$ formed

$$\text{(number of moles of } BaSO_4) = \frac{x}{233.40}$$

$$\text{(number of moles of } BaCl_2) = \frac{1.354}{208.25}$$

$$\frac{x}{233.40} = \frac{1.354}{208.25}$$

ARITHMETIC

$$x = (233.40)\left(\frac{1.354}{208.25}\right) = 1.518 \text{ grams of } BaSO_4$$

2. Ag^+ + Cl^- → $AgCl\downarrow$
   1 : 1 θ 1 } in moles
   107.87 : 35.453 θ 143.32 } in grams

We do *not* need to know whether the chloride came from $NaCl$, KCl, $CaCl_2$, or so on. For % Cl:

$$\% \text{ Cl} = \frac{(\text{grams of Cl})}{(\text{grams of sample})} \overset{\nearrow \text{unknown}}{\underset{\searrow \text{known}}{\times 100\%}}$$

SOLUTION
METHOD 1

Let x = number of grams of Cl that can produce 0.7476 grams of AgCl

$$\frac{x \text{ grams Cl}}{0.7476 \text{ grams AgCl}} = \frac{35.453 \text{ grams Cl}}{143.32 \text{ grams AgCl}}$$

ARITHMETIC

$$(x \text{ grams of Cl}) = (0.7476 \text{ ~~grams of AgCl~~})\left(\frac{35.453 \text{ grams of Cl}}{143.32 \text{ ~~grams of AgCl~~}}\right)$$

$$= 0.1849 \text{ grams of Cl}$$

$$\% \text{ Cl} = \frac{\text{grams of Cl}}{\text{grams of sample}} \times 100\% = \frac{0.1849}{0.2500} \times 100\% = 73.96\%$$

SOLUTION
METHOD 2

$$Ag^+ + Cl^- \to AgCl$$

(number of moles of Cl⁻)	**(number of moles of AgCl)**
1	1
2	2
3	3

(number of moles of Cl^-) = (number of moles of AgCl)

Let x = number of grams of Cl^- in the sample.

$$(\text{number of moles of } Cl^-) = \frac{x}{35.453}$$

$$\text{(number of moles of AgCl)} = \frac{0.7476}{143.32}$$

$$\frac{x}{35.453} = \frac{0.7476}{143.32}$$

ARITHMETIC

$$x = (35.453)\frac{(0.7476)}{(143.32)} = 0.1849 \text{ grams of Cl}$$

$$\% \text{ Cl} = \frac{\text{grams of Cl}}{\text{grams of sample}} \times 100\% = \frac{0.1849}{0.2500} \times 100\% = 73.96\%$$

3. $K^+ + ClO_4^- \rightleftharpoons KClO_4\downarrow$

   or

   $$\left.\begin{array}{ccc} K^+ & \rightarrow & KClO_4\downarrow \\ 1 & \ominus & 1 \\ 39.10 & \ominus & 138.55 \end{array}\right\} \begin{array}{l} \text{stoichiometric equation} \\ \text{in moles} \\ \text{in grams} \end{array}$$

   We do not need to know where the K came from. For % K,

   $$\% \text{ K} = \frac{\text{grams of K}}{\text{grams of sample}} \times 100\% \quad\begin{array}{l}\nearrow\text{unknown}\\ \searrow\text{known}\end{array}$$

SOLUTION
METHOD 1

Let x = number of grams of K that can produce 0.4240 grams of $KClO_4$.

$$\frac{x \text{ grams of K}}{0.4240 \text{ grams of } KClO_4} = \frac{39.10 \text{ grams of K}}{138.55 \text{ grams of } KClO_4}$$

ARITHMETIC

$$x \text{ grams K} = (0.4240 \text{ grams of } KClO_4)\left(\frac{39.10 \text{ grams of K}}{138.55 \text{ grams of } KClO_4}\right)$$

$$= 0.1196 \text{ grams of K}$$

$$\% \text{ K} = \frac{0.1196}{0.5742} \times 100\% = 20.83\%$$

SOLUTION
METHOD 2

$$K^+ \rightarrow KClO_4$$

(Number of moles of K)	(Number of moles of $KClO_4$)
1	1
2	2
3	3

$$(\text{number of moles of K}) = (\text{number of moles of } KClO_4)$$

Let x = number of grams of K in the sample.

$$(\text{number of moles of K}) = \frac{x}{39.10}$$

$$(\text{number of moles of } KClO_4) = \frac{0.4240}{138.55}$$

$$\frac{x}{39.10} = \frac{0.4240}{138.55}$$

ARITHMETIC

$$x = (39.10)\left(\frac{0.4240}{138.55}\right) = 0.1196 \text{ gram of K}$$

$$\% \text{ K} = \left(\frac{\text{grams of K}}{\text{grams of sample}}\right) \times 100\% = \frac{0.1196}{0.5742} \times 100\% = 20.83\%$$

4. $\left. \begin{array}{ccc} 2Mg & \rightarrow & Mg_2P_2O_7 \\ 2 & \theta & 1 \\ 48.62 & \theta & 229.57 \end{array} \right\}$ $\begin{array}{l} \text{stoichiometric equation} \\ \text{in moles} \\ \text{in grams} \end{array}$

SOLUTION
METHOD 1

Let x = number of grams of Mg that produce 0.4250 grams of $Mg_2P_2O_7$.

$$\frac{x \text{ grams of Mg}}{0.4250 \text{ grams of } Mg_2P_2O_7} = \frac{48.62 \text{ grams of Mg}}{222.57 \text{ grams of } Mg_2P_2O_7}$$

$$x \text{ grams of Mg} = (0.4250 \text{ grams of } Mg_2P_2O_7)\left(\frac{48.62 \text{ grams of Mg}}{222.57 \text{ grams of } Mg_2P_2O_7}\right)$$

$$= 0.09284 \text{ grams of Mg}$$

$$\% \text{ Mg} = \frac{\text{grams of Mg}}{\text{grams of } Mg_2P_2O_7} \times 100\% = \frac{0.09284}{0.6004} \times 100\% = 15.46\%$$

SOLUTION
METHOD 2

$$2Mg \rightarrow Mg_2P_2O_7$$

(Number of moles of Mg)	(Number of moles of $Mg_2P_2O_7$)
2	1
4	2
6	3

(Number of moles of Mg) = (2)(number of moles of $Mg_2P_2O_7$)

Let x = number of grams of Mg in the sample.

$$\text{(number of moles of Mg)} = \frac{x}{24.31}$$

$$\text{(number of moles of } Mg_2P_2O_7) = \frac{0.4250}{222.57}$$

$$\frac{x}{24.31} = (2)\left(\frac{0.4250}{222.57}\right)$$

ARITHMETIC

$$x = (24.31)(2)\left(\frac{0.4250}{222.57}\right) = 0.09284 \text{ gram of Mg}$$

$$\% \text{ Mg} = \frac{\text{(grams of Mg)}}{\text{(grams of sample)}} \times 100\% = \frac{0.09284}{0.6004} \times 100\% = 15.46\%$$

5. $2Fe_3O_4 \rightarrow 3Fe_2O_3$ } stoichiometric equation
        2   3 } in moles
    463.08   479.07 } in grams

SOLUTION
METHOD 1

Let x = number of grams of Fe_2O_3 that can be made from 1.0000 gram of Fe_3O_4.

$$\frac{x \text{ grams of } Fe_2O_3}{1.0000 \text{ grams of } Fe_3O_4} = \frac{479.07 \text{ grams of } Fe_2O_3}{463.08 \text{ grams of } Fe_3O_4}$$

ARITHMETIC

$$x \text{ grams of } Fe_2O_3 = (1.0000 \text{ grams of } Fe_3O_4)\left(\frac{479.07 \text{ grams of } Fe_2O_3}{463.08 \text{ grams of } Fe_3O_4}\right)$$

$$= 1.0345 \text{ grams of } Fe_2O_3$$

SOLUTION
METHOD 2

$$2Fe_3O_4 \rightarrow 3Fe_2O_3$$

(Number of moles of Fe_2O_3)	(Number of moles of Fe_3O_4)
3	2
6	4
9	6

$$(\text{number of moles of } Fe_2O_3) = \left(\frac{3}{2}\right) (\text{number of moles of } Fe_3O_4)$$

Let x = number of grams of Fe_2O_3.

$$(\text{number of moles of } Fe_2O_3) = \frac{x}{159.69}$$

$$(\text{number of moles of } Fe_3O_4) = \frac{1.0000}{231.54}$$

$$\frac{x}{159.69} = \left(\frac{3}{2}\right)\left(\frac{1.0000}{231.54}\right)$$

$$x = (159.69)\left(\frac{3}{2}\right)\left(\frac{1.0000}{231.54}\right) = 1.0345 \text{ grams of } Fe_2O_3$$

6.
$$\% \text{ NaCl} = \frac{(\text{grams of NaCl})\,\longleftarrow \text{unknown}}{(\text{grams of sample})\,\longleftarrow \text{known}} \times 100\%$$

$$\% \text{ KCl} = \frac{(\text{grams of KCl})\,\longleftarrow \text{unknown}}{(\text{grams of sample})\,\longleftarrow \text{known}} \times 100\%$$

Let x = grams NaCl and let y = grams KCl.

$$x + y = 0.1170 \leftarrow \text{ one equation with two unknowns}$$

To get another equation:

$$(\text{grams of AgCl from NaCl}) + (\text{grams of AgCl from KCl}) = 0.2500 \text{ gram}$$

$$(\text{number of moles of AgCl from NaCl}) = (\text{number of moles of NaCl})$$

$$\underbrace{\frac{(\text{grams of AgCl from NaCl})}{143.32}}_{\substack{\text{molecular weight} \\ \text{of AgCl}}} = \underbrace{\frac{\text{grams of NaCl}}{58.44}}_{\substack{\text{molecular weight} \\ \text{of NaCl}}}$$

(Remember what x equals.)

$$\frac{(\text{grams of AgCl from NaCl})}{143.32} = \frac{x}{58.44}$$

$$(\text{grams of AgCl from NaCl}) = \frac{143.32}{58.44}\,x$$

(number of moles of AgCl from KCl) = (number of moles of KCl)

$$\frac{\text{(grams of AgCl from KCl)}}{143.32} = \frac{\text{(grams of KCl)}}{74.56}$$

(Remember what y equals.)

$$\frac{\text{(grams of AgCl from KCl)}}{143.32} = \frac{y}{74.56}$$

$$\text{(grams of AgCl from KCl)} = \frac{143.32}{74.56} y$$

Two equations, two unknowns
$$\begin{cases} \dfrac{143.32}{58.44} x + \dfrac{143.32}{74.56} y = 0.2500 \leftarrow \text{the second equation} \\[2mm] \quad x + y = 0.1170 \leftarrow \text{the first equation. Remember?} \\[2mm] \quad\quad y = 0.1170 - x \end{cases}$$

ARITHMETIC

$$\frac{143.32}{58.44} x + \frac{143.32}{74.56} (0.1170 - x) = 0.2500$$

$$2.4524x + (1.9222)(0.1170 - x) = 0.2500$$

$$2.4524x + 0.2249 - 1.9222x = 0.2500$$

$$0.5302x = 0.0251$$

What we needed
$$\begin{cases} x = 0.0473 \text{ gram of NaCl} \\ y = 0.1170 - 0.0473 \\ y = 0.0697 \text{ gram of KCl} \end{cases}$$

$$\% \text{NaCl} = \frac{\text{(grams of NaCl)}}{\text{(grams of sample)}} \times 100\% = \frac{0.0473}{0.1170} \times 100\% = 40.43\%$$

$$\% \text{KCl} = \frac{\text{(grams of KCl)}}{\text{(grams of sample)}} \times 100\% = \frac{0.0697}{0.1170} \times 100\% = 59.57\%$$

7.
or
$$\text{moles}_1 = \text{moles}_2$$

$$M_1 v_1 = M_2 v_2$$

$$M_1 = 12 \text{ moles per liter}$$

$$v_1 = ?$$

$$M_2 = 1.0 \text{ mole per liter}$$

$$v_2 = 250. \text{ milliliters}$$

$$v_1 = \frac{M_2 v_2}{M_1}$$

ARITHMETIC

$$v_1 = \frac{(1.0)(250)}{12} = 21 \text{ milliliters}$$

Here 21 milliliters of 12M HCl are diluted to 250 milliliters with H_2O. Remember, though, to add the acid to the water, not the other way around.

8.

$$\text{grams}_1 = \text{grams}_2$$

$$\text{milligrams}_1 = \text{milligrams}_2$$

$$M_1 v_1 = M_2 v_2$$

$$M_1 = 50.0 \text{ milligrams per liter}$$

$$v_1 = ?$$

$$M_2 = 1.0 \text{ milligram per liter}$$

$$v_2 = 100. \text{ milliliters}$$

$$v_1 = \frac{M_2 v_2}{M_1}$$

ARITHMETIC

$$v_1 = \frac{(1.0)(100)}{50.0} = 2.0 \text{ milliliters}$$

Here 2.0 milliliters of 50.0 milligrams per liter of NaCl solution are diluted to 100. milliliters with H_2O.

9.

$$M_1 v_1 = M_2 v_2$$

$$M_1 = ?$$

$$v_1 = ?$$

$$M_2 = 0.100 \text{ mole per liter}$$

$$v_2 = 500 \text{ milliliters}$$

Find M_1:

ARITHMETIC

$$M_1 \quad \frac{\text{moles}}{\text{liter}} = \frac{42.0 \text{ grams per liter}}{208.25 \text{ grams per mole}}$$

Molecular weight of $BaCl_2$

$$M_1 = 0.202 \text{ mole per liter}$$

$$M_1 v_1 = M_2 v_2$$

$$v_1 = \frac{M_2 v_2}{M_1}$$

ARITHMETIC

$$v_1 = \frac{(0.100)(500.)}{(0.202)} = 248$$

Here 248 milliliters of 42 grams per liter of $BaCl_2$ are diluted to 500. milliliters with H_2O.

10. First, why use NaCl to make a solution 1000. milligrams per liter (1.000 gram per liter) in sodium? There are three reasons: (a) Metallic sodium is hard to get and weigh in a pure state; (b) metallic sodium reacts violently with H_2O; and (c) pure, weighable, stable NaCl is easy to get.

$$\frac{1.0000 \text{ gram}}{\text{liter}} \quad \text{means} \quad \frac{1.0000 \text{ gram}}{1000. \text{ milliliters}} \quad \text{or} \quad \frac{0.2500 \text{ gram}}{250.0 \text{ milliliters}}$$

$$(\text{moles of NaCl}) = (\text{moles of Na})$$

Let x = number of grams of NaCl needed.

$$(\text{moles of NaCl}) = \frac{x}{\underbrace{58.44}_{\substack{\text{molecular weight} \\ \text{of NaCl}}}}$$

$$(\text{moles of Na}) = \frac{0.2500}{\underbrace{22.9898}_{\substack{\text{atomic weight} \\ \text{of Na}}}}$$

$$\frac{x}{58.44} = \left(\frac{0.2500}{22.9898}\right)$$

ARITHMETIC

$$x = (58.44)\left(\frac{0.2500}{22.9898}\right)$$
$$= 0.6355 \text{ gram of NaCl}$$

Thus 0.6355 gram of NaCl is dissolved and diluted to 250. milliliters in a volumetric flask with distilled H_2O.

11.
$$M_1 v_1 = M_2 v_2$$
$$M_1 = ?$$
$$v_1 = ?$$
$$M_2 = 1.0 \text{ mole per liter}$$
$$v_2 = 1000. \text{ milliliters}$$

Find M_1:

ARITHMETIC STEPS

$$\frac{\text{number of grams of } H_3PO_4}{\text{gram of solution}}$$

$$= 0.85 \frac{\text{grams of } H_3PO_4}{\text{gram of solution}}$$

$$\frac{\text{number of grams of } H_3PO_4}{\text{milliliters of solution}}$$

$$= 0.85 \frac{\text{gram of } H_3PO_4}{\text{gram of solution}} \cdot 1.689 \frac{\text{gram of solution}}{\text{milliliters of solution}}$$

$$= 1.44 \frac{\text{grams of } H_3PO_4}{\text{milliliters of solution}}$$

$$\frac{\text{number of grams of } H_3PO_4}{\text{liter of solution}}$$

$$= 1.44 \frac{\text{grams of } H_3PO_4}{\text{milliliters of solution}} \cdot \frac{1000 \text{ milliliters}}{\text{liter}}$$

$$= 1440 \frac{\text{grams of } H_3PO_4}{\text{liter of solution}}$$

$$M_1 = \frac{\text{number of moles of } H_3PO_4}{\text{liter of solution}}$$

$$= 1440 \frac{\text{grams of } H_3PO_4}{\text{liter of solution}} \cdot \frac{1 \text{ mole of } H_3PO_4}{98.00 \text{ grams } H_3PO_4}$$

$$= 14.7 \frac{\text{moles of } H_3PO_4}{\text{liter of solution}}$$

$$M_1 v_1 = M_2 v_2$$

$$v_1 = \frac{M_2 v_2}{M_1}$$

ARITHMETIC

$$v_1 = \frac{(1.0)(1000)}{14.7} = 68 \text{ milliliters}$$

Here 68 milliliters of commercial phosphoric acid are diluted to 1000 milliliters with distilled H_2O.

12. $N_1 V_1 = N_2 V_2$

$N_1 = ?$

$V_1 = ?$

$N_2 = 0.120$ equivalents per liter

V_2 = 100 milliliters

Find N_1:

N_1 = (normality, N) = (molarity M)(number of protons donated or accepted)

$$N_1 = (0.500)(2) = 1.00 \text{ equivalent per liter}$$

$$v_1 = \frac{N_2 V_2}{N_1} = \frac{(.120)(100.)}{1.00} = 12.0 \text{ milliliters}$$

Here 12.0 milliliters of $0.500M$ $H_2C_2O_4$ are *cautiously* diluted to 100 milliliters with distilled H_2O.

13. $Ba(NO_3)_2 + Na_2SO_4 \rightleftharpoons BaSO_4\downarrow + 2NaNO_3$

or

$$Ba^{2+} + SO_4^{2-} \rightleftharpoons BaSO_4\downarrow$$

(Number of moles of Na_2SO_4)	(Number of moles of $BaSO_4$)
1	1
2	2
3	3

(number of moles of Na_2SO_4) = (number of moles of $BaSO_4$)

Let x = number of milliliters of Na_2SO_4 needed

(number of moles of Na_2SO_4) = (x milliliters) $\left(0.1 \dfrac{mole}{liter}\right) \left(\dfrac{1 \text{ liter}}{1000 \text{ milliliters}}\right)$

(number of moles of $Ba(NO_3)_2$)

$$= \frac{0.2200 \text{ grams of } Ba(NO_3)_2}{261.35 \dfrac{\text{grams of } Ba(NO_3)_2}{\text{moles of } Ba(NO_3)_2}}$$

$$(x)(0.10)\left(\frac{1}{1000}\right) = \left(\frac{0.2200}{261.38}\right)$$

ARITHMETIC

$$x = \frac{(0.2200)(1000)}{(261.35)(0.10)} = 8.4 \text{ milliliters}$$

14. $AgNO_3 + KI \rightleftharpoons AgI\downarrow + KNO_3$

or

$$Ag^+ + I^- \rightleftharpoons AgI\downarrow$$

(Number of moles of AgNO$_3$)	(Number of moles of KI)
1	1
2	2
3	3

(number of moles of AgNO$_3$) = (number of moles of KI)

Let x = number of milliliters of AgNO$_3$ required.

(number of moles of AgNO$_3$)

$$= (x \text{ milliliters}) \left(\frac{\text{number of moles}}{\text{liter}} \right) \left(\frac{1 \text{ liter}}{1000 \text{ milliliters}} \right)$$

$$\frac{\text{number of moles}}{\text{liter}} = \frac{60 \frac{\text{grams}}{\text{liter}}}{169.87 \frac{\text{grams}}{\text{mole}}} = 0.35 \frac{\text{mole}}{\text{liter}}$$

(number of moles of AgNO$_3$)

$$= (x \text{ milliliters}) \; 0.35 \left(\frac{\text{mole}}{\text{liter}} \right) \left(\frac{1 \text{ liter}}{1000 \text{ milliliters}} \right)$$

(number of moles of KI) $= \dfrac{0.4126 \text{ gram of KI}}{166.01 \frac{\text{gram of KI}}{\text{mole of KI}}}$

$$(x)(0.35) \left(\frac{1}{1000} \right) = \frac{0.4126}{166.01}$$

ARITHMETIC

$$x = \left(\frac{1000}{0.35} \right) \left(\frac{0.4126}{166.01} \right) = 7.1 \text{ milliliters}$$

15. % Cl in CaCl$_2$ $= \dfrac{(2)(35.453)}{110.99} \times 100\% = 63.89\%$

% Cl in MgCl$_2$ $= \dfrac{(2)(35.453)}{95.22} \times 100\% = 74.47\%$

Because the sample *could* be *all* MgCl$_2$, and because MgCl$_2$ has the higher % Cl, assume that the sample *is* all MgCl$_2$ to make sure that all the Cl$^-$ is precipitated as AgCl.

$$2 AgNO_3 + MgCl_2 \rightleftharpoons 2 AgCl\downarrow + Mg(NO_3)_2$$

(Number of moles of AgNO$_3$)	(Number of moles of MgCl$_2$)
2	1
4	2
6	3

(number of moles of $AgNO_3$) = (2)(number of moles of $MgCl_2$)

Let x = number of milliliters of $AgNO_3$ required.

$$\text{(number of moles of } AgNO_3) = (x \text{ milliliters}) \left(0.35 \frac{\text{mole}}{\text{liter}}\right) \left(\frac{1 \text{ liter}}{1000 \text{ milliliters}}\right)$$

$$\text{(number of moles of } MgCl_2) = \frac{0.4812 \text{ gram of } MgCl_2}{95.22 \frac{\text{grams of } MgCl_2}{\text{mole of } MgCl_2}}$$

$$(x)(0.35)\left(\frac{1}{1000}\right) = \frac{(2)(0.4812)}{(95.22)}$$

ARITHMETIC

$$x = \left(\frac{1000}{0.35}\right) \cdot \frac{(2)(0.4812)}{(95.22)} = 29 \text{ milliliters}$$

16. $HCl + NaOH \rightleftharpoons NaCl + H_2O$

$$H^+ + OH^- \rightleftharpoons H_2O$$

(Number of moles of HCl)	(Number of moles of NaOH)
1	1
2	2
3	3

(number of moles of HCl) = (number of moles of NaOH)

Let x = the molarity of HCl.

(number of moles of HCl)

$$= \left(x \frac{\text{moles}}{\text{liter}}\right) (25.00 \text{ milliliters}) \left(\frac{1 \text{ liter}}{1000 \text{ milliliters}}\right)$$

(number of moles of NaOH)

$$= \left(0.1123 \frac{\text{mole}}{\text{liter}}\right) (35.04 \text{ milliliters}) \left(\frac{1 \text{ liter}}{1000 \text{ milliliters}}\right)$$

$$(x)(25.00)\left(\frac{1}{1000}\right) = (0.1123)(35.04)\left(\frac{1}{1000}\right)$$

ARITHMETIC

$$x = \left(\frac{1}{25.00}\right)(0.1123)(35.04) = 0.1574 \text{ mole per liter}$$

17. $NaOH + KHC_8H_4O_4 \rightleftharpoons NaKC_8H_4O_4 + H_2O$

or

$$OH^- + HC_8H_4O_4^- \rightleftharpoons C_8H_4O_4^{2-} + H_2O$$

(Number of moles of NaOH)	(Number of moles of $KHC_8H_4O_4$)
1	1
2	2
3	3

(number of moles of NaOH) = (number of moles of $KHC_8H_4O_4$)

Let x = the molarity of NaOH.

$$(\text{number of moles of NaOH}) = \left(x\,\frac{\text{moles}}{\text{liter}}\right)(42.05\,\text{milliliters})\left(\frac{1\,\text{liter}}{1000.\,\text{milliliters}}\right)$$

$$(\text{number of moles of } KHC_8H_4O_4) = \left(\frac{0.7576\,\text{grams of } KHC_8H_4O_4}{204.23\,\dfrac{\text{grams of } KHC_8H_4O_4}{\text{mole of } KHC_8H_4O_4}}\right)$$

$$(x)(42.05)\left(\frac{1}{1000}\right) = \frac{0.7576}{204.23}$$

ARITHMETIC

$$x = \left(\frac{1000}{42.05}\right) \cdot \left(\frac{0.7576}{204.23}\right) = 0.08822\,\frac{\text{moles}}{\text{liter}}$$

18. $2HCl + Na_2CO_3 \rightleftharpoons 2NaCl + H_2CO_3$

or

$$2H^+ + CO_3^{2-} \rightleftharpoons H_2CO_3$$

(Number of moles of HCl)	(Number of moles of Na_2CO_3)
2	1
4	2
6	3

(number of moles of HCl) = (2)(number of moles of Na_2CO_3)

Let x = molarity of HCl.

$$(\text{number of moles of HCl}) = \left(x\,\frac{\text{moles}}{\text{liter}}\right)(40.36\,\text{milliliters})\left(\frac{1\,\text{liter}}{1000\,\text{milliliters}}\right)$$

$$(\text{number of moles of } Na_2CO_3) = \frac{(0.2345\,\text{grams of } Na_2CO_3)}{\left(105.99\,\dfrac{\text{grams of } Na_2CO_3}{\text{mole of } Na_2CO_3}\right)}$$

$$(x)(40.36)\left(\frac{1}{1000}\right) = (2)\left(\frac{0.2345}{105.99}\right)$$

$$x = \left(\frac{1000}{40.36}\right) \cdot \left(\frac{(2)\,(0.2345)}{(105.99)}\right) = 0.1096 \frac{\text{moles}}{\text{liter}}$$

19.

$$\%\ \text{Fe} = \frac{(\text{grams of Fe})}{(\text{grams of sample})} \times 100\%$$

(grams of Fe) — unknown
(grams of sample) — known

Find the number of grams of iron.

$$Cr_2O_7^{2-} + 6Fe^{2+} + 14H^+ \rightleftharpoons 2Cr^{3+} + 6Fe^{3+} + 7H_2O$$

(Number of Moles of Fe)	(Number of Moles of $Cr_2O_7^{2-}$) = (Number of Moles of $K_2Cr_2O_7$)
6	1
12	2
18	3

(number of moles of Fe) = (6)(number of moles of $K_2Cr_2O_7$)

Let x = (number of grams of Fe in the sample).

$$(\text{number of moles of Fe}) = \frac{x\ \text{grams of Fe}}{55.847\ \dfrac{\text{grams of Fe}}{\text{mole of Fe}}}$$

$$(\text{number of moles of } K_2Cr_2O_7) = (26.87\ \text{milliliters})\left(0.01524\ \frac{\text{moles}}{\text{liter}}\right)\left(\frac{1\ \text{liter}}{1000\ \text{milliliters}}\right)$$

$$\frac{x}{55.847} = (6)(26.87)(0.01524)\left(\frac{1}{1000}\right)$$

$$x = (55.847)(6)(26.87)(0.01524)\left(\frac{1}{1000}\right) = 0.1372\ \text{grams}$$

$$\%\ \text{Fe} = \frac{(\text{grams of Fe})}{(\text{grams of sample})} \times 100\%$$

$$= \frac{0.1372}{0.5285} \times 100\% = 25.96\%$$

20. (number of equivalents of analyte) = (number of equivalents of titrant)

(number of equivalents of Fe) = (number of equivalents of MnO_4^-)

$$= (N_{MnO_4^-})(\text{number of milliliters of } MnO_4^-)$$

$$\left(\frac{1\ \text{liter}}{1000\ \text{milliliters}}\right)$$

$$= (0.1012)(24.22)\left(\frac{1}{1000}\right)$$

$$= 2.451 \times 10^{-3} \text{ equivalent of Fe}$$

(number of equivalents) = (number of moles)(number of electrons in half-reaction)

$$\text{(number of moles)} = \frac{\text{(number of equivalents)}}{\text{(number of electrons in half-reaction)}}$$

Because

$$Fe^{3+} + e^- \rightleftharpoons Fe^{2+}$$

$$\text{(number of moles of Fe)} = \frac{\text{(number of equivalents of Fe)}}{1}$$

$$= 2.451 \times 10^{-3} \text{ moles of Fe}$$

Now, as before, find the number of grams of Fe.

Let x = (number of grams of Fe)

$$\frac{x \text{ grams of Fe}}{55.847 \frac{\text{grams of Fe}}{\text{mole of Fe}}} = \text{(number of moles of Fe)}$$

$$x = (55.847)(\text{number of moles of Fe})$$

$$= (55.847)(2.451 \times 10^{-3}) = 1.369 \times 10^{-1} \text{ gram of Fe}$$

$$\% \text{ Fe} = \frac{\text{(grams of Fe)}}{\text{(grams of sample)}} \times 100\% = \frac{1.369 \times 10^{-1}}{0.6282} \times 100\% = 21.79\% \text{ Fe}$$

21. The problem is easily solved by a simple proportioning approach. The question to ask is this: if there are 0.58 gram of tryptophan per 100 grams of protein, how many grams of protein are there for each 204 grams (1 mole) of tryptophan? If we assume at least 1 mole (204 grams) of tryptophan per mole of protein, the answer would be the minimum molecular weight of the protein.

Let x = number of grams of protein per 204 grams of tryptophan

$$\frac{x}{204} = \frac{100}{0.58}$$

ARITHMETIC

$$x = (204)\left(\frac{100}{0.58}\right) = 3.5 \times 10^4 \text{ grams per mole}$$

22. This problem involves coupled chemical reactions.

$$4I^- + 2Cu^{2+} \rightarrow 2CuI + I_2$$

$$I_2 + 2S_2O_3^{2-} \rightarrow S_4O_6^{2-} + 2I^-$$

There are two easy ways to solve the problem.

SOLUTION
METHOD 1

$$4I^- + 2Cu^{2+} \rightarrow 2CuI + I_2$$

$$I_2 + 2S_2O_3^{2-} \rightarrow S_4O_6^{2-} + 2I^-$$

(number of moles of Cu) = 2(number of moles of I_2)

(number of moles of I_2) = ½(number of moles $S_2O_3^{2-}$)

hence

2(number of moles I_2) = (number of moles $S_2O_3^{2-}$)

or

(number of moles of Cu) = (number of moles $S_2O_3^{2-}$)

The next method will yield the same results.

SOLUTION
METHOD 2

$$+ \begin{cases} 4I^- + 2Cu^{2+} \rightarrow 2CuI + I_2 \\ I_2 + 2S_2O_3^{2-} \rightarrow S_4O_6^{2-} + 2I^- \end{cases}$$

$$4I^- + 2Cu^{2+} + 2S_2O_3^{2-} \rightarrow 2CuI + S_4O_6^{2-} + 2I^-$$

(number of moles of Cu) = (number of moles of $S_2O_3^{2-}$)

Using the fact that,

(number of moles of Cu) = (number of moles of $S_2O_3^{2-}$)

Let x = the number of grams of Cu in a sample.

$$\text{(number of moles of Cu)} = \frac{x \text{ grams of Cu}}{63.54 \text{ grams of Cu}}$$
$$\text{mole of Cu}$$

$$\text{(number of moles of } S_2O_3^{2-}) = \left(0.1034 \frac{\text{moles of } S_2O_3^{2-}}{\text{liter}}\right)(40.74 \text{ milliliters})\left(\frac{(1 \text{ liter})}{1000 \text{ milliliters}}\right)$$

$$\frac{x}{63.54} = (0.1034)(40.74)\left(\frac{1}{1000}\right)$$

ARITHMETIC

$$x = (63.54)(0.1034)(40.74)\left(\frac{1}{1000}\right) = 0.2677 \text{ gm Cu}$$

$$\% \text{ Cu} = \frac{\text{(gms Cu)}}{\text{(gms sample)}} \times 100\%$$

$$= \frac{0.2677}{0.7627} \times 100\% = 35.10\%$$

chapter 2
Acid-Base Chemistry

≡ 2.1 GENERAL DISCUSSION ≡

Although there are many theories of acid-base chemistry, each has its own particular usefulness. The theory allowing the easiest mathematical treatment is called the Brønsted-Lowry theory and depends upon definitions of acids and bases in terms of proton transfer. An acid is defined as a proton donor; a base is defined as a proton acceptor.

Example: Acetic acid, CH_3COOH, can donate protons to water and other species. Hence, it is an acid. Water acts as a base in the following reaction:

$$CH_3COOH + H_2O \rightleftharpoons CH_3COO^- + H_3O^+$$

acid	base	the acetate ion	the hydronium ion
		(conjugate base)	(conjugate acid)

Ammonia, NH_3, can accept protons from water and other species. Hence it is a base. Water acts as an acid in the following reaction:

$$NH_3 + H_2O \rightleftharpoons NH_4^+ + OH^-$$

base	acid	the ammonium ion	the hydroxide ion
		(conjugate acid)	(conjugate base)

Water, with its ability to accept or donate protons, that is, its ability to act as either acid or base, is called an *amphiprotic species* or an *ampholyte*. There are many other ampholytic species besides water; their chemistry will be examined later.

Most often, the acid reaction of a species like acetic acid with water,

$$CH_3COOH + H_2O \rightleftharpoons CH_3COO^- + H_3O^+$$

is written in shortened form:

$$CH_3COOH \rightleftharpoons CH_3COO^- + H^+$$

The presence of water is understood. H^+, the symbol for a proton, stands for the hydronium ion, H_3O^+.

How may hydrogen ion concentrations be described quantitatively? An equilibrium constant must be written. Take the chemical reaction

$$aA + bB \rightleftharpoons cC + dD$$

The $\rightleftharpoons$ mark means that the reactants and products are at equilibrium so that, unless the system is disturbed, the net concentrations of products and reactants remain unchanged. A disturbance could be dilution, the addition of products or reactants to the reaction vessel, or a change in temperature. The reaction may be read in this way: "The number a moles of species A react with the number b moles of species B to yield c moles of species C and d moles of species D. Products and reactants are in equilibrium."

The equilibrium constant for the reaction is written this way when concentrations are small:

$$K = \frac{[C]^c \cdot [D]^d}{[A]^a \cdot [B]^b}$$

For any solution species, say X, [X] means the concentration of X, and only X, in moles per liter, at equilibrium. [X] is called the *equilibrium concentration* of X.

Now write the reaction and equilibrium constant for the acid reaction of acetic acid with water:

$$CH_3COOH + H_2O \rightleftharpoons CH_3COO^- + H_3O^+$$

$$K = \frac{[H_3O^+][CH_3COO^-]}{[CH_3COOH][H_2O]}$$

Because $[H_2O]$ is a constant quantity, about $55.5M$, in dilute aqueous solutions

$$K[H_2O] = constant$$

Let

$$K_a = K[H_2O] = \frac{[H_3O^+][CH_3COO^-]}{[CH_3COOH]}$$

or

$$K_a = \frac{[H^+][CH_3COO^-]}{[CH_3COOH]}$$

The constant K_a is called the *acid dissociation constant*, the *dissociation constant*, or the *ionization constant*.

Water, too, undergoes acid-base reactions. The one that *everyone ought to remember, together with its equilibrium constant*, is called the *autoprotolysis* (self-proton-splitting) or *autoionization* (self-ionization) reaction.

Long form: $\quad H_2O + H_2O \rightleftharpoons H_3O^+ + OH^- \quad K_w = [H_3O^+][OH^-]$

$[H_2O]$ is taken as constant when water is the solvent.

Shortened form: $\qquad H_2O \rightleftharpoons H^+ + OH^- \qquad K_w = [H^+][OH^-]$ $\qquad\qquad$ (2.1)

The constant, K_w, has a value that is easy to remember. At $25°C$, where most solution equilibrium constants are measured,

$$K_w = 1.00 \times 10^{-14}$$

There are important conclusions that can be drawn from the fact of the constancy of K_w: (1) if $[H^+]$ is known, then $[OH^-]$ can be calculated; and (2) if $[OH^-]$ is known, then $[H^+]$ can be calculated.

Examples: Calculate $[OH^-]$ when $[H^+] = 1.0, 1.0 \times 10^{-7}$, and $1.0 \times 10^{-14} M$.

$$K_w = [H^+][OH^-]$$

$$K_w = 1.00 \times 10^{-14}$$

$$[OH^-] = \frac{K_w}{[H^+]}$$

$$[OH^-] = \frac{1.00 \times 10^{-14}}{[H^+]}$$

where $[H^+] = 1.0M$,

$$[OH^-] = \frac{1.00 \times 10^{-14}}{1.0} = 1.0 \times 10^{-14} M$$

where $[H^+] = 1.0 \times 10^{-7} M$,

$$[OH^-] = \frac{1.00 \times 10^{-14}}{10^{-7}} = 1.0 \times 10^{-7} M$$

where $[H^+] = 1.0 \times 10^{-14} M$,

$$[OH^-] = \frac{1.00 \times 10^{-14}}{1.0 \times 10^{-14}} = 1.0 M$$

Examples: Calculate $[H^+]$ when $[OH^-] = 1.0, 1.0 \times 10^{-7}$, and $1.0 \times 10^{-14} M$.

$$K_w = [H^+][OH^-]$$

$$K_w = 1.00 \times 10^{-14}$$

$$[H^+] = \frac{K_w}{[OH^-]}$$

$$[H^+] = \frac{1.00 \times 10^{-14}}{[OH^-]}$$

where $[OH^-] = 1.0M$,

$$[H^+] = \frac{1.00 \times 10^{-14}}{1.0}$$

$$= 1.0 \times 10^{-14} M$$

where $[OH^-] = 1.0 \times 10^{-7} M$,

$$[H^+] = \frac{1.00 \times 10^{-14}}{1.0 \times 10^{-7}} = 1.0 \times 10^{-7} M$$

where $[OH^-] = 1.0 \times 10^{-14} M$,

$$[H^+] = \frac{1.00 \times 10^{-14}}{1.0 \times 10^{-14}} = 1.0 M$$

A range of $[H^+]$-values from $1M$ to $10^{-14} M$ covers almost all of the values of biochemical and analytical significance. It is awkward, though, to write down an exponential number every time some representation of a widely variable $[H^+]$-value is desired. It is also awkward to try to graph, say, $[H^+]$-values running from 10^{-3} to $10^{-12} M$, as they may run in a titration. Potentiometric instruments for measuring acidity respond logarithmically to hydrogen ion concentrations. The logarithmic representation of hydrogen ion concentration was proposed in 1909 by Sørensen, a biochemist working in Copenhagen. The symbol used is pH (from "puissance d'hydrogen" or hydrogen strength. Sørensen's article was written in French). The definition of pH given by Sørensen, and the one to be used throughout most of this book, is

$$pH = -\log[H^+] \tag{2.2}$$

or

$$pH = \log \frac{1}{[H^+]}$$

[Remember that $-\log x = \log (1/x)$.]

Example: Calculate the pH of a solution where $[H^+] = 5.0 \times 10^{-7} M$.

$$pH = -\log [H^+] = -\log (5.0 \times 10^{-7}) = -\log 5.0 - \log 10^{-7}$$
$$= -0.70 + 7.00 = 6.30$$

Example: Calculate the value of $[H^+]$ in a solution whose pH is measured as 4.70.

$$pH = -\log [H^+]$$
$$4.70 = -\log [H^+]$$
$$-4.70 = +\log [H^+]$$
$$0.30 - 5.00 = \log [H^+]$$
$$antilog (0.30 - 5.00) = [H^+]$$
$$[antilog(0.30)] \cdot [antilog(-5.00)] = [H^+]$$
$$2.0 \times 10^{-5} M = [H^+]$$

A handy relationship that is easily remembered may be derived very simply. A few definitions are in order. Let

$$p anything = -\log anything$$
$$pK_w = -\log K_w$$
$$pOH = -\log [OH^-]$$

Then

$$[H^+]\,[OH^-] \;=\; K_w \tag{2.1}$$

$$\log([H^+]\cdot[OH^-]) \;=\; \log K_w$$

$$\log[H^+] \;+\; \log[OH^-] \;=\; +\,\log K_w$$

$$-\log[H^+] \;-\; \log[OH^-] \;=\; -\log K_w$$

$$pH \;+\; pOH \;=\; pK_w$$

$$pK_w \;=\; -\log K_w \;=\; -\log(1.00 \times 10^{-14}) \;=\; 14.00$$

$$pH \;+\; pOH \;=\; 14.00 \tag{2.3}$$

$\equiv$ 2.2 STRONG ACIDS AND BASES $\equiv$

One is often faced with problems of the following kind:

¤ **(a)** Given that a solution is made by diluting 0.1000 mole of HCl to 1.000 liter with water, calculate its pH.

¤ **(b)** Given that a solution is made by diluting 0.1000 mole of CH_3COOH to 1.000 liter with water, calculate its pH.

The problems look alike. The algebra used in the solution of the first one is very simple; much more thought is required for the solution of the second. The reason for this disparity in ease of solution is that problem (a) concerns a *strong acid*, while problem (b) concerns a *weak acid*.

2.3 Strong Acids

A strong acid, such as HCl, is one that very readily gives away protons. In water, the reaction

$$HCl + H_2O \;\rightleftharpoons\; H_3O^+ + Cl^- \quad \text{(long form)}$$

or

$$HCl \;\rightleftharpoons\; H^+ + Cl^- \quad \text{(shortened form)}$$

is almost complete. The equilibrium lies far to the right, and the value of K_a

$$K_a \;=\; \frac{[H^+]\,[Cl^-]}{[HCl]}$$

is almost infinitely large, since the value of [HCl] is near zero. Strong acids are said to be completely ionized, or completely dissociated. They are not many in number; the more common ones are HCl, HBr, HI, $HClO_4$, and HNO_3.

A weak acid, such as CH_3COOH, acetic acid, is one that does not readily give away protons. In water, the reaction

Long form:

$$CH_3COOH + H_2O \rightleftharpoons H_3O^+ + CH_3COO^-$$

or

Shortened form:

$$CH_3COOH \rightleftharpoons H^+ + CH_3COO^-$$

does not proceed very far. The equilibrium lies far to the left, and the value of K_a

$$K_a = \frac{[H^+][CH_3COO^-]}{[CH_3COOH]} \cong 1.8 \times 10^{-5}$$

is very small, because $[H^+]$ and $[CH_3COO^-]$ are both small and $[CH_3COOH]$ is very large. Weak acids are said to be only partially ionized, or incompletely dissociated. They are also said to be weak electrolytes, for incomplete dissociation provides few ions to conduct electricity.

The terms *strong acid* and *concentrated acid* must not be confused with each other, nor must the terms *weak acid* and *dilute acid* be confused with each other. A concentrated acid is one made up of many moles per liter of either a strong or a weak acid. Examples of concentrated acids are:

HCl	CH_3COOH
10*M*	10*M*
concentrated strong acid	concentrated weak acid

A dilute acid is one made up of very few moles per liter of either a strong or a weak acid. Examples of dilute acid are

HCl	CH_3COOH
$10^{-4}M$	$10^{-4}M$
dilute strong acid	dilute weak acid

Figuring the pH of solutions of strong acids is easy. Given the analytical concentration of a strong acid, the process is almost trivial.

$$\text{analytical concentration of strong acid HCl} = C_{HCl}$$

C_{HCl} = number of moles per liter of HCl added to solution. This is the number of moles per liter of HCl that would be present in solution if *no* dissociation had occurred.

Two equilibria are important in the dissociation of HCl in H_2O.

$$HCl \rightleftharpoons H^+ + Cl^- \quad \text{(almost complete)}$$
$$H_2O \rightleftharpoons H^+ + OH^- \quad \text{(very small dissociation)}$$

Because of the one-to-one stoichiometry of the first reaction:

$$C_{HCl} = [Cl^-]$$
$$[Cl^-] = [H^+] \quad \text{from HCl}$$

There are two sources of hydrogen ion in the solution: the dissociation of HCl (usually the larger source) and the dissociation of H_2O (usually the smaller source by far).

$$[H^+] = (H^+) \text{ from HCl} + (H^+) \text{ from } H_2O$$

or

$$[H^+] = [Cl^-] + (H^+) \quad \text{from } H_2O$$

or

$$[H^+] = C_{HCl} + (H^+) \quad \text{from } H_2O$$

Because of the one-to-one stoichiometry of the dissociation of water:

$$(H^+) \text{ from } H_2O = [OH^-]$$

$$[H^+] = C_{HCl} + [OH^-]$$

$$K_w = [H^+][OH^-]$$

$$[OH^-] = \frac{K_w}{[H^+]}$$

$$[H^+] = C_{HCl} + \frac{K_w}{[H^+]} \tag{2.4}$$

Equation 2.4 is valid. It is also a quadratic equation, which is not always easy to solve. A method of successive approximations is easier to use.

First approximation: Let

$$[H^+]' = C_{HCl}$$

Second approximation:

$$[H^+]'' = C_{HCl} + \frac{K_w}{[H^+]'}$$

Third approximation:

$$[H^+]''' = C_{HCl} + \frac{K_w}{[H^+]''}$$

and so on.

Example: Calculate the pH of a solution where $C_{HCl} = 0.10M$.

First approximation:

$$[H^+]' = C_{HCl} = 1.0 \times 10^{-1}M$$

Second approximation:

$$[H^+]'' = C_{HCl} + \frac{K_w}{[H^+]'} = 1.0 \times 10^{-1} + \frac{1.00 \times 10^{-14}}{1.0 \times 10^{-1}} = 1.0 \times 10^{-1}$$

$$[H^+]' = [H^+]''$$

The approximations have converged.

$$pH = -\log [H^+] = -\log (1.0 \times 10^{-1}) = 1.00$$

Example: Calculate the pH of a solution $1.0 \times 10^{-7}M$ in HCl.

First approximation:

$$[H^+]' = C_{HCl}$$

$$[H^+]' = 1.0 \times 10^{-7} M$$

Second approximation:

$$[H^+]'' = C_{HCl} + \frac{K_w}{[H^+]'} = 1.0 \times 10^{-7} + \frac{1.00 \times 10^{-14}}{1.0 \times 10^{-7}}$$

$$= 1.0 \times 10^{-7} + 1.0 \times 10^{-7} M = 2.0 \times 10^{-7} M$$

Third approximation:

$$[H^+]''' = C_{HCl} + \frac{K_w}{[H^+]''} = 1.0 \times 10^{-7} + \frac{1.00 \times 10^{-14}}{2.0 \times 10^{-7}}$$

$$= 1.0 \times 10^{-7} + 5.0 \times 10^{-8} = 1.5 \times 10^{-7} M$$

Fourth approximation:

$$[H^+]'''' = C_{HCl} + \frac{K_w}{[H^+]'''} = 1.0 \times 10^{-7} + \frac{1.00 \times 10^{-14}}{1.5 \times 10^{-7}}$$

$$= 1.0 \times 10^{-7} + 7.0 \times 10^{-8} = 1.7 \times 10^{-7} M$$

Fifth approximation:

$$[H^+]^{v} = C_{HCl} + \frac{K_w}{[H^+]''''} = 1.0 \times 10^{-7} + \frac{1.00 \times 10^{-14}}{1.7 \times 10^{-7}}$$

$$= 1.0 \times 10^{-7} + 6.0 \times 10^{-8} = 1.6 \times 10^{-7} M$$

$[H^+]^{v}$ and $[H^+]''''$ are close. The answer is between 1.6×10^{-7} and $1.7 \times 10^{-7} M$.

The many approximations could have been replaced with a single quadratic equation. Generally, solving a quadratic is justified when: (a) many approximations must be done; or (b) when an absurd, that is, negative or imaginary, answer appears.

There is a general solution for all quadratic equations. The equation is first changed to the form of

$$ax^2 + bx + c = 0 \tag{2.5}$$

The terms a, b, and c are constants. The term x is the variable, or unknown.

$$x = \frac{-b \pm \sqrt{b^2 - 4ac}}{2a} \tag{2.6}$$

When a negative answer is not possible,

$$x = \frac{-b + \sqrt{b^2 - 4ac}}{2a}$$

$$[H^+] = C_{HCl} + \frac{K_w}{[H^+]} \tag{2.4}$$

$$[H^+]^2 = C_{HCl} \cdot [H^+] + K_w$$

$$[H^+]^2 - C_{HCl}[H^+] - K_w = 0$$

In the example just discussed,

$$a = 1$$

$$b = -C_{HCl} = -1.0 \times 10^{-7} M$$

$$c = -K_w = -1.00 \times 10^{-14}$$

$$x = [H^+]$$

$$[H^+] = \frac{C_{HCl} + [C^2_{HCl} - 4(-K_w)]^{1/2}}{2} = \frac{+ C_{HCl} + (C^2_{HCl} + 4K_w)^{1/2}}{2}$$

$$= \frac{1.0 \times 10^{-7} + [(1.0 \times 10^{-7})^2 + 4.00 \times 10^{-14}]^{1/2}}{2} \qquad (2.7)$$

$$= \frac{10^{-7} + (5.0 \times 10^{-14})^{1/2}}{2} = \frac{3.2 \times 10^{-7}}{2}$$

$$[H^+] = 1.6 \times 10^{-7}$$

$$pH = -\log[H^+] = -\log(1.6 \times 10^{-7}) = 6.80$$

2.4 Strong Bases

The hydroxide ion is the strongest base in water solutions. Solutions of the alkali hydroxides, NaOH, KOH, and LiOH are strong base solutions. The dissolution and dissociation of NaOH in water are almost complete:

$$NaOH \rightleftharpoons Na^+ + OH^-$$

Given the analytical concentration of a strong base, figuring pH is easy.

$$\text{analytical concentration of strong base NaOH} = C_{NaOH}$$

$$C_{NaOH} = \text{number of moles per liter of NaOH added to solution}$$

This is the number of moles per liter of NaOH that would be present in the solution if NaOH did *not* dissociate.

Consider the dissolution of NaOH in H_2O. Two equilibria are important.

$$NaOH \rightleftharpoons Na^+ + OH^- \quad \text{(almost complete)}$$
$$H_2O \rightleftharpoons H^+ + OH^- \quad \text{(very small dissociation)}$$

Because of the one-to-one stoichiometry of the first reaction,

$$C_{NaOH} = [Na^+]$$
$$[Na^+] = (OH^-) \quad \text{from NaOH}$$

There are two sources of hydroxide ion in the solution: the dissolution and dissociation of NaOH (usually the larger source) and the dissociation of H_2O (usually the smaller source by far).

$$[OH^-] = (OH^-) \text{ from NaOH} + (OH^-) \text{ from } H_2O$$

or

$$= [Na^+] + (OH^-) \text{ from } H_2O$$

or

$$= C_{NaOH} + (OH^-) \text{ from } H_2O$$

Because of the one-to-one stoichiometry of the dissociation of water:

$$(OH^-) \text{ from } H_2O = [H^+]$$

$$[OH^-] = C_{NaOH} + [H^+]$$

$$K_w = [H^+][OH^-]$$

$$[H^+] = \frac{K_w}{[OH^-]}$$

$$[OH^-] = C_{NaOH} + \frac{K_w}{[OH^-]} \tag{2.8}$$

Equation 2.8 is valid. It is also a quadratic equation, which is not always easy to solve. A method of successive approximations is easier to use.

First approximation: Let

$$[OH^-]' = C_{NaOH}$$

Second approximation:

$$[OH^-]'' = C_{NaOH} + \frac{K_w}{[OH^-]'}$$

Third approximation:

$$[OH^-]''' = C_{NaOH} + \frac{K_w}{[OH^-]''}$$

and so on.

Example: Calculate the pH of a solution where $C_{NaOH} = 0.10M$.

First approximation:

$$[OH^-]' = C_{NaOH}$$

$$= 1.0 \times 10^{-1}M$$

Second approximation:

$$[OH^-]'' = C_{NaOH} + \frac{K_w}{[OH^-]'}$$

$$= 1.0 \times 10^{-1} + \frac{1.00 \times 10^{-14}}{1.0 \times 10^{-1}}$$

$$= 1.0 \times 10^{-1}M$$

$$[OH^-]' = [OH^-]''$$

The approximations have converged.

$$pOH = -\log [OH^-] = -\log (1.0 \times 10^{-1}) = 1.00$$

$$pH + pOH = 14.00$$

$$pH = 14.00 - pOH = 14.00 - 1.00 = 13.00$$

Example: Calculate the pH of a solution $1.0 \times 10^{-7} M$ in NaOH.

First approximation:

$$[OH^-]' = C_{NaOH}$$

$$[OH^-]' = 1.0 \times 10^{-7} M$$

Second approximation:

$$[OH^-]'' = C_{NaOH} + \frac{K_w}{[OH^-]'} = 1.0 \times 10^{-7} + \frac{1.00 \times 10^{-14}}{1.0 \times 10^{-7}} = 2.0 \times 10^{-7} M$$

Third approximation:

$$[OH^-]''' = C_{NaOH} + \frac{K_w}{[OH^-]''} = 1.0 \times 10^{-7} + \frac{1.00 \times 10^{-14}}{2.0 \times 10^{-7}}$$

$$= 1.0 \times 10^{-7} + 5.0 \times 10^{-8} = 1.5 \times 10^{-7} M$$

Fourth approximation:

$$[OH^-]'''' = C_{NaOH} + \frac{K_w}{[OH^-]'''} = 1.0 \times 10^{-7} + \frac{1.00 \times 10^{-14}}{1.5 \times 10^{-7}}$$

$$= 1.0 \times 10^{-7} + 7.0 \times 10^{-8} = 1.7 \times 10^{-7} M$$

Fifth approximation:

$$[OH^-]^{\prime\prime\prime\prime\prime} = C_{NaOH} + \frac{K_w}{[OH^-]''''} = 1.0 \times 10^{-7} + \frac{1.00 \times 10^{-14}}{1.7 \times 10^{-7}}$$

$$= 1.0 \times 10^{-7} + 6.0 \times 10^{-8} = 1.6 \times 10^{-7} M$$

$[OH^-]^{\prime\prime\prime\prime\prime}$ and $[OH^-]''''$ are close. The answer is between 1.6×10^{-7} and $1.7 \times 10^{-7} M$.

The many approximations could have been replaced with a single quadratic equation.

$$[OH^-] = C_{NaOH} + \frac{K_w}{[OH^-]} \tag{2.8}$$

$$[OH^-]^2 = C_{NaOH} \cdot [OH^-] + K_w$$

$$[OH^-]^2 - C_{NaOH} \cdot [OH^-] - K_w = 0$$

$$[OH^-] = \frac{C_{NaOH} + [C^2_{NaOH} - 4(-K_w)]^{1/2}}{2}$$

$$[OH^-] = \frac{1.0 \times 10^{-7} + [(1.0 \times 10^{-7})^2 + (4.00 \times 10^{-14})]^{1/2}}{2}$$

$$[OH^-] \simeq 1.6 \times 10^{-7} M$$

$$pOH = -\log [OH^-] = -\log (1.6 \times 10^{-7}) = 6.80$$

$$pH + pOH = 14.00$$

$$pH = 14.00 - 6.80 = 7.20$$

1. Calculate $[H^+]$ in water solutions where $[OH^-]$ is
   (a) $1.00 \times 10^{-4}M$.
       Ans: *$1.00 \times 10^{-10}M$*
   (b) $2.00 \times 10^{-4}M$.
       Ans: *$5.00 \times 10^{-11}M$*
   (c) $5.00 \times 10^{-7}M$
       Ans: *$2.00 \times 10^{-8}M$*

2. Calculate pH in solutions when $[H^+]$ is
   (a) $1.0 \times 10^{-8}M$.
       Ans: *8.00*
   (b) $3.0 \times 10^{-8}M$.
       Ans: *7.52*
   (c) $2.9 \times 10^{-12}M$.
       Ans: *11.54*

3. Calculate pH and pOH in solutions where $[H^+]$ is
   (a) $5.0 \times 10^{-9}M$.
       Ans: *pH = 8.30; pOH = 5.70*
   (b) $2.3 \times 10^{-2}M$.
       Ans: *pH = 1.64; pOH = 12.36*
   (c) $6.1 \times 10^{-12}M$.
       Ans: *pH = 11.21; pOH = 2.79*

4. Calculate the value of $[H^+]$ in solutions where pH is
   (a) 5.00.
       Ans: *$1.0 \times 10^{-5}M$*
   (b) 4.30.
       Ans: *$5.0 \times 10^{-5}M$*
   (c) 12.80.
       Ans: *$1.6 \times 10^{-13}M$*

5. Calculate the pH of solutions:
   (a) $1.0 \times 10^{-1}M$ in $HClO_4$.
       Ans: *1.00*
   (b) $1.0 \times 10^{-4}M$ in HNO_3.
       Ans: *4.00*

6. Calculate the pH of a solution $1.0 \times 10^{-7}M$ in HI.
   Ans: *6.80*

7. Calculate the pH of solutions:
   (a) $1.0 \times 10^{-3}M$ in KOH.
       Ans: *11.00*

(b) 1.0 $\times$ $10^{-5}M$ in KOH.

     Ans: *9.00*

8. Calculate the pH of a solution made by dissolving 0.20 grams of HNO_3 in water, and diluting to a total volume of 500.0 milliliters.

    Ans: *2.20*

9. Calculate the pH of a solution made by diluting 10.0 milliliters of 0.094M NaOH to 500.0 milliliters with water.

    Ans: *11.28*

$\equiv$ 2.6 DETAILED SOLUTIONS TO PROBLEMS $\equiv$

1. For all three concentrations listed, one approach will work:

$$K_w = [H^+][OH^-]$$

$$[H^+] = \frac{K_w}{[OH^-]}$$

(a) $[H^+] = \dfrac{1.00 \times 10^{-14}}{1.00 \times 10^{-4}} = 1.00 \times 10^{-10}M$

(b) $[H^+] = \dfrac{1.00 \times 10^{-14}}{2.00 \times 10^{-4}} = 5.00 \times 10^{-11}M$

(c) $[H^+] = \dfrac{1.00 \times 10^{-14}}{5.00 \times 10^{-7}} = 2.00 \times 10^{-8}M$

2. For all three concentrations listed, one approach will work:

$$pH = -\log[H^+]$$

(a) $pH = -\log(1.0 \times 10^{-8}) = 8.00$

(b) $pH = -\log(3.0 \times 10^{-8}) = 7.52$

(c) $pH = -\log(2.9 \times 10^{-12}) = 11.54$

3. For all the concentrations listed, one approach will work:

$$\text{figure pH:} \quad pH = -\log[H^+]$$
$$pH + pOH = 14.00$$
$$pOH = 14.00 - pH$$

(a) $pH = -\log(5.0 \times 10^{-9}) = 8.30$

    $pOH = 14.00 - 8.30 = 5.70$

(b) $pH = -\log(2.3 \times 10^{-2}) = 1.64$

    $pOH = 14.00 - 1.64 = 12.36$

(c) $pH = -\log(6.1 \times 10^{-12}) = 11.21$

    $pOH = 14.00 - pH = 2.79$

4. For all three solutions, remember the definition of pH;

$$pH = -\log [H^+]$$
$$-pH = \log [H^+]$$
$$antilog(-pH) = [H^+]$$

(a)
$$5.00 = -\log [H^+]$$
$$-5.00 = \log [H^+]$$
$$antilog(-5.00) = [H^+]$$
$$1.0 \times 10^{-5} = [H^+]$$

(b)
$$4.30 = -\log [H^+]$$
$$-4.30 = \log [H^+]$$
$$+0.70-5.00 = \log [H^+]$$
$$antilog(+0.70 - 5.00) = [H^+]$$
$$[antilog(0.70)] \cdot [antilog(-5.00)] = [H^+]$$
$$5.0 \times 10^{-5} = [H^+]$$

(c)
$$12.80 = -\log [H^+]$$
$$-12.80 = \log [H^+]$$
$$+0.20 - 13.00 = \log[H^+]$$
$$antilog(0.20 - 13.00) = [H^+]$$
$$[antilog(0.20)] \cdot [antilog(-13.00)] = [H^+]$$
$$1.6 \times 10^{-13} = [H^+]$$

5. One approach will work for both solutions:

$$[H^+] = C_{acid} + \frac{K_w}{[H^+]}$$

First approximation:

$$[H^+]' = C_{acid}$$

Second approximation:

$$[H^+]'' = C_{acid} + \frac{K_w}{[H^+]'}$$

When the approximations converge:

$$pH = -\log[H^+]$$

(a) $[H^+]' = 1.0 \times 10^{-1}$

$$[H^+]'' = 1.0 \times 10^{-1} + \frac{1.0 \times 10^{-14}}{1.0 \times 10^{-1}}$$

$$= 1.0 \times 10^{-1} + 10^{-13} \qquad = 1.0 \times 10^{-1}$$

The approximations have converged.

$$pH = -\log (1.0 \times 10^{-1}) \qquad = 1.00$$

(b) $[H^+]' = 1.0 \times 10^{-4}$

$$[H^+]'' = 1.0 \times 10^{-4} + \frac{1.00 \times 10^{-14}}{1.0 \times 10^{-4}} = 1.0 \times 10^{-4} + \cancel{10^{-10}} = 1.0 \times 10^{-4}M$$

The approximations have converged.

$$pH = -\log(1.0 \times 10^{-4})$$
$$= \cancel{4}.00$$

6. This problem is best solved with a quadratic equation.

$$[H^+] = C_{HI} + \frac{K_w}{[H^+]}$$

$$[H^+]^2 = C_{HI} \cdot [H^+] + K_w$$

$$[H^+]^2 - C_{HI} \cdot [H^+] - K_w = 0$$

$$[H^+] = \frac{C_{HI} + (C_{HI}^2 + 4K_w)^{1/2}}{2}$$

$$= \frac{1.0 \times 10^{-7} + [(1.0 \times 10^{-7})^2 + 4.00 \times 10^{-14}]^{1/2}}{2} = 1.6 \times 10^{-7}M$$

$$pH = -\log(1.6 \times 10^{-7}) = 6.80$$

7. The same approach will work for both solutions:

$$[OH^-] = C_{KOH} + \frac{K_w}{[OH^-]}$$

First approximation:

$$[OH^-]' = C_{KOH}$$

Second approximation:

$$[OH^-]'' = [OH^-]' + \frac{K_w}{[OH^-]'}$$

When the approximations converge:

$$pOH = -\log[OH^-]$$
$$pH = 14.00 - pOH$$

(a) $[OH^-]' = 1.0 \times 10^{-3}M$

$$[OH^-]'' = 1.0 \times 10^{-3} + \frac{1.00 \times 10^{-14}}{1.0 \times 10^{-3}} = 1.0 \times 10^{-3} + \cancel{10^{-11}} = 1.0 \times 10^{-3}M$$

The approximations have converged.

$$pOH = -\log(1.0 \times 10^{-3}) = 3.00$$
$$pH = 14.00 - 3.00 = 11.00$$

(b) $[OH^-]' = 1.0 \times 10^{-5}M$

$$[OH^-]'' = 1.0 \times 10^{-5} + \frac{1.00 \times 10^{-14}}{1.0 \times 10^{-5}}$$

$$[OH^-]'' = 1.0 \times 10^{-5} + \cancel{10^{-9}}$$

$$[OH^-]'' = 1.0 \times 10^{-5}M$$

The approximations have converged.

$$pOH = -\log (1.0 \times 10^{-5}) = 5.00$$

$$pH = 14.00 - 5.00 = 9.00$$

8. First calculate C_{HNO_3}:

$$\text{(number of moles of HNO}_3) = \frac{\text{(number of grams of HNO}_3)}{\text{(molecular weight of HNO}_3)}$$

$$= \frac{0.20}{63.01} = 3.2 \times 10^{-3} \text{ moles}$$

$$C_{HNO_3} = \frac{\text{(number of moles of HNO}_3)}{\text{(number of liters of solution)}} = \frac{3.2 \times 10^{-3} \text{ moles}}{5.0000 \times 10^{-1} l} = 6.3 \times 10^{-3} M$$

Then, begin the approximations.

$$[H^+]' = C_{HNO_3}$$

$$[H^+]'' = C_{HNO_3} + \frac{K_w}{[H^+]'}$$

$$[H^+]' = 6.3 \times 10^{-3}$$

$$[H^+]'' = 6.3 \times 10^{-3} + \frac{1.00 \times 10^{-14}}{6.3 \times 10^{-3}} = 6.3 \times 10^{-3} + \cancel{1.6 \times 10^{-12}} = 6.3 \times 10^{-3} M$$

$$[H^+]' = [H^+]''$$

The approximations have converged.

$$pH = -\log (6.3 \times 10^{-3}) = 2.20$$

9. First, calculate C_{NaOH}, remembering from stoichiometry that:

$$M_1 v_1 = M_2 v_2$$

$M_1 = 0.094M$

$v_1 = 10.0$ milliliters

$M_2 = C_{NaOH} = ?$

$v_2 = 500.0$ milliliters

$$M_2 = \frac{M_1 v_1}{v_2}$$

$$C_{NaOH} = \frac{M_1 v_1}{v_2} = \frac{(0.094)(10.0)}{(500.0)} = 1.9 \times 10^{-3} M$$

Then, begin the approximations.

$$[OH^-]' = C_{NaOH}$$

$$[OH^-]'' = C_{NaOH} + \frac{K_w}{[OH^-]'}$$

$$[OH^-]' = 1.9 \times 10^{-3} M = 1.9 \times 10^{-3} + \frac{1.00 \times 10^{-14}}{1.9 \times 10^{-3}} = 1.9 \times 10^{-3} + \cancel{5.3 \times 10^{-12}}$$

$$[OH^-]' = [OH^-]''$$

The approximations have converged.

$$pOH = -\log(1.9 \times 10^{-3}) = 2.72$$

$$pH = 14.00 - 2.72 = 11.28$$

≡ 2.7 WEAK MONOPROTIC ACIDS ≡

Weak acids outnumber strong acids in the number of species isolated or synthesized. Many of them are capable of donating more than one proton to basic species; these are called polyprotic acids. We shall be concerned in this section, however, only with monoprotic weak acids, those with only one donatable proton per molecule of acid. (This one proton may be called the replaceable proton, or the ionizable proton.) Polyprotic acids will be dealt with in detail later.

By our earlier definition, a weak acid is one which does not easily donate protons. This means that it is not easily ionized in water. Acetic acid, CH_3COOH, is an example of a monoprotic weak acid.

Long form: $CH_3COOH + H_2O \rightleftharpoons H_3O^+ + CH_3COO^-$

Shortened form: $CH_3COOH \rightleftharpoons H^+ + CH_3COO^-$

$$K_a = \frac{[H^+][CH_3COO^-]}{[CH_3COOH]} \simeq 1.8 \times 10^{-5}$$

2.8 Weak Acids Alone in Solution

Let us represent a monoprotic weak acid as HA, where H is a proton and A an anion with a -1 charge. Suppose that its K_a is on the order of 10^{-5}:

$$HA \rightleftharpoons H^+ + A^- \qquad K_a = \frac{[H^+][A^-]}{[HA]}$$

A problem most often stated is this:
Given C_{HA}, the value of the analytical concentration of HA and K_a, and that the analytical concentration of the anion, C_{A^-}, is zero (no added anion), calculate pH.

C_{HA} = (analytical concentration of HA)

C_{HA} = (concentration of moles per liter of HA added to solution. This is what the concentration of HA would be if no dissociation occurred.)

C_{A^-} = (concentration in moles per liter of A^- added to solution. This is what the concentration of A^- would be if any A^- had been added as, say, NaA) = 0.

To solve the problem of determining $[H^+]$ in this solution, begin with K_a.

$$K_a = \frac{[H^+][A^-]}{[HA]}$$

$$[H^+] = K_a \frac{[HA]}{[A^-]} \qquad\qquad (2.9)$$

If K_a is known, only [HA] and [A⁻] are needed before [H⁺] can be calculated.

The equilibrium concentrations of HA and A⁻ are easily calculated. Because the stoichiometry of the dissociation of HA is one-to-one,

$$HA \rightleftharpoons H^+ + A^-$$

and

$$[HA] = C_{HA} - \text{(concentration dissociated)}$$
$$[A^-] = \text{(concentration dissociated)}$$
$$[HA] = C_{HA} - [A^-] \tag{2.10}$$

Equation 2.10 is sometimes called a mass balance equation.

Another equation, called a charge balance equation, is useful here. It is based on the fact that the sum of positive charges in a solution must equal the sum of the negative charges in that solution.

Positive ions	Negative ions
H⁺	OH⁻
	A⁻

$$[H^+] = [OH^-] + [A^-] \tag{2.11}$$

Equation 2.11 is a charge balance equation. [A⁻] is easily derived:

$$[A^-] = [H^+] - [OH^-]$$

Now, remember the mass balance equation (2.10):

$$[HA] = C_{HA} - [A^-]$$
$$[HA] = C_{HA} - ([H^+] - [OH^-])$$
$$[HA] = C_{HA} - [H^+] + [OH^-]$$

Now recall Equation 2.9.

$$[H^+] = K_a \frac{[HA]}{[A^-]}$$

Substitute

$$[H^+] = K_a \frac{(C_{HA} - [H^+] + [OH^-])}{([H^+] - [OH^-])}$$

$$K_w = [H^+][OH^-]$$

$$[OH^-] = \frac{K_w}{[H^+]}$$

$$[H^+] = K_a \frac{\left(C_{HA} - [H^+] + \dfrac{K_w}{[H^+]}\right)}{\left([H^+] - \dfrac{K_w}{[H^+]}\right)} \tag{2.12}$$

Now there is only one variable, [H⁺], in Eq. 2.12, a completely valid equation. We note that Eq. 2.12 is a cubic equation, harder to solve than a quadratic. The approximation process can begin at this point. Consider the terms [H⁺] and $K_w/[H^+]$.

$$\frac{K_w}{[H^+]} = [OH^-]$$

In an acid solution:

$$[H^+] \gg [OH^-]$$

$$[H^+] = K_a \frac{(C_{HA} - [H^+] + \cancel{\frac{K_w}{[H^+]}})}{([H^+] - \cancel{\frac{K_w}{[H^+]}})}$$

$$[H^+] = \frac{K_a(C_{HA} - [H^+])}{[H^+]}$$

$$[H^+]^2 = K_a(C_{HA} - [H^+]) \tag{2.13}$$

We can stop at Eq. 2.13, a quadratic. It is sometimes possible, however, to do a series of successive approximations and to get two values of $[H^+]$ that are close to each other. From Eq. 2.13,

$$[H^+] = [K_a (C_{HA} - [H^+])]^{1/2}$$

First approximation: Let

$$[H^+]' = (K_a C_{HA})^{1/2}$$

(In short, suppose that $C_{HA} \gg [H^+]$.)

Second approximation:

$$[H^+]'' = [K_a (C_{HA} - [H^+]')]^{1/2}$$

Third approximation:

$$[H^+]''' = [K_a (C_{HA} - [H^+]'')]^{1/2}$$

and so forth.

If the approximations converge after a couple of tries, the approach is worthwhile. Otherwise, a quadratic equation may be set up and solved.

Example: If

$$K_a = 1.0 \times 10^{-5}$$

and

$$C_{HA} = 1.0 \times 10^{-2} M$$

calculate the pH.

First approximation:

$$[H^+]' = \sqrt{K_a C_{HA}} = [(1.0 \times 10^{-5})(1.0 \times 10^{-2})]^{1/2} = 3.2 \times 10^{-4} M$$

Second approximation:

$$[H^+]'' = [K_a (C_{HA} - [H^+]')]^{1/2} = [(1.0 \times 10^{-5})(1.0 \times 10^{-2} - 3.2 \times 10^{-4})]^{1/2}$$
$$= 3.1 \times 10^{-4} M$$

Third approximation:

$$[H^+]''' = [K_a (C_{HA} - [H^+]')]^{1/2} = [(1.0 \times 10^{-5})(1.0 \times 10^{-2} - 3.1 \times 10^{-4})]^{1/2}$$
$$= 3.1 \times 10^{-4} M$$

$[H^+]''$ and $[H^+]'''$ are equal.

$$pH = -\log(3.1 \times 10^{-4}) \qquad = 3.51$$

(Remember that two or three significant figures are the most that can be had for K_a-values, and that pH can only be read to ± 0.02 pH units on ordinary apparatus.)

Example: If

$$K_a = 1.0 \times 10^{-5}$$

and

$$C_{HA} = 1.0 \times 10^{-5} M$$

calculate the pH.

First approximation:

$$[H^+]' = K_a\, C_{HA}$$

$$[H^+]' = [(1.0 \times 10^{-5})(1.0 \times 10^{-5})]^{\frac{1}{2}} = 1.0 \times 10^{-5} M$$

Second approximation:

$$[H^+]'' = [K_a\,(C_{HA} - [H^+]')]^{1/2} = [(1.0 \times 10^{-5})(1.0 \times 10^{-5} - 1.0 \times 10^{-5})]^{\frac{1}{2}}$$
$$= 0\, M \qquad \text{(which is absurd)}$$

A quadratic approach must be used.

$$[H^+] = [K_a(C_{HA} - [H^+])]^{1/2} \qquad\qquad (2.13)$$

$$[H^+]^2 = K_a C_{HA} - K_a[H^+]$$

$$[H^+]^2 + K_a[H^+] - K_a C_{HA} = 0$$

$$[H^+] = \frac{-K_a + (K_a^2 + 4K_a C_{HA})^{1/2}}{2}$$

$$= \frac{-1.0 \times 10^{-5} + [(1.0 \times 10^{-5})^2 + 4(1.0 \times 10^{-5})(1.0 \times 10^{-5})]^{1/2}}{2}$$

$$= 6.2 \times 10^{-6} M$$

$$pH = -\log(6.2 \times 10^{-6}) = 5.21$$

2.9 Buffers of Weak Monoprotic Acids

A buffer solution is one made by mixing a weak acid and its anion (or salt, since most anions are available as the sodium salts of the acid). Buffers resist changes in pH and are hence of importance in laboratory work and physiological processes.

Example: Suppose that a buffer is made from acetic acid, abbreviated HAc, and sodium acetate, abbreviated NaAc.

If a strong base, like NaOH, is added to the buffer, it reacts with the HAc, and the pH does not rise much.

$$OH^- + HAc \rightleftharpoons H_2O + Ac^-$$

If a strong base is added to water, the pH will rise noticeably.

If a strong acid, like HCl, is added to the buffer, the acetate ion, a weak base, will react with the protons, and the pH does not fall much.

$$H^+ + Ac^- \rightleftharpoons HAc$$

If a strong acid is added to water, the pH will fall noticeably.

The quantitative description of buffers is not difficult. Generally, one is asked to compute the pH of a solution made by dissolving a weak acid and its salt in water or to compute the concentrations of weak acid and anion needed to give a buffer of a certain pH. Methods of successive approximation can often be used.

As in the case of weak acid alone, equation 2.9 holds.

$$[H^+] = K_a \frac{[HA]}{[A^-]} \tag{2.9}$$

Terms for [HA] and [A$^-$] must be derived. Suppose that the buffer is made with a mixture of HA and its salt NaA.

C_{HA} = analytical concentration of HA

C_{HA} = number of moles of HA added per liter of solution

C_{A^-} = analytical concentration of A$^-$

C_{A^-} = number of moles of NaA added per liter of solution

Because HA and A$^-$ may be formed from both HA and A$^-$,

$$C_{HA} + C_{A^-} = [HA] + [A^-] \tag{2.14}$$

Equation 2.14 is the mass balance equation. Now, a charge balance equation can be written:

Positive ions	Negative ions
Na$^+$	A$^-$
H$^+$	OH$^-$

$$[Na^+] + [H^+] = [A^-] + [OH^-] \tag{2.15}$$

$$C_{A^-} = [Na^+]$$

$$C_{A^-} + [H^+] = [A^-] + [OH^-]$$

$$[A^-] = C_{A^-} + [H^+] - [OH^-] \tag{2.16}$$

This is an equation for [A$^-$]

Now rearrange the mass balance equation (2.14)

$$[HA] = C_{HA} + C_{A^-} - [A^-]$$

$$[HA] = C_{HA} + C_{A^-} - (C_{A^-} + [H^+] - [OH^-])$$

$$[HA] = C_{HA} - [H^+] + [OH^-] \tag{2.17}$$

Substitute equations 2.16 and 2.17 into equation 2.9.

$$[H^+] = K_a \left(\frac{C_{HA} - [H^+] + [OH^-]}{C_{A^-} + [H^+] - [OH^-]} \right)$$

$$K_w = [H^+][OH^-]$$

$$[OH^-] = \frac{K_w}{[H^+]}$$

$$[H^+] = K_a \left(\frac{C_{HA} - [H^+] + \dfrac{K_w}{[H^+]}}{C_{A^-} + [H^+] - \dfrac{K_w}{[H^+]}} \right) \tag{2.18}$$

Equation 2.18 is valid, but is exceedingly difficult to solve. A method of successive approximations is applied.

First approximation: Assume that

$$C_{HA} \gg \left(-[H^+] + \frac{K_w}{[H^+]} \right)$$

and that

$$C_{A^-} \gg \left([H^+] - \frac{K_w}{[H^+]} \right)$$

$$[H^+]' = K_a \frac{C_{HA}}{C_{A^-}}$$

Then determine whether the buffer is acidic or basic

$$[H^+]'[OH^-]' = K_w$$

$$[OH^-]' = \frac{K_w}{[H^+]'}$$

If $[H^+]' > [OH^-]'$ the buffer is acidic.

Second approximation:

$$[H^+]'' = K_a \left(\frac{C_{HA} - [H^+]'}{C_{A^-} + [H^+]'} \right)$$

Third approximation:

$$[H^+]''' = K_a \left(\frac{C_{HA} - [H^+]''}{C_{A^-} + [H^+]''} \right)$$

.
.
.

and so forth

If $[H^+]' < [OH^-]'$, the buffer is basic.

Second approximation:

$$[H^+]'' = K_a \frac{\left(C_{HA} + \dfrac{K_w}{[H^+]'} \right)}{\left(C_{A^-} - \dfrac{K_w}{[H^+]'} \right)}$$

Third approximation:

$$[H^+]''' = K_a \frac{\left(C_{HA} + \dfrac{K_w}{[H^+]''} \right)}{\left(C_{A^-} - \dfrac{K_w}{[H^+]''} \right)}$$

.
.
.

and so forth

If $[H^+]' = [OH^-]'$, we proceed this way:

Second approximation:

$$[H^+]'' = K_a \frac{\left(C_{HA} - [H^+]' + \dfrac{K_w}{[H^+]'}\right)}{\left(C_{A^-} + [H^+]' - \dfrac{K_w}{[H^+]'}\right)}$$

Third approximation:

$$[H^+]''' = K_a \frac{\left(C_{HA} - [H^+]'' + \dfrac{K_w}{[H^+]''}\right)}{\left(C_{A^-} + [H^+]'' - \dfrac{K_w}{[H^+]''}\right)}$$

and so forth.

The convergence of approximations is very slow when there are wide disparities in C_{HA} and C_{A^-}, and approximations may actually diverge wildly when K_a reaches a very low (ca. 10^{-8}) value. Overall, however, the approach is a valid one for *useful* ranges of buffer concentrations.

It is worth noting that the equation for the value of $[H^+]'$ even has a name, once it is changed into a slightly different form.

$$[H^+]' = K_a \frac{C_{HA}}{C_{A^-}}$$

$$\log [H^+]' = \log \left(K_a \frac{C_{HA}}{C_{A^-}}\right)$$

$$\log [H^+]' = \log K_a + \log \frac{C_{HA}}{C_{A^-}}$$

$$-\log [H^+]' = -\log K_a - \log \frac{C_{HA}}{C_{A^-}}$$

$$pH' = pK_a - \log \frac{C_{HA}}{C_{A^-}}$$

$$pH' = pK_a + \log \frac{C_{A^-}}{C_{HA}} \qquad (2.19)$$

Equation 2.19 is called the Henderson-Hasselbalch Equation.

Example: Calculate the pH of a buffer made from 0.010 mole of HA and 0.10 mole of NaA, dissolved in 1 liter of water, when $K_a = 1.0 \times 10^{-5}$.

First approximation:

$$[H^+]' = K_a \frac{C_{HA}}{C_{A^-}} = 1.0 \times 10^{-5} \frac{1.0 \times 10^{-2}}{1.0 \times 10^{-1}} = 1.0 \times 10^{-6}M$$

$$[OH^-]' = \frac{K_w}{[H^+]'} = \frac{1.00 \times 10^{-14}}{1.0 \times 10^{-6}} = 1.0 \times 10^{-8}M$$

$$[H^+]' > [OH^-]'$$

so that the buffer is *acidic*.

Second approximation:

$$[H^+]'' = K_a \frac{(C_{HA} - [H^+]')}{(C_{A^-} + [H^+]')} = 1.0 \times 10^{-5} \frac{(1.0 \times 10^{-2} - 1.0 \times 10^{-6})}{(1.0 \times 10^{-1} + 1.0 \times 10^{-6})}$$

$$= 1.0 \times 10^{-6} M$$

$$[H^+]' = [H^+]''$$

$$pH = -\log(1.0 \times 10^{-6}) = 6.00$$

Often, a student is asked to do something like this in the laboratory:

"Prepare 1.0 liter of a 0.10 molar acetic acid-acetate pH 4.74 buffer using acetic acid and sodium acetate." (For acetic acid $K_a = 1.8 \times 10^{-5}$.)

First, the "0.10 molar" means that

$$C_{HAc} + C_{Ac^-} = 0.10M$$

Next, calculate $[H^+]'$ from the pH.

$$pH = 4.74$$

$$[H^+]' = 10^{-4.74} = 1.8 \times 10^{-5} = K_a \frac{C_{HAc}}{C_{Ac^-}}$$

$$\frac{C_{HAc}}{C_{Ac^-}} = \frac{[H^+]'}{K_a}$$

$$\frac{C_{HAc}}{C_{Ac^-}} = \frac{\cancel{1.8 \times 10^{-5}}}{\cancel{1.8 \times 10^{-5}}}$$

$$\frac{C_{HAc}}{C_{Ac^-}} = 1.0$$

$$C_{HAc} = C_{Ac^-}$$

$$C_{HAc} + C_{Ac^-} = 0.10$$

$$C_{HAc} = 0.050M$$

$$C_{Ac^-} = 0.050M$$

Now, try a second approximation:

$$[H^+]'' = K_a \left(\frac{C_{HAc} - [H^+]'}{C_{Ac^-} + [H^+]'} \right) = 1.8 \times 10^{-5}$$

Thus a buffer made $0.050M$ in NaAc and $0.050M$ in HAc will exhibit a pH of 4.74. It is interesting to note that for monoprotic weak acids where $C_{HA} = C_{A^-}$, pH = pK_a.

If the buffer above were diluted to 2 liters, C_{HAc} would become $0.025M$ and C_{Ac^-} would become $0.025M$. The pH would remain constant! Why?

$$[H^+]' = K_a \frac{C_{HAc_1}}{C_{Ac_1^-}} = (1.8 \times 10^{-5}) \left(\frac{\cancel{0.025}}{\cancel{0.025}} \right) = 1.8 \times 10^{-5}$$

$$pH = -\log(1.8 \times 10^{-5}) = 4.74$$

Buffers may be greatly diluted before their pH values are changed. The more dilute a buffer is, though, the less able it is to withstand attack by a strong acid or a strong base. Its *capacity* is lowered with dilution, while the pH remains constant.

2.10 The Anions of Weak Acids Alone

When the salts of weak acids are dissolved in water, without any of the weak acids being added, the solutions produced are far from being neutral. They are basic. Take the anion of acetic acid as an example. Acetic acid does *not* readily give up a proton. This means that acetate ions can remove protons from water. Suppose that sodium acetate, NaAc, is dissolved in water. The reaction that makes the solution basic is this one:

$$Ac^- + H_2O \rightleftharpoons HAc + OH^-$$

The process is called hydrolysis (water-splitting). A constant for the reaction may be called the *hydrolysis constant*, K_{hy}, and written this way:

$$K_{hy} = \frac{[OH^-][HAc]}{[Ac^-]}$$

Values of K_{hy} are seldom tabulated in texts. It is possible, however, to calculate the value of K_{hy}, simply by multiplying K_{hy} through by 1.

$$\frac{[H^+]}{[H^+]} = 1$$

$$K_{hy} = \frac{[H^+][OH^-][HAc]}{[H^+][Ac^-]}$$

$$K_w = [H^+][OH^-]$$

$$K_a = \frac{[H^+][Ac^-]}{[HAc]}$$

$$K_{hy} = \frac{K_w}{K_a}$$

$$\frac{K_w}{K_a} = \frac{[OH^-][HAc]}{[Ac^-]}$$

If we are faced with the problem of determining $[OH^-]$, given K_w, K_a, and C_{Ac^-} (added as NaAc), the first thing to do is to rearrange the previous expression.

$$[OH^-] = \frac{K_w}{K_a}\frac{[Ac^-]}{[HAc]} \tag{2.20}$$

The values of $[Ac^-]$ and $[HAc]$ must be determined. The mass and charge balance equations must be written first.

$$C_{Ac^-} = [Ac^-] + [HAc] \quad \text{(mass balance)} \tag{2.21}$$

$$[HAc] = C_{Ac^-} - [Ac^-] \tag{2.22}$$

Positive ions	Negative ions
Na^+	Ac^-
H^+	OH^-

$$[Na^+] + [H^+] = [Ac^-] + [OH^-] \quad \text{(charge balance)} \tag{2.23}$$

$$C_{Ac^-} = [Na^+]$$

$$C_{Ac^-} + [H^+] = [Ac^-] + [OH^-]$$
$$[Ac^-] = C_{Ac^-} + [H^+] - [OH^-] \qquad (2.24)$$

Remember that $[HAc] = C_{Ac^-} - [Ac^-]$.

Substitute from Equation 2.24

$$[HAc] = C_{Ac^-} - (C_{Ac^-} + [H^+] - [OH^-])$$
$$[HAc] = [OH^-] - [H^+]$$

Substitute the new terms for $[Ac^-]$ and $[HAc]$ into Equation 2.20.

$$[OH^-] = \frac{K_w}{K_a} \left(\frac{C_{Ac^-} + [H^+] - [OH^-]}{[OH^-] - [H^+]} \right)$$

$$[OH^-] = \frac{K_w}{K_a} \left(\frac{C_{Ac^-} + \dfrac{K_w}{[OH^-]} - [OH^-]}{[OH^-] - \dfrac{K_w}{[OH^-]}} \right) \qquad (2.25)$$

Equation 2.25 is valid. It is also not very easy to solve. A process of successive approximations is useful once more. First, the solution is basic. This means that

$$[OH^-] \gg [H^+] \quad \text{or} \quad [OH^-] \gg \frac{K_w}{[OH^-]}$$

$$[OH^-] = \frac{K_w}{K_a} \left(\frac{C_{Ac^-} - [OH^-]}{[OH^-]} \right)$$

$$[OH^-]^2 = \frac{K_w}{K_a} (C_{Ac^-} - [OH^-])$$

or

$$[OH^-] = \left[\frac{K_w}{K_a} (C_{Ac^-} - [OH^-]) \right]^{1/2} \qquad (2.26)$$

Equation 2.26 may be solved as a quadratic or the process of successive approximations may be applied.

First approximation:

$$[OH^-]' = \left(\frac{K_w}{K_a} C_{Ac^-} \right)^{1/2}$$

Second approximation:

$$[OH^-]'' = \left[\frac{K_w}{K_a} (C_{Ac^-} - [OH^-]') \right]^{1/2}$$

Third approximation:

$$[OH^-]''' = \left[\frac{K_w}{K_a} (C_{Ac^-} - [OH^-]'') \right]^{1/2}$$

.
.
.

and so forth.

Example: Acetic acid, HAc, has a K_a value of 1.8×10^{-5}. Calculate the pH of a solution 0.10M in sodium acetate, NaAc.

First approximation:

$$[OH^-]' = \left(\frac{K_w}{K_a}C_{Ac^-}\right)^{1/2}$$

$$C_{Ac^-} = 0.10M$$

$$[OH^-]' = \left(\frac{1.00 \times 10^{-14}}{1.8 \times 10^{-5}} \times 0.10\right)^{1/2} = 7.4 \times 10^{-6}M$$

Second approximation:

$$[OH^-]'' = \left[\frac{K_w}{K_a}(C_{Ac^-} - [OH^-]')\right]^{1/2}$$

$$= \left[\left(\frac{1.00 \times 10^{-14}}{1.8 \times 10^{-5}}\right)(0.10 - 7.4 \times 10^{-6})\right]^{1/2} = 7.4 \times 10^{-6}M$$

$$[OH^-]' = [OH^-]''$$

$$pOH = -\log(7.4 \times 10^{-6}) = 5.13$$

$$pH = 14.00 - 5.13 = 8.87$$

2.11 Distribution Functions

Nowadays, it is easy to measure pH. A meter and electrode combination that is highly portable and relatively accurate can be purchased for about $100. The question arises: Given the pH, K_a, and an analytical concentration, can we easily determine the equilibrium concentrations of the weak acid and its anion? A few definitions and conditions need to be made first:

  ⊐ 1. Given: a weak acid, HA.

  ⊐ 2. Analytical concentration of acid and anion = C_A.

  ⊐ 3. Sum of equilibrium concentrations of all A-bearing species = C_A; [HA] + [A$^-$] = C_A. (It is *not* necessary that C_A = C_{A^-}, nor that C_A = C_{HA}.)

Then define

$$\alpha_0 = \frac{[HA]}{[HA] + [A^-]} = \frac{[HA]}{C_A} \tag{2.27}$$

and

$$\alpha_1 = \frac{[A^-]}{[HA] + [A^-]} = \frac{[A^-]}{C_A} \tag{2.28}$$

Finding [HA] and [A$^-$] is possible if C_A, α_0, and α_1 are known

$$[HA] = \alpha_0 C_A$$

$$[A^-] = \alpha_1 C_A$$

Also, α_0 and α_1 can be expressed in terms of $[H^+]$ (known), and K_a (known).

$$\alpha_0 = \frac{[HA]}{[HA] + [A^-]}$$

For HA,

$$HA \rightleftharpoons H^+ + A^-$$

$$K_a = \frac{[H^+][A^-]}{[HA]}$$

$$[A^-] = \frac{K_a[HA]}{[H^+]}$$

$$\alpha_0 = \frac{[HA]}{[HA] + \dfrac{K_a[HA]}{[H^+]}} = \frac{\cancel{[HA]}}{\cancel{[HA]}\left(1 + \dfrac{K_a}{[H^+]}\right)} = \frac{1}{1 + \dfrac{K_a}{[H^+]}}$$

$$= \frac{1}{\dfrac{[H^+]}{[H^+]} + \dfrac{K_a}{[H^+]}} = \frac{1}{\dfrac{[H^+] + K_a}{[H^+]}} = \frac{[H^+]}{[H^+] + K_a} \qquad (2.29)$$

and then α_1,

$$\alpha_1 = \frac{[A^-]}{[HA] + [A^-]}$$

$$[HA] = \frac{[H^+][A^-]}{K_a}$$

$$\alpha_1 = \frac{[A^-]}{\dfrac{[H^+][A^-]}{K_a} + [A^-]} = \frac{\cancel{[A^-]}}{\cancel{[A^-]}\left(\dfrac{[H^+]}{K_a} + 1\right)} = \frac{1}{\dfrac{[H^+] + K_a}{K_a}}$$

$$\alpha_1 = \frac{K_a}{[H^+] + K_a} \qquad (2.30)$$

Example: Benzoic acid,

 COOH

or HBz for short, has a K_a of 6.3×10^{-5}. Calculate the values of [HBz] and [Bz⁻] in a solution where the analytical concentration of benzoates is $0.10M$ and the pH is found to be 4.00.

At pH 4.00,

$$[H^+] = 1.0 \times 10^{-4}M$$

$$C_{Bz} = 0.10M$$

$$[HBz] = \alpha_0 C_{Bz}$$

$$\alpha_0 = \frac{[H^+]}{[H^+] + K_a}$$

$$\alpha_0 = \frac{1.0 \times 10^{-4}}{1.0 \times 10^{-4} + 6.3 \times 10^{-5}}$$

$$\alpha_0 = 0.61$$
$$[HBz] = (0.61)(0.10)$$
$$[HBz] = 0.061M$$
$$[Bz^-] = \alpha_1 \cdot C_{Bz}$$
$$\alpha_1 = \frac{K_a}{[H^+] + K_a} = \frac{6.3 \times 10^{-5}}{1.0 \times 10^{-4} + 6.3 \times 10^{-5}} = 0.39$$
$$[Bz^-] = (0.39)(0.10)$$
$$[Bz^-] = 0.039M$$

Note that
$$\alpha_0 + \alpha_1 = 1.00$$
$$[HBz] + [Bz^-] = 0.10 = C_{Bz}$$

≡ 2.12 PROBLEMS ≡

(K_a values may be found in Table 2.4 at the end of the chapter.)

1. Butyric acid, $CH_3CH_2CH_2COOH$, or HBu for short, smells vile and has a K_a of 1.48×10^{-5}. Calculate the pH of a solution $0.0010M$ in HBu.
   Ans: *pH = 3.94*

2. Nitrous acid, HNO_2, has a K_a of 5.1×10^{-4}. Calculate the pH of a solution $1.0 \times 10^{-4}M$ in HNO_2.
   Ans: *pH = 4.06*

3. Calculate the pH of a solution in which the analytical concentration of acetic acid is $0.030M$ and the analytical concentration of sodium acetate is $0.020M$. For acetic acid, $K_a = 1.8 \times 10^{-5}$.
   Ans: *pH = 4.57*

4. Calculate the pH of a solution made this way: 200. milliliters of the solution of Problem 3 are diluted to 500. milliliters.
   Ans: *pH = 4.57*

5. Calculate the pH of 200. milliliters of the solution of Problem 3 after the addition of 1.00×10^{-4} moles of solid NaOH. (Assume that the volume does not change.)
   Ans: *pH = 4.57*

6. Calculate the pH of 200. milliliters of the solution of Problem 3 after the addition of 1.0×10^{-4} moles of concentrated HCl. (Assume that the volume does not change.)
   Ans: *pH = 4.57*

7. A chemist with clogged sinuses desires to make a 1.00 liter of a 0.050 molar butyric acid-butyrate buffer of pH 5.00. How many moles of HBu and NaBu will be required to make this solution? ($K_a = 1.48 \times 10^{-5}$).
   Ans: *0.020 moles of HBu and 0.030 moles of NaBu*

8. Calculate the pH of a solution made by mixing 100 milliliters of 0.010M acetic acid with 100. milliliters of 0.0050M sodium hydroxide (K_a = 1.8 × 10^{-5}) (Assume that the volumes are additive.)

   Ans: *pH = 4.74*

9. Benzoic acid, 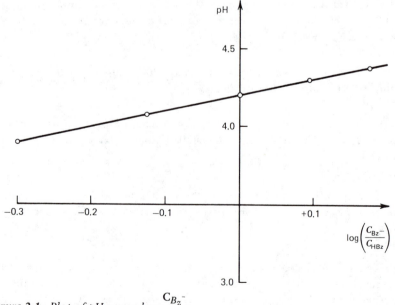COOH , or HBz for short, is a weak, monoprotic acid. A

   series of benzoic acid-sodium benzoate buffers is made up, and the pH of each is measured. Figure 2.1 is a plot of pH *versus* log(C_{Bz^-}/C_{HBz}). From Figure 2.1 using first approximations, determine the K_a for benzoic acid.

   $$C_{Bz^-} = \text{analytical concentration of sodium benzoate}$$

   $$C_{HBz} = \text{analytical concentration of benzoic acid}$$

   Ans: **$K_a = 6.3 \times 10^{-5}$**

10. Calculate the pH of a solution made by dissolving 0.020 moles of sodium benzoate in a little water and diluting this solution to 250. milliliters.

    Ans: *pH = 8.56*

11. 6.005 grams of acetic acid are dissolved in 900.0 milliliters of water. Then 100.0 milliliters of 1.000M NaOH are added to this solution. Calculate the pH of the solution after the addition of NaOH. (Assume that the volumes are additive.)

    Ans: *pH = 8.87*

Figure 2.1 *Plot of pH versus log* $\dfrac{C_{B_z^-}}{C_{HB_z}}$

12. Propionic acid, CH_3CH_2COOH, or HPro for short, has a K_a value of 1.3×10^{-5}. If a solution is made up of sodium propionate and propionic acid so that $C_{Pro} = [HPro] + [Pro^-] = 1.00 \times 10^{-2}M$, has a pH of 6.00, calculate [HPro] and [Pro$^-$].
Ans: *[HPro] = 7.1 × 10^{-4}M; [Pro$^-$] = 9.3 × 10^{-3}M*

$\equiv$ 2.13 DETAILED SOLUTIONS TO PROBLEMS $\equiv$

1. Try a method of successive approximations:

$$[H^+]' = (K_a C_{HBu})^{1/2} = [(1.48 \times 10^{-5})(1.0 \times 10^{-3})]^{1/2} = 1.2 \times 10^{-4}M$$

$$[H^+]'' = [K_a(C_{HBu} - [H^+]')]^{1/2} = [(1.48 \times 10^{-5})(1.0 \times 10^{-3} - 1.2 \times 10^{-4})]^{1/2}$$
$$= 1.1 \times 10^{-4}M$$

$$[H^+]''' = [K_a(C_{HBu} - [H^+]'')]^{1/2}$$
$$= [(1.48 \times 10^{-5}) \cdot (1.0 \times 10^{-3} - 1.1 \times 10^{-4})]^{1/2} = 1.1 \times 10^{-4}M$$

$$[H^+]'' = [H^+]'''$$

The approximations have converged.

$$pH = -\log(1.1 \times 10^{-4}) = 3.94$$

2. Try a method of successive approximations:

$$[H^+]' = (K_a C_{HNO_2})^{1/2} = [(5.1 \times 10^{-4})(1.0 \times 10^{-4})]^{1/2} = 2.2 \times 10^{-4}M$$

$$[H^+]'' = [(5.1 \times 10^{-4})(1.0 \times 10^{-4} - 2.2 \times 10^{-4})]^{1/2}$$
$$= [(5.1 \times 10^{-4})(-1.2 \times 10^{-4})]^{1/2}$$

This square root is an imaginary number. The method of successive approximations will not work here. Use a quadratic equation.

$$[H^+] = [K_a(C_{HNO_2} - [H^+])]^{1/2}$$

$$[H^+]^2 = K_a(C_{HNO_2} - [H^+])$$

$$[H^+]^2 = K_a C_{HNO_2} - K_a[H^+]$$

$$[H^+]^2 + K_a[H^+] - K_a C_{HNO_2} = 0$$

$$[H^+] = \frac{-K_a + (K_a^2 + 4K_a C_{HNO_2})^{1/2}}{2}$$

$$= \frac{-5.1 \times 10^{-4} + [(5.1 \times 10^{-4})^2 + 4(5.1 \times 10^{-4})(1.0 \times 10^{-4})]^{1/2}}{2}$$

$$= 8.6 \times 10^{-5}M$$

$$pH = -\log(8.6 \times 10^{-5}) = 4.06$$

3. This is a buffer solution:

$$[H^+]' = K_a \frac{C_{HAc}}{C_{Ac^-}} = 1.8 \times 10^{-5} \left(\frac{3.0 \times 10^{-2}}{2.0 \times 10^{-2}}\right) = 2.7 \times 10^{-5}M$$

$$[OH^-]' = \frac{K_w}{[H^+]'} = \frac{(1.00 \times 10^{-14})}{(2.7 \times 10^{-5})} = 3.7 \times 10^{-10}M$$

$$[H^+]' \gg [OH^-]'$$

so that the solution is decidedly *acidic*.

$$[H^+]'' = K_a \left(\frac{C_{HAc} - [H^+]'}{C_{Ac^-} + [H^+]'} \right) = 1.8 \times 10^{-5} \left(\frac{3.0 \times 10^{-2} - 2.7 \times 10^{-5}}{2.0 \times 10^{-2} + 2.7 \times 10^{-5}} \right)$$

$$= 2.7 \times 10^{-5} M$$

$$[H^+]' = [H^+]''$$

The approximations have converged:

$$pH = -\log(2.7 \times 10^{-5}) = 4.57$$

4. To get C_{HAc_2} and $C_{Ac_2^-}$ for this problem, remember that

$$M_1 v_1 = M_2 v_2$$

$$C_{HAc_1} v_1 = C_{HAc_2} v_2$$

$$C_{HAc_2} = C_{HAc_1} \frac{v_1}{v_2}$$

$$C_{HAc_2} = (3.0 \times 10^{-2}) \cdot \left(\frac{200.}{500.} \right) \qquad C_{HAc_1} = 3.0 \times 10^{-2} M$$

$$v_1 = 200. \text{ milliliters}$$

$$C_{HAc_2} = ?$$

$$v_2 = 500 \text{ milliliters}$$

$$C_{Ac_1^-} \cdot v_1 = C_{Ac_2^-} \cdot v_2$$

$$C_{Ac_2^-} = C_{Ac_1^-} \cdot \frac{v_1}{v_2}$$

$$C_{Ac_2^-} = (2.0 \times 10^{-2}) \left(\frac{200.}{500.} \right) \qquad C_{Ac_1^-} = 2.0 \times 10^{-2} M$$

$$v_1 = 200 \text{ milliliters}$$

$$C_{Ac_2^-} = ?$$

$$v_2 = 500. \text{ milliliters}$$

$$[H^+]' = K_a \frac{C_{HAc_2}}{C_{Ac_2^-}} = (1.8 \times 10^{-5}) \frac{(3.0 \times 10^{-2}) \, (200./500.)}{(2.0 \times 10^{-2}) \, (200./500.)} = 2.7 \times 10^{-5} M$$

$$[H^+]'' = K_a \left(\frac{C_{HAc_2} - [H^+]'}{C_{Ac_2^-} + [H^+]'} \right)$$

$$= (1.8 \times 10^{-5}) \cdot \frac{(3.0 \times 10^{-2})(200./500.) - 2.7 \times 10^{-5}}{(2.0 \times 10^{-2})(200./500.) + 2.7 \times 10^{-5}} = 2.7 \times 10^{-5} M$$

$$[H^+]' = [H^+]''$$

The approximations have converged: $[H^+]$ and the pH do not change much with reasonable dilution. This is true of most buffers.

$$pH = -\log(2.7 \times 10^{-5}) = 4.57$$

5. Here, stoichiometry must be combined with equilibrium calculations. The reac-

tion which is important is that between the added NaOH and the acetic acid. It goes very nearly to completion.

$$HAc + OH^- \rightarrow H_2O +' Ac^-$$

(number of moles of Ac^- formed) = (number of moles of OH^- added)

$$= 1.00 \times 10^{-4} \text{ moles}$$

(number of moles of HAc left)
    = (number of moles of HAc to start) - (number of moles of Ac^- formed)

$$= (3.0 \times 10^{-2})(2.00 \times 10^{-1}) - (1.00 \times 10^{-4}) = 5.9 \times 10^{-3} \text{ moles}$$

$$C_{HAc} = \frac{\text{(number of moles of HAc left)}}{\text{(volume of solution, liters)}} = \frac{5.9 \times 10^{-3}}{2.00 \times 10^{-1}} = 3.0 \times 10^{-2} M$$

(number of moles of Ac^- present)
    = (number of moles of Ac^- to start) + (number of moles of Ac^- formed)

$$= (2.00 \times 10^{-2})(2.00 \times 10^{-1}) + (1.00 \times 10^{-4}) = 4.1 \times 10^{-3} \text{ moles}$$

$$C_{Ac^-} = \frac{\text{(number of moles of } Ac^- \text{ present)}}{\text{(volume of solution, liters)}} = \frac{4.1 \times 10^{-3}}{2 \times 10^{-1}} = 2.0 \times 10^{-2} M$$

Now, the equilibrium approximations can begin

$$[H^+]' = K_a \frac{C_{HAc}}{C_{Ac^-}} = (1.8 \times 10^{-5})\left(\frac{3.0 \times 10^{-2}}{2.0 \times 10^{-2}}\right) = 2.7 \times 10^{-5} M$$

The buffer is decidedly acidic.

$$[H^+]'' = K_a \left(\frac{C_{HAc} - [H^+]'}{C_{Ac^-} + [H^+]'}\right)$$

$$= (1.8 \times 10^{-5})\left(\frac{3.0 \times 10^{-2} - 2.7 \times 10^{-5}}{2.0 \times 10^{-2} + 2.7 \times 10^{-5}}\right) = 2.7 \times 10^{-5} M$$

$$[H^+]' = [H^+]''$$

The approximations have converged. Note that $[H^+]$ and pH did not (within the limitations of *two* significant figures) change much with the addition of a little NaOH. The ability of a buffer to resist the pH change on the addition of strong acid or base is called its *buffer capacity*. The buffer capacity is dependent upon the analytical concentrations of acid and anion.

$$pH = -\log (2.7 \times 10^{-5}) = 4.57$$

6. Again, stoichiometry must be combined with equilibrium calculations. The reaction which is important is that between the added HCl (strong acid) and the acetate ion. It goes very nearly to completion.

$$H^+ + Ac^- \rightarrow HAc$$

(number of moles of HAc formed) = (number of moles of H^+ added)

(number of moles of HAc formed) = 1.00×10^{-4} moles

(number of moles of Ac^- left)
    = (number of moles of Ac^- to start) - (number of moles of HAc formed)

$$= (2.0 \times 10^{-2})(2.00 \times 10^{-1}) - 1.00 \times 10^{-4} = 3.9 \times 10^{-3} \text{ moles}$$

$$C_{Ac^-} = \frac{\text{(number of moles of Ac}^- \text{ left)}}{\text{(volume of solution, liters)}} = \frac{3.9 \times 10^{-3}}{2.00 \times 10^{-1}} = 2.0 \times 10^{-2} M$$

(number of moles of HAc present)

$$= \text{(number of moles of HAc to start)} + \text{(number of moles of HAc formed)}$$

$$= (3.0 \times 10^{-2})(2.00 \times 10^{-1}) + (1.00 \times 10^{-4}) = 6.1 \times 10^{-3} \text{ moles}$$

$$C_{HAc} = \frac{\text{(number of moles of HAc present)}}{\text{(volume of solution, liters)}} = \frac{6.1 \times 10^{-3}}{2.00 \times 10^{-1}} = 3.0 \times 10^{-2} M$$

Now, the equilibrium approximations can begin:

$$[H^+]' = K_a \frac{C_{HAc}}{C_{Ac^-}} = (1.8 \times 10^{-5}) \left(\frac{3.0 \times 10^{-2}}{2.0 \times 10^{-2}} \right) = 2.7 \times 10^{-5} M$$

The buffer is acidic

$$[H^+]'' = K_a \left(\frac{C_{HAc} - [H^+]'}{C_{Ac^-} + [H^+]'} \right)$$

$$= (1.8 \times 10^{-5}) \left(\frac{3.0 \times 10^{-2} - 2.7 \times 10^{-5}}{2.0 \times 10^{-2} + 2.7 \times 10^{-5}} \right) = 2.7 \times 10^{-5} M$$

$$[H^+]' = [H^+]''$$

The approximations have converged. Note that $[H^+]$ and the pH did not (within the limitation of *two* significant figures) change much with the addition of a little HCl.

$$pH = -\log (2.7 \times 10^{-5}) = 4.57$$

7. First, the "0.050 molar" means that

$$C_{HBu} + C_{Bu^-} = 0.05 M$$

Next, calculate $[H^+]'$ from pH

$$pH = 5.00$$

$$[H^+]' = 10^{-5.00} = 1.0 \times 10^{-5} = K_a \frac{C_{HBu}}{C_{Bu^-}}$$

$$\frac{C_{HBu}}{C_{Bu^-}} = \frac{[H^+]'}{K_a}$$

$$\frac{C_{HBu}}{C_{Bu^-}} = \frac{1.0 \times 10^{-5}}{1.48 \times 10^{-5}}$$

$$\frac{C_{HBu}}{C_{Bu^-}} = 6.8 \times 10^{-1}$$

$$C_{HBu} = 0.68 \, C_{Bu^-}$$

$$C_{HBu} + C_{Bu^-} = 0.050$$

$$0.68 \, C_{Bu^-} + C_{Bu^-} = 0.050$$

$$1.68 \, C_{Bu^-} = 0.050$$

$$C_{Bu^-} = \frac{0.050}{1.68} = 0.030M$$

$$C_{HBu} = 0.68 \, C_{Bu^-} = 0.020M$$

Now, try a second approximation:

$$[H^+]'' = K_a \left(\frac{C_{HBu} - [H^+]'}{C_{Bu^-} + [H^+]'}\right) = (1.48 \times 10^{-5}) \left(\frac{0.020 - 10^{-5}}{0.030 + 10^{-5}}\right) = 1.0 \times 10^{-5}M$$

The solution, then, is to be 0.020M in butyric acid 0.030M in sodium butyrate.

(number of moles of HBu) $= (C_{HBu}, M)$ (volume, liters) $= (0.020)(1.00)$

$$= 0.020 \text{ moles}$$

(number of moles of NaBu) $= (C_{Bu^-}, M)$ (volume, liters) $= (0.030) \cdot (1.00)$

$$= 0.030 \text{ moles}$$

8. Again, some stoichiometric calculations must be made before the equilibrium calculations. The important reaction is the one between NaOH and HAc. It goes very nearly to completion.

$$HAc + OH^- \rightarrow H_2O + Ac^-$$

(number of moles of Ac^- formed) = (number of moles of NaOH added)

$$= (5.0 \times 10^{-3})(1.00 \times 10^{-1})$$

$$= 5.0 \times 10^{-4} \text{ moles}$$

$$C_{Ac^-} = \frac{(\text{number of moles of } Ac^- \text{ formed})}{(total \text{ volume of solution, liters})} = \frac{5.0 \times 10^{-4}}{(1.00 \times 10^{-1} + 1.00 \times 10^{-1})}$$

$$= 2.5 \times 10^{-3}M$$

(number of moles of HAc left)

$$= (\text{number of moles of HAc to start}) - (\text{number of moles of } Ac^- \text{ formed})$$

$$= (1.0 \times 10^{-2})(1.00 \times 10^{-1}) - 5.0 \times 10^{-4} = 5.0 \times 10^{-4} \text{ moles}$$

$$C_{HAc} = \frac{(\text{number of moles of HAc formed})}{(total \text{ volume of solution, liters})}$$

$$= \frac{5.0 \times 10^{-4}}{(1.00 \times 10^{-1} + 1.00 \times 10^{-1})} = 2.5 \times 10^{-3}M$$

Now, the equilibrium approximations can begin.

$$[H^+]' = K_a \frac{C_{HAc}}{C_{Ac^-}} = (1.8 \times 10^{-5}) \left(\frac{2.5 \times 10^{-3}}{2.5 \times 10^{-3}}\right) = 1.8 \times 10^{-5}M$$

$$[H^+]'' = K_a \left(\frac{C_{HAc} - [H^+]'}{C_{Ac^-} + [H^+]'}\right)$$

$$= (1.8 \times 10^{-5}) \left(\frac{2.5 \times 10^{-3} - 1.8 \times 10^{-5}}{2.5 \times 10^{-3} + 1.8 \times 10^{-5}}\right) = 1.8 \times 10^{-5}M$$

$$pH = -\log(1.8 \times 10^{-5}) = 4.74$$

9. To a good first approximation,

$$[H^+] = K_a \frac{C_{HBz}}{C_{Bz^-}}$$

$$\log [H^+] = \log K_a + \log \frac{C_{HBz}}{C_{Bz^-}}$$

$$-\log [H^+] = -\log K_a - \log \frac{C_{HBz}}{C_{Bz^-}}$$

$$pH = pK_a - \log \frac{C_{HBz}}{C_{Bz^-}}$$

$$pH = pK_a + \log \frac{C_{Bz^-}}{C_{HBz}}$$

This is just the Henderson-Hasselbalch equation, where

$$C_{Bz^-} = C_{HBz}$$

$$\log \frac{C_{Bz^-}}{C_{HBz}} = \log 1 = 0$$

and

$$pH = pK_a$$

Refer to Figure 2.1, where $\log C_{Bz^-}/C_{HBz} = 0$.

$$pH = 4.20$$
$$pK_a = 4.20$$
$$-\log K_a = 4.20$$
$$K_a = 6.3 \times 10^{-5}$$

10. This problem involves the hydrolysis of the benzoate ion.

$$Bz^- + H_2O \rightleftharpoons HBz + OH^-$$

$$C_{Bz^-} = \frac{\text{(number of moles of NaBz)}}{\text{(number of liters of solution)}} = \frac{0.020}{0.250} = 8.0 \times 10^{-2} M$$

Now, the approximations can begin.

$$[OH^-]' = \left[\frac{K_w}{K_a} C_{Bz^-} \right]^{1/2} = \left[\left(\frac{1.00 \times 10^{-14}}{6.3 \times 10^{-5}} \right) (8.0 \times 10^{-2}) \right]^{1/2} = 3.6 \times 10^{-6} M$$

$$[OH^-]'' = \left[\frac{K_w}{K_a} (C_{Bz^-} - [OH^-]') \right]^{1/2}$$

$$= \left[\left(\frac{1.00 \times 10^{-14}}{6.3 \times 10^{-5}} \right) (8.0 \times 10^{-2} - 3.6 \times 10^{-6}) \right]^{1/2} = 3.6 \times 10^{-6} M$$

$$[OH^-]' = [OH^-]''$$

The approximations have converged.

$$pOH = -\log (3.6 \times 10^{-6})$$
$$pOH = 5.44$$
$$pH = 14.00 - 5.44 = 8.56$$

11. There is some stoichiometry to be done first. The important reaction is the one between NaOH and HAc. It goes very nearly to completion.

$$HAc + OH^- \rightarrow H_2O + Ac^-$$

(number of moles of Ac^- formed) = (number of moles of NaOH added)
$$= (1.000)(0.1000) = 0.1000 \text{ moles}$$

$$C_{Ac^-} = \frac{0.1000 \text{ moles}}{(0.900 + 0.100) \text{ liter}} = 0.100M$$

(number of moles of HAc to start) = $\dfrac{6.005 \text{ grams}}{60.05 \text{ grams per mole}} = 0.1000 \text{ mole}$

(number of moles of HAc left) = (number of moles of HAc to start) −
$$\text{(number of moles of } Ac^- \text{ formed)}$$

$$= 0.1000 - 0.1000 = 0.000 \text{ moles}$$

$$C_{HAc} = 0M$$

Thus we have sodium acetate alone in water. There is no excess HAc, nor is there excess OH^-. The hydrolysis of the acetate ion will determine the pH of the solution.

$$Ac^- + H_2O \rightarrow HAc + OH^-$$

Now the approximations can begin.

$$[OH^-]' = \left(\frac{K_w}{K_a} C_{Ac^-}\right)^{1/2} = \left[\left(\frac{1.00 \times 10^{-14}}{1.8 \times 10^{-5}}\right)(1.000 \times 10^{-1})\right]^{1/2} = 7.4 \times 10^{-6}M$$

$$[OH^-]'' = \left[\frac{K_w}{K_a}(C_{Ac^-} - [OH^-]')\right]^{1/2}$$

$$= \left[\frac{1.00 \times 10^{-14}}{1.8 \times 10^{-5}}(1.000 \times 10^{-1} - 7.4 \times 10^{-6})\right]^{1/2} = 7.4 \times 10^{-6}M$$

$$[OH^-]' = [OH^-]''$$

The approximations have converged.

$$pOH = -\log(7.4 \times 10^{-6}) = 5.13$$
$$pH = 14.00 - 5.13 = 8.87$$

12. Here, the α-functions can be used

$$\alpha_0 = \frac{[HPro]}{C_{Pro}}$$

$$[HPro] = \alpha_0 C_{Pro}$$

$$\alpha_0 = \frac{[H^+]}{[H^+] + K_a}$$

at pH 6.00,

$$[H^+] = 1.00 \times 10^{-6} M$$

$$\alpha_0 = \frac{1.00 \times 10^{-6} M}{1.00 \times 10^{-6} + 1.3 \times 10^{-5}}$$

$$\alpha_0 = 0.071$$

$$[HPro] = (0.071)(1.00 \times 10^{-2})$$

$$[HPro] = 7.1 \times 10^{-4} M$$

$$\alpha_1 = \frac{[Pro^-]}{C_{Pro}}$$

$$[Pro^-] = \alpha_1 C_{Pro}$$

$$\alpha_1 = \frac{K_a}{[H^+] + K_a}$$

$$\alpha_1 = \frac{1.3 \times 10^{-5}}{1.00 \times 10^{-6} + 1.3 \times 10^{-4}}$$

$$\alpha_1 = 0.93$$

$$[Pro^-] = (0.93)(1.00 \times 10^{-2})$$

$$[Pro^-] = 9.3 \times 10^{-3} M$$

≡ 2.14 WEAK POLYPROTIC ACIDS ≡

Weak monoprotic acids are certainly important. There are other weak acids that are of great physiological and chemical significance. These have more than one ionizable proton and are known as *polyprotic* acids. Acids with two ionizable protons are known as diprotic acids. An example of this class is carbonic acid, H_2CO_3.

$$H_2CO_3 \rightleftharpoons H^+ + HCO_3^- \qquad K_1 = \frac{[H^+][HCO_3^-]}{[H_2CO_3]} = 3.5 \times 10^{-7}$$

$$HCO_3^- \rightleftharpoons H^+ + CO_3^{2-} \qquad K_2 = \frac{[H^+][CO_3^{2-}]}{[HCO_3^-]} = 6.0 \times 10^{-11}$$

Diprotic acids ionize as does carbonic acid, in two stages, with the first ionization being stronger than the second.

Acids with three ionizable protons are known as triprotic acids. An example of this class is phosphoric acid, H_3PO_4.

$$H_3PO_4 \rightleftharpoons H^+ + H_2PO_4^- \qquad K_1 = \frac{[H^+][H_2PO_4^-]}{[H_3PO_4]} = 7.1 \times 10^{-3}$$

$$H_2PO_4^- \rightleftharpoons H^+ + HPO_4^{2-} \qquad K_2 = \frac{[H^+][HPO_4^{2-}]}{[H_2PO_4^-]} = 6.2 \times 10^{-8}$$

$$HPO_4^{2-} \rightleftharpoons H^+ + PO_4^{3-} \qquad K_3 = \frac{[H^+][PO_4^{3-}]}{[HPO_4^{2-}]} = 4.4 \times 10^{-13}$$

Triprotic acids generally ionize as does phosphoric acid, in three stages, with the first ionization being stronger than the second, which is stronger than the third.

Tetraprotic acids, such as ethylenediamine tetraacetic acid, homocystine, and cystine, have four protons and ionize in four stages. The first ionization is the strongest; the last is the weakest.

Some amino acids, such as glycine, a diprotic acid, go through two stages of ionization. Their charge arrangement, however, is slightly different from that of an inorganic acid like H_2CO_3, or an organic acid like $H_2C_2O_4$. The following reaction sequence illustrates this point.

Glycine:

$$
\begin{array}{ccc}
\begin{array}{c} \text{H}+ \\ | \\ \text{H}-\text{N}-\text{H} \\ | \\ \text{H}-\text{C}-\text{H} \\ | \\ \text{C}=\text{O} \\ | \\ \text{OH} \end{array}
&
\underset{\rightleftharpoons}{K_1}
&
\begin{array}{c} \text{H}+ \\ | \\ \text{H}-\text{N}-\text{H} \\ | \\ \text{H}-\text{C}-\text{H} \\ | \\ \text{C}=\text{O} \\ | \\ \text{O}- \end{array}
\end{array}
$$

$$
\begin{array}{ccc}
\underset{\rightleftharpoons}{K_2}
&
\begin{array}{c} \text{H}-\text{N}-\text{H} \\ | \\ \text{H}-\text{C}-\text{H} \\ | \\ \text{C}=\text{O} \\ | \\ \text{O}- \end{array}
\end{array}
$$

Diprotic acid or cation or acid	Monovalent anion or zwitterion or ampholyte	Divalent anion or anion or unprotonated anion or base

Carbonic acid:

$$ H_2CO_3 \quad \underset{\rightleftharpoons}{K_1} \quad HCO_3^- \quad \underset{\rightleftharpoons}{K_2} \quad CO_3^{2-} $$

acid	ampholyte	unprotonated anion or base

Any old diprotic acid:

$$ H_2A \quad \underset{\rightleftharpoons}{K_1} \quad HA^- \quad \underset{\rightleftharpoons}{K_2} \quad A^{2-} $$

acid	ampholyte	unprotonated anion or base

Terms such as "zwitterion," then, need confuse no one.

2.15 Distribution Functions

Just looking at the number of ionizations that polyprotic acids undergo, one can easily get the feeling that finding the pH of a solution of, say, carbonate species could be a very difficult problem. It really is not if one uses the right approximations and some distribution functions known as α-functions. Alpha-functions have already been derived for weak monoprotic acids. The functions will be more useful in the case of weak polyprotic acids. Their derivation is straightforward.

Take the case of diprotic acid H_2A first.

$$ H_2A \rightleftharpoons H^+ + HA^- \quad K_1 = \frac{[H^+][HA^-]}{[H_2A]} $$

$$ HA^- \rightleftharpoons H^+ + A^{2-} \quad K_2 = \frac{[H^+][A^{2-}]}{[HA^-]} $$

Define the analytical concentration of all A-bearing species, C_A:

$$C_A = [H_2A] + [HA^-] + [A^{2-}]$$

Define three α-functions:

$$\alpha_0 = \frac{[H_2A]}{C_A} = \frac{[H_2A]}{[H_2A] + [HA^-] + [A^{2-}]} \qquad (2.31)$$

$$\alpha_1 = \frac{[HA^-]}{C_A} = \frac{[HA^-]}{[H_2A] + [HA^-] + [A^{2-}]} \qquad (2.32)$$

$$\alpha_2 = \frac{[A^{2-}]}{C_A} = \frac{[A^{2-}]}{[H_2A] + [HA^-] + [A^{2-}]} \qquad (2.33)$$

Note that brackets signify an equilibrium concentration. The term $[H_2A]$ is the number of moles of H_2A (*not* HA^- nor A^{2-}) present per liter of solution.

If $[H^+]$, K_1, and K_2 are known, the values of α_0, α_1, and α_2 may be calculated. It remains to get α_0, α_1, and α_2 into terms of $[H^+]$, K_1, and K_2. Let us start with α_0 as defined in Equation 2.31. First, make every term in Equation 2.31 a function of $[H_2A]$, K_1, K_2, and/or $[H^+]$. $[H_2A]$ already meets this condition. Now try $[HA^-]$

$$K_1 = \frac{[H^+][HA^-]}{[H_2A]}$$

$$[HA^-] = \frac{K_1[H_2A]}{[H^+]}$$

Last, try $[A^{2-}]$

$$K_2 = \frac{[H^+][A^{2-}]}{[HA^-]}$$

$$[A^{2-}] = \frac{K_2[HA^-]}{[H^+]}$$

But

$$[HA^-] = \frac{K_1[H_2A]}{[H^+]}$$

$$[A^{2-}] = \frac{K_1 K_2[H_2A]}{[H^+]^2}$$

$$\alpha_0 = \frac{[H_2A]}{[H_2A] + \dfrac{K_1[H_2A]}{[H^+]} + \dfrac{K_1 K_2[H_2A]}{[H^+]^2}}$$

$$\alpha_0 = \frac{\cancel{[H_2A]}}{\cancel{[H_2A]}\left(1 + \dfrac{K_1}{[H^+]} + \dfrac{K_1 K_2}{[H^+]^2}\right)}$$

Add the fractions in the denominator.

$$\alpha_0 = \frac{1}{\dfrac{[H^+]^2 + [H^+]K_1 + K_1 K_2}{[H^+]^2}}$$

$$\alpha_0 = \frac{[H^+]^2}{[H^+]^2 + [H^+]K_1 + K_1K_2} \tag{2.34}$$

Now we can proceed to α_1 as defined in Equation 2.32. Make every term a function of $[HA^-]$, K_1, K_2, and/or $[H^+]$, just as was done for every term in Equation 2.31.

$$\alpha_1 = \frac{[HA^-]}{\dfrac{[H^+][HA^-]}{K_1} + [HA^-] + \dfrac{K_2[HA^-]}{[H^+]}}$$

$$\alpha_1 = \frac{[\cancel{HA^-}]}{[\cancel{HA^-}]\left(\dfrac{[H^+]}{K_1} + 1 + \dfrac{K_2}{[H^+]}\right)}$$

Add the fractions in the denominator.

$$\alpha_1 = \frac{1}{\dfrac{[H^+]^2 + [H^+]K_1 + K_1K_2}{[H^+]K_1}}$$

$$\alpha_1 = \frac{[H^+]K_1}{[H^+]^2 + [H^+]K_1 + K_1K_2} \tag{2.35}$$

At last, we can evaluate α_2 as defined in Equation 2.33. Make every term a function of $[A^{2-}]$, K_1, K_2, and/or $[H^+]$.

$$\alpha_2 = \frac{[A^{2-}]}{\dfrac{[H^+]^2[A^{2-}]}{K_1K_2} + \dfrac{[H^+][A^{2-}]}{K_2} + [A^{2-}]}$$

$$\alpha_2 = \frac{K_1K_2}{[H^+]^2 + [H^+]K_1 + K_1K_2}$$

Now assemble α_0, α_1, and α_2

$$\alpha_0 = \frac{[H_2A]}{C_A} = \frac{[H_2A]}{[H_2A] + [HA^-] + [A^{2-}]} = \frac{[H^+]^2}{[H^+]^2 + [H^+]K_1 + K_1K_2}$$

$$\alpha_1 = \frac{[HA^-]}{C_A} = \frac{[HA^-]}{[H_2A] + [HA^-] + [A^{2-}]} = \frac{[H^+]K_1}{[H^+]^2 + [H^+]K_1 + K_1K_2}$$

$$\alpha_2 = \frac{[A^{2-}]}{C_A} = \frac{[A^{2-}]}{[H_2A] + [HA^-] + [A^{2-}]} = \frac{K_1K_2}{[H^+]^2 + [H^+]K_1 + K_1K_2}$$

Note how each α-fraction has the same denominator, and how each denominator term appears in the numerator of either α_0, α_1, or α_2. Do not, however equate $[H^+]^2$ and $[H_2A]$, $[H^+]K_1$ and $[HA^-]$, or K_1K_2 and $[A^{2-}]$. In a word

$$\left.\begin{array}{c} [H^+]^2 \neq [H_2A] \\[2mm] [H^+]K_1 \neq [HA^-] \\[2mm] K_1K_2 \neq [A^{2-}] \end{array}\right\} \quad \text{most of the time}$$

Equilibrium concentrations can only be calculated when both the α-fractions *and* the analytical concentration are known.

$$[H_2A] = \alpha_0\, C_A$$

$$[HA^-] = \alpha_1\, C_A$$

$$[A^{2-}] = \alpha_2\, C_A$$

What about the α-fractions for triprotic acids? Consider triprotic acid H_3Z.

$$H_3Z \rightleftharpoons H^+ + H_2Z^- \qquad K_1 = \frac{[H^+]\,[H_2Z^-]}{[H_3Z]}$$

$$H_2Z^- \rightleftharpoons H^+ + HZ^{2-} \qquad K_2 = \frac{[H^+]\,[HZ^{2-}]}{[H_2Z^-]}$$

$$HZ^{2-} \rightleftharpoons H^+ + Z^{3-} \qquad K_3 = \frac{[H^+]\,[Z^{3-}]}{[HZ^{2-}]}$$

Let

$$C_Z = [H_3Z] + [H_2Z^-] + [HZ^{2-}] + [Z^{3-}]$$

We can write each α-function in terms of $[H^+]$, K_1, K_2, and K_3. The derivations are similar to those just shown for the α-functions of diprotic acids.

$$\alpha_0 = \frac{[H_3Z]}{C_Z} = \frac{[H^+]^3}{[H^+]^3 + [H^+]^2 K_1 + [H^+] K_1 K_2 + K_1 K_2 K_3} \qquad (2.37)$$

$$\alpha_1 = \frac{[H_2Z^-]}{C_Z} = \frac{[H^+]^2 K_1}{[H^+]^3 + [H^+]^2 K_1 + [H^+] K_1 K_2 + K_1 K_2 K_3} \qquad (2.38)$$

$$\alpha_2 = \frac{[HZ^{2-}]}{C_Z} = \frac{[H^+] K_1 K_2}{[H^+]^3 + [H^+]^2 K_1 + [H^+] K_1 K_2 + K_1 K_2 K_3} \qquad (2.39)$$

$$\alpha_3 = \frac{[Z^{3-}]}{C_Z} = \frac{K_1 K_2 K_3}{[H^+]^3 + [H^+]^2 K_1 + [H^+] K_1 K_2 + K_1 K_2 K_3} \qquad (2.40)$$

The distribution (or α) functions just derived can be used to determine which of the many species that a weak acid can generate predominates at a given pH value. Tables of α-values, and graphs of some of these values are found at the end of the chapter.

Example: Which of the succinate-bearing species predominates at pH 2.0? At pH 3.0? At pH 6.0? Use Table 2.11 on page 00.

At pH 2.0, α_0 is largest. Hence H_2Su predominates.
At pH 3.0, α_0 is largest, and H_2Su predominates, followed by HSu^-.
At pH 6.0, α_2 is largest. Hence the base, Su^{2-} predominates.

Example: If a solution is made by dissolving 0.10 mole of H_3PO_4 in water, adjusting the pH to 2.5, and diluting the solution to 1.00 liter, keeping pH constant, what are the values of $[H_3PO_4]$, $[H_2PO_4^-]$, $[HPO_4^{2-}]$, and $[PO_4^{3-}]$? Use Table 2.12 on page 00.

$$[H_3PO_4] = \alpha_0\, C_{PO_4} = (0.31)(0.10) = 0.031M$$

$$[H_2PO_4^-] = \alpha_1\, C_{PO_4} = (0.69)(0.10) = 0.069M$$

$$[HPO_4^{2-}] = \alpha_2\, C_{PO_4} = (1.4 \times 10^{-5})(0.10) = 1.4 \times 10^{-6}M$$

$$[PO_4^{3-}] = \alpha_3 C_{PO_4} = (\sim0)(0.10) \simeq 0M$$

At this low pH $[H_3PO_4]$ and $[H_2PO_4^-]$ predominate. There is very little HPO_4^{2-}, and almost no PO_4^{3-}

2.16 Solutions of Weak Polyprotic Acids in Water

If pH and an analytical concentration are specified, it is easy, using α-functions, to find equilibrium concentrations of the various species to which a weak polyprotic acid can give rise. Another sort of problem frequently addressed is this: Given an analytical concentration of a weak diprotic acid, H_2A, and the values of K_1 and K_2, find the pH of the solution. Two approaches may be taken to solve the problem. One is intuitive, the other more rigorous. Both yield the same result: A weak diprotic acid, placed in water (or a weak triprotic acid, for that matter), behaves in much the same fashion as a weak monoprotic acid.

First, the intuitive approach will be taken. Carbonic acid is a typical diprotic weak acid.

$$H_2CO_3 \rightleftharpoons H^+ + HCO_3^- \quad K_1 = 3.5 \times 10^{-7}$$

$$HCO_3^- \rightleftharpoons H^+ + CO_3^{2-} \quad K_2 = 6.0 \times 10^{-11}$$

Note that K_1 is much larger (by a factor of between 5000 and 6000) than K_2. Thus the dissociation of HCO_3^- is a negligible source of hydrogen ions. The dissociation of H_2CO_3 is the primary source of hydrogen ions in the solution. Therefore, H_2CO_3, placed in water, behaves like a weak monoprotic acid. The approximations for figuring the value of $[H^+]$ are like those for figuring $[H^+]$ when a weak monoprotic acid, like acetic acid, is placed in water.

First approximation:

$$[H^+]' = (K_1 C_{H_2CO_3})^{1/2} \tag{2.41}$$

Second approximation:

$$[H^+]'' = (K_1(C_{H_2CO_3} - [H^+]'))^{1/2}$$

Third approximation:

$$[H^+]''' = (K_1(C_{H_2CO_3} - [H^+]''))^{1/2}$$

and so forth.

The intuitive approach can be accepted without question, particularly if you are in a hurry. The other approach is more rigorous and more intellectually satisfying.

Suppose that a weak diprotic acid, H_2A, is placed in water in known analytical concentration, C_{H_2A}. These equilibria are important:

$$H_2A \rightleftharpoons H^+ + HA^- \quad K_1 = \frac{[H^+][HA^-]}{[H_2A]}$$

$$HA^- \rightleftharpoons H^+ + A^{2-} \quad K_2 = \frac{[H^+][A^{2-}]}{[HA^-]}$$

$$H_2O \rightleftharpoons H^+ + OH^- \quad K_w = [H^+][OH^-]$$

The charge balance equation can be written.

$$\text{Positive ion} \qquad\qquad \text{Negative ions}$$

$$H^+ \qquad\qquad\qquad HA^-, A^{2-}, OH^-$$

$$[H^+] = [HA^-] + 2[A^{2-}] + [OH^-]$$

Now the α-functions are introduced.

$$C_A = C_{H_2A}$$

$$[HA^-] = \alpha_1 C_{H_2A}$$

$$[A^{2-}] = \alpha_2 C_{H_2A}$$

$$[H^+] = \alpha_1 C_{H_2A} + 2\alpha_2 C_{H_2A} + [OH^-] \qquad\qquad (2.42)$$

Approximations are begun: In an acid solution, $[OH^-]$ is small.

$$[H^+] \cong \alpha_1 \cdot C_{H_2A} + 2\alpha_2 C_{H_2A}$$

The equation is not as handy to work with as it could be. At low pH values that would be encountered in solutions of various weak acids in water, $\alpha_1 \gg \alpha_2$.

$$\alpha_1 C_{H_2A} \cong \alpha_1 C_{H_2A} + 2\alpha_2 C_{H_2A}$$

Then

$$[H^+] \cong \alpha_1 C_{H_2A}$$

$$[H^+] = \left[\frac{K_1[H^+]}{K_1K_2 + K_1[H^+] + [H^+]^2} \right] \cdot C_{H_2A}$$

At low pH values,

$$[H^+]^2 \gg K_1[H^+] + K_1K_2$$

$$[H^+] = \frac{K_1[H^+]}{[H^+]^2} \cdot C_{H_2A}$$

$$[H^+] = \frac{K_1}{[H^+]} C_{H_2A}$$

$$[H^+]^2 = K_1 C_{H_2A}$$

First approximation:

$$[H^+]' = (K_1 C_{H_2A})^{1/2} \qquad\qquad (2.43)$$

Second approximation:
Substitute $[H^+]'$ into Eq. 2.42.

$$\alpha_1' = \frac{[H^+]'K_1}{([H^+]')^2 + [H^+]'K_1 + K_1K_2}$$

$$\alpha_2' = \frac{K_1K_2}{([H^+]')^2 + [H^+]'K_1 + K_1K_2}$$

$$[H^+]'' = \alpha_1'C_{H_2A} + 2\alpha_2'C_{H_2A} + \frac{K_w}{[H^+]'}$$

Third approximation:
Substitute $[H^+]''$ into Eq. 2.42.

$$\alpha_1'' = \frac{[H^+]''K_1}{([H^+]'')^2 + [H^+]''K_1 + K_1K_2}$$

$$\alpha_2'' = \frac{K_1K_2}{([H^+]'')^2 + [H^+]''K_1 + K_1K_2}$$

$$[H^+]''' = \alpha_1''C_{H_2A} + 2\alpha_2''C_{H_2A} + \frac{K_w}{[H^+]''}$$

and so forth.

Example: What is the pH of a 0.010M solution of weak acid H_2A if $K_1 = 1.0 \times 10^{-3}$ and $K_2 = 1.0 \times 10^{-4}$?

First approximation:

$$[H^+]' = (K_1C_{H_2A})^{1/2} \qquad (2.43)$$

$$[H^+]' = ((1.0 \times 10^{-3})(1.0 \times 10^{-2}))^{1/2}$$

$$[H^+]' = 3.2 \times 10^{-3}M$$

$$pH = 2.50$$

Second approximation:

$$\alpha_1' = \frac{[H^+]'K_1}{([H^+]')^2 + [H^+]'K_1 + K_1K_2}$$

$$= \frac{(3.2 \times 10^{-3}) \cdot (1.0 \times 10^{-3})}{(3.2 \times 10^{-3})^2 + (3.2 \times 10^{-3})(1.0 \times 10^{-3}) + (1.0 \times 10^{-3})(1.0 \times 10^{-4})}$$

$$= 0.24$$

$$\alpha_2' = \frac{K_1K_2}{([H^+]')^2 + [H^+]'K_1 + K_1K_2}$$

$$= \frac{(1.0 \times 10^{-3})(1.0 \times 10^{-4})}{(3.2 \times 10^{-3})^2 + (3.2 \times 10^{-3})(1.0 \times 10^{-3}) + (1.0 \times 10^{-3})(1.0 \times 10^{-4})}$$

$$= 7.5 \times 10^{-3}$$

$$[H^+]'' = \alpha_1'C_{H_2A} + 2\alpha_2'C_{H_2A} + \frac{K_w}{[H^+]'}$$

$$= (0.24)(1.0 \times 10^{-2}) + (2)(7.5 \times 10^{-3})(1.0 \times 10^{-2}) + \frac{1.0 \times 10^{-14}}{(3.2 \times 10^{-3})}$$

$$= 2.5 \times 10^{-3}M$$

$$pH = 2.60$$

Third approximation:

$$\alpha_1'' = \frac{[H^+]''K_1}{([H^+]'')^2 + [H^+]''K_1 + K_1K_2}$$

$$= \frac{(2.5 \times 10^{-3})(1.0 \times 10^{-3})}{(2.5 \times 10^{-3})^2 + (2.5 \times 10^{-3})(1.0 \times 10^{-3}) + (1.0 \times 10^{-3})(1.0 \times 10^{-4})}$$

$$= 0.28$$

$$\alpha_2'' = \frac{K_1 K_2}{([H^+]'')^2 + [H^+]'' K_1 + K_1 K_2}$$

$$= \frac{(1.0 \times 10^{-3})(1.0 \times 10^{-4})}{(2.5 \times 10^{-3})^2 + (2.5 \times 10^{-3})(1.0 \times 10^{-3}) + (1.0 \times 10^{-3})(1.0 \times 10^{-4})}$$

$$= 0.011$$

$$[H^+]''' = \alpha_1'' \cdot C_{H_2A} + 2\alpha_2'' C_{H_2A} + \frac{K_w}{[H^+]''}$$

$$= (0.28)(1.0 \times 10^{-2}) + (2)(0.011)(1.0 \times 10^{-2}) + \frac{1.00 \times 10^{-14}}{2.5 \times 10^{-3}}$$

$$= 3.0 \times 10^{-3} M$$

$$pH = 2.52$$

The values of pH converge within a value of about ±0.02 pH unit after further approximation.

Weak triprotic acids can be treated in the same way as weak diprotic acids. In the case of weak triprotic acid H_3Z,

First approximation:

$$[H^+]' = (K_1 C_{H_3Z})^{1/2}$$

Second approximation:

$$\alpha_1' = \frac{([H^+]')^2 K_1}{([H^+]')^3 + ([H^+]')^2 K_1 + [H^+]' \cdot K_1 K_2 + K_1 K_2 K_3}$$

$$\alpha_2' = \frac{[H^+]' K_1 K_2}{([H^+]')^3 + ([H^+]')^2 K_1 + [H^+]' \cdot K_1 K_2 + K_1 K_2 K_3}$$

$$\alpha_3' = \frac{K_1 K_2 K_3}{([H^+]')^3 + ([H^+]')^2 K_1 + [H^+]' K_1 K_2 + K_1 K_2 K_3}$$

$$[H^+]'' = \alpha_1' C_{H_3Z} + 2\alpha_2' C_{H_3Z} + 3\alpha_3' C_{H_3Z} + \frac{K_w}{[H^+]'}$$

Third approximation:

$$\alpha_1'' = \frac{([H^+]'')^2 \cdot K_1}{([H^+]'')^3 + ([H^+]'')^2 \cdot K_1 + [H^+]'' K_1 K_2 + K_1 K_2 K_3}$$

$$\alpha_2'' = \frac{[H^+]'' \cdot K_1 K_2}{([H^+]'')^3 + ([H^+]'')^2 K_1 + [H^+]'' \cdot K_1 K_2 + K_1 K_2 K_3}$$

$$\alpha_3'' = \frac{K_1 K_2 K_3}{([H^+]'')^3 + ([H^+]'')^2 \cdot K_1 + [H^+]'' \cdot K_1 K_2 + K_1 K_2 K_3}$$

$$[H^+]''' = \alpha_1''C_{H_2A} + 2\alpha_2''C_{H_2A} + 3\alpha_3''C_{H_2A} + \frac{K_w}{[H^+]''}$$

and so forth.

2.17 Buffers of Weak Polyprotic Acids

Weak polyprotic acids and their subspecies provide more controlled pH ranges than do weak monoprotic acids. Take, for comparison, the buffers of acetic acid, carbonic acid, and phosphoric acid.

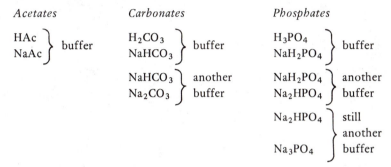

The only buffers that can be made in the acetate system are mixtures of acetic acid and the acetate ion. Acetic acid serves as the acid, buffering the effect of added strong base. Acetate ion serves as the base, buffering the effect of added strong acid.

By contrast, the carbonate system offers two separate buffers. The first is the carbonic acid and bicarbonate mixture. Carbonic acid serves as the acid, buffering the effect of added strong base. Bicarbonate ion, an ampholyte, serves as the base, buffering the effect of added strong acid. In the other buffer bicarbonate ion and carbonate ion are mixed. Bicarbonate ion, an ampholyte, here serves as the acid, buffering the effect of added strong base. Carbonate ion serves as the base, buffering the effect of added strong acid.

The phosphate system offers three separate buffers. The first is the phosphoric acid and dihydrogen phosphate mixture. Phosphoric acid serves as the acid, buffering the effect of added strong base. Dihydrogen phosphate ion, an ampholyte, serves as the base, buffering the effect of added strong acid. In the next buffer, dihydrogen phosphate ion and hydrogen phosphate ion are mixed. Dihydrogen phosphate ion, an ampholyte, serves as the acid here, buffering the effect of added strong base. Hydrogen phosphate ion, also an ampholyte, serves as the base here, buffering the effect of added strong acid. In the very last buffer, hydrogen phosphate ion and phosphate ion are mixed. Hydrogen phosphate ion, still an ampholyte, serves as the acid here, buffering the effect of added strong base. Unprotonated phosphate ion serves as the base, buffering the effect of added strong acid.

The calculations connected with the buffer systems of polyprotic acids are no more complex than those for the buffers of monoprotic acids. Consider the buffers of weak diprotic acid H_2A.

> **Buffer 1** / A mixture of H_2A and $NaHA$, with both C_{H_2A} and C_{HA^-} known.

$$K_1 = \frac{[H^+][HA^-]}{[H_2A]}$$

$$[H^+] = K_1 \frac{[H_2A]}{[HA^-]}$$

From this, a rigorous equation may be developed:

$$[H^+] = K_1 \frac{C_{H_2A} - [H^+] + \dfrac{K_w}{[H^+]}}{C_{HA^-} + [H^+] - \dfrac{K_w}{[H^+]}}$$

Approximations may be begun:

First approximation:

$$[H^+]' = K_1 \frac{C_{H_2A}}{C_{HA^-}} \qquad \text{and from this,} \qquad pH = pK_1 + \log \frac{C_{HA^-}}{C_{H_2A}}$$

$$[OH^-]' = \frac{K_w}{[H^+]'}$$

If $[H^+]' > [OH^-]'$

Second approximation:

$$[H^+]'' = K_1 \frac{(C_{H_2A} - [H^+]')}{(C_{HA^-} + [H^+]')}$$

Third approximation:

$$[H^+]''' = K_1 \frac{(C_{H_2A} - [H^+]'')}{(C_{HA^-} + [H^+]'')}$$

If $[OH^-]' > [H^+]'$

Second approximation:

$$[H^+]'' = K_1 \frac{(C_{H_2A} + [OH^-]')}{(C_{HA^-} - [OH^-]')}$$

Third approximation:

$$[H^+]''' = K_1 \frac{(C_{H_2A} + [OH^-]'')}{(C_{HA^-} - [OH^-]'')}$$

and so forth.

> **Buffer 2** / A mixture of $NaHA$ and Na_2A, both with C_{HA^-} and $C_{A^{2-}}$ known:

$$K_2 = \frac{[H^+][A^{2-}]}{[HA^-]}$$

$$[H^+] = K_2 \frac{[HA^-]}{[A^{2-}]}$$

From this, a rigorous equation may be developed:

$$[H^+] = K_2 \left(\frac{C_{HA^-} - [H^+] + [OH^-]}{C_{A^{2-}} + [H^+] - [OH^-]} \right)$$

Approximations may be begun:

First approximation:

$$[H^+]' = K_2 \frac{C_{HA^-}}{C_{A^{2-}}} \quad \text{and from this,} \quad pH = pK_2 + \log \frac{C_{A^{2-}}}{C_{HA^-}}$$

$$[OH^-]' = \frac{K_w}{[H^+]'}$$

If $[H^+]' > [OH^-]'$

Second approximation:

$$[H^+]'' = K_2 \left(\frac{C_{HA^-} - [H^+]'}{C_{A^{2-}} + [H^+]'} \right)$$

Third approximation:

$$[H^+]''' = K_2 \left(\frac{C_{HA^-} - [H^+]''}{C_{A^{2-}} + [H^+]''} \right)$$

If $[OH^-]' > [H^+]'$

Second approximation:

$$[H^+]'' = K_2 \left(\frac{C_{HA^-} + [OH^-]'}{C_{A^{2-}} - [OH^-]'} \right)$$

Third approximation:

$$[H^+]''' = K_2 \left(\frac{C_{HA^-} + [OH^-]''}{C_{A^{2-}} - [OH^-]''} \right)$$

Example: Calculate the pH of a solution 0.0050M in $NaHCO_3$ and 0.0050M in Na_2CO_3. For carbonic acid, $K_1 = 3.5 \times 10^{-7}$ and $K_2 = 6.0 \times 10^{-11}$.

$$C_{HCO_3^-} = 0.005M$$

$$C_{CO_3^{2-}} = 0.005M$$

First approximation:

$$[H^+]' = K_2 \frac{C_{HCO_3^-}}{C_{CO_3^{2-}}} = (6.0 \times 10^{-11}) \left(\frac{5.0 \times 10^{-3}}{5.0 \times 10^{-3}} \right) = 6.0 \times 10^{-11} M$$

$$[OH^-]' = \frac{K_w}{[H^+]'} = \frac{1.00 \times 10^{-14}}{6.0 \times 10^{-11}} = 1.7 \times 10^{-4} M$$

$$[OH^-]' > [H^+]'$$

Second approximation:

$$[H^+]'' = K_2 \left(\frac{C_{HCO_3^-} + [OH^-]'}{C_{CO_3^{2-}} - [OH^-]'} \right)$$

$$= (6.0 \times 10^{-11}) \left(\frac{5 \times 10^{-3} + 1.7 \times 10^{-4}}{5 \times 10^{-3} - 1.7 \times 10^{-4}} \right) = 6.4 \times 10^{-11} M$$

$$[OH^-]'' = \frac{K_w}{[H^+]''} = \frac{1.00 \times 10^{-14}}{6.4 \times 10^{-11}} = 1.6 \times 10^{-4} M$$

Third approximation:

$$[H^+]''' = K_2 \left(\frac{C_{HCO_3^-} + [OH^-]''}{C_{CO_3^{2-}} - [OH^-]''} \right)$$

$$= (6.0 \times 10^{-11}) \left(\frac{5 \times 10^{-3} + 1.6 \times 10^{-4}}{5 \times 10^{-3} - 1.6 \times 10^{-4}} \right) = 6.4 \times 10^{-11} M$$

$$[H^+]'' = [H^+]'''$$

The approximations have converged.

$$pH = -\log (6.4 \times 10^{-11}) = 10.19$$

Next, consider the buffers of weak triprotic acid H_3Z.

⯈ Buffer 1 / A mixture of H_3Z and NaH_2Z, with both C_{H_3Z} and $C_{H_2Z^-}$ known.

$$K_1 = \frac{[H^+][H_2Z^-]}{[H_3Z]}$$

$$[H^+] = K_1 \frac{[H_3Z]}{[H_2Z^-]}$$

From this, a rigorous equation may be developed:

$$[H^+] = K_1 \frac{(C_{H_3Z} - [H^+] + [OH^-])}{(C_{H_2Z^-} + [H^+] - [OH^-])}$$

Approximations may be begun.

First approximation:

$$[H^+]' = K_1 \frac{C_{H_3Z}}{C_{H_2Z^-}} \quad \text{and, from this,} \quad pH = pK_1 + \log \frac{C_{H_2Z^-}}{C_{H_3Z}}$$

$$[OH^-]' = \frac{K_w}{[H^+]'}$$

If $[H^+]' > [OH^-]'$ If $[OH^-]' > [H^+]'$

Second approximation: Second approximation:

$$[H^+]'' = K_1 \left(\frac{C_{H_3Z} - [H^+]'}{C_{H_2Z^-} + [H^+]'} \right) \quad [H^+]'' = K_1 \left(\frac{C_{H_3Z} + [OH^-]'}{C_{H_2Z^-} - [OH^-]'} \right)$$

Third approximation: Third approximation:

$$[H^+]''' = K_1 \left(\frac{C_{H_3Z} - [H^+]''}{C_{H_2Z^-} + [H^+]''} \right) \quad [H^+]''' = K_1 \left(\frac{C_{H_3Z} + [OH^-]''}{C_{H_2Z^-} - [OH^-]''} \right)$$

and so forth.

⯈ Buffer 2 / A mixture of NaH_2Z and Na_2HZ, with both $C_{H_2Z^-}$ and $C_{HZ^{2-}}$ known.

$$K_2 = \frac{[H^+][HZ^{2-}]}{[H_2Z^-]}$$

$$[H^+] = K_2 \frac{[H_2Z^-]}{[HZ^{2-}]}$$

From this, a rigorous equation may be developed:

$$[H^+] = K_2 \left(\frac{C_{H_2Z^-} - [H^+] + [OH^-]}{C_{HZ^{2-}} + [H^+] - [OH^-]} \right)$$

142 / Acid-Base Chemistry

Approximations may be begun.

First approximation:

$$[H^+]' = K_2 \frac{C_{H_2Z^-}}{C_{HZ^{2-}}} \quad \text{and, from this,} \quad pH = pK_2 + \log \frac{C_{HZ^{2-}}}{C_{H_2Z^-}}$$

$$[OH^-]' = \frac{K_w}{[H^+]'}$$

If $[H^+]' > [OH^-]'$ If $[OH^-]' > [H^+]'$

Second approximation: Second approximation:

$$[H^+]'' = K_2 \left(\frac{C_{H_2Z^-} - [H^+]'}{C_{HZ^{2-}} + [H^+]'}\right) \qquad [H^+]'' = K_2 \left(\frac{C_{H_2Z^-} + [OH^-]'}{C_{HZ^{2-}} - [OH^-]'}\right)$$

Third approximation: Third approximation:

$$[H^+]''' = K_2 \left(\frac{C_{H_2Z^-} - [H^+]''}{C_{HZ^{2-}} + [H^+]''}\right) \qquad [H^+]''' = K_2 \left(\frac{C_{H_2Z^-} + [OH^-]''}{C_{HZ^{2-}} - [OH^-]''}\right)$$

and so forth.

➤ Buffer 3 / A mixture of Na_2HZ and Na_3Z, with both $C_{HZ^{2-}}$ and $C_{Z^{3-}}$ known.

$$K_3 = \frac{[H^+][Z^{3-}]}{[HZ^{2-}]}$$

$$[H^+] = K_3 \frac{[HZ^{2-}]}{[Z^{3-}]}$$

From this, a rigorous equation can be developed:

$$[H^+] = K_3 \left(\frac{C_{HZ^{2-}} - [H^+] + [OH^-]}{C_{Z^{3-}} + [H^+] - [OH^-]}\right)$$

First approximation:

$$[H^+]' = K_3 \frac{C_{HZ^{2-}}}{C_{Z^{3-}}} \quad \text{and from this,} \quad pH = pK_3 + \log \frac{C_{Z^{3-}}}{C_{HZ^{2-}}}$$

$$[OH^-]' = \frac{K_w}{[H^+]'}$$

If $[H^+]' > [OH^-]'$ If $[OH^-]' > [H^+]'$

Second approximation: Second approximation:

$$[H^+]'' = K_3 \frac{(C_{HZ^{2-}} - [H^+]')}{(C_{Z^{3-}} + [H^+]')} \qquad [H^+]'' = K_3 \frac{(C_{HZ^{2-}} + [OH^-]')}{(C_{Z^{3-}} - [OH^-]')}$$

Third approximation: Third approximation:

$$[H^+]''' = K_3 \frac{(C_{HZ^{2-}} - [H^+]'')}{(C_{Z^{3-}} + [H^+]'')} \qquad [H^+]''' = K_3 \frac{(C_{HZ^{2-}} + [OH^-]'')}{(C_{Z^{3-}} - [OH^-]'')}$$

and so forth.

Example: Calculate the pH of a solution 0.050M in H_3PO_4 and 0.050M in NaH_2PO_4.

$$K_1 = 7.1 \times 10^{-3} \qquad C_{H_3PO_4} = 0.050M$$

$$K_2 = 6.2 \times 10^{-8} \qquad C_{H_2PO_4^-} = 0.050M$$

$$K_3 = 4.4 \times 10^{-13}$$

First approximation:

$$[H^+]' = K_1 \frac{C_{H_3PO_4}}{C_{H_2PO_4^-}} = (7.1 \times 10^{-3})\left(\frac{5.0 \times 10^{-2}}{5.0 \times 10^{-2}}\right) = 7.1 \times 10^{-3}M$$

$$[OH^-]' = \frac{K_w}{[H^+]'} = \frac{1.00 \times 10^{-14}}{7.1 \times 10^{-3}} = 1.4 \times 10^{-12}M$$

$$[H^+]' \gg [OH^-]'$$

Second approximation:

$$[H^+]'' = K_1 \left(\frac{C_{H_3PO_4} - [H^+]'}{C_{H_2PO_4^-} + [H^+]'}\right)$$

$$= 7.1 \times 10^{-3} \left(\frac{5.0 \times 10^{-2} - 7.1 \times 10^{-3}}{5.0 \times 10^{-2} + 7.1 \times 10^{-3}}\right) = 5.3 \times 10^{-3}M$$

Third approximation:

$$[H^+]''' = K_1 \left(\frac{C_{H_3PO_4} - [H^+]''}{C_{H_2PO_4^-} + [H^+]''}\right)$$

$$= (7.1 \times 10^{-3})\left(\frac{5.0 \times 10^{-2} - 5.3 \times 10^{-3}}{5.0 \times 10^{-2} + 5.3 \times 10^{-3}}\right) = 5.7 \times 10^{-3}M$$

$$pH''' = 2.24$$

Fourth approximation:

$$[H^+]'''' = K_1 \left(\frac{C_{H_3PO_4} - [H^+]'''}{C_{H_2PO_4^-} + [H^+]'''}\right)$$

$$= (7.1 \times 10^{-3})\left(\frac{5.0 \times 10^{-2} - 5.7 \times 10^{-3}}{5.0 \times 10^{-2} + 5.7 \times 10^{-3}}\right)$$

$$= 5.6 \times 10^{-3}M$$

$$pH'''' = 2.25$$

It is a good idea to be able to prepare buffers of a given pH. A problem like this is often faced: "Suppose that you have available H_3PO_4, NaH_2PO_4, Na_2HPO_4, Na_3PO_4, and distilled water. How would you prepare 1.00 liter of a buffer of pH 7.21 whose total concentration of phosphates is 0.100M?"

First what two phosphate-bearing species ought to be used? Near a pH of 7.2, the values of α_1 and α_2 are high, while α_0 and α_3 are near zero (Table 2.1 and Figure 2.2). Thus, at pH 7.2, $H_2PO_4^-$ and HPO_4^{2-} are the dominant phosphate species.

DISTRIBUTION DIAGRAM FOR PHOSPHORIC ACID

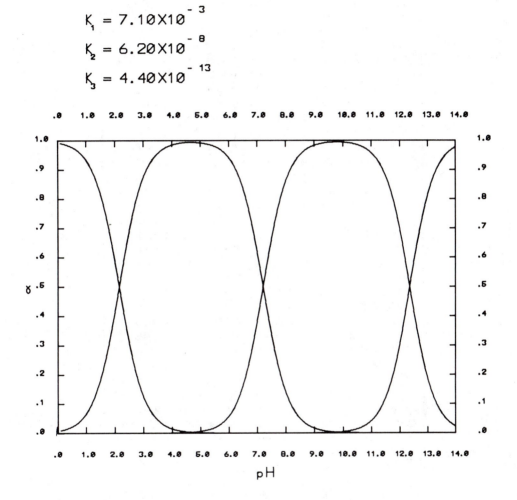

$$K_1 = 7.10 \times 10^{-3}$$
$$K_2 = 6.20 \times 10^{-8}$$
$$K_3 = 4.40 \times 10^{-13}$$

Figure 2.2 *Distribution Diagram for Phosphoric Acid.*

Table 2.1
α VALUES FOR PHOSPHORIC ACIDS

$$K_1 = 7.1E - 3$$
$$K_2 = 6.2E - 8$$
$$K_3 = 4.4E - 13$$

pH	Alpha(0)	Alpha(1)	Alpha(2)	Alpha(3)
0.50	0.978E + 00	0.220E - 01	0.431E - 08	0.599E - 20
1.00	0.934E + 00	0.663E - 01	0.411E - 07	0.181E - 18
1.50	0.817E + 00	0.183E + 00	0.359E - 06	0.500E - 17
2.00	0.585E + 00	0.415E + 00	0.257E - 05	0.113E - 15
2.50	0.308E + 00	0.692E + 00	0.136E - 04	0.189E - 14
3.00	0.123E + 00	0.876E + 00	0.543E - 04	0.239E - 13
3.50	0.426E - 01	0.957E + 00	0.188E - 03	0.261E - 12
4.00	0.139E - 01	0.986E + 00	0.611E - 03	0.269E - 11
4.50	0.443E - 02	0.994E + 00	0.195E - 02	0.271E - 10
5.00	0.140E - 02	0.992E + 00	0.615E - 02	0.271E - 09
5.50	0.437E - 03	0.980E + 00	0.192E - 01	0.267E - 08
6.00	0.133E - 03	0.941E + 00	0.584E - 01	0.257E - 07
6.50	0.372E - 04	0.836E + 00	0.164E + 00	0.228E - 06
7.00	0.869E - 05	0.617E + 00	0.383E + 00	0.168E - 05
7.50	0.150E - 05	0.338E + 00	0.662E + 00	0.921E - 05
8.00	0.196E - 06	0.139E + 00	0.861E + 00	0.379E - 04
8.50	0.216E - 07	0.485E - 01	0.951E + 00	0.132E - 03
9.00	0.223E - 08	0.159E - 01	0.984E + 00	0.433E - 03
9.50	0.226E - 09	0.507E - 02	0.994E + 00	0.138E - 02
10.00	0.226E - 10	0.160E - 02	0.994E + 00	0.437E - 02
10.50	0.224E - 11	0.503E - 03	0.986E + 00	0.137E - 01
11.00	0.218E - 12	0.154E - 03	0.958E + 00	0.421E - 01
11.50	0.199E - 13	0.448E - 04	0.878E + 00	0.122E + 00
12.00	0.158E - 14	0.112E - 04	0.694E + 00	0.306E + 00
12.50	0.950E - 16	0.213E - 05	0.418E + 00	0.582E + 00
13.00	0.421E - 17	0.299E - 06	0.185E + 00	0.815E + 00
13.50	0.152E - 18	0.342E - 07	0.671E - 01	0.933E + 00
14.00	0.505E - 20	0.358E - 08	0.222E - 01	0.978E + 00

$$C_{PO_4} = 0.10M$$

$$C_{PO_4} = C_{H_2PO_4^-} + C_{HPO_4^{2-}}$$

at pH 7.21,

$$[H^+] = 6.2 \times 10^{-8}M$$

First approximation:

$$[H^+]' = K_2 \frac{C_{H_2PO_4^-}}{C_{HPO_4^{2-}}}$$

$$(6.2 \times 10^{-8}) = (6.2 \times 10^{-8}) \frac{C_{H_2PO_4^-}}{C_{HPO_4^{2-}}}$$

$$C_{H_2PO_4^-} = C_{HPO_4^{2-}}$$

$$0.10 = C_{H_2PO_4^-} + C_{HPO_4^{2-}}$$

$$C_{H_2PO_4^-} = 0.050M$$

$$C_{HPO_4^{2-}} = 0.050M$$

$$[OH^-]' = \frac{K_w}{[H^+]'} = \frac{1.00 \times 10^{-14}}{6.2 \times 10^{-8}} = 1.6 \times 10^{-7}$$

The buffer is basic, but just barely. The general expression is

$$[H^+] = K_2 \frac{(C_{H_2PO_4^-} - [H^+] + [OH^-])}{(C_{HPO_4^{2-}} + [H^+] - [OH^-])}$$

We can make a second approximation:

$$[H^+]'' = K_2 \frac{(C_{H_2PO_4^-}) - [H^+]' + [OH^-]')}{(C_{HPO_4^{2-}} + [H^+]' - [OH^-]')}$$

$$= (6.2 \times 10^{-8}) \frac{(0.050 - 6.2 \times 10^{-8} + 1.6 \times 10^{-7})}{(0.050 + 6.2 \times 10^{-8} - 1.6 \times 10^{-7})} = 6.2 \times 10^{-8}M$$

$[H^+]'$ and $[H^+]''$ are identical.

To make the buffer, we dissolve 0.05 mole of NaH_2PO_4 and 0.05 mole of Na_2HPO_4 in the same beaker and then dilute the solution to 1.00 liter.

2.18   The Unprotonated Anions of Weak Acids

If a salt such as Na_2CO_3 or Na_3PO_4 is placed in water, the resulting solution will be basic. The reason is that such unprotonated anions act as bases in water. The process is called hydrolysis, and is in every way similar to the hydrolysis reaction of the anion of a monoprotic acid.

NaAc in water: $Ac^- + H_2O \rightleftharpoons HAc + OH^-$ $K_{hy} = \dfrac{[OH^-][HAc]}{[Ac^-]}$

Na_2CO_3 in water: $CO_3^{2-} + H_2O \rightleftharpoons HCO_3^- + OH^-$ $K_{hy} = \dfrac{[OH^-][HCO_3^-]}{[CO_3^{2-}]}$

Na_3PO_4 in water: $PO_4^{3-} + H_2O \rightleftharpoons HPO_4^{2-} + OH^-$ $K_{hy} = \dfrac{[OH^-][HPO_4^{2-}]}{[PO_4^{3-}]}$

The hydrolysis constants, as we have chosen to call them, are not often found in tables but may be evaluated from K_w and acid dissociation constants, just as are the hydrolysis constants for the anions of weak monoprotic acids.

For Na_2CO_3 in H_2O,

$$K_{hy} = \frac{[OH^-][HCO_3^-]}{[CO_3^{2-}]}$$

$$K_{hy} = \frac{[H^+]}{[H^+]} \cdot \frac{[OH^-][HCO_3^-]}{[CO_3^{2-}]}$$

$$K_w = [H^+][OH^-]$$

For H_2CO_3,

$$K_2 = \frac{[H^+][CO_3^{2-}]}{[HCO_3^-]}$$

$$K_{hy} = \frac{K_w}{K_2}$$

For Na_3PO_4 in H_2O,

$$K_{hy} = \frac{[OH^-][HPO_4^{2-}]}{[PO_4^{3-}]}$$

$$K_{hy} = \frac{[H^+]}{[H^+]} \cdot \frac{[OH^-][HPO_4^{2-}]}{[PO_4^{3-}]}$$

$$K_w = [H^+][OH^-]$$

For H_3PO_4,

$$K_3 = \frac{[H^+][PO_4^{3-}]}{[HPO_4^{2-}]}$$

$$K_{hy} = \frac{K_w}{K_3}$$

For the most part, the hydrolyses of the monoprotonated anions, ampholytes such as HCO_3^- and HPO_4^{2-}, are unimportant when the principal species in solution are CO_3^{2-} and PO_4^{3-}. The hydrolyses of anions of polyprotic acids can be treated much as the hydrolyses of anions of monoprotic acids. A general equation for either the salt Na_2A or Na_3Z can be derived.

For Na_2A,

$$[OH^-] = \left(\frac{K_w}{K_2} (C_A{}^{2-} - [OH^-]) \right)^{1/2}$$

For Na_3Z,

$$[OH^-] = \left(\frac{K_w}{K_3} (C_Z{}^{3-} - [OH^-]) \right)^{1/2}$$

Either quadratics may be solved, or a series of approximations made. The approximations are similar in form to those for anions of weak monoprotic acids. The approximations will be shown for the case of Na_2A.

First approximation:

$$[OH^-]' = \left(\frac{K_w}{K_2} C_A{}^{2-} \right)^{1/2}$$

Second approximation:

$$[OH^-]'' = \left[\frac{K_w}{K_2} (C_A{}^{2-} - [OH^-]') \right]^{1/2}$$

Third approximation:

$$[OH^-]''' = \left[\frac{K_w}{K_2}(C_A{}^{2-} - [OH^-]'')\right]^{1/2}$$

and so forth.

Example: Calculate the pH of a solution 0.10M in Na_2CO_3.

$$[OH^-]' = \left(\frac{K_w}{K_2}C_{CO_3^{2-}}\right)^{1/2} = \left[\frac{(1.00 \times 10^{-14})}{(6.0 \times 10^{-11})} \cdot (1.0 \times 10^{-1})\right]^{1/2} = 4.1 \times 10^{-3}M$$

$$[OH^-]'' = \left[\frac{K_w}{K_2}(C_{CO_3^{2-}} - [OH^-]')\right]^{1/2}$$

$$= \left[\left(\frac{1.00 \times 10^{-14}}{6.0 \times 10^{-11}}\right)(1.0 \times 10^{-1} - 4.1 \times 10^{-3})\right]^{1/2} = 4.1 \times 10^{-3}M$$

$$[OH^-]' = [OH^-]''$$

The approximations have converged.

$$pOH = -\log(4.1 \times 10^{-3}) = 2.39$$

$$pH = 14.00 - 2.39 = 11.61$$

2.19 Ampholytes

Solutions of weak polyprotic acids alone in water, buffer systems of weak polyprotic acids, and solutions of the unprotonated anions of weak acids all behave much like their counterparts in weak monoprotic acid systems. No monoprotic acid, however, can give rise to species such as ampholytes, those protonated anions or zwitterions which can act either as acids or as bases.

CARBONIC ACID SYSTEM

PHOSPHORIC ACID SYSTEM

Ampholyte

HCO_3^-
As acid:
$$HCO_3^- \rightleftharpoons H^+ + CO_3^{2-}$$
As base:
$$HCO_3^- + H^+ \rightleftharpoons H_2CO_3$$

Ampholytes

$H_2PO_4^-$
As acid:
$$H_2PO_4^- \rightleftharpoons H^+ + HPO_4^{2-}$$
As base:
$$H_2PO_4^- + H^+ \rightleftharpoons H_3PO_4$$

HPO_4^{2-}
As acid:
$$HPO_4^{2-} \rightleftharpoons H^+ + PO_4^{3-}$$
As base:
$$HPO_4^{2-} + H^+ \rightleftharpoons H_2PO_4^-$$

If we are given a typical problem, for example, to calculate the pH of a solution of $NaHCO_3$, where $C_{HCO_3^-}$ is known, where do we begin?

Consider a solution of NaHA, a salt of H_2A, where C_{HA^-}, the analytical concentration of the ampholyte, is known. The reactions involving the release or consumption of hydrogen ions are as follows.

$$HA^- \rightleftharpoons H^+ + A^{2-} \qquad K_2 = \frac{[H^+][A^{2-}]}{[HA^-]}$$

$$HA^- + H^+ \rightleftharpoons H_2A \qquad \frac{1}{K_1} = \frac{[H_2A]}{[H^+][HA^-]}$$

$$H_2O \rightleftharpoons H^+ + OH^- \qquad K_w = [H^+][OH^-]$$

The equilibrium concentration of hydrogen ion, $[H^+]$, is found by simple chemical bookkeeping.

$$[H^+] = [H^+ \text{ released by } HA^-] - [H^+ \text{ consumed by } HA^-] + [H^+ \text{ released by } H_2O]$$

Because all three reactions exhibit one-to-one stoichiometry:

$$[H^+] = [A^{2-}] - [H_2A] + [OH^-] \qquad (2.44)$$

Now, each term on the right is to be put into terms of $[H^+]$, $[HA^-]$, K_1, K_2, and/or K_w.

$$K_2 = \frac{[H^+][A^{2-}]}{[HA^-]}$$

$$[A^{2-}] = \frac{K_2[HA^-]}{[H^+]}$$

$$\frac{1}{K_1} = \frac{[H_2A]}{[H^+][HA^-]}$$

$$[H_2A] = \frac{[H^+][HA^-]}{K_1}$$

$$K_w = [H^+][OH^-]$$

$$[OH^-] = \frac{K_w}{[H^+]}$$

We substitute the new terms into Equation 2.44,

$$[H^+] = \frac{K_2[HA^-]}{[H^+]} - \frac{[H^+][HA^-]}{K_1} + \frac{K_w}{[H^+]}$$

clear the fractions,

$$K_1[H^+]^2 = K_1K_2[HA^-] - [H^+]^2[HA^-] + K_1K_w$$

and regroup by placing the $[H^+]^2$ terms on the left.

$$K_1[H^+]^2 + [HA^-][H^+]^2 = K_1K_2[HA^-] + K_1K_w$$

$$(K_1 + [HA^-])[H^+]^2 = K_1K_2[HA^-] + K_1K_w$$

$$[H^+]^2 = \frac{K_1K_2[HA^-] + K_1K_w}{K_1 + [HA^-]}$$

$$[H^+] = \left(\frac{K_1 K_2 [HA^-] + K_1 K_w}{K_1 + [HA^-]}\right)^{1/2}$$

or

$$[H^+] = \left(\underbrace{\frac{K_1 K_2 [HA^-]}{K_1 + [HA^-]}}_{\text{term 1}} + \underbrace{\frac{K_1 K_w}{K_1 + [HA^-]}}_{\text{term 2}}\right)^{1/2}$$

Consider two limiting cases:

    ¤   **1.** $[HA^-]$ becomes vanishingly small ($\rightarrow 0$).

    ¤   **2.** $[HA^-]$ becomes extremely large ($\rightarrow \infty$) (for our purposes, "large" can mean 0.01 to 0.1M or so).

≫ First Case / Term 1 will vanish:

$$\frac{K_1 K_2 [HA^-]}{K_1 + [HA^-]} = \frac{K_1 K_2 (0)}{K_1 + 0} = 0$$

Term 2 will not vanish:

$$\frac{K_1 K_w}{K_1 + [HA^-]} = \frac{K_1 K_w}{K_1 + 0} = K_w$$

$$[H^+] = (K_w)^{1/2}$$

This is what $[H^+]$ would be in a pure H_2O solution, which is the case as $[HA^-] \rightarrow 0$.

≫ Second Case / Term 1 will not vanish as $[HA^-] \rightarrow \infty$.

$$\frac{K_1 K_2 [HA^-]}{K_1 + [HA^-]} = \frac{K_1 K_2}{\dfrac{K_1 + [HA^-]}{[HA^-]}} = \frac{K_1 K_2}{\dfrac{K_1}{[HA^-]} + 1} = \frac{K_1 K_2}{\dfrac{K_1}{\infty} + 1} = \frac{K_1 K_2}{0 + 1} = K_1 K_2$$

Term 2 will vanish as $[HA^-] \rightarrow \infty$

$$\frac{K_1 K_w}{K_1 + [HA^-]} = \frac{K_1 K_w}{K_1 + \infty} = \frac{K_1 K_w}{\infty} = 0$$

and

$$[H^+] = (K_1 K_2)^{1/2} \qquad (2.45)$$

This is a familiar formula and works where $C_{HA^-} \simeq 0.01$ to 0.1M. It is interesting that $[H^+]$ is not dependent on C_{HA^-} so that a fivefold dilution of, say 0.1M $NaHCO_3$ should cause no appreciable change in pH. The familiar formula,

$$pH = \frac{pK_1 + pK_2}{2}$$

is easily derived.

$$[H^+] = (K_1 K_1)^{1/2}$$

$$\log [H^+] = \log (K_1 K_2)^{1/2} = \tfrac{1}{2} \log (K_1 K_2) = \tfrac{1}{2}(\log K_1 + \log K_2)$$

$$-\log [H^+] = \tfrac{1}{2}(-\log K_1 - \log K_2)$$

$$pH = \frac{1}{2}(pK_1 + pK_2)$$

$$pH = \frac{pK_1 + pK_2}{2} \tag{2.46}$$

For the ampholytes of such triprotic acids as H_3Z, similar relationships may be derived: Given $C_{H_2Z^-}$, and that $[H_2Z^-]$ is large,

$$[H^+] = (K_1K_2)^{1/2}$$

$$pH = \frac{pK_1 + pK_2}{2}$$

given $C_{HZ^{2-}}$ and that $[HZ^{2-}]$ is large,

$$[H^+] = (K_2K_3)^{1/2}$$

$$pH = \frac{pK_2 + pK_3}{2}$$

For the most part, the simple formulas given here work rather well. If it does become necessary to make approximations, a good one to start with is that

$$C_{HA^-} = [HA^-]$$

Then the complete expression becomes

$$[H^+] = \left(\frac{K_1K_2\,C_{HA^-}}{K_1 + C_{HA^-}} + \frac{K_1K_w}{C_{HA^-}} \right)^{1/2}$$

If $C_{HA^-} \gg K_1$,

$$[H^+] = \left(K_1K_2 + \frac{K_1K_w}{C_{HA^-}} \right)^{1/2}$$

The expression $(K_1K_2)^{1/2}$ works very well where $[HA^-]$ is large, as in titrations, and the pH is below 8. If the pH is greater than 8, the expression

$$[H^+] = \left(K_1K_2 + \frac{K_1K_w}{C_{HA^-}} \right)^{1/2}$$

is often valid.

Example: Calculate the pH of a solution 0.10M in $NaHCO_3$ ($K_1 = 3.5 \times 10^{-7}, K_2 = 6.0 \times 10^{-11}$).

$$[H^+]' = (K_1K_2)^{1/2} = [(3.5 \times 10^{-7})(6.0 \times 10^{-11})]^{1/2} = 4.6 \times 10^{-9}$$

$$pH = -\log(4.6 \times 10^{-9}) = 8.34$$

Because the pH > 8, another calculation is needed.

$$[H^+]'' = \left(K_1K_2 + \frac{K_1K_w}{C_{HCO_3^-}} \right)^{1/2}$$

$$[H^+]'' = \left((3.5 \times 10^{-7})(6.0 \times 10^{-11}) + \frac{(3.5 \times 10^{-7})(1.00 \times 10^{-14})}{1.0 \times 10^{-1}} \right)^{1/2}$$

$$[H^+]'' = 4.6 \times 10^{-9}$$

$$pH = 8.34$$

The second calculation was a good precaution.

Example: Calculate the pH of a solution $1.0 \times 10^{-3}M$ in sodium hydrogen malonate, NaHMal. For malonic acid, H_2Mal, $K_1 = 1.51 \times 10^{-3}$ and $K_2 = 2.2 \times 10^{-6}$.

$$[H^+]' = (K_1 K_2)^{1/2} = [(1.51 \times 10^{-3})(2.2 \times 10^{-6})]^{1/2} = 5.76 \times 10^{-5}M$$

$$pH = -\log(5.76 \times 10^{-5}) = 4.24$$

This tells us that the solution is acidic. Another calculation must be made because K_1 and C_{HMal^-} are close.

$$[H^+]'' = \left(\frac{K_1 K_2\, C_{HMal^-}}{K_1 + C_{HMal^-}} \right)^{1/2}$$

$$= \left(\frac{(1.51 \times 10^{-3})(2.2 \times 10^{-6})(1.0 \times 10^{-3})}{1.51 \times 10^{-3} + 1.00 \times 10^{-3}} \right)^{1/2} = 3.64 \times 10^{-5}M$$

$$pH = -\log(3.64 \times 10^{-5}) = 4.44$$

≡ 2.20 PROBLEMS ≡

1. Consider a malonate solution, where $C_{Mal} = 0.010M$. Calculate $[H_2Mal]$, $[HMal^-]$, and $[Mal^{2-}]$ at:

	[H₂Mal], M	[HMal⁻], M	[Mal²⁻], M
(a) pH 2.00			
Ans:	8.7×10^{-3}	1.3×10^{-3}	2.9×10^{-7}
(b) pH 4.00			
Ans:	6.1×10^{-4}	9.2×10^{-3}	2.0×10^{-4}
(c) pH 5.50			
Ans:	1.2×10^{-5}	5.9×10^{-3}	4.1×10^{-3}

$K_1 = 1.5 \times 10^{-3}$ *and* $K_2 = 2.2 \times 10^{-6}$.

2. Prove that, at pH 2.82, $[H_2Mal] = [HMal^-]$ in any malonate solution. What would you call this solution?
   Ans: *See detailed solutions. It is a buffer.*

3. It was asserted in the preceding section that, for triprotic acid H_3Z,

$$\alpha_0 = \frac{[H^+]^3}{[H^+]^3 + [H^+]^2 K_1 + [H^+]K_1 K_2 + K_1 K_2 K_3}.$$

Prove this.
Ans: *See detailed solutions.*

4. Calculate the pH of a solution made $0.10M$ in 8-hydroxyquinoline, H_2Qu if $K_1 = 1.07 \times 10^{-5}$ and $K_2 = 6.0 \times 10^{-11}$.
Ans: *pH = 3.00*

5. Calculate the pH of a solution $0.10M$ in H_3PO_4 ($K_1 = 7.1 \times 10^{-3}$, $K_2 = 6.2 \times 10^{-8}$, and $K_3 = 4.4 \times 10^{-13}$).
Ans: *pH = between 1.57 and 1.64*

6. Calculate the pH of a solution made by mixing 0.010 mole of NaH_2AsO_4 and 0.020 mole of Na_2HAsO_4 and adjusting the volume to 1.00 liter. For arsenic acid, $K_1 = 6.0 \times 10^{-3}$, $K_2 = 1.05 \times 10^{-7}$, and $K_3 = 4.0 \times 10^{-12}$.
Ans: *pH = 7.28*

7. Calculate the pH of 200. milliliters of the solution in Problem 6 when 0.00010 mole of concentrated HCl are added, with no change in volume.
Ans: *pH = 7.25*

8. Calculate the pH of 200. milliliters of the solution in Problem 6 when 0.00010 mole of concentrated NaOH are added, with no change in volume.
Ans: *pH = 7.31*

9. Calculate the pH of a solution made by adding 50. milliliters of $0.10M$ NaOH to 100. milliliters of $0.10M$ H_3PO_4.
Ans: *pH = 2.28*

10. Calculate the pH of a solution made by adding 150. milliliters of $0.10M$ NaOH to 100 milliliters of $0.10M$ H_3PO_4.
Ans: *pH = 7.21*

11. Calculate the pH of a solution made by adding 0.050 mole of sodium oxalate, $Na_2C_2O_4$, to 1.00 liter of water ($K_1 = 5.6 \times 10^{-2}$ and $K_2 = 6.2 \times 10^{-5}$).
Ans: *pH = 8.45*

12. Calculate the pH of a solution made by adding 200. milliliters of $0.10M$ NaOH to 100. milliliters of $0.10M$ tartaric acid, H_2Tar. ($K_1 = 2.0 \times 10^{-3}$ and $K_2 = 8.9 \times 10^{-5}$).
Ans: *pH = 8.29*

13. Calculate the pH of a solution of $0.10M$ in sodium hydrogen salicylate, NaHSal. ($K_2 = 1.07 \times 10^{-3}$ and $K_2 = 1.0 \times 10^{-13}$.)
Ans: *pH = 8.00*

14. Calculate the pH of a solution made by adding 100. milliliters of $0.10M$ NaOH to 100. milliliters of $0.10M$ H_3PO_4.
Ans: *pH = 4.68*

15. Why might potassium acid phthalate, KHPh, be regarded as a buffer? See the α-diagram for phthalic acid ($K_1 = 8.0 \times 10^{-4}$ and $K_2 = 4.0 \times 10^{-6}$).
Ans: *See Figure 2.3 and detailed answers.*

$\equiv$ 2.21 DETAILED SOLUTIONS TO PROBLEMS $\equiv$

1. For each pH involved, these relationships hold:

$$[H_2Mal] = \alpha_0 C_{Mal}$$

$$[HMal^-] = \alpha_1 C_{Mal}$$

$$[Mal^{2-}] = \alpha_2 C_{Mal}$$

$K_1 = 1.51 \times 10^{-3}$ and $K_2 = 2.2 \times 10^{-6}$.

(a) At pH 2.00, $[H^+] = 1.00 \times 10^{-2}M$.

$$[H_2Mal] = \alpha_0 C_{Mal} = \left[\frac{[H^+]^2}{[H^+]^2 + [H^+]K_1 + K_1K_2}\right] C_{Mal}$$

$$= \left[\frac{(1.0 \times 10^{-2})^2}{(1.0 \times 10^{-2})^2 + (1.0 \times 10^{-2})(1.51 \times 10^{-3}) + (1.51 \times 10^{-3})(2.2 \times 10^{-6})}\right] \cdot$$

$$(1.0 \times 10^{-2})$$

$$= 8.7 \times 10^{-3}M$$

$$[HMal^-] = \alpha_1 C_{Mal} = \left[\frac{[H^+]K_1}{[H^+]^2 + [H^+]K_1 + K_1K_2}\right] C_{Mal}$$

$$= \left[\frac{(1.0 \times 10^{-2})(1.51 \times 10^{-3})}{(1.0 \times 10^{-2})^2 + (1.0 \times 10^{-2})(1.5 \times 10^{-3}) + (1.51 \times 10^{-3})(2.2 \times 10^{-6})}\right] \cdot$$

$$(1.0 \times 10^{-2})$$

$$= 1.3 \times 10^{-3}M$$

$$[Mal^{2-}] = \alpha_2 C_{Mal} = \left[\frac{K_1K_2}{[H^+]^2 + [H^+]K_1 + K_1K_2}\right] C_{Mal}$$

$$= \left[\frac{(1.51 \times 10^{-3})(2.2 \times 10^{-6})}{(1.0 \times 10^{-2})^2 + (1.0 \times 10^{-2})(1.51 \times 10^{-3}) + (1.51 \times 10^{-3})(2.2 \times 10^{-6})}\right] \cdot$$

$$(1.0 \times 10^{-2})$$

$$= 2.9 \times 10^{-7}M$$

(b) Figuring out α_0, α_1, and α_2 is all right—once. If a table is available, such as Table 2.2, α_0, α_1, and α_2 can be read from it at pH 4.00.

$[H_2Mal] = \alpha_0 C_{Mal} = (6.09 \times 10^{-2})(1.00 \times 10^{-2}) = 6.1 \times 10^{-4}M$

$[HMal^-] = \alpha_1 C_{Mal} = (9.19 \times 10^{-1})(1.00 \times 10^{-2}) = 9.2 \times 10^{-3}M$

$[Mal^{2-}] = \alpha_2 C_{Mal} = (2.02 \times 10^{-2})(1.00 \times 10^{-2}) = 2.0 \times 10^{-4}M$

(c) At pH 5.50, Table 2.2 can be used again.

$[H_2Mal] = \alpha_0 C_{Mal} = (1.23 \times 10^{-3})(1.00 \times 10^{-2}) = 1.2 \times 10^{-5}M$

$[HMal^-] = \alpha_1 C_{Mal} = (5.89 \times 10^{-1})(1.00 \times 10^{-2}) = 5.9 \times 10^{-3}M$

$[Mal^{2-}] = \alpha_2 C_{Mal} = (4.10 \times 10^{-1})(1.00 \times 10^{-2}) = 4.1 \times 10^{-3}M$

Table 2.2
α-VALUES FOR MALONIC ACID
$$K_1 = 1.51E - 3$$
$$K_2 = 2.2E - 6$$

pH	α_0	α_1	α_2
0.50	0.995E + 00	0.475E − 02	0.331E − 07
1.00	0.985E + 00	0.149E − 01	0.327E − 06
1.50	0.954E + 00	0.456E − 01	0.317E − 05
2.00	0.869E + 00	0.131E + 00	0.289E − 04
2.50	0.677E + 00	0.323E + 00	0.225E − 03
3.00	0.398E + 00	0.601E + 00	0.132E − 02
3.50	0.172E + 00	0.822E + 00	0.572E − 02
4.00	0.609E − 01	0.919E + 00	0.202E − 01
4.50	0.192E − 01	0.917E + 00	0.638E − 01
5.00	0.540E − 02	0.815E + 00	0.179E + 00
5.50	0.123E − 02	0.589E + 00	0.410E + 00
6.00	0.207E − 03	0.312E + 00	0.687E + 00
6.50	0.263E − 04	0.126E + 00	0.874E + 00
7.00	0.288E − 05	0.435E − 01	0.957E + 00
7.50	0.297E − 06	0.142E − 01	0.986E + 00
8.00	0.300E − 07	0.452E − 02	0.995E + 00
8.50	0.301E − 08	0.144E − 02	0.999E + 00
9.00	0.301E − 09	0.454E − 03	0.100E + 01
9.50	0.301E − 10	0.144E − 03	0.100E + 01
10.00	0.301E − 11	0.455E − 04	0.100E + 01
10.50	0.301E − 12	0.144E − 04	0.100E + 01
11.00	0.301E − 13	0.455E − 05	0.100E + 01
11.50	0.301E − 14	0.144E − 05	0.100E + 01
12.00	0.301E − 15	0.455E − 06	0.100E + 01
12.50	0.301E − 16	0.144E − 06	0.100E + 01
13.00	0.301E − 17	0.455E − 07	0.100E + 01
13.50	0.301E − 18	0.144E − 07	0.100E + 01
14.00	0.301E − 19	0.455E − 08	0.100E + 01

2. At pH 2.82, $[H^+] = 1.5 \times 10^{-3}$, and $\alpha_0 = \alpha_1$. Why?

$$\alpha_0 = \frac{[H^+]^2}{[H^+]^2 + [H^+]K_1 + K_1K_2}$$

$$= \frac{(1.5 \times 10^{-3})^2}{(1.5 \times 10^{-3})^2 + (1.5 \times 10^{-3})(1.51 \times 10^{-3}) + (1.51 \times 10^{-3})(2.2 \times 10^{-6})}$$

$$\alpha_1 = \frac{[H^+]K_1}{(1.5 \times 10^{-3})^2 + (1.5 \times 10^{-3})(1.51 \times 10^{-3}) + (1.51 \times 10^{-3})(2.2 \times 10^{-6})}$$

$$= \frac{(1.5 \times 10^{-3})^2}{(1.5 \times 10^{-3})^2 + (1.5 \times 10^{-3})(1.51 \times 10^{-3}) + (1.51 \times 10^{-3})(2.2 \times 10^{-6})}$$

Therefore,

$$\alpha_0 = \alpha_1$$

$$\alpha_0 = \frac{[H_2Mal]}{C_{Mal}}$$

$$\alpha_1 = \frac{[HMal^-]}{C_{Mal}}$$

$$\frac{[H_2Mal]}{C_{Mal}} = \frac{[HMal^-]}{C_{Mal}}$$

and

$$[H_2Mal] = [HMal^-]$$

The solution is a *buffer*.

3. $H_3Z \rightleftharpoons H^+ + H_2Z^-$ $K_1 = \dfrac{[H^+][H_2Z^-]}{[H_3Z]}$

   $H_2Z^- \rightleftharpoons H^+ + HZ^{2-}$ $K_2 = \dfrac{[H^+][HZ^{2-}]}{[H_2Z^-]}$

   $HZ^{2-} \rightleftharpoons H^+ + Z^{3-}$ $K_3 = \dfrac{[H^+][Z^{3-}]}{[HZ^{2-}]}$

$$\alpha_0 = \frac{[H_3Z]}{[H_3Z] + [H_2Z^-] + [HZ^{2-}] + [Z^{3-}]}$$

Put $[H_3Z]$, $[H_2Z^-]$, $[HZ^{2-}]$, and $[Z^{3-}]$ all in terms of $[H_3Z]$, $[H^+]$, K_1, K_2 and/or K_3. $[H_3Z]$ can be left as it is.

From K_1 $[H_2Z^-] = \dfrac{K_1[H_3Z]}{[H^+]}$

From K_2 $[HZ^{2-}] = \dfrac{K_2[H_2Z^-]}{[H^+]}$

But $[H_2Z^-] = \dfrac{K_1[H_3Z]}{[H^+]}$

   $[HZ^{2-}] = \dfrac{K_1K_2[H_3Z]}{[H^+]^2}$

From K_3: $[Z^{3-}] = \dfrac{K_3[HZ^{2-}]}{[H^+]}$

But $[HZ^{2-}] = \dfrac{K_1K_2[H_3Z]}{[H^+]^2}$

   $[Z^{3-}] = \dfrac{K_1K_2K_3[H_3Z]}{[H^+]^3}$

Now we obtain the following results:

$$\alpha_0 = \frac{[H_3Z]}{[H_3Z] + \dfrac{K_1[H_3Z]}{[H^+]} + \dfrac{K_1K_2[H_3Z]}{[H^+]^2} + \dfrac{K_1K_2K_3[H_3Z]}{[H^+]^3}}$$

$$\alpha_0 = \frac{[H_3Z]}{[H_3Z]\left(1 + \dfrac{K_1}{[H^+]} + \dfrac{K_1K_2}{[H^+]^2} + \dfrac{K_1K_2K_3}{[H^+]^3}\right)}$$

$$\alpha_0 = \frac{1}{\dfrac{[H^+]^3 + [H^+]^2 K_1 + [H^+]K_1K_2 + K_1K_2K_3}{[H^+]^3}}$$

$$\alpha_0 = \frac{[H^+]^3}{[H^+]^3 + [H^+]^2 K_1 + [H^+]K_1K_2 + K_1K_2K_3}$$

4. Try a method of approximations.

First approximation:

$$[H^+]' = (K_1 C_{H_2Qu})^{1/2} = [(1.07 \times 10^{-5})(1.0 \times 10^{-1})]^{1/2} = 1.0 \times 10^{-3} M$$

$$\alpha_1' = \frac{[H^+]'K_1}{([H^+]')^2 + [H^+]'K_1 + K_1K_2}$$

$$= \frac{(1.0 \times 10^{-3})(1.07 \times 10^{-5})}{(1.0 \times 10^{-3})^2 + (1.0 \times 10^{-3})(1.07 \times 10^{-5}) + (1.07 \times 10^{-5})(6.0 \times 10^{-11})}$$

$$= 0.010$$

$$\alpha_2' = \frac{K_1K_2}{([H^+]')^2 + [H^+]'K_1 + K_1K_2}$$

$$= \frac{(1.07 \times 10^{-5})(6.0 \times 10^{-11})}{(1.0 \times 10^{-3})^2 + (1.0 \times 10^{-3})(1.07 \times 10^{-5}) + (1.07 \times 10^{-5})(6.0 \times 10^{-11})}$$

$$\simeq 0.$$

Second approximation:

$$[H^+]'' = \alpha_1' C_{H_2Qu} + 2\alpha_2' C_{H_2Qu} + \frac{K_w}{[H^+]'}$$

$$= (0.010)(1.0 \times 10^{-1}) + (2)(0)(1.0 \times 10^{-1}) + \frac{1.00 \times 10^{-14}}{(1.0 \times 10^{-3})} = 1.0 \times 10^{-3} M$$

$[H^+]'$ and $[H^+]''$ are the same.

$$pH = -\log(1.0 \times 10^{-3})$$

$$pH = 3.00$$

5. From charge balance:

$$[H^+] = [H_2PO_4^-] + 2[HPO_4^{2-}] + 3[PO_4^{3-}] + [OH^-]$$

$$C_{PO_4} = C_{H_3PO_4}$$

$$[H_2PO_4^-] = \alpha_1 C_{H_3PO_4}$$

$$[HPO_4^{2-}] = \alpha_2 C_{H_3PO_4}$$

$$[PO_4^{3-}] = \alpha_3 C_{H_3PO_4}$$

$$[OH^-] = \frac{K_w}{[H^+]}$$

$$[H^+] = \alpha_1 C_{H_3PO_4} + 2\alpha_2 C_{H_3PO_4} + 3\alpha_3 C_{H_3PO_4} + \frac{K_w}{[H^+]}$$

Now, approximations may be used:

$$[H^+]' = (K_1 C_{H_3PO_4})^{1/2} = [(7.1 \times 10^{-3})(1.0 \times 10^{-1})]^{1/2} = 2.7 \times 10^{-2} M$$

$$pH' = -\log(2.7 \times 10^{-2}) = 1.57$$

Using the α-plot for H_3PO_4, Figure 2.2, at pH 1.57, we obtain

$$\alpha_1' = 0.23$$

$$\alpha_2' = 0$$

$$\alpha_3' = 0$$

$$[H^+]'' = \alpha_1' C_{H_3PO_4} + 2\alpha_2' C_{H_3PO_4} + 3\alpha_3' C_{H_3PO_4} + \frac{K_w}{[H^+]'}$$

$$= (0.23)(1.0 \times 10^{-1}) + (2)(0)(1 \times 10^{-1}) + (3)(0)(1 \times 10^{-1})$$

$$+ \frac{1.00 \times 10^{-14}}{2.45 \times 10^{-2}}$$

$$= 2.3 \times 10^{-2} M$$

$$pH'' = 1.64$$

The pH lies between 1.57 and 1.64. Further approximations could be made, but the accuracy we could get from the graph probably makes them useless. To get very accurate approximations, we should have to compute α_1, α_2, and α_3 many times—a wearying business.

6. This is a buffer solution.

$$K_2 = \frac{[H^+][HAsO_4^{2-}]}{[H_2AsO_4^-]}$$

$$[H^+] = K_2 \frac{[H_2AsO_4^-]}{[HAsO_4^{2-}]}$$

Now, we can begin approximating

$$[H^+]' = K_2 \frac{C_{H_2AsO_4^-}}{C_{HAsO_4^{2-}}}$$

$$C_{H_2AsO_4^-} = 0.010M$$

$$C_{HAsO_4^{2-}} = 0.020M$$

$$[H^+]' = (1.05 \times 10^{-7}) \frac{0.010}{0.020} = 5.2 \times 10^{-8}M$$

$$[OH^-]' = \frac{K_w}{[H^+]'} = \frac{1.00 \times 10^{-14}}{5.25 \times 10^{-8}} = 2.0 \times 10^{-7}M$$

The solution is just barely basic, so a second approximation could include both $[H^+]'$ and $[OH^-]'$.

$$[H^+]'' = K_2 \frac{(C_{H_2AsO_4^-} - [H^+]' + [OH^-]')}{(C_{HAsO_4^{2-}} + [H^+]' - [OH^-]')}$$

$$= (1.05 \times 10^{-7}) \cdot \frac{(1.0 \times 10^{-2} - 5.2 \times 10^{-8} + 2.0 \times 10^{-7})}{(2.0 \times 10^{-2} + 5.2 \times 10^{-8} - 2.0 \times 10^{-7})} = 5.2 \times 10^{-8}M$$

Now $[H^+]'$ and $[H^+]''$ are equal. The approximations have converged.

$$pH = -\log(5.2 \times 10^{-8}) = 7.28$$

7. Here, it is necessary to do a little stoichiometry before the equilibrium calculations. The important reaction is that between the added HCl (strong acid) and the hydrogen arsenate ($HAsO_4^{2-}$) ion. It goes very nearly to completion.

$$H^+ + HAsO_4^{2-} \rightarrow H_2AsO_4^-$$

(number of moles of $H_2AsO_4^-$ formed)
$$= \text{(number of moles of } H^+ \text{ added)}$$

(number of moles of $H_2AsO_4^-$ formed)
$$= 1.0 \times 10^{-4} \text{ moles}$$

(number of moles of $HAsO_4^{2-}$ left)
$$= \text{(number of moles of } HAsO_4^{2-} \text{ to start)} -$$
$$\text{(number of moles of } H_2AsO_4^- \text{ formed)}$$
$$= (2.0 \times 10^{-2})(2.00 \times 10^{-1}) - 1.0 \times 10^{-4} = 3.9 \times 10^{-3} \text{ moles}$$

$$C_{HAsO_4^{2-}} = \frac{\text{(number of moles of } HAsO_4^{2-} \text{ left)}}{\text{(volume of solution, liters)}} = \frac{3.9 \times 10^{-3}}{2.00 \times 10^{-1}}$$

(number of moles of $H_2AsO_4^-$ present)
$$= \text{(number of moles of } H_2AsO_4^- \text{ to start)} +$$
$$\text{(number of moles of } H_2AsO_4^- \text{ formed)}$$
$$= (1.0 \times 10^{-2})(2.00 \times 10^{-1}) + 1.0 \times 10^{-4} = 2.1 \times 10^{-3}$$

$$C_{H_2AsO_4^-} = \frac{\text{(number of moles of } H_2AsO_4^- \text{ present)}}{\text{(volume of solution, liters)}} = \frac{2.1 \times 10^{-3}}{2.00 \times 10^{-1}}$$

Now, the equilibrium approximations can begin.

First approximation:

$$[H^+]' = K_2 \frac{C_{H_2AsO_4^-}}{C_{HAsO_4^{2-}}} = 1.05 \times 10^{-7} \cdot \left(\frac{\frac{2.1 \times 10^{-3}}{2.00 \times 10^{-1}}}{\frac{3.9 \times 10^{-3}}{2.00 \times 10^{-1}}}\right) = 5.6 \times 10^{-8} M$$

$$[OH^-]' = \frac{K_w}{[H^+]'} = \frac{(1.00 \times 10^{-14})}{(5.6 \times 10^{-8})} = 1.8 \times 10^{-7} M$$

Second approximation:

$$[H^+]'' = K_2 \left(\frac{C_{H_2AsO_4^-} - [H^+]' + [OH^-]'}{C_{HAsO_4^{2-}} + [H^+]' - [OH^-]'}\right)$$

$$= (1.05 \times 10^{-7})\left(\frac{\frac{2.1 \times 10^{-3}}{2.00 \times 10^{-1}} - 5.6 \times 10^{-8} + 1.8 \times 10^{-7}}{\frac{3.9 \times 10^{-3}}{2.00 \times 10^{-1}} + 5.6 \times 10^{-8} - 1.8 \times 10^{-7}}\right)$$

$$= 5.6 \times 10^{-8} M$$

$$[H^+]' = [H^+]''$$

The approximations have converged.

$$pH = -\log(5.6 \times 10^{-8}) = 7.25$$

8. Some stoichiometry must be done before the equilibrium calculations. The important reaction is that between the added NaOH (strong base) and the dihydrogen arsenate ($H_2AsO_4^-$) ion. It goes very nearly to completion.

$$OH^- + H_2AsO_4^- \rightarrow HAsO_4^{2-} + H_2O$$

(number of moles of $HAsO_4^{2-}$ formed)

$$= \text{(number of moles of } OH^- \text{ added)}$$

(number of moles of $HAsO_4^{2-}$ formed)

$$= 1.0 \times 10^{-4} \text{ moles}$$

(number of moles of $H_2AsO_4^-$ left) = (number of moles of $H_2AsO_4^-$ to start) −
(number of moles of $HAsO_4^{2-}$ formed)

(number of moles of $H_2AsO_4^-$ left) = $(1.0 \times 10^{-2})(2.00 \times 10^{-1}) - 1.0 \times 10^{-4}$

(number of moles of $H_2AsO_4^-$ left) = 1.9×10^{-3} moles

$$C_{H_2AsO_4^-} = \frac{\text{(number of moles of } H_2AsO_4^- \text{ left)}}{\text{(volume of solution, liters)}}$$

$$C_{H_2AsO_4^-} = \frac{1.9 \times 10^{-3}}{2.00 \times 10^{-1}}$$

(number of moles of $HAsO_4^{2-}$ present)

$$= \text{(number of moles of } HAsO_4^{2-} \text{ to start)} +$$
$$\text{(number of moles of } HAsO_4^{2-} \text{ formed)}$$

$$= (2.0 \times 10^{-2})(2.00 \times 10^{-1}) + 1.0 \times 10^{-4}$$

$$= 4.1 \times 10^{-3} \text{ mole}$$

$$C_{HAsO_4^{2-}} = \frac{\text{(number of moles of } HAsO_4^{2-} \text{ present)}}{\text{(volume of solution, liters)}}$$

$$= \frac{4.1 \times 10^{-3}}{2.00 \times 10^{-1}}$$

Now, equilibrium approximations can begin.

First approximation:

$$[H^+]' = K_2 \frac{C_{H_2AsO_4^-}}{C_{HAsO_4^{2-}}} = (1.05 \times 10^{-7}) \cdot \left(\frac{\dfrac{1.9 \times 10^{-3}}{2.00 \times 10^{-1}}}{\dfrac{4.1 \times 10^{-3}}{2.00 \times 10^{-1}}} \right) = 4.9 \times 10^{-8} M$$

$$[OH^-]' = \frac{K_w}{[H^+]'} = \frac{(1.00 \times 10^{-14})}{(4.9 \times 10^{-8})} = 2.1 \times 10^{-7} M$$

Second approximation:

$$[H^+]'' = K_2 \left(\frac{C_{H_2AsO_4^-} - [H^+]' + [OH^-]'}{C_{HAsO_4^{2-}} + [H^+]' - [OH^-]'} \right)$$

$$= (1.05 \times 10^{-7}) \left(\frac{\dfrac{1.9 \times 10^{-3}}{2.00 \times 10^{-1}} - 4.9 \times 10^{-8} + 2.1 \times 10^{-7}}{\dfrac{4.1 \times 10^{-3}}{2.00 \times 10^{-1}} + 4.9 \times 10^{-8} - 2.1 \times 10^{-7}} \right)$$

$$= 4.9 \times 10^{-8} M$$

$$[H^+]' = [H^+]''$$

The approximations have converged.

$$pH = -\log (4.9 \times 10^{-8}) = 7.31$$

9. Again, some stoichiometry must be done before the equilibrium calculations can be made. The important reaction is that of OH^- with H_3PO_4. It goes very nearly to completion.

$$H_3PO_4 + OH^- \rightarrow H_2PO_4^- + H_2O$$

(number of moles of $H_2PO_4^-$ formed)

$\qquad$ = (number of moles of OH^- added)

$\qquad$ = $(1.0 \times 10^{-1})(5.0 \times 10^{-2})$ = 5.0×10^{-3} moles

(number of moles of H_3PO_4 left)

$\qquad$ = (number of moles of H_3PO_4 to start) −
$\qquad$ (number of moles of $H_2PO_4^-$ formed)

$\qquad$ = $(1.00 \times 10^{-1})(1.0 \times 10^{-1})$ − 5.0×10^{-3} = 5.0×10^{-3} moles

$$C_{H_3PO_4} = \frac{\text{(number of moles of } H_3PO_4 \text{ left)}}{\text{(volume of solution, liters)}}$$

$$= \frac{(5 \times 10^{-3})}{(1 \times 10^{-1} + 5 \times 10^{-2})} = 3.3 \times 10^{-2} M$$

(number of moles of $H_2PO_4^-$ present)

$\qquad$ = (number of moles of $H_2PO_4^-$ formed) = 5.0×10^{-3} moles

$$C_{H_2PO_4^-} = \frac{\text{(number of moles of } H_2PO_4^- \text{ present)}}{\text{(volume of solution, liters)}}$$

$$= \frac{(5.0 \times 10^{-3})}{(1.00 \times 10^{-1} + 5.0 \times 10^{-2})} = 3.3 \times 10^{-2} M$$

Now the equilibrium approximations can begin. This solution is a *buffer*.

First approximation:

$$[H^+]' = K_1 \frac{C_{H_3PO_4}}{C_{H_2PO_4^-}} = (7.1 \times 10^{-3}) \left(\frac{3.3 \times 10^{-2} M}{3.3 \times 10^{-2} M} \right) = 7.1 \times 10^{-3} M$$

(It is an *acid* buffer.)

Second approximation:

$$[H^+]'' = K_1 \cdot \left(\frac{C_{H_3PO_4} - [H^+]'}{C_{H_2PO_4^-} + [H^+]'} \right)$$

$$= (7.1 \times 10^{-3}) \left(\frac{3.3 \times 10^{-2} - 7.1 \times 10^{-3}}{3.3 \times 10^{-2} + 7.1 \times 10^{-3}} \right) = 4.6 \times 10^{-3} M$$

Third approximation:

$$[H^+]''' = K_1 \left(\frac{C_{H_3PO_4} - [H^+]''}{C_{H_2PO_4^-} + [H^+]''} \right)$$

$$= (7.1 \times 10^{-3}) \cdot \left(\frac{3.3 \times 10^{-2} - 4.6 \times 10^{-3}}{3.3 \times 10^{-2} + 4.6 \times 10^{-3}} \right) = 5.4 \times 10^{-3} M$$

Fourth approximation:

$$[H^+]'''' = K_1 \cdot \left(\frac{C_{H_3PO_4} - [H^+]'''}{C_{H_2PO_4^-} + [H^+]'''} \right)$$

$$= 7.1 \times 10^{-3} \left(\frac{3.3 \times 10^{-2} - 5.4 \times 10^{-3}}{3.3 \times 10^{-2} + 5.4 \times 10^{-3}} \right) = 5.1 \times 10^{-3} M$$

Fifth approximation:

$$[H^+]^{TAL} = K_1 \cdot \left(\frac{C_{H_3PO_4} - [H^+]''''}{C_{H_2PO_4^-} + [H^+]''''} \right)$$

$$= (7.1 \times 10^{-3}) \cdot \left(\frac{3.3 \times 10^{-2} - 5.1 \times 10^{-3}}{3.3 \times 10^{-2} + 5.1 \times 10^{-3}} \right) = 5.2 \times 10^{-3} M$$

It is probably safe to say that $[H^+] = 5.2 \times 10^{-3} M$

$$pH = -\log (5.2 \times 10^{-3}) = 2.28$$

10. Here, *two* reactions occur:

$$\begin{array}{ccccc} H_3PO_4 & + & OH^- & \rightarrow & H_2PO_4^- + H_2O \\ 10^{-2} \text{ moles} & & 10^{-2} \text{ moles} & & 10^{-2} \text{ moles} \end{array}$$

First, as shown above, *all* of the H_3PO_4 is converted to $H_2PO_4^-$ by 1.0×10^{-2} mole of NaOH. But 150. milliliters of 0.10M NaOH were added to the solution. That means that 1.5×10^{-2} mole of NaOH was added. So, after 1.0×10^{-2} mole was used up converting H_3PO_4 to $H_2PO_4^-$, 0.5×10^{-2} mole of NaOH was left. This reacted with the $H_2PO_4^-$.

$$\begin{array}{ccccc} H_2PO_4^- & + & OH^- & \rightarrow & HPO_4^{2-} + H_2O \\ 0.5 \times 10^{-2} \text{ mole} & & 0.5 \times 10^{-2} \text{ mole} & & 0.5 \times 10^{-2} \text{ mole} \end{array}$$

(number of moles of HPO_4^{2-} formed) $= 0.50 \times 10^{-2}$ mole

$$C_{HPO_4^{2-}} = \frac{\text{(number of moles of } HPO_4^{2-} \text{ formed)}}{\text{(total volume of solution, liters)}}$$

$$= \frac{(0.50 \times 10^{-2})}{(1.00 \times 10^{-1} + 1.50 \times 10^{-1})} = 2.0 \times 10^{-2} M$$

(number of moles of $H_2PO_4^-$ left) $=$ (number of moles of $H_2PO_4^-$ to start) $-$
$=$ (number of moles of HPO_4^{2-} formed)

$$= 1.0 \times 10^{-2} - 0.50 \times 10^{-2} = 0.5 \times 10^{-2} \text{ mole}$$

$$C_{H_2PO_4^-} = \frac{\text{(number of moles of } H_2PO_4^- \text{ left)}}{\text{(total volume of solution, liters)}}$$

$$= \frac{(0.5 \times 10^{-2})}{(1.00 \times 10^{-1} + 1.50 \times 10^{-1})} = 2 \times 10^{-2} M$$

This solution is a *buffer*. The equilibrium approximations may be begun now.

First approximation:

$$[H^+]' = K_2 \frac{C_{H_2PO_4^-}}{C_{HPO_4^{2-}}} = (6.2 \times 10^{-8}) \left(\frac{2 \times 10^{-2}}{2.0 \times 10^{-2}} \right) = 6.2 \times 10^{-8} M$$

$$[OH^-]' = \frac{K_w}{[H^+]'} = \frac{(1.00 \times 10^{-14})}{(6.2 \times 10^{-8})} = 1.6 \times 10^{-7} M$$

Second approximation:

$$[H^+]'' = K_2 \cdot \frac{(C_{H_2PO_4^-} - [H^+]' + [OH^-]')}{(C_{HPO_4^{2-}} + [H^+]' - [OH^-]')}$$

$$= (6.2 \times 10^{-8}) \cdot \left(\frac{2 \times 10^{-2} - 6.2 \times 10^{-8} + 1.6 \times 10^{-7}}{2.0 \times 10^{-2} + 6.2 \times 10^{-8} - 1.6 \times 10^{-7}} \right)$$

$$= 6.2 \times 10^{-8} M$$

$$[H^+]' = [H^+]''$$

The approximations have converged.

$$pH = -\log(6.2 \times 10^{-8}) = 7.21$$

11. This is a case of hydrolysis.

$$C_2O_4^{2-} + H_2O \rightleftharpoons HC_2O_4^- + OH^-$$

First approximation:

$$[OH^-]' = \left(\frac{K_w}{K_2} C_{C_2O_4^{2-}} \right)^{1/2} = \left[\frac{(1.00 \times 10^{-14})}{(6.2 \times 10^{-5})} \cdot (5.0 \times 10^{-2}) \right]^{1/2} = 2.8 \times 10^{-6} M$$

Second approximation:

$$[OH^-]'' = \left[\frac{K_w}{K_2} (C_{C_2O_4^{2-}} - [OH^-]') \right]^{1/2}$$

$$= \left[\left(\frac{1.00 \times 10^{-14}}{6.2 \times 10^{-5}} \right) (5.0 \times 10^{-2} - 2.8 \times 10^{-6}) \right]^{1/2} = 2.8 \times 10^{-6} M$$

$$[OH^-]' = [OH^-]''$$

The approximations have converged.

$$pOH = -\log(2.8 \times 10^{-6}) = 5.55$$

$$pH = 14.00 - 5.55 = 8.45$$

12. Two reactions take place here. The amount of NaOH added is 2×10^{-2} mole. The amount of H_2Tar to start with is 1×10^{-2} mole.

$$\begin{array}{ccccc}
H_2Tar & + & OH^- & \rightarrow & HTar^- & + H_2O \\
0.01 \text{ mole} & & 0.01 \text{ mole} & & 0.01 \text{ mole} &
\end{array}$$

$$\begin{array}{ccccc}
HTar^- & + & OH^- & \rightarrow & Tar^{2-} & + H_2O \\
0.01 \text{ mole} & & 0.01 \text{ mole} & & 0.01 \text{ mole} &
\end{array}$$

Now, since all the H_2Tar and $HTar^-$ are used up, just the Tar^{2-} is left. This hydrolyzes.

$$Tar^{2-} + H_2O \rightleftharpoons HTar^- + OH^-$$

$$C_{Tar^{2-}} = \frac{0.010 \text{ mole}}{0.300 \text{ liter}} = 3.3 \times 10^{-2} M$$

Now the approximations can begin.

First approximation:

$$[OH^-]' = \left(\frac{K_w}{K_2} C_{Tar^{2-}}\right)^{1/2} = \left[\left(\frac{1.00 \times 10^{-14}}{8.9 \times 10^{-5}}\right)(3.3 \times 10^{-2})\right]^{1/2} = 1.9 \times 10^{-6} M$$

Second approximation:

$$[OH^-]'' = \left[\frac{K_w}{K_2}(C_{Tar^{2-}} - [OH^-]')\right]^{1/2}$$

$$= \left[\left(\frac{1.00 \times 10^{-14}}{8.9 \times 10^{-5}}\right)(3.3 \times 10^{-2} - 1.9 \times 10^{-6})\right]^{1/2} = 1.9 \times 10^{-6} M$$

$$[OH^-]' = [OH^-]''$$

The approximations have converged.

$$pOH = -\log(1.9 \times 10^{-6}) = 5.72$$

$$pH = 14.00 - 5.72 = 8.28$$

13. Hydrogen salicylate, $HSal^-$, is an ampholyte. We use a first approximation because $C_{HSal^-} = 0.10 M$.

$$[H^+]' = (K_1 K_2)^{1/2} = [(1.07 \times 10^{-3})(1.0 \times 10^{-13})]^{1/2} = 1.0 \times 10^{-8} M$$

$$pH = -\log(1.0 \times 10^{-8})$$

$$pH = 8.00$$

Because $C_{HSal^-} = 0.10$, and because $pH = 8.00$, no further approximations are needed.

14. The reaction between H_3PO_4 and NaOH is

$$\begin{array}{ccccc} H_3PO_4 & + & OH^- & \rightarrow & H_2PO_4^- + H_2O \\ 10^{-2} \text{ mole} & & 10^{-2} \text{ mole} & & 10^{-2} \text{ mole} \end{array}$$

The solution consists now of water and the ampholyte $H_2PO_4^-$. All the H_3PO_4 reacts with all the OH^-.

(number of moles of $H_2PO_4^-$ formed) $= 1.0 \times 10^{-2}$ mole

$$C_{H_2PO_4^-} = \frac{\text{(number of moles of } H_2PO_4^- \text{ formed)}}{\text{(volume of solution, liters)}}$$

$$= \frac{1.0 \times 10^{-2}}{(1.00 \times 10^{-1} + 1.00 \times 10^{-1})} = 5.0 \times 10^{-2} \text{ mole}$$

We use a first approximation.

$$[H^+]' = (K_1 K_2)^{1/2} = [(7.1 \times 10^{-3})(6.2 \times 10^{-8})]^{1/2} = 2.1 \times 10^{-5} M$$

$$pH = -\log(2.1 \times 10^{-5}) = 4.68$$

15. First, if $[H^+] = (K_1 K_2)^{1/2}$, dilution would not affect the pH much. Second look at the peak of the α_1 ($[HPh^-]/C_{Ph}$) curve (Figure 2.3). Even a fair-sized change in α_1, which would mean a decrease in $[HPh^-]$ and an increase in either $[H_2Ph]$ or $[Ph^-]$, does not correspond to a large pH shift. A change in α_1 could result from the addition of either a strong acid or a strong base. The KHPh resists pH change and is hence called a buffer.

DISTRIBUTION DIAGRAM FOR PHTHALIC ACID

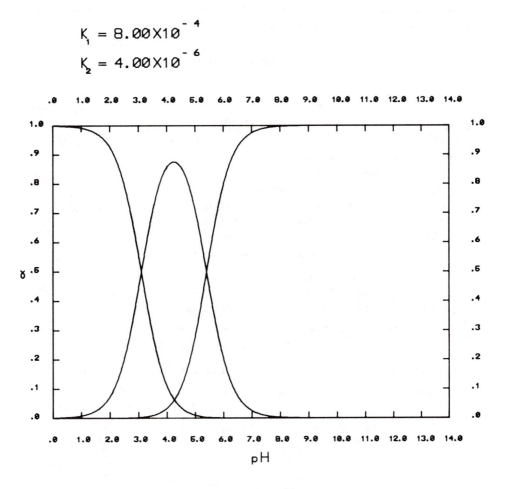

$$K_1 = 8.00 \times 10^{-4}$$

$$K_2 = 4.00 \times 10^{-6}$$

Figure 2.3 *Distribution Diagram for Phthalic Acid.*

≡ 2.22 THE TITRATION OF WEAK ACIDS ≡

Titrations of weak acids serve at least three purposes: (1) assay, (2) determination of dissociation constants, and (3) the identification of unknown compounds. In this section, ways of predicting the shapes of titration curves will be examined. Prediction of curve shapes for monoprotic and well-behaved polyprotic acids by simple methods will be discussed. Ill-behaved polyprotic acids have curve shapes that are predictable when more subtle means are applied; these means will be treated.

In the titration of a weak acid, strong base, usually NaOH of $0.1M$ (or $0.1N$, which is the same thing for NaOH) is added to the titration vessel from a buret, and is called the "titrant." The weak acid, in the titration vessel, often a beaker, is usually about $0.1M$ to $0.01M$. It is called the "analyte" or "titrate." The solution in the titration vessel is kept stirred. The pH of the solution is measured with a combination of a pH-sensitive glass electrode, a reference electrode, and a pH meter.

2.23 Titration of Weak Monoprotic Acids with Strong Bases

Suppose, for the sake of argument, that the titration vessel contains 100.00 milliliters of $0.1000M$ HA ($K_a = 1.00 \times 10^{-5}$). Suppose, further, that the buret contains $0.1000M$ NaOH. These numbers are chosen for convenience. There are four chemical cases to discuss in predicting the shape of the titration curve.

In case 1, no NaOH has been added to the solution. $C_{HA} = 0.1000M$, and $C_{A^-} = 0M$. Here, the weak acid, HA, is alone in solution.

In case 2, some NaOH has been added to the titration vessel but not enough to react with all the HA. The reaction between the strong base, OH^- from NaOH, and the weak acid, HA, is

$$HA + OH^- \rightleftharpoons A^- + H_2O$$

Its equilibrium lies far to the right. Because the stoichiometry is one-to-one, it is possible to say that for each millimole of NaOH added, 1 millimole of A^- ion is formed and 1 millimole of HA is lost. (A millimole is 10^{-3} mole.)

$$\text{(number of millimoles of X)} = \text{(number of milliliters of X)} \cdot$$
$$\text{(molarity, } M, \text{ of X)}$$

$$\text{(number of millimoles of } A^- \text{ formed)} = \text{(number of millimoles of } OH^- \text{ added)}$$

$$\text{(number of millimoles of HA left)} = \text{(number of millimoles of HA to start)} - $$
$$\text{(number of millimoles of } A^- \text{ formed)}$$

The solutions found in the titration vessel in case 2 are *buffer solutions* because they contain large, finite analytical concentrations of both weak acid HA and its anion A^-.

$$C_{HA} > 0$$

$$C_{A^-} > 0$$

Simple approximations often apply when between 10 and 90 percent of the HA has been reacted.

In case 3, just enough NaOH has been added to the titration vessel so that the endpoint is reached.

$$(\text{number of millimoles of HA left}) = 0$$

$$(\text{number of millimoles of } A^- \text{ formed}) = (\text{number of millimoles of } OH^- \text{ added})$$

$$C_{HA} = 0$$

$$C_{A^-} > 0$$

Because the analytical concentration of HA is zero, while the analytical concentration of A^- is large, the hydrolysis of A^- determines the pH of the solution.

$$A^- + H_2O \rightleftharpoons HA + OH^-$$

In case 4, past the endpoint, there is excess strong base in solution. This excess strong base determines the value of $[OH^-]$.

$$[OH^-] = \frac{(\text{number of millimoles of } OH^- \text{ excess})}{(\text{number of milliliters of solution})}$$

It is usually sufficient to use first approximations in all four cases for calculating the pH of the solution in the titration vessel. Concentrations of the pH-determining species are generally between 0.1 and 0.01M. Let us calculate now the pH-values of the solution in the titration vessel when 0, 25.00, 50.00, 90.00, 100.00, 125.00, and 150.00 milliliters of 0.1000M NaOH have been added.

A. 0 milliliters of NaOH added. This is case 1.

$$[H^+]' = (K_a C_{HA})^{1/2} \qquad\qquad K_a = 1.00 \times 10^{-5}M$$

$$C_{HA} = 1.000 \times 10^{-1}M$$

$$= [(1.00 \times 10^{-5})(1.000 \times 10^{-1})]^{1/2} = 1.00 \times 10^{-3}M$$

$$pH = -\log(1.00 \times 10^{-3}) = 3.00$$

(Most pH meters can be read to only two decimal places.)

B. 25.00 milliliters of NaOH added. This is case 2, the case of the buffer region, where between 10 and 90% of the HA has been reacted.

$$(\text{number of millimoles of } A^- \text{ formed}) = (\text{number of millimoles of } OH^- \text{ added})$$

$$= (25.00)(0.1000)$$

$$= 2.500 \text{ millimoles}$$

$$C_{A^-} = \frac{(\text{number of millimoles of } A^- \text{ formed})}{(\text{number of milliliters of solution})}$$

$$C_{A^-} = \frac{2.500}{125.00}$$

(number of millimoles of HA left)

= (number of millimoles of HA to start) −
(number of millimoles of A⁻ formed)

= (100.00)(0.1000) − (25.00)(0.1000) = 10.00 − 2.500 = 7.50 millimoles

$$C_{HA} = \frac{\text{(number of millimoles of HA left)}}{\text{(number of milliliters of solution)}} = \frac{7.50}{125.00}$$

In a buffer,

$$[H^+]' = K_a \frac{C_{HA}}{C_{A^-}} = (1.00 \times 10^{-5}) \left(\frac{\frac{7.50}{125}}{\frac{2.500}{125}} \right) = 3.00 \times 10^{-5} M$$

$$pH = -\log (3.00 \times 10^{-5}) = 4.52$$

C. 50.00 milliliters NaOH added. This is *still* case 2, the case of the buffer region.

(number of millimoles of A⁻ formed)

= (number of millimoles of OH⁻ added)

= (50.00)(0.1000) = 5.000 millimoles

$$C_{A^-} = \frac{\text{(number of millimoles of A}^-\text{ formed)}}{\text{(number of milliliters of solution)}}$$

$$C_{A^-} = \frac{5.000}{150.00}$$

(number of millimoles of HA left)

= (number of millimoles of HA to start)−
(number of millimoles A⁻ formed)

= (100.00)(0.1000) − (50.00)(0.1000) = 5.00 millimoles

$$C_{HA} = \frac{\text{(number of millimoles of HA left)}}{\text{(number of milliliters of solution)}} = \frac{5.00}{150.00}$$

In a buffer,

$$[H^+]' = K_a \frac{C_{HA}}{C_{A^-}} = (1.00 \times 10^{-5}) \left(\frac{\frac{5.00}{150.00}}{\frac{5.00}{150.00}} \right) = 1.00 \times 10^{-5} M$$

(It is interesting to note that, at the midway point in the titration of a weak acid, $[H^+]' = K_a$.)

$$pH = -\log (1.00 \times 10^{-5}) = 5.00$$

D. 90.00 milliliters of NaOH added. This is *still* case 2, the case of the buffer region.

(number of millimoles of A⁻ formed)

= (number of millimoles of OH⁻ added)

$$= (90.00)(0.1000) = 9.000 \text{ millimoles}$$

$$c_{A^-} = \frac{\text{(number of millimoles of } A^- \text{ formed)}}{\text{(number of milliliters of solution)}}$$

$$c_{A^-} = \frac{9.000}{190.00}$$

(number of millimoles of HA left)
= (number of millimoles of HA to start) –
(number of millimoles of A^- formed

$$= (10.00)(0.1000) - (90.00)(0.1000) = 1.000 \text{ millimoles}$$

$$c_{HA} = \frac{\text{(number of millimoles of HA left)}}{\text{(number of milliliters of solution)}} = \frac{1.000}{190.00}$$

In a buffer,

$$[H^+]' = K_a \frac{c_{HA}}{c_{A^-}} = (1.00 \times 10^{-5}) \left(\frac{\frac{1.000}{190.00}}{\frac{9.000}{190.00}} \right) = 1.11 \times 10^{-6}M$$

$$pH = -\log(1.11 \times 10^{-6})$$

$$pH = 5.95$$

E. 100.00 milliliters of NaOH added. Case 3 applies here. The hydrolysis of A^- determines pH.

$$c_{HA} = 0$$

(number of millimoles of A^- formed)

= (number of millimoles of OH^- added)

= (100.00)(0.1000) = 10.00 millimoles

$$c_{A^-} = \frac{\text{(number of millimoles of } A^- \text{ formed)}}{\text{(number of milliliters of solution)}} = \frac{10.00}{200.00} = 0.05000M$$

When the hydrolysis of A^- determines the pH,

$$[OH^-]' = \left(\frac{K_w}{K_a} c_{A^-} \right)^{1/2} = \left[\left(\frac{1.00 \times 10^{-14}}{1.00 \times 10^{-5}} \right) (5.000 \times 10^{-2}) \right]^{1/2} = 7.07 \times 10^{-6}M$$

$$pOH = -\log(7.07 \times 10^{-6}) = 5.15$$

$$pH = 14.00 - 5.15 = 8.85$$

(Note that the pH at the endpoint *need not* be 7.00.)

F. 125.00 milliliters of NaOH added. This is case 4; there is excess strong base in the solution. From the stoichiometry of the reaction

$$HA + OH^- \rightleftharpoons A^- + H_2O$$

it can be concluded that, *after the endpoint*

(number of millimoles of OH^- excess)

= (number of millimoles of OH^- added) −
(number of millimoles of HA to start)

= (125.00)(0.1000) − (100.00)(0.1000) = 2.50 millimoles

$$[OH^-] = \frac{(\text{number of millimoles of } OH^- \text{ excess})}{(\text{number of milliliters of solution})}$$

$$[OH^-] = \frac{2.50}{225.00}$$

$$[OH^-] = 1.11 \times 10^{-2}M$$

$$pOH = -\log(1.1 \times 10^{-2}) = 1.95$$

$$pH = 14.00 - 1.95 = 12.05$$

G. 150.00 milliliters of NaOH added. This is still case 4. There is a greater excess of strong base in the solution.

(number of millimoles of OH^- excess)

= (number of millimoles of OH^- added) −
(number of millimoles of HA to start)

= (150.00)(0.1000) − (100.00)(0.1000) = 5.00 millimoles

$$[OH^-] = \frac{(\text{number of millimoles of } OH^- \text{ excess})}{(\text{number of milliliters of solution})} = \frac{5.00}{250.00} = 2.00 \times 10^{-2}M$$

$$pOH = -\log(2.00 \times 10^{-2}) = 1.70$$

$$pH = 14.00 - 1.70 = 12.30$$

The curve that would be described by these (and other) points was shown in Figure 2.4.

2.24 Titration of Weak, Well-Behaved Polyprotic Acids with Strong Bases

The titration curves of some polyprotic acids may be calculated very simply. Such acids will be designated "well-behaved" polyprotic acids. If dissociation constants are different by factors of 100 or more, endpoints at low pH-values are not blurred, simple approximations may be used, and the acids are called "well-behaved."

The apparatus used in the titration of weak polyprotic acids is the same as that used for weak monoprotic acids. The titrant is a strong base, usually NaOH, of known molarity around 0.1M.

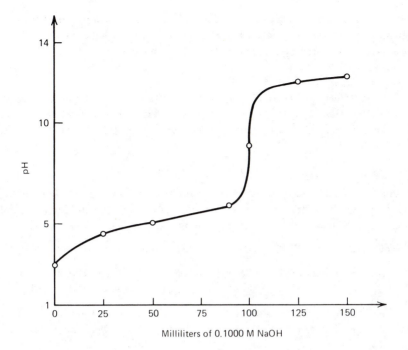

Figure 2.4 *Titration of 0.1000M HA ($K_a = 1.00 \times 10^{-5}$) by 0.1000M NaOH. Initial Analyte Volume 100.00 ml 0.1000M HA.*

Suppose that a titration vessel contains 100.00 milliliters of a solution 0.1000M in H_2B ($K_1 = 1.00 \times 10^{-4}$, $K_2 = 1.00 \times 10^{-7}$). The titrant used is 0.1000M NaOH.

In the case of the *mono*protic weak acid, HA, there is one reaction of titrant with analyte

$$HA + OH^- \rightleftharpoons A^- + H_2O$$

There can be only one reaction because there is only one proton for HA to give away. The diprotic acid, H_2B, on the other hand, has two protons to give away. There are two reactions of H_2B analyte with NaOH titrant. These reactions occur successively, not simultaneously. Only when the first reaction is complete does the second one begin. The first reaction is

$$H_2B + OH^- \rightleftharpoons HB^- + H_2O$$

When the first reaction is complete, that is, when enough OH^- has been added to convert all the H_2B to HB^-, the first endpoint is reached. Only when all the H_2B is converted to HB^-, that is, only then the first endpoint is reached, can the second reaction begin. The second reaction is

$$HB^- + OH^- \rightleftharpoons B^{2-} + H_2O$$

When the second reaction is complete, that is, when enough OH^- has been added to convert all the HB^- to B^{2-}, the second endpoint is reached. After the second endpoint is reached, the addition of more OH^- to the analyte solution simply raises the pH, because there is no protonated species like H_2B or HB^- to react with the added OH^-. There are six chemical cases to discuss in predicting the shape of the titration curve.

In case 1 no NaOH has been added to the solution; $C_{H_2B} = 0.1000M$, $C_{HB^-} = 0$, and $C_{B^{2-}} = 0$. Here the weak acid, H_2B, is alone in solution.

In case 2 some NaOH has been added to the titration vessel but not enough to react all the H_2B to HB^-. Case 2 deals with the region of the titration curve before the first endpoint. The reaction between the strong base (OH^- from NaOH) and the weak, diprotic acid H_2B, is

$$H_2B + OH^- \rightleftharpoons HB^- + H_2O$$

Its equilibrium lies far to the right. Because the reaction stoichiometry is one-to-one, it is possible to say that for each millimole of NaOH added, 1 millimole of the ampholyte, HB^-, is formed and 1 millimole of the diprotic weak acid, H_2B, is lost.

(number of millimoles of HB^- formed) = (number of millimoles of OH^- added)

(number of millimoles of H_2B left) = (number of millimoles of H_2B to start) − (number of millimoles of OH^- added)

The solutions found in the titration vessel in case 2 are *buffer solutions*, because they contain large, finite analytical concentrations of both weak acid H_2B and ampholyte HB^-.

$$C_{H_2B} > 0$$

$$C_{HB^-} > 0$$

$$C_{B^{2-}} = 0$$

Simple approximations often apply when between 10 and 90% of the H_2B has been converted to HB^-.

In case 3, just enough NaOH has been added to the titration vessel so that all of the H_2B is converted to HB^-, but so that no HB^- is converted to B^{2-}. In other words, the first endpoint has been reached. Case 3 is just the case of the first endpoint.

(number of millimoles of H_2B left) = 0

(number of millimoles of HB^- formed) = (number of millimoles of OH^- added)

(number of millimoles of B^{2-} formed) = 0

$$C_{H_2B} = 0$$

$$C_{HB^-} > 0$$

$$C_{B^{2-}} = 0$$

Because the analytical concentration of HB^- is large, while the analytical concentrations of H_2B and B^{2-} are zero, the solution in the titration vessel at the first endpoint is an ampholyte solution.

In case 4, past the first endpoint, but before the second endpoint, enough NaOH has been added to convert some but not all of the HB^- to B^{2-}. The reaction between the strong base (OH^- from NaOH) and the ampholyte, HB^-, is

$$HB^- + OH^- \rightleftharpoons B^{2-} + H_2O$$

Its equilibrium lies far to the right. Because the reaction stoichiometry is one-to-one, it is possible to say that for each millimole of NaOH added past the first endpoint, 1 millimole of the unprotonated anion, B^{2-}, is formed, and 1 millimole of the ampholyte, HB^-, is lost.

(number of millimoles of B^{2-} formed)

       = (number of millimoles of OH^- added past first endpoint)

(number of millimoles of HB^- left)

       = (number of millimoles of HB^- to start) –
       (number of millimoles of OH^- added past first endpoint)

(number of millimoles of HB^- to start)

       = (number of millimoles of OH^- added to first endpoint)

(number of millimoles of HB^- left)

       = (number of millimoles of OH^- added to first endpoint) –
       (number of millimoles of OH^- added past first endpoint)

The solutions found in the titration vessel in case 4 are *buffer solutions* because they contain large, finite analytical concentrations of both ampholyte HB^- and unprotonated anion B^{2-}.

$$C_{H_2B} = 0$$

$$C_{HB^-} > 0$$

$$C_{B^{2-}} > 0$$

Simple approximations often apply when between 10 and 90% of the HB^- has been converted to B^{2-}.

In case 5, just enough NaOH has been added to the titration vessel so that all of the HB^- is converted to B^{2-}, but so that there is no excess OH^- in solution. Case 5 is just the case of the second endpoint.

(number of millimoles of H_2B left)

       = 0

(number of millimoles of HB^- left)

       = 0

(number of millimoles of B^{2-} formed)

       = (number of millimoles of OH^- added past first endpoint)

$$C_{H_2B} = 0$$

$$C_{HB^-} = 0$$

$$C_{B^{2-}} > 0$$

Because the analytical concentrations of H_2B and HB^- are zero, and because the analytical concentration of B^{2-} is large, the hydrolysis of B^{2-} determines the pH of the solution.

$$B^{2-} + H_2O \rightleftharpoons HB^- + OH^-$$

In case 6, past the second endpoint, there is excess strong base in solution. This excess strong base determines the value of $[OH^-]$.

$$[OH^-] = \frac{\text{(number of millimoles of } OH^- \text{ excess)}}{\text{(number of milliliters of solution)}}$$

With well-behaved polyprotic acids, it is usually sufficient to use first approximations in all six cases for calculating the pH of the solution in the titration vessel. Concentrations of the pH-determining species are generally between 0.1 and 0.01M. Let us now calculate the pH values of the analyte solution when 0, 50.00, 100.00, 150.00, 200.00, and 250.00 milliliters of 0.1000M NaOH are added to the titration vessel.

The solution in the titration vessel before the start of the titration consists of 100.00 milliliters of 0.1000M H_2B ($C_{H_2B} = 0.1000M$).

$$K_1 = 1.00 \times 10^{-4}$$

$$K_2 = 1.00 \times 10^{-7}$$

A. 0 milliliters of NaOH are added. This is case 1.

$$[H^+]' = (K_1 C_{H_2B})^{1/2} = [(1.00 \times 10^{-4})(0.1000)]^{1/2} = 3.16 \times 10^{-3} M$$

$$pH = -\log (3.16 \times 10^{-3}) = 2.50$$

B. 50.00 milliliters of NaOH are added. This is case 2, the case of the buffer region.

(number of millimoles of HB^- formed)
$$= \text{(number of millimoles of } OH^- \text{ added)}$$
$$= (50.00)(0.1000) = 5.000 \text{ millimoles}$$
$$= \frac{\text{(number of millimoles of } HB^- \text{ formed)}}{\text{(number of milliliters of solution)}}$$

$$C_{HB^-} = \frac{5.000}{150.00}$$

(number of millimoles of H_2B left)

$$= \text{(number of millimoles of } H_2B \text{ to start)} -$$
$$\text{(number of millimoles of } HB^- \text{ formed)}$$

$$= (100.00)(0.1000) - (50.00)(0.1000) = 5.000 \text{ millimoles}$$

$$C_{H_2B} = \frac{\text{(number of millimoles of } H_2B \text{ left)}}{\text{(number of milliliters of solution)}}$$

$$C_{H_2B} = \frac{5.000}{150.00}$$

In a buffer of H_2B and HB^-,

$$[H^+]' = K_1 \frac{C_{H_2B}}{C_{HB^-}}$$

$$= (1.00 \times 10^{-4}) \left(\frac{\frac{5.000}{150.00}}{\frac{5.000}{150.00}} \right) = 1.00 \times 10^{-4} M$$

Note that, midway between 0 milliliter of NaOH added, and the first endpoint in the titration of this well-behaved weak diprotic acid,

$$[H^+]' = K_1$$

$$pH = -\log (1.00 \times 10^{-4}) = 4.00$$

C. 100 milliliters of NaOH added. Case 3 applies here. The ampholyte, HB^-, determines pH.

$$C_{H_2B} = 0$$

$$C_{B^{2-}} = 0$$

(number of millimoles of HB^- formed) = (number of millimoles of OH^- added)

$$= (100.00)(0.1000) = 10.00 \text{ millimoles}$$

$$C_{HB^-} = \frac{\text{(number of millimoles of } HB^- \text{ formed)}}{\text{(number of milliliters of solution)}}$$

$$= \frac{10.00}{200.00} = 5.000 \times 10^{-2} M$$

In this concentration range, a first approximation will probably do.

$$[H^+]' = (K_1 K_2)^{1/2} = [(1.00 \times 10^{-4})(1.00 \times 10^{-7})]^{1/2} = 3.16 \times 10^{-6} M$$

$$pH = -\log (3.16 \times 10^{-6}) = 5.50$$

D. 150.00 milliliters of NaOH added. Case 4 applies here. The solution is a buffer of HB^- and B^{2-}.

(number of millimoles of B^{2-} formed)

= (number of millimoles of OH^- added past first endpoint)

= (50.00)(0.1000) = 5.000 millimoles

$$C_{B^{2-}} = \frac{\text{(number of millimoles of } B^{2-} \text{ formed)}}{\text{(number of milliliters of solution)}} = \frac{5.000}{250.00}$$

(number of millimoles of HB^- left)

= (number of millimoles of OH^- added to first endpoint) – (number of millimoles of OH^- added past first endpoint)

= (100.00)(0.1000) – (50.00)(0.1000) = 5.000 millimoles

$$C_{HB^-} = \frac{\text{(number of millimoles of HB}^- \text{ left)}}{\text{(number of milliliters of solution)}} = \frac{5.000}{250.00}$$

In a buffer of HB^- and B^{2-},

$$[H^+]' = K_2 \frac{C_{HB^-}}{C_{B^{2-}}} = (1.00 \times 10^{-7}) \frac{\left(\frac{5.000}{250.00}\right)}{\left(\frac{5.000}{250.00}\right)} = 1.00 \times 10^{-7} M$$

Note that, midway between the first and second endpoints in the titration of this well-behaved weak diprotic acid,

$$[H^+]' = K_2$$

$$pH = -\log(1.00 \times 10^{-7}) = 7.00$$

E. 200 milliliters of NaOH added. Case 5 applies here. Hydrolysis of the unprotonated anion, B^{2-}, determines the pH.

(number of millimoles of B^{2-} formed)
  = (number of millimoles of OH^- added past first endpoint)

  = (100.00)(0.1000) = 10.00 millimoles

$$C_{B^{2-}} = \frac{\text{(number of millimoles of B}^{2-} \text{ formed)}}{\text{(number of milliliters of solution)}} = \frac{10.00}{300.00} = 3.333 \times 10^{-2} M$$

When hydrolysis of B^{2-} determines pH,

$$[OH^-]' = \left(\frac{K_w}{K_2} C_{B^{2-}}\right)^{1/2} = \left[\left(\frac{1.00 \times 10^{-14}}{1.00 \times 10^{-7}}\right)(3.333 \times 10^{-2})\right]^{1/2} = 5.77 \times 10^{-5} M$$

$$pOH = -\log(5.77 \times 10^{-5}) = 4.24$$

$$pH = 14.00 - 4.24 = 9.76$$

F. 250.00 milliliters of NaOH added. Case 6 applies here. There is excess strong base in the solution.
   From the stoichiometry of the two reactions:

$$H_2B + OH^- \rightleftharpoons HB^- + H_2O$$

$$HB^- + OH^- \rightleftharpoons B^{2-} + H_2O$$

it can be concluded that, *after the second endpoint:*

(number of millimoles of OH^- excess)
     = (number of millimoles of OH^- added) –
       (2)(number of millimoles of H_2B to start)

     = (250.00)(0.1000) – (2)(100.00)(0.1000) = 5.000 millimoles

$$[OH^-] = \frac{\text{(number of millimoles of OH}^- \text{ excess)}}{\text{(number of milliliters of solution)}} = \frac{5.000}{350.00} = 1.428 \times 10^{-2} M$$

$$pOH = -\log(1.428 \times 10^{-2}) = 1.84$$

$$pH = 14.00 - 1.84 = 12.16$$

The complete titration curve would look somewhat like the one shown in Figure 2.5.

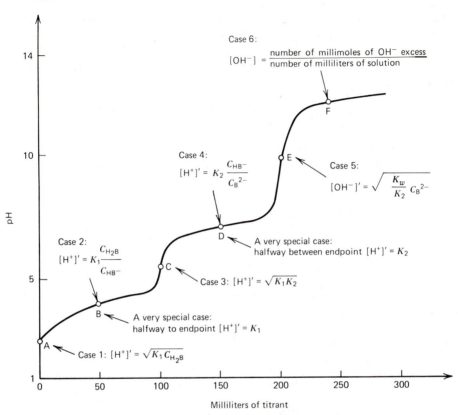

Figure 2.5 *Titration of 0.1000M H_2B ($K_1 = 1.00 \times 10^{-4}$, $K_2 = 1.00 \times 10^{-7}$) by 0.1000 M NaOH. Initial Analyte Volume 100.00 ml 0.1000 M H_2B.*

2.25 Titration of Any Polyprotic Acids (Well- or Ill-Behaved) with Strong Bases

Not all polyprotic acids are well-behaved. A great many display values of successive ionization constants that are not separated by factors of 100. When dissociation constants are close together, the simple approximations that serve to calculate titration curve points for well-behaved weak acids begin to break down. It is difficult, in these cases, to calculate $[H^+]$ in terms of analytical concentrations and successive ionization constants. What is to be done?

What is described here is a numerical approach. The approach involves letting $[H^+]$

serve as an independent variable and calculating the volume of titrant that must be added to a given analyte solution to reach that value of $[H^+]$. The approach used with well-behaved polyprotic acids is opposite to this: A volume of titrant is specified and a value of $[H^+]$ then calculated.

The derivation of a complete, and rather bulky, expression to calculate the needed volume of titrant to reach a given $[H^+]$-value is given here.*

First, suppose that an analyte solution consists of a known volume of weak acid H_2B. The total equilibrium hydrogen ion concentration, $[H^+]$, is given by the expression:

$$[H^+] = +(H^+) \frac{from}{H_2B} - (H^+) \frac{reacted\ into}{H_2O\ by\ base} + (H^+) \frac{from}{H_2O}$$

Define the following terms:

C_b = concentration of strong base added (within analyte) and then converted to H_2O

$$[OH^-] = (H^+) \frac{from}{H_2O}$$

$$(H^+) \frac{from}{H_2B} = C_{H/H_2B}$$

Then, using the defined terms, we obtain

$$[H^+] = C_{H/H_2B} + [OH^-] - C_b \tag{2.47}$$

By simple chemical bookkeeping,

$$C_{H/H_2B} = [HB^-] + 2[B^{2-}]$$

Substituting into Equation 2.47 gives

$$[H^+] = [HB^-] + 2[B^{2-}] + [OH^-] - C_b$$

$$[OH^-] = \frac{K_w}{[H^+]}$$

$$[H^+] = [HB^-] + 2[B^{2-}] + \frac{K_w}{[H^+]} - C_b$$

Let C_B be the sum of the concentrations of all B-containing species. Remember the alpha-functions.

$$[HB^-] = \alpha_1 C_B$$

$$[B^{2-}] = \alpha_2 C_B$$

$$[H^+] = \alpha_1 C_B + 2\alpha_2 C_B + \frac{K_w}{[H^+]} - C_b \tag{2.48}$$

* Courtesy of Professors John P. Walters and John C. Wright, of the Department of Chemistry, University of Wisconsin.

We define some more terms:

M = molarity of base in the buret

x = volume of base added from buret

V = initial volume of H_2B analyte solution—before addition of base

$V + x$ = analyte solution volume in titration vessel at any time during titration

$C_{H_2B}^0$ = initial analytical concentration of H_2B in analyte solution

When no base has been added,

$$C_B = C_{H_2B}^0$$

At any point in the titration,

$$C_B = C_{H_2B}^0 \left(\frac{V}{V + x} \right)$$

$$C_b = M \left(\frac{x}{V + x} \right)$$

Then we rearrange Equation 2.48.

$$C_b = \alpha_1 C_B + 2\alpha_2 C_B + \frac{K_w}{[H^+]} - [H^+]$$

Now, the new terms can be substituted:

$$M \left(\frac{x}{V + x} \right) = \alpha_1 C_{H_2B}^0 \left(\frac{V}{V + x} \right) + 2\alpha_2 C_{H_2B}^0 \left(\frac{V}{V + x} \right) + \frac{K_w}{[H^+]} - [H^+]$$

$$Mx = \alpha_1 C_{H_2B}^0 V + 2\alpha_2 C_{H_2B}^0 V + \frac{K_w}{[H^+]} (V + x) - [H^+](V + x)$$

$$Mx - \frac{K_w}{[H^+]} (V + x) + [H^+](V + x) = \alpha_1 C_{H_2B}^0 V + 2\alpha_2 C_{H_2B}^0 V$$

$$Mx - \frac{K_w}{[H^+]} V - \frac{K_w}{[H^+]} x + [H^+] V + [H^+] x = \alpha_1 C_{H_2B}^0 V + 2\alpha_2 C_{H_2B}^0 V$$

$$Mx - \frac{K_w}{[H^+]} x + [H^+] x = \alpha_1 C_{H_2B}^0 V + 2\alpha_2 C_{H_2B}^0 V + \frac{K_w}{[H^+]} V - [H^+] V$$

$$\left(M - \frac{K_w}{[H^+]} + [H^+] \right) x = \left(\alpha_1 C_{H_2B}^0 + 2\alpha_2 C_{H_2B}^0 + \frac{K_w}{[H^+]} - [H^+] \right) V$$

$$x = \frac{\left(\alpha_1 C_{H_2B}^0 + 2\alpha_2 C_{H_2B}^0 + \frac{K_w}{[H^+]} - [H^+] \right) V}{\left(M - \frac{K_w}{[H^+]} + [H^+] \right)} \tag{2.49}$$

Equation 2.49 is a complete (and very bulky) expression for calculating points on a titration curve before the last endpoint. Given $[H^+]$, $C_{H_2B}^0$, M, and V, plus either K-values or tables of alpha-values, we can calculate the volume of strong base needed to reach the specified $[H^+]$-value. Under certain conditions, the expression can be simplified.

Suppose that

$$\alpha_1 C_{H_2B}^0 + 2\alpha_2 C_{H_2B}^0 > [H^+]$$

$$M > [H^+]$$

$$[H^+] > [OH^-] = \frac{K_w}{[H^+]}$$

Equation 2.49 becomes

$$x \simeq \frac{(\alpha_1 C_{H_2B}^0 + 2\alpha_2 C_{H_2B}^0)V}{M}$$

But if

$$M = C_{H_2B}^0 = 0.1000M$$

$$x \simeq \frac{(\alpha_1 + 2\alpha_2)(0.1000)V}{(0.1000)}$$

or

$$x \simeq (\alpha_1 + 2\alpha_2)V \qquad (2.50)$$

This approximate expression is valid at intermediate regions of the curve.

Example: Draw a titration curve for the titration of 50.2 milliliters of 0.0617M succinic acid,

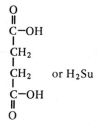

or H_2Su

Table 2.3
THEORETICAL TITRATION CURVE FOR
SUCCINIC ACID[a]

Chemistry 221   Professor J. Walters
Fall Semester 1973   Experiment No. 6
Prepared for Margaret Clark

Acid Constants: Experimental Data:
K_1 = 6.46E - 05   Acid concentration   = .0617
K_2 = 3.31E - 06   Acid volume   = 50.2
                       NaOH concentration = 0.0954

pH	α_1	$2\alpha_2$	$\dfrac{(\alpha_1 + 2\alpha_2) \cdot V \cdot C_{H_2B}^0}{M}$	x, ml of NaOH
2.00	.0064	.0000	.21	−4.57
2.25	.0114	.0000	.37	−2.45

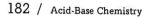

Table 2.3 continued

pH	α_1	$2\alpha_2$	$\dfrac{(\alpha_1 + 2\alpha_2) \cdot V \cdot C_{H_2B}^0}{M}$	x, ml of NaOH
2.50	.0200	.0000	.65	-.98
2.75	.0350	.0001	1.14	.20
3.00	.0606	.0002	1.98	1.44
3.25	.1029	.0006	3.38	3.07
3.50	.1693	.0018	5.61	5.43
3.75	.2651	.0049	8.93	8.82
4.00	.3873	.0128	13.41	13.34
4.25	.5182	.0305	18.80	18.76
4.50	.6272	.0657	24.63	24.60
4.75	.6842	.1274	30.49	30.47
5.00	.6729	.2228	36.32	36.31
5.25	.5967	.3514	42.19	42.18
5.50	.4771	.4996	47.93	47.92
5.75	.3461	.6444	53.08	53.08
6.00	.2311	.7653	57.20	57.20
6.25	.1450	.8537	60.14	60.14
6.50	.0871	.9124	62.08	62.08
6.75	.0510	.9489	63.27	63.27
7.00	.0293	.9706	63.98	63.98
7.25	.0167	.9833	64.39	64.39
7.50	.0095	.9905	64.63	64.63
7.75	.0053	.9947	64.76	64.76
8.00	.0030	.9970	64.84	64.84
8.25	.0017	.9983	64.88	64.88
8.50	.0010	.9990	64.90	64.91
8.75	.0005	.9995	64.92	64.92
9.00	.0003	.9997	64.92	64.94
9.25	.0002	.9998	64.93	64.95
9.50	.0001	.9999	64.93	64.97
9.75	.0001	.9999	64.93	65.00
10.00	.0000	1.0000	64.93	65.05
10.25	.0000	1.0000	64.93	65.15
10.50	.0000	1.0000	64.93	65.32
10.75	.0000	1.0000	64.93	65.62
11.00	.0000	1.0000	64.93	66.15
11.25	.0000	1.0000	64.93	67.12
11.50	.0000	1.0000	64.93	68.88

[a] *Courtesy of Professor John P. Walters, Department of Chemistry, University of Wisconsin.*

for short. ($K_1 = 6.46 \times 10^{-5}$ and $K_2 = 3.31 \times 10^{-6}$.) The titrant is $0.0954M$ NaOH. Alpha-tables or curves would be useful here. Table 2.3 is made up of data generated by the rigorous formula just derived. The negative values for milliliters of NaOH can be ignored; they simply mean that an equivalent amount of $0.0954M$ strong acid would have to be added to reach the low pH-values (2.00, 2.25, and 2.50). The data are plotted in Figure 2.6. Because K_1 and K_2 are close together, the first endpoint is obscured and only the second endpoint shows.

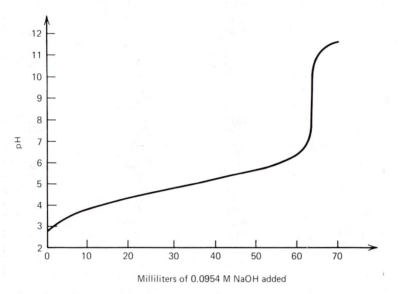

Milliliters of 0.0954 M NaOH added

Figure 2.6 *Plot of data from Table 2.3. Courtesy of Prof. John P. Walters, Dep't. of Chemistry, University of Wisconsin.*

≡ 2.26 PROBLEMS ≡

If titration stoichiometry presents difficulties, review the text and problems in "Stoichiometry," Section 1.11 of Chapter 1.

1. A solution is prepared by diluting 6.005 grams of acetic acid (CH_3COOH) to 1.000 liter with H_2O. A 50.00-milliliter aliquot is taken and titrated with $0.1000M$ NaOH. Calculate the pH when 0.00, 20.00, 25.00, 30.00, 50.00, 75.00, and 100.00 milliliters of NaOH are added. ($K_a = 1.8 \times 10^{-5}$.)
   Ans: *2.87, 4.57, 4.74, 4.92, 8.72, 12.30, and 12.52*

2. All that is known of a weak acid, HA, is that it is monoprotic and crystalline at room temperature. A sample, weighing 0.5000 gram is dissolved in 50.00 milliliters of H_2O and titrated with $0.1000M$ NaOH. The following data are taken from the titration curve:

Milliliters of NaOH	pH
0.00	2.65
20.47	4.21
(endpoint) 40.94	8.43

From these data, calculate the molecular weight of the acid, and, using first approximations, its K_a-value.

Ans: *Molecular weight = 122.1 grams per mole; K_a = 6.2 × 10^{-5}*

3. Phosphoric acid, H_3PO_4, is a relatively well-behaved triprotic weak acid. If a sample of $0.1000M$ H_3PO_4 is titrated with $0.1000M$ NaOH, estimate the pH at these points:

   (a) Halfway between the initial point and the first endpoint.

   Ans: *2.15*

   (b) At the first endpoint.

   Ans: *4.68*

   (c) Halfway between the first endpoint and the second endpoint.

   Ans: *7.21*

   (d) At the second endpoint.

   Ans: *9.77*

   (e) Why might it be hard to define the titration curve after the second endpoint?

   Ans: *See detailed solutions.*

   $(K_1 = 7.1 × 10^{-3}, K_2 = 6.2 × 10^{-8}, \text{and } K_3 = 4.4 × 10^{-13}.)$

4. An unknown weak acid, H_2B, is titrated. The acid is crystalline; 0.3658 gram is dissolved in 50.2 milliliters of water, and the sample is titrated with $0.0954M$ NaOH. The curve seen suggests that this weak acid is not badly behaved. Using first approximations, and the following data,* calculate the molecular weight, K_1, and K_2 for the acid.

Milliliters of OH⁻	pH
18.42	2.85
(First endpoint) 36.83	4.26
55.24	5.66
(Second endpoint) 73.66	~8.5

Ans: *Molecular weight = 104. grams per mole; K_1 = 1.4 × 10^{-3}, and K_2 = 2.2 × 10^{-6}*

5. Table 2.4 shows the number of milliliters of $0.1000M$ NaOH that have to be added to 50.00 milliliters of 0.1000 oxalic acid ($K_1 = 5.6 × 10^{-2}$; $K_2 = 6.2 × 10^{-5}$) to produce the pH values shown. (Each number in the last column can be rounded to two decimal places.)

* Courtesy of Professor John P. Walters, Department of Chemistry, University of Wisconsin.

(a) Complete the empty places in the table using the α-values for oxalic acid (Table 2.5).

(b) Plot the data from Table 2.4. On the same graph plot $(\alpha_1 + 2\alpha_2)V$ as you calculated it above.

(c) Mark the spots where the two curves differ.

(d) Looking at your titration curves, give a simple but quantitative explanation of why the two curves differ. Do one or two scratch calculations to prove your point.

Table 2.4
TITRATION OF OXALIC ACID[a]

pH	α_1	α_2	$(\alpha_1 + 2\alpha_2)V$	Milliliter base to be added to H_2A (Eq. 2.49)
0				−45.213
0.5				−36.179
1.0				−16.017
1.5				12.330
2.0				34.296
2.5				45.274
3.0				51.034
3.5				57.582
4.0				68.941
4.5				83.054
5.0				93.039
5.5				97.569
6.0				99.205
6.5				99.746
7.0				99.919
7.5				99.975
8.0				99.993
8.5				100.002
9.0				100.014
9.5				100.047
10.0				100.150
10.5				100.476
11.0				101.515
11.5				104.898
12.0				116.667
12.5				169.371
13.0				1.5×10^{14}
13.5				∞
14.0				∞

[a] Courtesy of Professor John P. Walters, Department of Chemistry, University of Wisconsin.

(e) Using your common sense and knowledge of stoichiometry for titration reactions, calculate the pH of the above solution after the addition of 200 milliliters of 0.1000M NaOH. Mark this on your graph.

Ans: *12.60*

Table 2.5
α-VALUES FOR OXALIC ACID.
$K_1 = 5.6E - 2; K_2 = 6.2E - 5.$

pH	α_o	α_1	α_2
0.50	0.850E + 00	0.150E + 00	0.295E - 04
1.00	0.641E + 00	0.359E + 00	0.223E - 03
1.50	0.360E + 00	0.638E + 00	0.125E - 02
2.00	0.151E + 00	0.844E + 00	0.523E - 02
2.50	0.525E - 01	0.929E + 00	0.182E - 01
3.00	0.165E - 01	0.926E + 00	0.574E - 01
3.50	0.470E - 02	0.832E + 00	0.163E + 00
4.00	0.110E - 02	0.617E + 00	0.382E + 00
4.50	0.191E - 03	0.338E + 00	0.662E + 00
5.00	0.248E - 04	0.139E + 00	0.861E + 00
5.50	0.274E - 05	0.485E - 01	0.951E + 00
6.00	0.283E - 06	0.159E - 01	0.984E + 00
6.50	0.287E - 07	0.507E - 02	0.995E + 00
7.00	0.288E - 08	0.161E - 02	0.998E + 00
7.50	0.288E - 09	0.510E - 03	0.999E + 00
8.00	0.288E - 10	0.161E - 03	0.100E + 01
8.50	0.288E - 11	0.510E - 04	0.100E + 01
9.00	0.288E - 12	0.161E - 04	0.100E + 01
9.50	0.288E - 13	0.510E - 05	0.100E + 01
10.00	0.288E - 14	0.161E - 05	0.100E + 01
10.50	0.288E - 15	0.510E - 06	0.100E + 01
11.00	0.288E - 16	0.161E - 06	0.100E + 01
11.50	0.288E - 17	0.510E - 07	0.100E + 01
12.00	0.288E - 18	0.161E - 07	0.100E + 01
12.50	0.288E - 19	0.510E - 08	0.100E + 01
13.00	0.288E - 20	0.161E - 08	0.100E + 01
13.50	0.288E - 21	0.510E - 09	0.100E + 01
14.00	0.288E - 22	0.161E - 09	0.100E + 01

6. Prove that for any polyprotic acid, the equation exactly describing the titration curve is

$$x = \frac{\left(\alpha_1 C^0 + 2\alpha_2 C^0 + 3\alpha_3 C^0 + \cdots + n\alpha_n C^0 + \dfrac{K_w}{[H^+]}\right) V}{\left(M - \dfrac{K_w}{[H^+]} + [H^+]\right)}$$

C^0 = initial concentration of acid in the titration vessel.

7. A sample consists of NaH_2PO_4 and H_3PO_4. Its volume is 25.00 milliliters. It is titrated with $0.1000M$ NaOH. The first endpoint appears when 10.00 milliliters of NaOH have been added. The second endpoint appears when 26.00 milliliters of NaOH have been added. (molecular weight of H_3PO_4 = 98.00 gram/mole; molecular weight of NaH_2PO_4 = 120.00 gram/mole).

(a) How many millimoles of H_3PO_4 were present?
Ans: *1.000*

(b) How many grams of H_3PO_4 were present?
Ans: *0.09800*

(c) How many millimoles of NaH_2PO_4 were present?
Ans: *0.6000*

(d) How many grams of NaH_2PO_4 were present?
Ans: *0.0720*

≡ 2.27 DETAILED SOLUTIONS TO PROBLEMS ≡

1. Abbreviate CH_3COOH as HAc. First, calculate C^0_{HAc} (before the addition of base).

$$\text{(number of moles of HAc)} = \frac{\text{(number of grams of HAc)}}{\text{(molecular weight of HAc)}}$$

$$= \frac{6.005 \text{ grams}}{60.05 \text{ gram/mole}} = 0.1000 \text{ moles}$$

$$C^0_{HAc} = \frac{\text{(number of moles of HAc)}}{\text{(number of liters of solution)}}$$

$$= \frac{0.1000}{1.000} = 0.1000M$$

(a) At 0.00 milliliter of NaOH added (case 1):

$$[H^+]' = (K_a C_{HAc})^{1/2}$$

$$= [(1.8 \times 10^{-5})(0.1000)]^{1/2} = 1.3 \times 10^{-3}M$$

$$pH = -\log(1.3 \times 10^{-3}) = 2.87$$

(b) At 20.00 milliliters of NaOH added (case 2):

(number of millimoles of Ac^- formed)

$$= \text{(number of millimoles of } OH^- \text{ added)}$$

$$= (20.00)(0.1000) = 2.000 \text{ millimoles}$$

$$C_{Ac^-} = \frac{\text{(number of millimoles of } Ac^- \text{ formed)}}{\text{(number of milliliters of solution)}} = \frac{2.000}{70.00}$$

(number of millimoles of HAc left)

$$= \text{(number of millimoles of HAc to start)} -$$
$$\text{(number of millimoles of Ac}^- \text{ formed)}$$

$$= (50.00)(0.1000) - (20.00)(0.1000) = 3.000 \text{ millimoles}$$

$$C_{HAc} = \frac{\text{(number of millimoles of HAc left)}}{\text{(number of milliliters of solution)}} = \frac{3.000}{70.00}$$

$$[H^+]' = K_a \frac{C_{HAc}}{C_{Ac^-}} = (1.8 \times 10^{-5}) \frac{\left(\frac{3.000}{70.00}\right)}{\left(\frac{2.000}{70.00}\right)} = 2.7 \times 10^{-5} M$$

$$pH = -\log(2.7 \times 10^{-5}) = 4.57$$

(c) At 25.00 milliliters of NaOH added (case 2):

(number of millimoles of Ac$^-$ formed)
$$= \text{(number of millimoles of OH}^- \text{ added)}$$

$$= (25.00)(0.1000) = 2.500 \text{ millimoles}$$

$$C_{Ac^-} = \frac{\text{(number of millimoles of Ac}^- \text{ formed)}}{\text{(number of milliliters of solution)}} = \frac{2.500}{75.00}$$

(number of millimoles of HAc left)
$$= \text{(number of millimoles of HAc to start)} -$$
$$\text{(number of millimoles of Ac}^- \text{ formed)}$$

$$= (50.00)(0.1000) - (25.00)(0.1000) = 2.500 \text{ millimoles}$$

$$C_{HAc} = \frac{\text{(number of millimoles of HAc left)}}{\text{(number of milliliters of solution)}} = \frac{2.500}{75.00}$$

$$[H^+]' = K_a \frac{C_{HAc}}{C_{Ac^-}} = (1.8 \times 10^{-5}) \frac{\left(\frac{2.500}{75.00}\right)}{\left(\frac{2.500}{75.00}\right)}$$

$$= 1.8 \times 10^{-5} M \quad \left(\begin{array}{l}\text{This is the special} \\ \text{case where } [H^+]' = K_a.\end{array}\right)$$

$$pH = -\log 1.8 \times 10^{-5} = 4.74$$

(d) At 30.00 milliliters of NaOH added (case 2):

(number of millimoles of Ac$^-$ formed)
$$= \text{(number of millimoles of OH}^- \text{ added)}$$

$$= (30.00)(0.1000) = 3.000 \text{ millimoles}$$

$$C_{Ac^-} = \frac{\text{(number of millimoles of Ac}^- \text{ formed)}}{\text{(number of milliliters of solution)}} = \frac{3.000}{80.00}$$

(number of millimoles of HAc left)
$$= \text{(number of millimoles of HAc to start)} -$$
$$\text{(number of millimoles of Ac}^- \text{ formed)}$$

$$= (50.00)(0.1000) - (30.00)(0.1000) = 2.000 \text{ millimoles}$$

$$C_{HAc} = \frac{(\text{number of millimoles of HAc left})}{(\text{number of milliliters of solution})} = \frac{2.000}{80.00}$$

$$[H^+]' = K_a \frac{C_{HAc}}{C_{Ac^-}} = (1.8 \times 10^{-5}) \frac{\left(\dfrac{2.000}{80.00}\right)}{\left(\dfrac{3.000}{80.00}\right)} = 1.2 \times 10^{-5} M$$

$$pH = -\log(1.2 \times 10^{-5}) = 4.92$$

Note that it is plain that the region between 0.00 milliliter of NaOH added and the endpoint is a buffer region. From 20.00 milliliters of NaOH added to 30.00 milliliters of NaOH added, the pH only changes from 4.57 to 4.92.

(e) At 50.00 milliliters of NaOH added (case 3): This is the endpoint, and hydrolysis of the acetate ion determines the pH.

$$C_{HAc} = 0$$

(number of millimoles of Ac^- formed)

$$= (\text{number of millimoles of } OH^- \text{ added})$$

$$= (50.00)(0.1000) = 5.000 \text{ millimoles}$$

$$C_{Ac^-} = \frac{(\text{number of millimoles of } Ac^- \text{ formed})}{(\text{number of milliliters of solution})} = \frac{5.000}{100.00}$$

$$= 5.000 \times 10^{-2} M$$

$$[OH^-]' = \left(\frac{K_w}{K_a} C_{Ac^-}\right)^{1/2} = \left[\left(\frac{1.00 \times 10^{-14}}{1.8 \times 10^{-5}}\right)(5.000 \times 10^{-2})\right]^{1/2}$$

$$= 5.3 \times 10^{-6} M$$

$$pOH = -\log(5.3 \times 10^{-6}) = 5.28$$

$$pH = 14.00 - 5.28 = 8.72$$

(f) At 75.00 milliliters of NaOH added (case 4):

(number of millimoles of OH^- excess)

$$= (\text{number of millimoles of } OH^- \text{ added}) -$$
(number of millimoles of HAc to start)

$$= (75.00)(0.1000) - (50.00)(0.1000) = 2.500 \text{ millimoles}$$

$$[OH^-] = \frac{(\text{number of millimoles of } OH^- \text{ excess})}{(\text{number of milliliters of solution})} = \frac{2.500}{125.00}$$

$$= 2.000 \times 10^{-2} M$$

$$pOH = -\log(2.000 \times 10^{-2}) = 1.70$$

$$pH = 14.00 - 1.70 = 12.30$$

(g) At 100.00 milliliters of NaOH added (case 4):

(number of millimoles of OH⁻ excess)

$$= \text{(number of millimoles of OH}^- \text{ added)} -$$
$$\text{(number of millimoles of HAc to start)}$$

$$= (100.00)(0.1000) - (50.00)(0.1000) = 5.00 \text{ millimoles}$$

$$[OH^-] = \frac{\text{(number of millimoles of OH}^- \text{ excess)}}{\text{(number of milliliters of solution)}} = \frac{5.000}{150.00}$$

$$= 3.333 \times 10^{-2} M$$

$$pOH = -\log (3.33 \times 10^{-2}) = +1.48$$

$$pH = 14.00 - 1.48 = 12.52$$

2. $HA + OH^- \rightleftharpoons A^- + H_2O$

At the endpoint,

$$\text{(number of moles of HA)} = \text{(number of moles of NaOH)}$$

$$\frac{\text{(number of grams of HA)}}{\text{(molecular weight of HA)}} = \left(\frac{\text{number of milliliters of NaOH}}{1000}\right)(M_{NaOH})$$

$$\text{(molecular weight of HA)} = \frac{\text{(number of grams of HA)}(1000)}{\text{(number of milliliters of NaOH)}(M_{NaOH})}$$

$$= \frac{(0.5000)(1000)}{(40.94)(0.1000)} = 122.1 \text{ gram/mole}$$

To find K_a with a first approximation, note that 20.47 milliliters is exactly half of 40.94 milliliters

$$[H^+]' = K_a \frac{C_{HA}}{C_{A^-}}$$

$$C_{HA} = C_{A^-}$$

$$[H^+]' = K_a$$

$$pH = 4.21$$

$$-\log [H^+] = 4.21$$

$$[H^+] = 6.2 \times 10^{-5} M$$

$$K_a = 6.2 \times 10^{-5} M$$

3. Use first approximations. Refer to Figure 2.7.
   (a) Here, a buffer of H_3PO_4 and $H_2PO_4^-$ is present.

$$C_{H_3PO_4} = C_{H_2PO_4^-}$$

$$[H^+]' = K_1 \frac{C_{H_3PO_4}}{C_{H_2PO_4^-}} = K_1 = 7.1 \times 10^{-3} M$$

$$pH = -\log (7.1 \times 10^{-3}) = 2.15$$

   (b) Here, the ampholyte, $H_2PO_4^-$, is present.

$$[H^+]' = (K_1 K_2)^{1/2} = [(7.1 \times 10^{-3})(6.2 \times 10^{-8})]^{1/2} = 2.1 \times 10^{-5} M$$

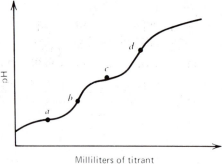

Somewhere out here is the third endpoint

Figure 2.7 *Titration of* H_3PO_4 *by NaOH.*

$$pH = -\log (2.1 \times 10^{-5}) = 4.68$$

(c) Here, a buffer of $H_2PO_4^-$ and HPO_4^{2-} is present.

$$C_{H_2PO_4^-} = C_{HPO_4^{2-}}$$

$$[H^+]' = K_2 \frac{C_{H_2PO_4^-}}{C_{HPO_4^{2-}}} = 6.2 \times 10^{-8}M$$

$$pH = -\log (6.2 \times 10^{-8}) = 7.21$$

(d) The ampholyte, HPO_4^{2-} is present here.

$$[H^+]' = (K_2K_3)^{1/2} = [(6.2 \times 10^{-8})(4.4 \times 10^{-13})]^{1/2} = 1.7 \times 10^{-10}M$$

(Here, if one were striving for accuracy, further approximations would have to be made for the pH > 7.)

$$pH = -\log (1.7 \times 10^{-10}) = 9.77$$

(e) HPO_4^{2-} ($K_3 = 4.4 \times 10^{-13}$) is not really a much stronger acid than H_2O ($K_w = 1.00 \times 10^{-14}$). Adding strong base to an HPO_4^{2-} solution is a bit like adding strong base to water.

4. At the first endpoint, this reaction is complete:

$$H_2B + OH^- \rightleftharpoons HB^- + H_2O$$

Thus

(number of moles of H_2B) = (number of moles of NaOH)

$$\frac{\text{(number of grams of } H_2B)}{\text{(molecular weight of } H_2B)} = \left(\frac{\text{number of milliliters of NaOH}}{1000} \right) (M_{NaOH})$$

$$\text{(molecular weight of } H_2B) = \frac{\text{(number of grams of } H_2B)(1000)}{\text{(number of milliliters of NaOH)}(M_{NaOH})}$$

$$= \frac{(0.3658)(1000)}{(36.83)(0.0954)} = 104. \text{ gram/mole}$$

192 / Acid-Base Chemistry

To find K_1, from first approximations, note that 18.42 milliliters is just halfway to the first endpoint.

$$C_{H_2B} = C_{HB^-}$$

$$[H^+]' = K_1 \frac{C_{H_2B}}{C_{HB^-}} = K_1$$

$$pH = 2.85$$

$$-\log [H^+] = 2.85$$

$$[H^+] = 1.4 \times 10^{-3} M$$

$$K_1 = 1.4 \times 10^{-3}$$

To find K_2, from first approximations, note that 55.24 is just halfway between the first and second endpoints.

$$C_{HB^-} = C_{B^{2-}}$$

$$[H^+]' = K_2 \frac{C_{HB^-}}{C_{B^{2-}}} = K_2$$

$$pH = 5.66$$

$$-\log [H^+] = 5.66$$

$$[H^+] = 2.2 \times 10^{-6} M$$

$$K_2 = 2.2 \times 10^{-6}$$

5. (a) See Table 2.6.
   (b) and (c) See Figure 2.8.
   (d) Explanation of why the two curves differ. In all regions of the titration curve, the volume of base added is exactly related to the pH by Equation (2.49)

$$x = \left(\frac{\alpha_1 C_{H_2A}^0 + 2\alpha_2 C_{H_2A}^0 + \dfrac{K_w}{[H^+]} - [H^+]}{M - \dfrac{K_w}{[H^+]} + [H^+]} \right) V \qquad (2.49)$$

When the $[H^+]$ is large (0.1 to 0.001M), Equation 2.49 reduces to

$$x = \left(\frac{\alpha_2 C_{H_2A}^0 + 2\alpha_2 C_{H_2A}^0 - [H^+]}{M + [H^+]} \right) V$$

The $[H^+]$ terms in the above equation should not be neglected. If they are, then the expression

$$x = \frac{\alpha_1 C_{H_2A}^0 + 2\alpha_2 C_{H_2A}^0}{M} V$$

results, but it is in error from the actual titration curve.

The same reasoning applies when the $[H^+]$ is small ($< 10^{-8}$); however, the important term then is $[OH^-]$ or $K_w/[H^+]$.

Table 2.6
OXALIC ACID TITRATION DATA FOR PROBLEM 5

pH	α_1	α_2	$(\alpha_1 + 2\alpha_2)V$ (Eq. 2.50)	Milliliter base to be added to H_2A (Eq. 2.49)
0				−45.213
0.5	0.15	3.0×10^{-5}	7.5	−36.179
1.0	0.36	2.2×10^{-4}	18.	−16.017
1.5	0.64	1.2×10^{-3}	32.	12.330
2.0	0.84	5.2×10^{-3}	42.	34.296
2.5	0.93	1.8×10^{-2}	48.	45.274
3.0	0.93	5.7×10^{-2}	52.	51.034
3.5	0.83	0.16	58.	57.582
4.0	0.62	0.38	69.	68.941
4.5	0.34	0.66	83.	83.054
5.0	0.14	0.86	93.	93.039
5.5	4.8×10^{-2}	0.95	97.	97.569
6.0	1.6×10^{-2}	0.98	99.	99.205
6.5	5.1×10^{-3}	1.0	100.	99.746
7.0	1.6×10^{-3}	1.0	100.	99.919
7.5	5.1×10^{-4}	1.0	100.	99.975
8.0	1.6×10^{-4}	1.0	100.	99.993
8.5	5.1×10^{-5}	1.0	100.	100.002
9.0	1.6×10^{-5}	1.0	100.	100.014
9.5	5.1×10^{-6}	1.0	100.	100.047
10.0	1.6×10^{-6}	1.0	100.	100.150
10.5	5.1×10^{-7}	1.0	100.	100.476
11.0	1.6×10^{-7}	1.0	100.	101.515
11.5	5.1×10^{-8}	1.0	100.	104.898
12.0	1.6×10^{-8}	1.0	100.	116.667
12.5	5.1×10^{-9}	1.0	100.	169.371
13.0	1.6×10^{-9}	1.0	100.	1.5×10^{14}
13.5	5.0×10^{-10}	1.0	100.	∞
14.0	1.6×10^{-10}	1.0	100.	∞

The only scratch calculations that are required are those necessary to show that if $[H^+] > 0.001$,
then

$$[H^+] \gg \frac{K_w}{[H^+]} \quad \text{and} \quad \sim \alpha_1 C^0_{H_2A} + 2\alpha_2 C^0_{H_2A}$$

or if $[H^+] < 10^{-8}$,
then

$$\frac{K_w}{[H^+]} \gg [H^+] \quad \text{and} \quad \sim \alpha_1 C^0_{H_2A} + 2\alpha_2 C^0_{H_2A}$$

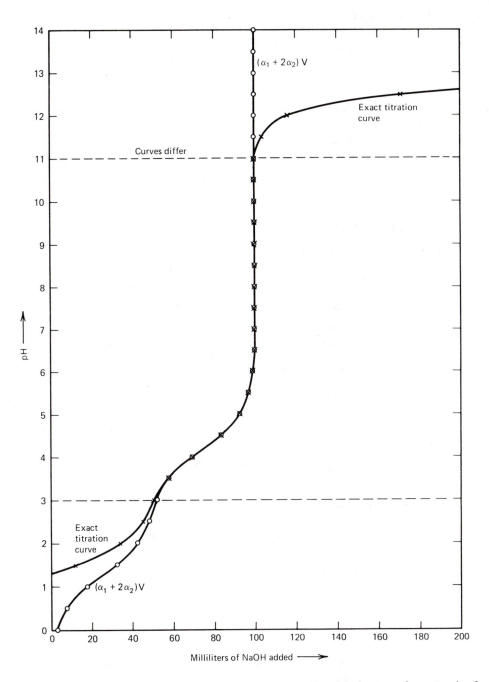

Figure 2.8 *Plot of Data from Table 2.6. Courtesy of Prof. John P. Walters, Dep't of Chemistry, University of Wisconsin.*

It is clear that $C_{H_2A}^0$ and M are important. Only if

$$C_{H_2A}^0 = M$$

is

$$x = (\alpha_1 + 2\alpha_2)V$$

valid.

(e) Here, there is excess strong base in solution. Remember (or look up) case 6 for the titration.

From the stoichiometry of the two reactions (calling oxalic acid, $H_2C_2O_4$ simply H_2Ox):

$$H_2Ox + OH^- \rightleftharpoons HOx^- + H_2O$$

$$HOx^- + OH^- \rightleftharpoons Ox^{2-} + H_2O$$

it can be concluded that *after the second endpoint,*

(number of millimoles of OH^- excess)

$$= \text{(number of millimoles of } OH^- \text{ added)} -$$
$$\text{(2)(number of millimoles of } H_2Ox \text{ to start)}$$

$$= (200.00)(0.1000) - (2)(50.00)(0.1000) = 10.00 \text{ millimoles}$$

$$[OH^-] = \frac{\text{(number of millimoles of } OH^- \text{ excess)}}{\text{(number of milliliters of solution)}} = \frac{10.00}{250.00}$$

$$= 4.000 \times 10^{-2} M$$

$$pOH = -\log(4.000 \times 10^{-2}) = 1.40$$
$$pH = 14.00 - 1.40 = 12.60$$

6. Refer to the earlier derivation for the definition of symbols. Let the weak acid be H_nB.

$$C_{H/H_nB} = (H^+) \quad \text{from } H_nB$$
$$[H^+] = C_{H/H_nB} + [OH^-] - C_b$$

By simple chemical bookkeeping,

$$C_{H/H_nB} = [H_{n-1}B^-] + 2[H_{n-2}B^{2-}] + 3[H_{n-3}B^{3-}] +$$
$$\cdots + n[B^{n-}]$$

$$[OH^-] = \frac{K_w}{[H^+]}$$

$$[H^+] = [H_{n-1}B^-] + 2[H_{n-2}B^{2-}] + 3[H_{n-3}B^{3-}] +$$
$$\cdots + n[B^{n-}] + \frac{K_w}{[H^+]} - C_b$$

Let C_B be the sum of the concentrations of all the B-containing species. Remember the alpha-functions:

$$[H_{n-1}B^-] = \alpha_1 C_B$$

$$[H_{n-2}B^{2-}] = \alpha_2 C_B$$

$$[H_{n-3}B^{3-}] = \alpha_3 C_B$$

$$\vdots \qquad \vdots$$

$$[B^{n-}] = \alpha_n C_B$$

$$[H^+] = \alpha_1 C_B + 2\alpha_2 C_B + 3\alpha_3 C_B + \cdots + n\alpha_n C_B + \frac{K_w}{[H^+]} - C_b$$

$$C_b = \alpha_1 C_B + 2\alpha_2 C_B + 3\alpha_3 C_B + \cdots + n\alpha_n C_B + \frac{K_w}{[H^+]} - [H^+]$$

When M, x, V, and C^0 are defined (C^0 is the initial analytical concentration of H_nB in analyte solution):

$$M\left(\frac{x}{V+x}\right) = \alpha_1 C^0\left(\frac{V}{V+x}\right) + 2\alpha_2 C^0\left(\frac{V}{V+x}\right) + 3\alpha_3 C^0\left(\frac{V}{V+x}\right) +$$

$$\cdots + n\alpha_n C^0\left(\frac{V}{V+x}\right) + \frac{K_w}{[H^+]} - [H^+]$$

By following the same steps that were used in the earlier derivation (you can find them in the text), you obtain

$$x = \frac{\left(\alpha_1 C^0 + 2\alpha_2 C^0 + 3\alpha_3 C^0 + \cdots + n\alpha_n C^0 + \dfrac{K_w}{[H^+]} - [H^+] \; V\right)}{M - \dfrac{K_w}{[H^+]} + [H^+]}$$

7. Consider the titration curve in Figure 2.9

(number of millimoles of H_3PO_4)

$\qquad$ = (number of millimoles of NaOH to the first endpoint)

$\qquad$ = (number of millimoles of $H_2PO_4^-$ from H_3PO_4)

(number of millimoles of $H_2PO_4^-$ from H_3PO_4) +
(number of millimoles of NaH_2PO_4)

$\qquad$ = (number of millimoles of NaOH from the first to the second endpoint)

(number of millimoles of H_3PO_4) +
(number of millimoles of NaH_2PO_4)

$\qquad$ = (number of millimoles of NaOH from the first to the second endpoint)

(number of millimoles of NaH_2PO_4)

$\qquad$ = (number of millimoles of NaOH from the first to the second endpoint) –
(number of millimoles of H_3PO_4)

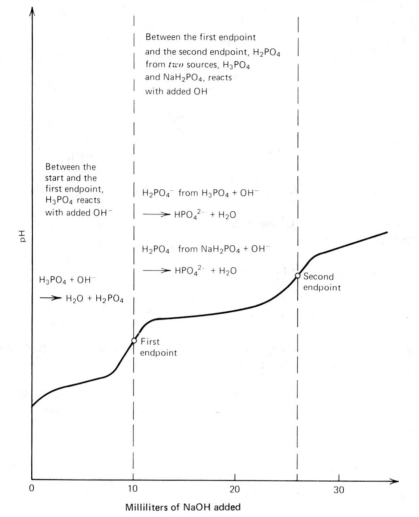

Figure 2.9 *Titration of phosphate mixutre in Problem 7.*

(a) (number of millimoles of H_3PO_4)

$= $ (number of millimoles of NaOH to the first endpoint)

$= (10.00)(0.1000) = 1.000$ millimoles

(b) (number of grams of H_3PO_4)

$$= \frac{(\text{number of millimoles of } H_3PO_4)(\text{molecular weight of } H_3PO_4)}{(1000 \text{ millimoles/mole})}$$

$$= \left(\frac{1.000}{1000}\right)(98.00) = 0.09800 \text{ gram}$$

(c) (number of millimoles of NaH_2PO_4)

198 / Acid-Base Chemistry

= (number of millimoles of NaOH from the first to the second endpoint) – (number of millimoles of H_3PO_4)

= $(26.00 - 10.00)(0.1000) - 1.000 = 0.600$ millimoles

(d) (number of grams of NaH_2PO_4)

$$= \left(\frac{\text{number of millimoles of } NaH_2PO_4}{1000 \text{ millimoles/mole}}\right) \text{(molecular weight of } NaH_2PO_4\text{)}$$

$$= \left(\frac{0.600}{1000}\right)(120.01) = 0.0720 \text{ gram}$$

≡ 2.28 WEAK BASES ≡

Weak bases are roughly analogous to weak acids. We shall be concerned here with weak bases that can accept one proton; these are called weak monoacidic bases. (Weak bases such as CO_3^{2-} and ampholytes such as HCO_3^- have already been treated.) Ammonia, NH_3, is an example of just such a monoacidic weak base.

$$NH_3 + H_2O \rightleftharpoons NH_4^+ + OH^-$$

$$K_b = \frac{[OH^-][NH_4^+]}{[NH_3]} = 1.8 \times 10^{-5}$$

2.29 Weak Bases Alone in Solution

Let us represent a weak monoacidic base as B. Suppose that its K_b is on the order of 10^{-5}.

$$B + H_2O \rightleftharpoons BH^+ + OH^- \qquad K_b = \frac{[OH^-][BH^+]}{[B]} = 1.00 \times 10^{-5}$$

A problem is often stated like this:

Given the values of the analytical concentration of B, C_B, and K_b, and that the analytical concentration of the protonated cation, C_{BH^+}, is zero (no added cation), calculate pH.

C_B = analytical concentration of B

C_B = concentration in moles per liter of B added to the solution. This is what the concentration of B would be if no reaction with water occurred.

C_{BH^+} = analytical concentration of BH^+ = 0

C_{BH^+} = concentration in moles per liter of BH^+ added to solution. This is what the concentration of BH^+ would be if any BH^+ had been added as, say BHCl (analogous to NH_4Cl) = 0.

We begin with K_b.

$$K_b = \frac{[OH^-][BH^+]}{[B]}$$

$$[OH^-] = K_b \frac{[B]}{[BH^+]} \qquad (2.51)$$

If K_b is known (from a table in some reference work), only [B] and [BH$^+$] are needed before [OH$^-$] can be calculated.

The derivation is similar to that used for weak acids alone in solution and is given somewhat sketchily.

Mass balance:

$$C_B = [B] + [BH^+] \qquad (2.52)$$

Charge balance:

$$[OH^-] = [BH^+] + [H^+] \qquad (2.53)$$

$$[BH^+] = [OH^-] - [H^+]$$

We rearrange Equation 2.52 and substitute from Equation 2.53:

$$[B] = C_B - [BH^+]$$

$$[B] = C_B - [OH^-] + [H^+]$$

Using Equation 2.51, we obtain

$$[OH^-] = K_b \frac{[B]}{[BH^+]}$$

We substitute the new terms for [BH$^+$] and [B] into Equation 2.51:

$$[OH^-] = K_b \left(\frac{C_B - [OH^-] + [H^+]}{[OH^-] - [H^+]} \right)$$

$$[H^+] = \frac{K_w}{[OH^-]}$$

$$[OH^-] = K_b \left(\frac{C_B - [OH^-] + \dfrac{K_w}{[OH^-]}}{[OH^-] - \dfrac{K_w}{[OH^-]}} \right) \qquad (2.54)$$

In a basic solution,

$$[OH^-] \gg [H^+]$$

$$[OH^-] \gg \frac{K_w}{[OH^-]}$$

Equation 2.54 becomes

$$[OH^-] = K_b \frac{(C_B - [OH^-])}{[OH^-]}$$

$$[OH^-]^2 = K_b (C_B - [OH^-]) \qquad (2.55)$$

We can either solve Equation 2.55 as a quadratic equation or do successive approximations. The approximation method is simple.

$$[OH^-] = [K_b(C_B - [OH^-])]^{1/2}$$

First approximation:

$$[OH^-]' = (K_b C_B)^{1/2}$$

Suppose that $C_B \gg [OH^-]$.

Second approximation:
$$[OH^-]'' = [K_b(C_B - [OH^-]')]^{1/2}$$
Third approximation:
$$[OH^-]''' = [K_b(C_B - [OH^-]'')]^{1/2}$$
and so forth.

If the approximations converge quickly, then the approach pays off.

Example: If $K_b = 1.00 \times 10^{-5}$ and $C_B = 1.00 \times 10^{-2}M$, calculate pH.

First approximation:
$$[OH^-]' = (K_bC_B)^{1/2} = [(1.00 \times 10^{-5})(1.00 \times 10^{-2})]^{1/2} = 3.16 \times 10^{-4}M$$
Second approximation:
$$[OH^-]'' = [K_b(C_B - [OH^-]')]^{1/2}$$
$$= [(1.00 \times 10^{-5})(1.00 \times 10^{-2} - 3.16 \times 10^{-4})]^{1/2} = 3.11 \times 10^{-4}M$$
Third approximation:
$$[OH^-]''' = [K_b(C_B - [OH^-]'')]^{1/2}$$
$$= [(1.00 \times 10^{-5})(1.00 \times 10^{-2} - 3.11 \times 10^{-4})]^{1/2} = 3.11 \times 10^{-4}M$$
The approximations have converged.
$$pOH = -\log(3.11 \times 10^{-4}) = 3.51$$
$$pH = 14.00 - 3.51 = 10.49$$

2.30 Buffers of Weak Monoacidic Bases

Mixtures of weak monoacidic bases and their cations (added often as salts) are buffer solutions, resisting pH change.

Example: A buffer may be made from ammonia, NH_3, and ammonium chloride, NH_4Cl.

Any strong base added reacts with the NH_4^+.

$$NH_4^+ + OH^- \rightleftharpoons NH_3 + H_2O$$

Any strong acid added reacts with the NH_3.

$$NH_3 + H^+ \rightleftharpoons NH_4^+$$

The quantitative description of buffers is easy. Given analytical concentrations of the base, C_B, and its protonated cation, C_{BH^+} (added as BHCl), we can find the pH of a buffer. The derivation that follows parallels that given for buffers of weak acids. Equation 2.51 holds:

$$[OH^-] = K_b \cdot \frac{[B]}{[BH^+]} \qquad (2.51)$$

Mass balance:

$$C_B + C_{BH^+} = [B] + [BH^+] \qquad (2.56)$$

Charge balance:

$$[OH^-] + [Cl^-] = [BH^+] + [H^+] \tag{2.57}$$

We find $[BH^+]$,

$$C_{BH^+} = [Cl^-]$$

substitute into Eq. 2.57, and rearrange.

$$[OH^-] + C_{BH^+} = [BH^+] + [H^+]$$

$$[BH^+] = C_{BH^+} + [OH^-] - [H^+] \tag{2.58}$$

By rearranging Eq. 2.56, we find $[B]$.

$$C_B = [B] + [BH^+] - C_{BH^+}$$

$$C_B = [B] + \cancel{C_{BH^+}} + [OH^-] - [H^+] - \cancel{C_{BH^+}}$$

$$C_B = [B] + [OH^-] - [H^+]$$

$$[B] = C_B - [OH^-] + [H^+] \tag{2.59}$$

Substituting Eqs. 2.58 and 2.59 into Eq. 2.51 yields

$$[OH^-] = K_b \left(\frac{C_B - [OH^-] + [H^+]}{C_{BH^+} + [OH^-] - [H^+]} \right)$$

$$[H^+] = \frac{K_w}{[OH^-]}$$

$$[OH^-] = K_b \frac{\left(C_B - [OH^-] + \dfrac{K_w}{[OH^-]} \right)}{\left(C_{BH^+} + [OH^-] - \dfrac{K_w}{[OH^-]} \right)} \tag{2.60}$$

An approximation method, starting with Eq. 2.60 and simple suppositions, is used to calculate $[OH^-]$.

Simple suppositions:

$$\left(C_B \gg -[OH^-] + \frac{K_w}{[OH^-]} \right)$$

$$\left(C_{BH^+} \gg [OH^-] - \frac{K_w}{[OH^-]} \right)$$

First approximation:

$$[OH^-]' = K_b \frac{C_B}{C_{BH^+}}$$

Determine whether the buffer is basic (most likely) or acidic (unlikely).

$$[H^+]' = \frac{K_w}{[OH^-]'}$$

If $[OH^-]' \gg [H^+]'$,   If $[OH^-]' \gg [H^+]'$,
the buffer is basic   the buffer is acidic

Second approximation:

$$[OH^-]'' = K_b \left(\frac{C_B - [OH^-]'}{C_{BH^+} + [OH^-]'} \right) \qquad [OH^-]'' = K_b \left(\frac{C_B + [H^+]'}{C_{BH^+} - [H^+]'} \right)$$

Third approximation:

$$[OH^-]''' = K_b \left(\frac{C_B - [OH^-]''}{C_{BH^+} + [OH^-]''}\right) \qquad [OH^-]''' = K_b \left(\frac{C_B + [H^+]''}{C_{BH^+} - [H^+]''}\right)$$

$$\vdots \qquad\qquad\qquad\qquad\qquad\qquad \vdots$$

and so forth   and so forth

Example: Calculate the pH of a buffer made from 0.010 mole of B and 0.10 mole of BHCl dissolved in 1 liter of water, when $K_b = 1.00 \times 10^{-5}$.

First approximation:

$$[OH^-]' = K_b \frac{C_B}{C_{BH^+}} = (1.00 \times 10^{-5}) \left(\frac{1.0 \times 10^{-2}}{1.0 \times 10^{-1}}\right) = 1.0 \times 10^{-6} M$$

$$[H^+]' = \frac{K_w}{[OH^-]'} = \left(\frac{1.00 \times 10^{-14}}{1.0 \times 10^{-6}}\right) = 1.0 \times 10^{-8} M$$

$$[OH^-]' \gg [H^+]'$$

The buffer is basic.

Second approximation:

$$[OH^-]'' = K_b \left(\frac{C_B - [OH^-]'}{C_{BH^+} + [OH^-]'}\right)$$

$$= (1.00 \times 10^{-5}) \cdot \left(\frac{1.0 \times 10^{-2} - 1.0 \times 10^{-6}}{1.0 \times 10^{-1} + 1.0 \times 10^{-6}}\right) = 1.0 \times 10^{-6} M$$

The approximations have converged.

$$pOH = -\log(1.0 \times 10^{-6}) = 6.00$$

$$pH = 14.00 - 6.00 = 8.00$$

2.31   The Protonated Cations of Weak Bases

Solutions of ammonium chloride are acidic. The reason is that the ammonium ion, NH_4^+, is a weak acid.

Long form:

$$NH_4^+ + H_2O \rightleftharpoons NH_3 + H_3O^+$$

or

Short form:

$$NH_4^+ \rightleftharpoons NH_3 + H^+$$

Suppose that BHCl, the chloride salt of base B, is dissolved in water. The pH of the resultant solution can be evaluated if C_{BH^+}, K_b, and K_w are known. The reaction of BH^+ is

$$BH^+ \rightleftharpoons B + H^+$$

By analogy to the hydrolysis of the anion of a weak acid, the reaction can be called hydrolysis and a hydrolysis constant written:

$$K_{hy} = \frac{[H^+][B]}{[BH^+]}$$

To evaluate K_{hy}, multiply the expression through by 1.

$$K_{hy} = \frac{[OH^-][H^+][B]}{[OH^-][BH^+]}$$

$$K_{hy} = \frac{K_w}{K_b}$$

$$\frac{K_w}{K_b} = \frac{[H^+][B]}{[BH^+]}$$

Given K_w, K_b, and C_{BH^+}, it is possible to evaluate $[H^+]$.

$$[H^+] = \frac{K_w}{K_b} \cdot \frac{[BH^+]}{[B]} \qquad (2.61)$$

Suppose that the BH^+ is added as BHCl.

Mass balance:

$$C_{BH^+} = [BH^+] + [B] \qquad (2.62)$$

Charge balance:

$$[BH^+] + [H^+] = [Cl^-] + [OH^-] \qquad (2.63)$$

$$C_{BH^+} = [Cl^-]$$

We substitute into Eq. 2.63:

$$[BH^+] + [H^+] = C_{BH^+} + [OH^-]$$

and rearrange

$$[BH^+] = C_{BH^+} - [H^+] + [OH^-] \qquad (2.64)$$

Then we rearrange Eq. 2.62,

$$[B] = C_{BH^+} - [BH^+]$$

substitute from Eq. 2.64,

$$[B] = C_{BH^+} - (C_{BH^+} - [H^+] + [OH^-])$$

and

$$[B] = [H^+] - [OH^-] \qquad (2.65)$$

substitute Eqs. 2.64 and 2.65 into Eq. 2.61.

$$[H^+] = \frac{K_w}{K_b} \cdot \left(\frac{C_{BH^+} - [H^+] + [OH^-]}{[H^+] - [OH^-]} \right) \qquad (2.66)$$

In an acidic solution, $[H^+] \gg [OH^-]$. Equation 2.66 becomes

$$[H^+] = \frac{K_w}{K_b} \cdot \left(\frac{C_{BH^+} - [H^+]}{[H^+]} \right)$$

$$[H^+]^2 = \frac{K_w}{K_b} \cdot (C_{BH^+} - [H^+]) \qquad (2.67)$$

Equation 2.67 can be rigorously solved (quadratic) or a series of approximations can be used.

$$[H^+] = \left[\frac{K_w}{K_b} (C_{BH^+} - [H^+]) \right]^{1/2}$$

First approximation:

$$[H^+]' = \left(\frac{K_w}{K_b} \cdot C_{BH^+} \right)^{1/2}$$

Second approximation:

$$[H^+]'' = \left[\frac{K_w}{K_b} (C_{BH^+} - [H^+]') \right]^{1/2}$$

Third approximation:

$$[H^+]''' = \left[\frac{K_w}{K_b} (C_{BH^+} - [H^+]'') \right]^{1/2}$$

and so forth.

Example: If a weak base B has a K_b of 1.00×10^{-5}, calculate the pH of a solution 0.100M in the salt BHCl.

First approximation:

$$[H^+]' = \left(\frac{K_w}{K_b} C_{BH^+} \right)^{1/2} = \left[\left(\frac{1.00 \times 10^{-14}}{1.00 \times 10^{-5}} \right) \cdot (1.00 \times 10^{-1}) \right]^{1/2}$$

$$= 1.00 \times 10^{-5} M$$

Second approximation:

$$[H^+]'' = \left[\frac{K_w}{K_b} (C_{BH^+} - [H^+]') \right]^{1/2}$$

$$= \left[\left(\frac{1.00 \times 10^{-14}}{1.00 \times 10^{-5}} \right) \cdot (1.00 \times 10^{-1} - 1.00 \times 10^{-5}) \right]^{1/2}$$

$$= 1.00 \times 10^{-5} M$$

The approximations have converged.

$$pH = -\log (1.00 \times 10^{-5}) = 5.00$$

2.32 The Titration of Weak Bases by Strong Acids

Strong acids may be used to titrate weak bases. The description of the titration curve of a weak base is easy, once determining the pH values of weak bases alone, buffers, protonated cations alone, and excess strong acids is mastered. Suppose that 100 milliliters of 0.1000M weak base B ($K_b = 1.00 \times 10^{-5}$) are titrated by 0.1000M HCl. The titration curve will look like the one shown in Figure 2.10. The titration curve may be described very simply in terms of four cases.

In case 1, no strong acid has been added to the solution and the weak base is alone: $C_B = 0.1000M$; $C_{BH^+} = 0.0000M$.

$$[OH^-]' = (K_b C_B)^{1/2} = [(1.00 \times 10^{-5})(1.000 \times 10^{-1})]^{1/2} = 1.00 \times 10^{-3} M$$

$$pOH = -\log (1.00 \times 10^{-3}) = 3.00$$

$$pH = 14.00 - 3.00 = 11.00$$

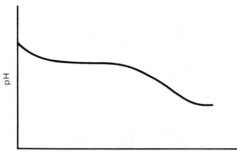

Figure 2.10 *Curve for titration of monoacidic weak base by strong acid.*

In case 2, some strong acid has been added to the titrant but not enough to react all of the base B to BH^+.

$$H^+ + B \rightleftharpoons BH^+$$

Both B and BH^+ are present in the solution in large, finite concentrations. The solution is a buffer.

$$C_B > 0; \; C_{BH^+} > 0$$

The approximation

$$[OH^-]' = K_b \frac{C_B}{C_{BH^+}}$$

may be used, when between 10 and 90% of the base has been titrated. Let us calculate the pH of the solution when 50.00 milliliters of 0.1000M HCl have been added.

(number of millimoles of BH^+ formed) = (number of millimoles of HCl added)

$$= (50.00)(0.1000) = 5.000 \text{ millimoles}$$

$$C_{BH^+} = \frac{\text{(number of millimoles of } BH^+ \text{ formed)}}{\text{(number of milliliters of solution)}} = \frac{5.000}{150.00}$$

(number of millimoles of B left)

$$= \text{(number of millimoles of B to start)} - \text{(number of millimoles of } BH^+ \text{ formed)}$$

$$= (100.00)(0.1000) - 5.000 = 5.000 \text{ millimoles}$$

$$C_B = \frac{\text{(number of millimoles of B left)}}{\text{(number of milliliters of solution)}} = \frac{5.000}{150.00}$$

$$[OH^-]' = K_b \frac{C_B}{C_{BH^+}} = (1.00 \times 10^{-5}) \cdot \left(\frac{\frac{5.000}{150.00}}{\frac{5.000}{150.00}} \right) = 1.00 \times 10^{-5} M$$

$$pOH = -\log (1.00 \times 10^{-5}) = 5.00$$
$$pH = 14.00 - 5.00 = 9.00$$

In case 3, the exact endpoint of the titration, just enough strong acid has been added to react with all the base. The solution at the endpoint contains a great deal of the protonated cation, BH^+, and no B is left; $C_B = 0$. The approximation

$$[H^+]' = \left(\frac{K_w}{K_b} C_{BH^+}\right)^{1/2}$$

is probably valid. The endpoint is reached when just 100.00 milliliters of HCl have been added.

(number of millimoles of BH^+ formed) = (number of millimoles of HCl added)

$$= (100.00)(0.1000) = 10.00 \text{ millimoles}$$

$$C_{BH^+} = \frac{\text{(number of millimoles of } BH^+ \text{ formed)}}{\text{(number of milliliters of solution)}}$$

$$= \frac{10.00}{200.00} = 5.000 \times 10^{-2} M$$

$$[H^+]' = \left(\frac{K_w}{K_b} C_{BH^+}\right)^{1/2} = \left[\left(\frac{1.00 \times 10^{-14}}{1.00 \times 10^{-5}}\right)(5.000 \times 10^{-2})\right]^{1/2} = 7.07 \times 10^{-6} M$$

$$pH = -\log (7.07 \times 10^{-6}) = 5.15$$

In case 4, all of the B has been reacted to BH^+ and excess strong acid has been added. Consider the point at which 150.00 milliliters of strong acid have been added.

(number of millimoles of HCl excess)

$$= \text{(number of millimoles of HCl added)} - \text{(number of millimoles of B to start)}$$

$$= (150.00)(0.1000) - (100.00)(0.1000) = 5.000 \text{ millimoles}$$

$$[H^+] = \frac{\text{(number of millimoles of HCl excess)}}{\text{(number of milliliters of solution)}} = \frac{5.000}{250.00} = 2.000 \times 10^{-2} M$$

$$pH = -\log (2.000 \times 10^{-2} M) = 1.70$$

≡ 2.33 PROBLEMS ≡

(See Table 2.8 at the end of the chapter for K_b-values.)

1. Calculate the pH of a solution of goopyl amine (a monoacidic weak base, Gup, $K_b = 4.0 \times 10^{-4}$) that is 0.1000M in Gup.
   Ans: **11.79**

2. Calculate the pH of a $1.00 \times 10^{-4} M$ solution of goopyl amine. See Problem 1 for K_b.
   Ans: **9.92**

3. Calculate the pH of a solution 0.0010M in NH_3 and 0.0020M in NH_4Cl. $K_b = 1.8 \times 10^{-5}$ for NH_3.
   Ans: **8.96**

4. Calculate the pH of a solution made by adding 20.00 milliliters of 0.1000M HCl to 50.00 milliliters of 0.1000M NH_3.

   Ans: *9.43*

5. Calculate the pH of a solution that is 0.010M in NH_4Cl.

   Ans: *5.62*

6. Goopyl amine is not terribly volatile. Calculate the pH at these points during the titration of 25.00 milliliters of 0.1000M Gup by 0.1000M HCl: (a) 0.00 milliliters of HCl added, (b) 12.50 milliliters of HCl added, (c) 25.00 milliliters of HCl added, and (d) 40.00 milliliters of HCl added. See Problem 1 for K_b.

   Ans: (a) *11.79*, (b) *10.60*, (c) *5.95*, and (d) *1.64*

≡ 2.34 DETAILED SOLUTIONS TO PROBLEMS ≡

1. Try a method of successive approximations.

$$[OH^-]' = (K_b\, C_{Gup})^{1/2} = [(4.0 \times 10^{-4})(1.00 \times 10^{-1})]^{1/2} = 6.3 \times 10^{-3} M$$

$$[OH^-]'' = [K_b\, (C_{Gup} - [OH^-]')]^{1/2}$$

$$= [(4.0 \times 10^{-4})(1.00 \times 10^{-1} - 6.3 \times 10^{-3})]^{1/2} = 6.1 \times 10^{-3} M$$

$$[OH^-]''' = [K_b\, (C_{Gup} - [OH^-]'')]^{1/2}$$

$$= [(4.0 \times 10^{-4})(1.00 \times 10^{-1} - 6.1 \times 10^{-3})]^{1/2} = 6.1 \times 10^{-3} M$$

$$pOH = -\log (6.1 \times 10^{-3}) = 2.21$$

$$pH = 14.00 - 2.21 = 11.79$$

2. Try a method of successive approximations.

$$[OH^-]' = (K_b \cdot C_{Gup})^{1/2} = [(4.0 \times 10^{-4})(1.00 \times 10^{-4})]^{1/2} = 2.0 \times 10^{-4} M$$

$$[OH^-]'' = [K_b(C_{Gup} - [OH^-]')]^{1/2}$$

$$= [(4.0 \times 10^{-4})(1.00 \times 10^{-4} - 2.0 \times 10^{-4})]^{1/2} = \text{an imaginary number}$$

A quadratic equation must be solved.

$$[OH^-]^2 = K_b(C_{Gup} - [OH^-])$$

$$[OH^-]^2 = K_b\, C_{Gup} - K_b\, [OH^-]$$

$$[OH^-]^2 + K_b\, [OH^-] - K_b\, C_{Gup} = 0$$

$$[OH^-] = \frac{-K_b + (K_b^2 + 4K_b\, C_{Gup})^{1/2}}{2}$$

$$= \frac{-4.0 \times 10^{-4} + [(4.0 \times 10^{-4})^2 + (4)(4.0 \times 10^{-4})(1.00 \times 10^{-4})]^{1/2}}{2}$$

$$= 8.3 \times 10^{-5} M$$

$$pOH = -\log(8.3 \times 10^{-5}) = 4.08$$

$$pH = 14.00 - 4.08 = 9.92$$

3. This is a buffer. Try a method of successive approximations first.

$$[OH^-]' = K_b \frac{C_{NH_3}}{C_{NH_4^+}} = (1.8 \times 10^{-5})\frac{0.0010}{0.0020} = 9.0 \times 10^{-6}M$$

$$[H^+]' = \frac{K_w}{[OH^-]'} = \frac{(1.00 \times 10^{-14})}{(9.0 \times 10^{-6})} = 1.1 \times 10^{-9}M$$

$$[OH^-]' > [H^+]'$$

The buffer is basic.

$$[OH^-]'' = K_b \frac{(C_{NH_3} - [OH^-]')}{(C_{NH_4^+} + [OH^-]')}$$

$$= (1.8 \times 10^{-5})\frac{(1.00 \times 10^{-3} - 9.0 \times 10^{-6})}{(2.00 \times 10^{-3} + 9.0 \times 10^{-6})} = 9.0 \times 10^{-6}M$$

$$pOH = -\log(9.0 \times 10^{-6}M) = 5.04$$

$$pH = 14.00 - 5.04 = 8.96$$

4. $H^+ + NH_3 \rightleftharpoons NH_4^+$

(number of millimoles of NH_4^+ formed)

$$= \text{(number of millimoles of } H^+ \text{ added)}$$

$$= (20.00)(0.1000) = 2.000 \text{ millimoles}$$

$$C_{NH_4^+} = \frac{\text{(number of millimoles of } NH_4^+ \text{ formed)}}{\text{(number of milliliters of solution)}} = \frac{2.000}{70.00}$$

(number of millimoles of NH_3 left) = (number of millimoles of NH_3 to start) −

(number of millimoles of NH_4^+ formed)

$$= (50.00)(0.1000) - 2.000 = 3.000 \text{ millimoles}$$

$$C_{NH_3} = \frac{\text{(number of millimoles of } NH_3 \text{ left)}}{\text{(number of milliliters of solution)}} = \frac{3.000}{70.00}$$

This solution is truly a buffer, for $C_{NH_4^+} > 0$ and $C_{NH_3} > 0$. Try a method of approximations first.

$$[OH^-]' = K_b \frac{C_{NH_3}}{C_{NH_4^+}} = (1.8 \times 10^{-5})\frac{\dfrac{3.000}{70.00}}{\dfrac{2.000}{70.00}} = 2.7 \times 10^{-5}M$$

(The buffer is *basic*.)

$$[OH^-]'' = K_b \left(\frac{C_{NH_3} - [OH^-]'}{C_{NH_4^+} + [OH^-]'}\right)$$

$$= (1.8 \times 10^{-5}) \left(\frac{\dfrac{3.000}{70.00} - 2.7 \times 10^{-5}}{\dfrac{2.000}{70.00} + 2.7 \times 10^{-5}} \right) = 2.7 \times 10^{-5} M$$

$$pOH = -\log (2.7 \times 10^{-5}) = 4.57$$

$$pH = 14.00 - 4.57 = 9.43$$

5.
$$C_{NH_4^+} = 1.0 \times 10^{-2} M$$

$$C_{NH_3} = 0$$

This is clearly a case where the reaction of interest is the hydrolysis of NH_4^+. We try a method of approximations.

$$[H^+]' = \left(\frac{K_w}{K_b} C_{NH_4^+} \right)^{1/2} = \left[\left(\frac{1.00 \times 10^{-14}}{1.8 \times 10^{-5}} \right)(1.0 \times 10^{-2}) \right]^{1/2}$$

$$= 2.4 \times 10^{-6} M$$

$$[H^+]'' = \left[\frac{K_w}{K_b} (C_{NH_4^+} - [H^+]') \right]^{1/2}$$

$$= \left[\left(\frac{1.00 \times 10^{-14}}{1.8 \times 10^{-5}} \right) \cdot (1.0 \times 10^{-2} - 2.4 \times 10^{-6}) \right]^{1/2}$$

$$= 2.4 \times 10^{-6} M$$

$$pH = -\log (2.4 \times 10^{-6}) = 5.62$$

6. Use first approximations here.

   (a) Case 1: Goopyl amine alone.

   $$[OH^-]' = (K_b C_{Gup})^{1/2}$$

   $$[(4.0 \times 10^{-4})(1.00 \times 10^{-1})]^{1/2}$$

   $$= 6.1 \times 10^{-3} M$$

   $$pOH = -\log (6.1 \times 10^{-3}) = 2.21$$

   $$pH = 14.00 - 2.21 = 11.79$$

   (b) Case 2: This is a buffer, and $[OH^-]' = K_b$. Why?

   $$Gup + H^+ \rightleftharpoons GupH^+$$

   (number of millimoles of $GupH^+$ formed)

   $\qquad$ = (number of millimoles of HCl added)

   $\qquad$ = (12.50)(0.1000) = 1.250 millimoles

   $$C_{GupH^+} = \frac{\text{(number of millimoles of } GupH^+ \text{ formed)}}{\text{(number of milliliters of solution)}} = \frac{1.250}{37.50}$$

(number of millimoles of Gup left)

$$= \text{(number of millimoles of Gup to start)} -$$
$$\text{(number of millimoles of GupH}^+ \text{ formed)}$$

$$= (25.00)(0.1000) - 1.250 = 1.250 \text{ millimoles}$$

$$C_{Gup} = \frac{\text{(number of millimoles of Gup left)}}{\text{(number of milliliters of solution)}} = \frac{1.250}{37.50}$$

$$[OH^-]' = K_b \frac{C_{Gup}}{C_{GupH^+}} = (4.0 \times 10^{-4}) \frac{\cancel{1.250}}{\cancel{37.50}} = 4.0 \times 10^{-4} M$$

$$pOH = -\log (4.0 \times 10^{-4}) = 3.40$$

$$pH = 14.00 - 3.40 = 10.60$$

(c) Case 3: This is the endpoint of the titration.

$$C_{GupH^+} > 0, \quad \text{but } C_{Gup} = 0$$

Why?

(number of millimoles of GupH$^+$ formed)

$$= \text{(number of millimoles of HCl added)}$$

$$= (25.00)(0.1000) = 2.500 \text{ millimoles}$$

$$C_{GupH^+} = \frac{\text{(number of millimoles of GupH}^+ \text{ formed)}}{\text{(number of milliliters of solution)}} = \frac{2.500}{50.00}$$

(number of millimoles of Gup left)

$$= \text{(number of millimoles of Gup to start)} -$$
$$\text{(number of millimoles of GupH}^+ \text{ formed)}$$

$$= (25.00)(0.1000) - 2.500 = 0.0000$$

$$C_{Gup} = \frac{\text{(number of millimoles of Gup left)}}{\text{(number of milliliters of solution)}}$$

$$= \frac{0.000}{50.00} = 0.000 M$$

The hydrolysis of GupH$^+$ (the goopium ion) is of interest here.

$$GupH^+ \rightleftharpoons Gup + H^+$$

$$[H^+]' = \left(\frac{K_w}{K_b} C_{GupH^+} \right)^{1/2} = \left[\left(\frac{1.00 \times 10^{-14}}{4.0 \times 10^{-4}} \right) \frac{(2.500)}{(50.00)} \right]^{1/2} = 1.12 \times 10^{-6} M$$

$$pH = -\log (1.12 \times 10^{-6}) = 5.95$$

(d) Here HCl is in excess and determines the pH.

(number of millimoles of HCl excess)

$$= \text{(number of millimoles of HCl added)} - \text{(number of millimoles of Gup to start)}$$

$$= (40.00)(0.1000) - (25.00)(0.1000)$$

$$= 1.500 \text{ millimoles}$$

$$[H^+] = \frac{\text{(number of millimoles of HCl excess)}}{\text{(number of milliliters of solution)}}$$

$$= \frac{1.500}{65.00} = 2.308 \times 10^{-2} M$$

$$pH = -\log(2.308 \times 10^{-2}) = 1.64$$

Table 2.7
DISSOCIATION (OR IONIZATION) CONSTANTS FOR WEAK ACIDS AT 25°C. FOR H_2O: $K_w = 1.00 \times 10^{-14}$.

Monoprotic Acids	
Acetic acid, CH_3COOH	1.8×10^{-5}
Propionic acid, CH_3CH_2COOH	1.3×10^{-5}
Butyric acid, $CH_3CH_2CH_2COOH$	1.48×10^{-5}
Benzoic acid,	6.3×10^{-5}
Lactic acid, $CH_3\overset{OH}{\underset{}{CH}}-\overset{O}{\underset{}{C}}-OH$	1.4×10^{-4}
Nitrous acid, HNO_2	5.1×10^{-4}
Hydrofluoric acid, HF	6.7×10^{-4}

Diprotic Acids	K_1	K_2
Carbonic, H_2CO_3	3.5×10^{-7}	6.0×10^{-11}
Hydrogen sulfide, H_2S	1.02×10^{-7}	1.21×10^{-15}
Oxalic, $HO\overset{O}{\underset{}{C}}-\overset{O}{\underset{}{C}}OH$	5.6×10^{-2}	6.2×10^{-5}
Malonic, $HO\overset{O}{\underset{}{C}}-CH_2-\overset{O}{\underset{}{C}}OH$	1.51×10^{-3}	2.2×10^{-6}
Succinic, $HO\overset{O}{\underset{}{C}}-(CH_2)_2-\overset{O}{\underset{}{C}}OH$	6.46×10^{-5}	3.31×10^{-6}
Phthalic,	8.0×10^{-4}	4.0×10^{-6}

Table 2.7 continued

	Diprotic Acids K_1	K_2
8-Hydroxyquinoline	1.07×10^{-5}	6×10^{-11}
Salicyclic	1.07×10^{-3}	1.0×10^{-13}
Tartaric $HO\overset{O}{\overset{\|}{C}}-\underset{\underset{OH}{\|}}{CH}-\underset{\underset{OH}{\|}}{CH}-\overset{O}{\overset{\|}{C}}-OH$	2.0×10^{-3}	8.9×10^{-5}

	Triprotic Acids K_1	K_2	K_3
Arsenic, H_3AsO_4	6.0×10^{-3}	1.05×10^{-7}	4.0×10^{-12}
Phosphoric, H_3PO_4	7.1×10^{-3}	6.2×10^{-8}	4.4×10^{-13}
Citric, $HO\overset{O}{\overset{\|}{C}}-CH_2-\underset{\underset{\overset{\|}{C}-OH}{\underset{\|}{\overset{OH}{\|}}}}{C}-CH_2\overset{O}{\overset{\|}{C}}OH$	8.3×10^{-4}	1.74×10^{-5}	4.0×10^{-7}

	A Tetraprotic Acid K_1	K_2	K_3	K_4
	6.6×10^{-3}	1.86×10^{-3}	6.3×10^{-7}	1.0×10^{-10}

Table 2.8
K_b-VALUES FOR SELECTED WEAK BASES AT 25°C

Ammonia, NH_3	1.8×10^{-5}
Ethylamine, $CH_3CH_2NH_2$	4.3×10^{-4}
Hydroxylamine, $HONH_2$	9.1×10^{-9}
Pyridine,	1.7×10^{-9}
Aniline	4.0×10^{-10}

Table 2.9
α-VALUES FOR CARBONIC ACID.
$$K_1 = 3.5E - 07$$
$$K_2 = 6.0E - 11$$

pH	α (0)	α (1)	α (2)
0.50	0.100E + 01	0.111E − 05	0.210E − 15
1.00	0.100E + 01	0.350E − 05	0.210E − 14
1.50	0.100E + 01	0.111E − 04	0.210E − 13
2.00	0.100E + 01	0.350E − 04	0.210E − 12
2.50	0.100E + 01	0.111E − 03	0.210E − 11
3.00	0.100E + 01	0.350E − 03	0.210E − 10
3.50	0.999E + 00	0.111E − 02	0.210E − 09
4.00	0.997E + 00	0.349E − 02	0.209E − 08
4.50	0.989E + 00	0.109E − 01	0.208E − 07
5.00	0.966E + 00	0.338E − 01	0.203E − 06
5.50	0.900E + 00	0.997E − 01	0.189E − 05
6.00	0.741E + 00	0.259E + 00	0.156E − 04
6.50	0.475E + 00	0.525E + 00	0.997E − 04
7.00	0.222E + 00	0.777E + 00	0.466E − 03
7.50	0.827E − 01	0.916E + 00	0.174E − 02
8.00	0.276E − 01	0.967E + 00	0.580E − 02
8.50	0.879E − 02	0.973E + 00	0.185E − 01
9.00	0.269E − 02	0.941E + 00	0.565E − 01
9.50	0.759E − 03	0.840E + 00	0.159E + 00
10.00	0.179E − 03	0.625E + 00	0.375E + 00
10.50	0.312E − 04	0.345E + 00	0.655E + 00
11.00	0.408E − 05	0.143E + 00	0.857E + 00
11.50	0.452E − 06	0.501E − 01	0.950E + 00
12.00	0.468E − 07	0.164E − 01	0.984E + 00
12.50	0.474E − 08	0.524E − 02	0.995E + 00
13.00	0.475E − 09	0.166E − 02	0.998E + 00
13.50	0.476E − 10	0.527E − 03	0.999E + 00
14.00	0.476E − 11	0.167E − 03	0.100E + 01

Table 2.10
α-VALUES FOR 8-HYDROXYQUINOLINE
$$K_1 = 1.07E - 5$$
$$K_2 = 6.0E - 11$$

pH	α (0)	α (1)	α (2)
0.50	0.100E + 01	0.338E – 04	0.642E – 14
1.00	0.100E + 01	0.107E – 03	0.642E – 13
1.50	0.100E + 01	0.338E – 03	0.642E – 12
2.00	0.999E + 00	0.107E – 02	0.641E – 11
2.50	0.997E + 00	0.337E – 02	0.640E – 10
3.00	0.989E + 00	0.106E – 01	0.635E – 09
3.50	0.967E + 00	0.327E – 01	0.621E – 08
4.00	0.903E + 00	0.967E – 01	0.580E – 07
4.50	0.747E + 00	0.253E + 00	0.480E – 06
5.00	0.483E + 00	0.517E + 00	0.310E – 05
5.50	0.228E + 00	0.772E + 00	0.146E – 04
6.00	0.855E – 01	0.914E + 00	0.549E – 04
6.50	0.287E – 01	0.971E + 00	0.184E – 03
7.00	0.925E – 02	0.990E + 00	0.594E – 03
7.50	0.294E – 02	0.995E + 00	0.189E – 02
8.00	0.928E – 03	0.993E + 00	0.596E – 02
8.50	0.290E – 03	0.981E + 00	0.186E – 01
9.00	0.882E – 04	0.943E + 00	0.566E – 01
9.50	0.248E – 04	0.841E + 00	0.159E + 00
10.00	0.584E – 05	0.625E + 00	0.375E + 00
10.50	0.102E – 05	0.345E + 00	0.655E + 00
11.00	0.134E – 06	0.143E + 00	0.857E + 00
11.50	0.148E – 07	0.501E – 01	0.950E + 00
12.00	0.153E – 08	0.164E – 01	0.984E + 00
12.50	0.155E – 09	0.524E – 02	0.995E + 00
13.00	0.156E – 10	0.166E – 02	0.998E + 00
13.50	0.156E – 11	0.527E – 03	0.999E + 00
14.00	0.156E – 12	0.167E – 03	0.100E + 01

Table 2.11
α-VALUES FOR PHTHALIC ACID
$$K_1 = 8.0E - 4$$
$$K_2 = 4.0E - 6$$

pH	α (0)	α (1)	α (2)
0.50	0.997E + 00	0.252E – 02	0.319E – 07
1.00	0.992E + 00	0.794E – 02	0.317E – 06
1.50	0.975E + 00	0.247E – 01	0.312E – 05
2.00	0.926E + 00	0.741E – 01	0.296E – 04
2.50	0.798E + 00	0.202E + 00	0.255E – 03
3.00	0.555E + 00	0.444E + 00	0.177E – 02
3.50	0.281E + 00	0.710E + 00	0.898E – 02
4.00	0.107E + 00	0.858E + 00	0.343E – 01
4.50	0.339E – 01	0.858E + 00	0.108E + 00
5.00	0.885E – 02	0.708E + 00	0.283E + 00
5.50	0.174E – 02	0.441E + 00	0.558E + 00
6.00	0.250E – 03	0.200E + 00	0.800E + 00
6.50	0.290E – 04	0.733E – 01	0.927E + 00
7.00	0.305E – 05	0.244E – 01	0.976E + 00
7.50	0.310E – 06	0.784E – 02	0.992E + 00
8.00	0.312E – 07	0.249E – 02	0.998E + 00
8.50	0.312E – 08	0.790E – 03	0.999E + 00
9.00	0.312E – 09	0.250E – 03	0.100E + 01
9.50	0.312E – 10	0.791E – 04	0.100E + 01
10.00	0.312E – 11	0.250E – 04	0.100E + 01
10.50	0.312E – 12	0.791E – 05	0.100E + 01
11.00	0.312E – 13	0.250E – 05	0.100E + 01
11.50	0.312E – 14	0.791E – 06	0.100E + 01
12.00	0.312E – 15	0.250E – 06	0.100E + 01
12.50	0.313E – 16	0.791E – 07	0.100E + 01
13.00	0.312E – 17	0.250E – 07	0.100E + 01
13.50	0.313E – 18	0.791E – 08	0.100E + 01
14.00	0.313E – 19	0.250E – 08	0.100E + 01

Table 2.12
α-VALUES FOR SUCCINIC ACID
$$K_1 = 6.46E - 05$$
$$K_2 = 3.31E - 06$$

pH	α (0)	α (1)	α (2)
0.50	0.100E + 01	0.204E − 03	0.214E − 08
1.00	0.999E + 00	0.646E − 03	0.214E − 07
1.50	0.998E + 00	0.204E − 02	0.213E − 06
2.00	0.994E + 00	0.642E − 02	0.212E − 05
2.50	0.980E + 00	0.200E − 01	0.210E − 04
3.00	0.939E + 00	0.607E − 01	0.201E − 03
3.50	0.829E + 00	0.169E + 00	0.177E − 02
4.00	0.600E + 00	0.387E + 00	0.128E − 01
4.50	0.307E + 00	0.627E + 00	0.657E − 01
5.00	0.104E + 00	0.673E + 00	0.223E + 00
5.50	0.234E − 01	0.477E + 00	0.499E + 00
6.00	0.358E − 02	0.231E + 00	0.765E + 00
6.50	0.427E − 03	0.872E − 01	0.912E + 00
7.00	0.454E − 04	0.293E − 01	0.971E + 00
7.50	0.463E − 05	0.946E − 02	0.991E + 00
8.00	0.466E − 06	0.301E − 02	0.997E + 00
8.50	0.467E − 07	0.954E − 03	0.999E + 00
9.00	0.468E − 08	0.302E − 03	0.100E + 01
9.50	0.468E − 09	0.955E − 04	0.100E + 01
10.00	0.468E − 10	0.302E − 04	0.100E + 01
10.50	0.468E − 11	0.955E − 05	0.100E + 01
11.00	0.468E − 12	0.302E − 05	0.100E + 01
11.50	0.468E − 13	0.955E − 06	0.100E + 01
12.00	0.468E − 14	0.302E − 06	0.100E + 01
12.50	0.468E − 15	0.955E − 07	0.100E + 01
13.00	0.468E − 16	0.302E − 07	0.100E + 01
13.50	0.468E − 17	0.955E − 08	0.100E + 01
14.00	0.468E − 18	0.302E − 08	0.100E + 01

chapter 3
The Control
of Solubility

≡ 3.1 INTRODUCTION ≡

Very often, before a measurement can be taken in analytical laboratory practice, a *separation* of the species on which the measurement is to be made from other species present must be done. The simplest example that comes to mind is separation by precipitation, the precipitate being weighed. Here is a description of such a separation as it was done in the 1860s. In the assay of gold-silver bricks, a button of metal, an alloy of gold and silver, was isolated. Then, in the words of Mark Twain:

> He (the assayer) has to separate the gold from the silver now. The button is hammered out flat and thin, put in the furnace, and kept for some time at a red heat; after cooling it off it is rolled up like a quill and heated in a glass vessel containing nitric acid; the acid dissolves the silver and leaves the gold pure and ready to be weighed on its own merits. Then saltwater is poured into the vessel containing the dissolved silver and the silver returns to palpable form again and sinks to the bottom. Nothing now remains but to weigh it. . .*

* Mark Twain, *Roughing It*, Chapter 36.

218

Separation science, or technology, if you will, is quite old. It not only includes precipitation, but also a number of other processes. Our concern here will be with the oldest of the processes, precipitation. Separation by precipitation depends upon the solubilities of the species separated. The control of solubility allows the control of precipitation, and hence of the separation of chemical species.

Solubilities may be controlled through the intelligent manipulation of a number of chemical processes. The solubility of each and every precipitate may not be controlled by each of the processes listed, but at least one of the processes controls the solubility of each precipitate. The processes are

¤ 1. The common ion effect, or the decrease in solubility of a precipitate with increasing excess of one of the ions used in its formation.

¤ 2. Differential solubility, in which separation is made possible through competing chemical reactions of the species to be separated with the same precipitating anion or cation.

¤ 3. Complex formation with a ligand, either a neutral species or an anion, which is not the anion of the precipitate. This increases solubility.

¤ 4. Weak acid formation, in which hydrogen ions from the solution surrounding the precipitate form weak acids with the anion of the precipitate and increase the solubility of the precipitate.

The first two processes are often studied at length in freshman chemistry, but the last two seldom are. Consequently, the last two will receive the most attention here.

Before any of the processes listed can be examined in detail, the reader must know something of solubilities and the simple quantities known as solubility product constants, or K_{sp}'s.

Solubility of a precipitate will most often be given the designation S in these pages. It will be in units of moles per liter of solution. Sometimes, solubility, thus defined, is called molar solubility. Solubility, or molar solubility, is simply the number of moles of a precipitate that will dissolve in a liter of a given solution. Such a solution is said to be saturated when so many moles of precipitate as can dissolve in it, have dissolved. A saturated solution is a solution at equilibrium.

The solubility product constant, or K_{sp}, is simply an equilibrium constant. Consider the equilibrium constant for the dissociation of $BaSO_4$. The dissolution reaction is just

$$BaSO_{4(s)} \rightleftharpoons Ba^{2+} + SO_4^{2-}$$

From the general equation just written

$$K = \frac{[Ba^{2+}] \cdot [SO_4^{2-}]}{[BaSO_{4(s)}]}$$

or

$$K \cdot [BaSO_{4(s)}] = [Ba^{2+}] \cdot [SO_4^{2-}]$$

Because $BaSO_4$ is a pure solid, the quantity $[BaSO_{4(s)}]$ may be taken as a constant quantity. A simple argument for this assertion might run like this: (1) The pure solid is a separate phase from the solution; (2) the density, or concentration, of $BaSO_4$ in the pure solid is a constant; and (3) the quantity $[BaSO_{4(s)}]$ is thus a constant. Thus the quantity

$$K \cdot [BaSO_{4(s)}] = \text{a constant}$$

or

$$K \cdot [BaSO_{4(s)}] = K_{sp}$$

$$K_{sp} = [Ba^{2+}] \cdot [SO_4^{2-}]$$

The term K_{sp} is called the solubility product constant. If a precipitate of $BaSO_4$ is mixed with pure water and allowed to come to equilibrium, and if the only reaction going on is

$$BaSO_{4(s)} \rightleftharpoons Ba^{2+} + SO_4^{2}$$

then the solubility of the $BaSO_4$ precipitate is easily defined:

S = solubility

S = (number of moles of precipitate that dissolve per liter of solution)

Because the stoichiometry of dissolution is one-to-one,

(number of moles of precipitate = (number of moles of Ba^{2+} per liter
that dissolve per liter of solution)   of solution)

or

$$S = [Ba^{2+}]$$

Also

(number of moles of precipitate = (number of moles of SO_4^{2-} per liter
that dissolve per liter of solution)   of solution)

or

$$S = [SO_4^{2-}]$$

$$K_{sp} = [Ba^{2+}] \cdot [SO_4^{2-}]$$

or

$$K_{sp} = (S) \cdot (S)$$

or

$$K_{sp} = S^2$$

In this simple system, if solubility, S, is known, the value of K_{sp} can be found.

Example: Careful measurement reveals the solubility of $BaSO_4$ in pure H_2O to be $1.04 \times 10^{-5} M$. Calculate the K_{sp} of $BaSO_4$.

As we have already argued,

$$K_{sp} = S^2$$

$$S = 1.04 \times 10^{-5}$$

$$K_{sp} = (1.04 \times 10^{-5})^2 = 1.08 \times 10^{-10}$$

Conversely, in the simple systems such as $BaSO_4$ in H_2O, if K_{sp} is known, the solubility can be found.

Example: The K_{sp} of cuprous iodide, CuI, is 5.1×10^{-12}. Calculate the solubility of CuI in water. If this is the only reaction

$$CuI \rightleftharpoons Cu^+ + I^-$$

then

$$K_{sp} = [Cu^+][I^-]$$

If

$$S = \text{solubility} = [Cu^+] = [I^-]$$

then

$$K_{sp} = [Cu^+][I^-] = (S)(S) = S^2$$

and

$$S = (K_{sp})^{1/2} = (5.1 \times 10^{-12})^{1/2} = 2.3 \times 10^{-6} M$$

Not all chemical compounds exhibit one-to-one stoichiometry, so it would be unreasonable to expect all precipitates to do so. Take Ag_2CrO_4, silver chromate, as an example. If Ag_2CrO_4 is placed into pure water and allowed to come to equilibrium, and if the only reaction is the dissociation of solid Ag_2CrO_4,

$$Ag_2CrO_4 \rightleftharpoons 2\,Ag^+ + CrO_4^{2-}$$

$$K_{sp} = [Ag^+]^2 \cdot [CrO_4^{2-}]$$

then the relationship between the solubility and K_{sp} is not hard to derive.

$$S = \text{solubility}$$

$$S = (\text{number of moles of precipitate that dissolve per liter of solution})$$

Because there are 2 moles of silver in every mole of silver chromate

(2) (number of moles of precipitate = (number of moles of Ag^+ per liter that dissolve per liter of solution)   of solution)

or

$$2S = [Ag^+]$$

Because there is 1 mole of chromate ion in every mole of silver chromate

(number of moles of precipitate = (number of moles of CrO_4^{2-} per liter of that dissolve per liter of solution)   solution)

or

$$S = [CrO_4^{2-}]$$

$$K_{sp} = [Ag^+]^2 \cdot [CrO_4^{2-}]$$

$$= (2S)^2 \cdot (S) = 4S^3$$

If the solubility, S, is known, the value of K_{sp} can be found. Conversely, if K_{sp} is known, the value of solubility can be found

$$S = \left(\frac{K_{sp}}{4}\right)^{1/3}$$

$\equiv$ 3.2 THE COMMON ION EFFECT $\equiv$

Thus far, we have considered the simplest kind of precipitation system-precipitates that are mixed with pure water and then come to equilibrium. In these simple systems, only the dissociation equilibrium has been of importance. It was early noted that the solubility of precipitates often decreases with the addition of an ion common to the precipitates. In the case of barium sulfate, if we have a saturated solution of $BaSO_4$, made by mixing an excess of solid $BaSO_4$ with water and allowing the system to come to equilibrium, we can decrease the solubility of the precipitate by the addition of $BaCl_2$ or Na_2SO_4. Why?

First, both $BaCl_2$ and Na_2SO_4 are highly soluble salts. They both dissociate completely on being dissolved. $BaCl_2$ dissociates into Ba^{2+} and Cl^- ions; Na_2SO_4 dissociates into Na^+ and SO_4^{2-} ions. The Cl^- or Na^+ ions need be of no concern to us. Excess Ba^{2+} or SO_4^{2-} ions are of concern. Either can depress solubility. In qualitative terms, the equilibrium equation for dissolution is

$$BaSO_4 \rightleftharpoons Ba^{2+} + SO_4^{2-}$$

The addition of Ba^{2+} ions, according to Le Chatelier's principle, forces the equilibrium to the left, diminishing solubility. The addition of SO_4^{2-} ions would have the same effect.

How much would solubility be depressed? The following example can provide some answers.

Example: Calculate the solubility of $BaSO_4$: (a) in H_2O; (b) in $0.00100M$ Na_2SO_4; (c) in $0.0100M$ Na_2SO_4; and (d) in $0.100M$ Na_2SO_4, where $K_{sp} = 1.08 \times 10^{-10}$.

(a) $BaSO_4 \rightleftharpoons Ba^{2+} + SO_4^{2-}$ $K_{sp} = [Ba^{2+}][SO_4^{2-}]$

solubility $= S$

$$S = [Ba^{2+}] = [SO_4^{2-}]$$

$$K_{sp} = [Ba^{2+}][SO_4^{2-}]$$

$$K_{sp} = S^2$$

$$S = (K_{sp})^{1/2} = (1.08 \times 10^{-10})^{1/2} = 1.04 \times 10^{-5}M$$

(b) In $0.00100M$ Na_2SO_4. This is a trifle more difficult.

$$BaSO_4 \rightleftharpoons Ba^{2+} + SO_4^{2-}   K_{sp} = [Ba^{2+}][SO_4^{2-}]$$

solubility $= S$

$$S = [Ba^{2+}]$$

But $S \neq [SO_4^{2-}]$. Why?

There are two sources of sulfate ions, the $BaSO_4$ precipitate and the Na_2SO_4 solution.

$$\text{(concentration of } SO_4^{2-} \text{ from } BaSO_4) = (SO_4^{2-})_{ppt}$$

$$(SO_4^{2-})_{ppt} = [Ba^{2+}]$$

$$(SO_4^{2-})_{ppt} = S$$

$$\text{(concentration of } SO_4^{2-} \text{ from } Na_2SO_4) = 0.00100M$$

$$[SO_4^{2-}] = \text{(concentration of } SO_4^{2-} \text{ from } BaSO_4) +$$
$$\text{(concentration of } SO_4^{2-} \text{ from } Na_2SO_4)$$

$$[SO_4^{2-}] = (SO_4^{2-})_{ppt} + 0.00100$$

$$[SO_4^{2-}] = S + 1.00 \times 10^{-3}$$

$$K_{sp} = [Ba^{2+}] [SO_4^{2-}]$$

$$K_{sp} = (S)(S + 1.00 \times 10^{-3})$$

This equation could be solved rigorously for S, but the solution takes a little time—perhaps 5 minutes. It is easier to go to a system of successive approximations.

$$S = \frac{K_{sp}}{1.00 \times 10^{-3} + S}$$

Suppose that $10^{-3} \gg S$. Then a good first approximation for solubility, called S', would be

First approximation:

$$S' = \frac{K_{sp}}{1.00 \times 10^{-3}}$$

The second approximation would involve setting S' back into the full expression for solubility.

Second approximation:

$$S'' = \frac{K_{sp}}{1.00 \times 10^{-3} + S'}$$

If the first and second approximations do not converge, that is if S' and S'' do not reach the desired closeness to each other (10, 5, or 1%, or whatever you like), a third approximation can be made.

Third approximation:

$$S''' = \frac{K_{sp}}{1.00 \times 10^{-3} + S''}$$

We continue approximating, or iterating, until the desired convergence occurs. With the $BaSO_4$ and $0.00100M$ Na_2SO_4 case:

$$S' = \frac{K_{sp}}{1.00 \times 10^{-3}} = \frac{1.08 \times 10^{-10}}{1.00 \times 10^{-3}} = 1.08 \times 10^{-7}M$$

$$S'' = \frac{1.08 \times 10^{-10}}{1.00 \times 10^{-3} + 1.08 \times 10^{-7}} = 1.08 \times 10^{-7}M$$

(c) In 0.0100M Na$_2$SO$_4$: Using the reasoning from part b,

$$S = \frac{K_{sp}}{1.00 \times 10^{-2} + S}$$

$$S' = \frac{K_{sp}}{1.00 \times 10^{-2}} = \frac{1.08 \times 10^{-10}}{1.00 \times 10^{-2}} = 1.08 \times 10^{-8}M$$

$$S'' = \frac{1.08 \times 10^{-10}}{1.00 \times 10^{-2} + S'}$$

$$= \frac{1.08 \times 10^{-10}}{1.00 \times 10^{-2} + 1.08 \times 10^{-8}} = 1.08 \times 10^{-8}M$$

(d) In 0.100M Na$_2$SO$_4$: Using the reasoning from part b, we find that

$$S = \frac{K_{sp}}{1.00 \times 10^{-1} + S}$$

$$S' = \frac{K_{sp}}{1.00 \times 10^{-1}} = \frac{1.08 \times 10^{-10}}{1.00 \times 10^{-1}} = 1.08 \times 10^{-9}M$$

$$S'' = \frac{1.08 \times 10^{-10}}{1.00 \times 10^{-1} + S'}$$

$$= \frac{1.08 \times 10^{-10}}{1.00 \times 10^{-1} + 1.08 \times 10^{-9}} = 1.08 \times 10^{-9}M$$

In this simple case, where only the dissociation equilibrium is important, solubility does indeed decrease as a concentration of common ion is increased.

Solution	Solubility, M
H$_2$O	1.04×10^{-5}
$10^{-3}M$ Na$_2$SO$_4$	1.08×10^{-7}
$10^{-2}M$ Na$_2$SO$_4$	1.08×10^{-8}
$10^{-1}M$ Na$_2$SO$_4$	1.08×10^{-9}

≡ 3.3 DIFFERENTIAL SOLUBILITY ≡

Sometimes it is possible to separate two metal ions that form sparingly soluble precipitates (which may be called "insoluble" in handbooks) from one another, by precipitating first one and then the other with the same anion. Let us take an example of strontium and barium.

Example: A solution is $1.0 \times 10^{-3}M$ in Ba(NO$_3$)$_2$ and Sr(NO$_3$)$_2$. Given that the nitrate salts are soluble, and that the only important equilibria of chromate salts are

$$BaCrO_4 \rightleftharpoons Ba^{2+} + CrO_4^{2-} \qquad K_{sp} = 2.4 \times 10^{-10}$$

$$SrCrO_4 \rightleftharpoons Sr^{2+} + CrO_4^{2-} \qquad K_{sp} = 3.6 \times 10^{-5}$$

let us answer some questions about their separation. Suppose that solid Na_2CrO_4 (soluble) is added very slowly to the solution. Remember that a precipitate begins to form whenever the product of ionic concentrations equals or exceeds the K_{sp}. In short, $BaCrO_4$ will form when

$$[Ba^{2+}] [CrO_4^{2-}] \gtreqless K_{sp}$$

and $SrCrO_4$ will form when

$$[Sr^{2+}] [CrO_4^{2-}] \gtreqless K_{sp}$$

(a) At what value of $[CrO_4^{2-}]$ will $BaCrO_4$ begin to form?
(b) At what value of $[CrO_4^{2-}]$ will $SrCrO_4$ begin to form?
(c) Will $BaCrO_4$ or $SrCrO_4$ form first?
(d) What will be the concentration of the first ion to precipitate when the second precipitate begins to form?

(a) If $[Ba^{2+}] [CrO_4^{2-}] \gtreqless K_{sp}$, $BaCrO_4$ will form.

$$[CrO_4^{2-}] = \frac{K_{sp}}{[Ba^{2+}]}$$

$$[Ba^{2+}] = 1.0 \times 10^{-3} M$$

$$[CrO_4^{2-}] = \frac{2.4 \times 10^{-10}}{1.0 \times 10^{-3}} = 2.4 \times 10^{-7} M$$

when $BaCrO_4$ begins to form.

(b) If $[Sr^{2+}] [CrO_4^{2-}] \gtreqless K_{sp}$, $SrCrO_4$ will form.

$$[CrO_4^{2-}] = \frac{K_{sp}}{[Sr^{2+}]}$$

$$[Sr^{2+}] = 1.0 \times 10^{-3} M$$

$$[CrO_4^{2-}] = \frac{3.6 \times 10^{-5}}{1.0 \times 10^{-3}} = 3.6 \times 10^{-2} M$$

when $SrCrO_4$ begins to form.

(c) $BaCrO_4$ forms first, because when $[CrO_4^{2-}]$ is only $2.4 \times 10^{-7} M$, it begins to form, while $[CrO_4^{2-}]$ must rise to $3.6 \times 10^{-2} M$ for $SrCrO_4$ to form.

(d) When the second precipitate, $SrCrO_4$, begins to form, $[CrO_4^{2-}] = 3.6 \times 10^{-2} M$. What is $[Ba^{2+}]$?

$$[Ba^{2+}] [CrO_4^{2-}] = K_{sp}$$

$$[Ba^{2+}] = \frac{K_{sp}}{[CrO_4^{2-}]} = \frac{2.4 \times 10^{-10}}{3.6 \times 10^{-2}} = 6.7 \times 10^{-9} M$$

Thus the separation of Ba^{2+} from Sr^{2+} is possible. It might be best, after the $BaCrO_4$ is filtered, to precipitate the Sr^{2+} with something other than CrO_4^{2-}, for the K_{sp} of $SrCrO_4$ is quite high.

$\equiv$ 3.4 PROBLEMS $\equiv$

(Use Table 3.1 for K_{sp} values.)

Table 3.1
SELECTED SOLUBILITY PRODUCT
CONSTANTS AT 25°C

Substance	K_{sp}
$Ag_2C_2O_4$	3.5×10^{-11}
Ag_2CrO_4	1.31×10^{-12}
$AgCl$	1.78×10^{-10}
$AgBr$	5.25×10^{-13}
AgI	9.0×10^{-17}
Ag_2S	2.0×10^{-49}
$AgSCN$	1.00×10^{-12}
Ag_3PO_4	1.3×10^{-20}
$BaCrO_4$	2.4×10^{-10}
$BaSO_4$	1.08×10^{-10}
$BiPO_4$	1.3×10^{-23}
$CaCO_3$	7.9×10^{-9}
CaC_2O_4	2.6×10^{-9}
CaF_2	4.0×10^{-11}
$Ca_3(PO_4)_2$	2.0×10^{-29}
CdS	7.8×10^{-27}
CuI	5.1×10^{-12}
PbC_2O_4	4.8×10^{-10}
$PbCrO_4$	1.8×10^{-14}
$PbCO_3$	3.3×10^{-14}
$PbSO_4$	1.6×10^{-10}
PbI_2	7.1×10^{-9}
$Pb_3(AsO_4)_2$	4.1×10^{-36}
$Pb(IO_3)_2$	2.6×10^{-13}
$SrCO_3$	1.1×10^{-10}
$SrCrO_4$	3.6×10^{-5}
$SrSO_4$	3.8×10^{-7}
SrF_2	2.8×10^{-9}

1. Assuming that dissociation equilibria are the only ones involved in the dissolution of the following precipitates, write the K_{sp} of each as a function of its solubility.

(a) AgSCN.

Ans: $K_{sp} = S^2$

(b) PbCrO$_4$.

Ans: $K_{sp} = S^2$

(c) PbI$_2$.

Ans: $K_{sp} = 4S^3$

(d) PbClF.

Ans: $K_{sp} = S^3$

(e) Ce$_2$S$_3$.

Ans: $K_{sp} = 108S^5$

2. Assuming that dissociation equilibria are the only ones involved in the dissociation of the following precipitates, calculate the molar solubility of each in water.

(a) PbSO$_4$.

Ans: $1.3 \times 10^{-5}M$

(b) Ag$_2$CrO$_4$.

Ans: $6.89 \times 10^{-5}M$

(c) Pb(IO$_3$)$_2$.

Ans: $4.0 \times 10^{-5}M$

(d) Pb$_3$(AsO$_4$)$_2$.

Ans: $3.3 \times 10^{-8}M$

3. Calculate the solubility of SrSO$_4$ ($K_{sp} = 3.8 \times 10^{-7}$) under the following conditions, assuming that only the dissociation equilibrium is important.

(a) Pure H$_2$O.

Ans: $6.2 \times 10^{-4}M$

(b) 0.0010M Sr(NO$_3$)$_2$.

Ans: $2.9 \times 10^{-4}M$

(c) 0.010M Sr(NO$_3$)$_2$.

Ans: $3.8 \times 10^{-5}M$

(d) 0.10M Sr(NO$_3$)$_2$.

Ans: $3.8 \times 10^{-6}M$

4. Calculate the solubility of Ag$_2$CrO$_4$ ($K_{sp} = 1.31 \times 10^{-12}$) under the following conditions, assuming that only the dissociation equilibrium is important:

(a) Pure H$_2$O.

Ans: $6.89 \times 10^{-5}M$

(b) 0.0100M AgNO$_3$.

Ans: $1.31 \times 10^{-8}M$

(c) 0.0100M Na$_2$CrO$_4$.

Ans: $5.72 \times 10^{-6}M$

5. Lead sulfate has a K_{sp} of 2.0×10^{-8}. If 5.0×10^{-3} moles of Pb(NO$_3$)$_2$ are treated with H$_2$SO$_4$, the final volume of solution being 100 milliliters, calculate the

number of moles of Pb^{2+} remaining in the solution:

(a) After the addition of 5.0×10^{-3} moles of H_2SO_4.
   Ans: *1.4 $\times$ 10^{-5} moles*

(b) After the addition of 5.5×10^{-3} moles of H_2SO_4.
   Ans: *4 $\times$ 10^{-7} moles*

6. A solution is $0.10M$ in $Sr(NO_3)_2$ and $0.050M$ in Ca^{2+}. Solid Na_2CO_3 is added to the solution. Assume that only the dissociation equilibria are important.

(a) Which carbonate precipitates first? What is the value of $[CO_3^{2-}]$ when this happens?
   Ans: *$SrCO_3$; [CO$_3^{2-}$] = 4.1 $\times$ 10^{-9}M*

(b) What is the concentration of $[Sr^{2+}]$ in the solution when $CaCO_3$ first begins to form?
   Ans: *2.6 $\times$ 10^{-3}M*

$\equiv$ 3.5 DETAILED SOLUTIONS TO PROBLEMS $\equiv$

1. (a) $AgSCN \rightleftharpoons Ag^+ + SCN^-$

   S = solubility

   S = (number of moles of precipitate that dissolve per liter of solution)

   Because there is 1 mole of silver in every mole of silver thiocyanate:

   (number of moles of precipitate   = (number of moles of Ag^+ per
   that dissolve per liter of solution)   liter of solution)
   or
   $$S = [Ag^+]$$

   Because there is 1 mole of thiocyanate in every mole of silver thiocyanate,

   (number of moles of precipitate   = (number of moles of SCN^-
   that dissolve per liter of solution)   per liter of solution)
   or
   $$S = [SCN^-]$$
   $$K_{sp} = [Ag^+] \cdot [SCN^-] = (S)(S) = S^2$$

(b) $PbCrO_4 \rightleftharpoons Pb^{2+} + CrO_4^{2-}$

   S = solubility

   S = (number of moles of precipitate that dissolve per liter of solution)

   Because there is 1 mole of lead in every mole of lead chromate,

   (number of moles of precipitate   = (number of moles of Pb^{2+}
   that dissolve per liter of solution)   per liter of solution)

or
$$S = [Pb^{2+}]$$

Because there is 1 mole of chromate in every mole of lead chromate,

(number of moles of precipitate = (number of moles of CrO_4^{2-}
that dissolve per liter of solution)   per liter of solution)

or
$$S = [CrO_4^{2-}]$$

$$K_{sp} = [Pb^{2+}][CrO_4^{2-}] = (S)(S) = S^2$$

(c) $PbI_2 \rightleftharpoons Pb^{2+} + 2I^-$

S = solubility

S = (number of moles of precipitate that dissolve
      per liter of solution)

Because there is 1 mole of lead in every mole of lead iodide,

(number of moles of precipitate = (number of moles of Pb^{2+}
that dissolve per liter of solution)   per liter of solution)

or
$$S = [Pb^{2+}]$$

Because there are 2 moles of iodide in every mole of lead iodide,

(2)(number of moles of precipitate = (number of moles of I^-
that dissolve per liter of solution)   per liter of solution)

or
$$2S = [I^-]$$

$$K_{sp} = [Pb^{2+}][I^-]^2$$

$$K_{sp} = (S)(2S)^2 \quad \text{(be sure to square this term)} = 4S^3$$

(d) $PbClF \rightleftharpoons Pb^{2+} + Cl^- + F^-$

S = solubility

S = (number of moles of precipitate that dissolve
      per liter of solution)

Because there is 1 mole of lead in every mole of PbClF,

(number of moles of precipitate = (number of moles of Pb^{2+}
that dissolve per liter of solution)   per liter of solution)

or
$$S = [Pb^{2+}]$$

Because there is 1 mole of chloride in every mole of PbClF,

(number of moles of precipitate = (number of moles of Cl^- per
that dissolve per liter of solution)   liter of solution)

or

$$S = [Cl^-]$$

Because there is 1 mole of fluoride in every mole of PbClF,

(number of moles of precipitate = (number of moles of F⁻
that dissolve per liter of solution)   per liter of solution)

or

$$S = [F^-]$$

$$K_{sp} = [Pb^{2+}] \cdot [Cl^-] \cdot [F^-] = (S) \cdot (S) \cdot (S) = S^3$$

(e) $Ce_2S_3 \rightleftharpoons 2Ce^{3+} + 3S^{2-}$

S (*not* S^{2-}) = solubility

S = (number of moles of precipitate that dissolve
per liter of solution)

Because there are 2 moles of cerium in every mole of cerous sulfide,

(2) (number of moles of precipitate = (number of moles of Ce^{3+}
that dissolve per liter of solution)   per liter of solution)

or

$$2S = [Ce^{3+}]$$

Because there are three moles of sulfide in every mole of cerous sulfide,

(3) (number of moles of precipitate = (number of moles of S^{2-}
that dissolve per liter of solution)   per liter of solution)

or

$$3S = [S^{2-}]$$

$$K_{sp} = [Ce^{3+}]^2 \cdot [S^{2-}]^3 = (2S)^2 \cdot (3S)^3 = (4S^2) \cdot (27S^3) = 108S^5$$

2. (a) $PbSO_4 \rightleftharpoons Pb^{2+} + SO_4^{2-}$

$$K_{sp} = [Pb^{2+}] \cdot [SO_4^{2-}]$$

$$S = \text{solubility}$$

$$[Pb^{2+}] = S$$

$$[SO_4^{2-}] = S$$

$$K_{sp} = (S)(S) = S^2$$

$$S = (K_{sp})^{1/2} = (1.6 \times 10^{-10})^{1/2} = 1.3 \times 10^{-5} M$$

(b) $Ag_2CrO_4 \rightleftharpoons 2Ag^+ + CrO_4^{2-}$

$$K_{sp} = [Ag^+]^2 \cdot [CrO_4^{2-}]$$

$$S = \text{solubility}$$

$$[Ag^+] = 2S$$

$$[CrO_4^{2-}] = S$$

$$K_{sp} = (2S)^2 \cdot (S) = 4S^3$$

$$S = \left(\frac{K_{sp}}{4}\right)^{1/3} = \left(\frac{1.31 \times 10^{-12}}{4}\right)^{1/3} = 6.89 \times 10^{-5} M$$

(c) $Pb(IO_3)_2 \rightleftharpoons Pb^{2+} + 2IO_3^-$

$$K_{sp} = [Pb^{2+}] \cdot [IO_3^-]^2$$

$$S = \text{solubility}$$

$$[Pb^{2+}] = S$$

$$[IO_3^-] = 2S$$

$$K_{sp} = (S)(2S)^2 = 4S^3$$

$$S = \left(\frac{K_{sp}}{4}\right)^{1/3} = \left(\frac{2.6 \times 10^{-13}}{4}\right)^{1/3} = 4.0 \times 10^{-5} M$$

(d) $Pb_3(AsO_4)_2 \rightleftharpoons 3Pb^{2+} + 2AsO_4^{3-}$

$$K_{sp} = [Pb^{2+}]^3 \cdot [AsO_4^{3-}]^2$$

$$S = \text{solubility}$$

$$[Pb^{2+}] = 3S$$

$$[AsO_4^{3-}] = 2S$$

$$K_{sp} = (3S)^3 \cdot (2S)^2 = (27S^3) \cdot (4S^2) = 108\,S^5$$

$$S = \left(\frac{K_{sp}}{108}\right)^{1/5} = \left(\frac{4.1 \times 10^{-36}}{108}\right)^{1/5} = 3.3 \times 10^{-8} M$$

3. (a) $SrSO_4 \rightleftharpoons Sr^{2+} + SO_4^{2-}$

$$K_{sp} = [Sr^{2+}] \cdot [SO_4^{2-}]$$

$$S = \text{solubility}$$

$$[Sr^{2+}] = S$$

$$[SO_4^{2-}] = S$$

$$K_{sp} = (S)(S) = S^2$$

$$S = (K_{sp})^{1/2} = (3.8 \times 10^{-7})^{1/2} = 6.2 \times 10^{-4} M$$

(b) $S = \text{solubility}$

$$[SO_4^{2-}] = S$$

But $[Sr^{2+}] \neq S$. Why?

There are two sources of strontium ions, the $SrSO_4$ precipitate and the $Sr(NO_3)_2$ solution.

$$\text{(concentration of } Sr^{2+} \text{ from } SrSO_4) = (Sr^{2+}) \text{ ppt}$$

$$(Sr^{2+}) \text{ ppt } = [SO_4^{2-}] = S$$

$$\text{concentration of } Sr^{2+} \text{ from } Sr(NO_3)_2 = 0.0010M$$

$$\begin{aligned} [Sr^{2+}] &= \text{(concentration of } Sr^{2+} \text{ from } SrSO_4) \\ &\quad + \text{(concentration of } Sr^{2+} \text{ from } Sr(NO_3)_2) \end{aligned}$$

$$= (Sr^{2+}) \text{ ppt } + 0.0010 = S + 1.0 \times 10^{-3}$$

$$K_{sp} = [Sr^{2+}] \cdot [SO_4^{2-}]$$

$$K_{sp} = (S + 1.0 \times 10^{-3})(S)$$

$$S = \frac{K_{sp}}{(1.0 \times 10^{-3} + S)}$$

First approximation:

$$S' = \frac{K_{sp}}{1.0 \times 10^{-3}} = \frac{3.8 \times 10^{-7}}{1.0 \times 10^{-3}} = 3.8 \times 10^{-4} M$$

Second approximation:

$$S'' = \frac{K_{sp}}{1.0 \times 10^{-3} + S'} = \frac{3.8 \times 10^{-7}}{1.0 \times 10^{-3} + 3.8 \times 10^{-4}} = 2.8 \times 10^{-4} M$$

S' and S'' are *not* equal.

Third approximation:

$$S''' = \frac{K_{sp}}{1.0 \times 10^{-3} + S''} = \frac{3.8 \times 10^{-7}}{1.0 \times 10^{-3} + 2.8 \times 10^{-4}} = 3.0 \times 10^{-4} M$$

Fourth approximation:

$$S'''' = \frac{K_{sp}}{1.0 \times 10^{-3} + S'''} = \frac{3.8 \times 10^{-7}}{10 \times 10^{-3} + 3.0 \times 10^{-4}} = 2.9 \times 10^{-4} M$$

Fifth approximation:

$$S''''' = \frac{K_{sp}}{1.0 \times 10^{-3} + S''''} = \frac{3.8 \times 10^{-7}}{1.0 \times 10^{-3} + 2.9 \times 10^{-4}} = 2.9 \times 10^{-4} M$$

S''''' and S'''' are equal.

(c) From the kind of reasoning used in part b:

$$S = \frac{K_{sp}}{10^{-2} + S}$$

First approximation:

$$S' = \frac{K_{sp}}{1.0 \times 10^{-2}} = \frac{3.8 \times 10^{-7}}{1.0 \times 10^{-2}} = 3.8 \times 10^{-5} M$$

Second approximation:

$$S'' = \frac{K_{sp}}{1.0 \times 10^{-2} + S'} = \frac{3.8 \times 10^{-7}}{1.0 \times 10^{-2} + 3.8 \times 10^{-5}} = 3.8 \times 10^{-5} M$$

S' and S'' are equal.

(d) Again, from the sort of reasoning used in part b:

$$S = \frac{K_{sp}}{1.0 \times 10^{-1} + S}$$

First approximation:

$$S' = \frac{3.8 \times 10^{-7}}{1.0 \times 10^{-1}} = 3.8 \times 10^{-6} M$$

Second approximation:

$$S'' = \frac{K_{sp}}{1.0 \times 10^{-1} + S'} \quad \frac{3.8 \times 10^{-7}}{1.0 \times 10^{-1} + 3.8 \times 10^{-6}} = 3.8 \times 10^{-6} M$$

Again, S' and S'' converged in two approximations.

4. (a) Remember problem 2 (b)? From it,

$$S = \left(\frac{K_{sp}}{4}\right)^{1/3} = \left(\frac{1.31 \times 10^{-12}}{4}\right)^{1/3} = 6.89 \times 10^{-5} M$$

(b) $Ag_2CrO_4 \rightleftharpoons 2Ag^+ + CrO_4^{2-}$

$$S = \text{solubility}$$

$$[CrO_4^{2-}] = S$$

But $[Ag^+] \neq 2S$. Why?
There are two sources of silver ions, the Ag_2CrO_4 precipitate and the $AgNO_3$ solution.

$$(\text{concentration of } Ag^+ \text{ from } Ag_2CrO_4) = (Ag^+) \text{ ppt}$$

$$(Ag^+) \text{ ppt} = 2[CrO_4^{2-}] = 2S$$

$$(\text{concentration of } Ag^+ \text{ from } AgNO_3) = 0.0100M$$

$$[Ag^+] = (\text{concentration of } Ag^+ \text{ from } Ag_2CrO_4)$$
$$+ (\text{concentration of } Ag^+ \text{ from } AgNO_3)$$

$$= (Ag^+) \text{ ppt} + 0.0100 = 2S + 1.00 \times 10^{-2}$$

$$K_{sp} = [Ag^+]^2 \cdot [CrO_4^{2-}] = (2S + 1.00 \times 10^{-2})^2 \cdot (S)$$

$$S = \frac{K_{sp}}{(2S + 1.00 \times 10^{-2})^2}$$

First approximation:

$$S' = \frac{K_{sp}}{(1.00 \times 10^{-2})^2} = \frac{1.31 \times 10^{-12}}{(1.00 \times 10^{-2})^2} = 1.31 \times 10^{-8} M$$

Second approximation:

$$S'' = \frac{K_{sp}}{(2S' + 10^{-2})^2} = \frac{1.31 \times 10^{-12}}{[(2)(1.31 \times 10^{-8}) + 1.00 \times 10^{-2}]^2} = 1.31 \times 10^{-8} M$$

Note the rapid convergence of S' and S''.

(c) $Ag_2CrO_4 \rightleftharpoons 2Ag^+ + CrO_4^{2-}$

$$S = \text{solubility}$$
$$[Ag^+] = 2S$$

But $[CrO_4^{2-}] \neq S$. Why?

There are two sources of chromate ions, the Ag_2CrO_4 precipitate and the Na_2CrO_4 solution.

$$(\text{concentration of } CrO_4^{2-} \text{ from } Ag_2CrO_4) = (CrO_4^{2-}) \text{ ppt}$$

$$(CrO_4^{2-}) \text{ ppt} = \frac{1}{2}[Ag^+] = (\frac{1}{2})(2S) = S$$

$$(\text{concentration of } CrO_4^{2-} \text{ from } Na_2CrO_4) = 0.0100 M$$

$$[CrO_4^{2-}] = (\text{concentration of } CrO_4^{2-} \text{ from } Ag_2CrO_4)$$
$$+ (\text{concentration of } CrO_4^{2-} \text{ from } Na_2CrO_4)$$

$$= (CrO_4^{2-}) \text{ ppt} + 0.0100 = S + 1.00 \times 10^{-2}$$

$$K_{sp} = [Ag^+]^2 \cdot [CrO_4^{2-}] = (2S)^2 \cdot (S + 1.00 \times 10^{-2})$$

$$4S^2 = \frac{K_{sp}}{S + 1.00 \times 10^{-2}}$$

$$S = \left(\frac{K_{sp}}{4S + 4.00 \times 10^{-2}}\right)^{1/2}$$

First approximation:

$$S' = \left(\frac{K_{sp}}{4.00 \times 10^{-2}}\right)^{1/2} = \left(\frac{1.31 \times 10^{-12}}{4.00 \times 10^{-2}}\right)^{1/2} = 5.72 \times 10^{-6} M$$

Second approximation:

$$S'' = \left(\frac{K_{sp}}{4S' + 4.00 \times 10^{-2}}\right)^{1/2} = \left(\frac{1.31 \times 10^{-12}}{(4)(5.72 \times 10^{-6}) + 4.00 \times 10^{-2}}\right)^{1/2}$$

$$= 5.72 \times 10^{-6} M$$

Note how S' and S'' converge. Also note the tremendous effect a $10^{-2} M$ solution of $AgNO_3$ has, in comparison with the effect of $10^{-2} M$ Na_2CrO_4.

5. This is a common-ion problem similar to Problems 3 and 4.

$$PbSO_4 \rightleftharpoons Pb^{2+} + SO_4^{2-}$$
$$1 \rightleftharpoons 1 : 1$$

(a) If the solution initially contains 5.0×10^{-3} moles of Pb^{2+}, and 5.0×10^{-3} moles of H_2SO_4 are added, then 5.0×10^{-3} moles of SO_4^{2-} have been added and there is no excess Pb^{2+} or SO_4^{2-}. These ions will exist in solution only because of dissolution of the precipitate. At this point, there is no common-ion excess and the problem is one of simply calculating the solubility of $PbSO_4$.

$$K_{sp} = [Pb^{2+}] \cdot [SO_4^{2-}]$$
$$[Pb^{2+}] = S$$
$$[SO_4^{2-}] = S$$
$$K_{sp} = S^2$$
$$S = (2.0 \times 10^{-8})^{1/2} = 1.4 \times 10^{-4} M$$

We need to know the *amount* of lead remaining in solution. The above answer is a concentration (moles per liter). Since the solution volume is 100. milliliters, or 0.100 liter, the amount is 1.4×10^{-5} moles.

(b) When 5.5×10^{-3} moles have been added, there is an excess of 0.5×10^{-3} moles of the sulfate ion present in a volume of 0.100 liter

$$\frac{5 \times 10^{-4} \text{ mole}}{1.00 \times 10^{-1} \text{ liter}} = 5 \times 10^{-3} M = [SO_4^{2-}]$$

We make the approximation in the above calculation that

$$[SO_4^{2-}] = S + 0.005 \cong 0.005$$

This is based on an assumption that 5×10^{-3} is, in fact, substantially greater than S.

$$K_{sp} = [Pb^{2+}] \cdot [SO_4^{2-}] = S (0.005)$$

$$S = \frac{2.0 \times 10^{-8}}{5 \times 10^{-3}} = 4 \times 10^{-6} M$$

We may test the assumption.

$$5 \times 10^{-3} \gg 4 \times 10^{-6}$$

We need to know the amount of lead remaining in solution

$$4 \times 10^{-6} \frac{\text{mole}}{\text{liter}} \times 0.100 \text{ liter} = 4 \times 10^{-7} \text{ mole}$$

6. This problem is the basis of a separation by precipitation, which we call differential precipitation. It is based on the fact that, for a constant concentration of cation, the compound whose K_{sp} is numerically smallest will precipitate first, that is, it will require less anion to meet the numerical balance required by the solubility product constant.

(a) For $SrCO_3$,

$$K_{sp} = [Sr^{2+}][CO_3^{2-}]$$

$$[CO_3^{2-}] = \frac{K_{sp}}{[Sr^{2+}]} = \frac{4.1 \times 10^{-10}}{(1.0 \times 10^{-1})} = 4.1 \times 10^{-9}M$$

For $CaCO_3$,

$$K_{sp} = [Ca^{2+}][CO_3^{2-}]$$

$$[CO_3^{2-}] = \frac{K_{sp}}{[Ca^{2+}]} = \frac{7.9 \times 10^{-9}}{5.0 \times 10^{-2}} = 1.6 \times 10^{-7}M$$

$SrCO_3$ will precipitate first.

(b) We must first calculate the $[CO_3^{2-}]$ that will be free in solution when $CaCO_3$ begins to precipitate, since this will act as a common ion to the (already precipitated) $SrCO_3$, repressing its solubility and thus lowering the $[Sr^{2+}]$ in solution.

When $CaCO_3$ *first* begins to precipitate,

$$[CO_3^{2-}] = 1.6 \times 10^{-7}M$$

Knowing the $[CO_3^{2-}]$, we can solve the K_{sp} expression for $SrCO_3$ to determine how much $[Sr^{2+}]$ can exist in solution with this much CO_3^{2-} (i.e., a common-ion situation).

$$K_{sp} = [Sr^{2+}] \cdot [CO_3^{2-}]$$

$$4.1 \times 10^{-10} = [Sr^{2+}] \cdot (1.6 \times 10^{-7})$$

$$[Sr^{2+}] = \frac{4.1 \times 10^{-10}}{1.6 \times 10^{-7}} = 2.6 \times 10^{-3}M$$

Much of the original amount of Sr^{2+} is still in solution when the $CaCO_3$ precipitates, that is, the separation of the two cations is not complete. The solubility products are too close in numerical magnitude.

≡ 3.6 COMPLEX FORMATION WITH A LIGAND ≡

As everyone knows, AgCl will promptly dissolve in a concentrated solution of aqueous ammonia (designated $NH_3(aq)$ or "ammonium hydroxide, NH_4OH").* Thus the study of the complexation of a metal with a ligand that is not the anion of the precipitate is well worth investigation.

The term *ligand* has been used here. It would be good to give examples of ligands, which may be either anions or molecular species.

The ligand is NH_3, a neutral species:

$$\underset{\text{metal ion}}{Ag^+} + \underset{\text{ligand}}{NH_3} \rightleftharpoons \underset{\text{complex}}{AgNH_3^+}$$

---

* *It is not very smart to dissolve any silver salt in NH_3 and allow the solution to dry, for silver azide, formerly known as "fulminating silver," may be formed thereby and may explode with some violence.*

$$AgNH_3^+ + NH_3 \rightleftharpoons Ag(NH_3)_2^+$$

$$\text{complex} \quad\quad \text{ligand} \quad\quad \text{complex}$$

The ligand is Cl^-, an anion:

$$Fe^{3+} + Cl^- \rightleftharpoons FeCl^{2+}$$

$$\text{metal ion} \quad \text{ligand} \quad \text{complex}$$

$$FeCl^{2+} + Cl^- \rightleftharpoons FeCl_2^+$$

$$\text{complex} \quad\quad \text{ligand} \quad \text{complex}$$

$$FeCl_2^+ + Cl^- \rightleftharpoons FeCl_3$$

$$\text{complex} \quad\quad \text{ligand} \quad \text{complex}$$

$$FeCl_3 + Cl^- \rightleftharpoons FeCl_4^-$$

$$\text{complex} \quad \text{ligand} \quad \text{complex}$$

If ligands compete with precipitate anions for the metal ions of the precipitate, there is bound to be an enhancement of solubility. How much enhancement? This depends upon the K_{sp} of the precipitate, the equilibrium concentration of the ligand, and the values of equilibrium constants for the successive complexation reactions. It does not appear to be an easy question to answer, but a careful, systematic approach involving some rather simple fractions designated as β_0 will make answering "how much enhancement?" straightforward.

In order to treat the enhancement of solubility of ligands, let us take the imaginary species MA. Suppose that $MA_{(solution)}$, and such complexes as MA_2^-, MA_3^{2-}, and so forth, do not exist. Suppose, instead, that ligand X can form a series of complex ions with the M^+ cation. Here is how the reactions might appear (k_1, k_2, and so forth, are called successive formation constants).

$$MA \rightleftharpoons M^+ + A^- \qquad\qquad K_{sp} = [M^+] \cdot [A^-] \qquad (3.1)$$

$$M^+ + X \rightleftharpoons MX^+ \qquad\qquad k_1 = \frac{[MX^+]}{[M^+] \cdot [X]} \qquad (3.2)$$

$$MX^+ + X \rightleftharpoons MX_2^+ \qquad\qquad k_2 = \frac{[MX_2^+]}{[MX^+] \cdot [X]} \qquad (3.3)$$

$$MX_2^+ + X \rightleftharpoons MX_3^+ \qquad\qquad k_3 = \frac{[MX_3^+]}{[MX_2^+] \cdot [X]} \qquad (3.4)$$

$$\vdots \qquad\qquad\qquad\qquad \vdots$$

$$MX_{n-1}^+ + X \rightleftharpoons MX_n^+ \qquad\qquad k_n = \frac{[MX_n^+]}{[MX_{n-1}^+] \cdot [X]} \qquad (3.5)$$

The solubility of the precipitate is easily defined. Because the precipitate is one-to-one, and because of the absence of surplus A^- or higher M^+ - A^- complexes, the solubility is simply equal to the equilibrium concentration of A^-.

$$S = [A^-] \qquad (3.6)$$

To write $S = [M^+]$ is *not* correct. There are many species containing M^+ in this solution. The solubility of the precipitate is equal to the sum of the concentrations of all the M-bearing species in solution.

$$S = [M^+] + [MX^+] + [MX_2^+] + [MX_3^+] + \cdots + [MX_n^+] \qquad (3.7)$$

Conveniently, we can also define the analytical concentration of M, C_M, as the sum of the concentrations of all the M-bearing species in solution:

$$C_M = [M^+] + [MX^+] + [MX_2^+] + [MX_3^+] + \cdots + [MX_n^+] \qquad (3.8)$$

$$S = C_M \qquad (3.9)$$

We also define a new term, β_0, which will come in very handy.

$$\beta_0 = \frac{[M^+]}{C_M} \qquad (3.10)$$

In this case,

$$\beta_0 = \frac{[M^+]}{S} \qquad (3.11)$$

We rearrange Equation 3.11.

$$[M^+] = \beta_0 \cdot S \qquad (3.12)$$

Then we substitute Equations 3.6 and 3.12 into Equation 3.1.

$$K_{sp} = (\beta_0 S)(S) = \beta_0 S^2$$

$$S = \left(\frac{K_{sp}}{\beta_0}\right)^{1/2} \qquad (3.13)$$

There is only one catch to this argument: β_0's numerical value is not given, nor has any way to evaluate it been given. To solve this problem, we recall Equation 3.10:

$$\beta_0 = \frac{[M^+]}{C_M} \qquad (3.10)$$

and substitute from Equation 3.8:

$$\beta_0 = \frac{[M^+]}{[M^+] + [MX^+] + [MX_2^+] + [MX_3^+] + \cdots + [MX_n^+]} \qquad (3.14)$$

To evaluate β_0, if $[X]$ and k_1 through k_n are known, is not hard to do. All terms in the expression must be expressed as functions of $[M^+]$, and/or $[X]$ and k_1 through k_n. The term $[M^+]$ needs no further work. We consider $[MX^+]$ and rearrange Equation 3.2.

$$[MX^+] = k_1 [M^+] [X] \qquad (3.15)$$

Next, turning to $[MX_2^+]$, we rearrange Equation 3.3.

$$[MX_2^+] = k_2 [MX^+] [X]$$

and substitute Equation 3.15.

$$[MX_2^+] = k_1 k_2 [M^+] [X]^2 \qquad (3.16)$$

Then we derive a term for $[MX_3^+]$. Rearrange Equation 3.4.

$$[MX_3^+] = k_3 [MX_2^+] [X]$$

We substitute Equation 3.16 so that

$$[MX_3^+] = k_1 k_2 k_3 [M^+] [X]^3 \qquad (3.17)$$

By similar reasoning,

$$[MX_n^+] = k_1 k_2 k_3 \cdots k_n [M^+] [X]^n \qquad (3.18)$$

We then substitute Equations 3.15 through 3.18 into Equation 3.14.

$$\beta_0 = \frac{[M^+]}{[M^+] + k_1 \cdot [M^+] \cdot [X] + k_1 k_2 \cdot [M^+] \cdot [X]^2 + k_1 k_2 k_3 \cdot [M^+] \cdot [X]^3 + \cdots + k_1 k_2 k_3 \cdots k_n [M^+] \cdot [X]^n}$$

$$\beta_0 = \frac{\cancel{[M^+]}}{\cancel{[M^+]}(1 + k_1[X] + k_1 k_2 [X]^2 + k_1 k_2 k_3 [X]^3 + \cdots + k_1 k_2 k_3 \cdots k_n [X]^n)}$$

$$\beta_0 = \frac{1}{1 + k_1[X] + k_1 k_2 [X]^2 + k_1 k_2 k_3 [X]^3 + \cdots + k_1 k_2 k_3 \cdots k_n [X]^n} \qquad (3.19)$$

Thus, if $[X]$ and k_1 through k_n are known, β_0 can be evaluated and used in Equation 3.13.

$$S = \left(\frac{K_{sp}}{\beta_0} \right)^{1/2} \qquad (3.13)$$

A numerical example would prove useful here. Consider AgBr, a precipitate less soluble than AgCl. The important equilibria are:

$$AgBr_{(solid)} \rightleftharpoons Ag^+ + Br^- \qquad K_{sp} = 5.25 \times 10^{-13}$$
$$Ag^+ + NH_3 \rightleftharpoons AgNH_3^+ \qquad k_1 = 1585$$
$$AgNH_3^+ + NH_3 \rightleftharpoons Ag(NH_3)_2^+ \qquad k_2 = 6761$$

What is the solubility of AgBr in a solution where the equilibrium concentration of NH_3 is $0.100M$?

Equations 3.1 through 3.19 can be applied here. To begin with, we define solubility, S.

$$S = \text{solubility}$$
$$S = [Br^-]$$
$$S = [Ag^+] + [AgNH_3^+] + [Ag(NH_3)_2^+]$$
$$S = C_{Ag}$$
$$\beta_0 = \frac{[Ag^+]}{C_{Ag}}$$
$$\beta_0 = \frac{[Ag^+]}{S}$$
$$[Ag^+] = \beta_0 S$$
$$K_{sp} = [Ag^+][Br^-] = (\beta_0 S)(S) = \beta_0 \cdot S^2$$
$$S = \left(\frac{K_{sp}}{\beta_0} \right)^{1/2} \qquad (3.13)$$

Then β_0 is evaluated.

$$\beta_0 = \frac{1}{1 + k_1[NH_3] + k_1 k_2 [NH_3]^2}$$
$$= \frac{1}{1 + (1585)(1.00 \times 10^{-1}) + (1585)(6761)(1.00 \times 10^{-1})^2}$$

$$= \frac{1}{1 + 158 + 1.07 \times 10^5} = \frac{1}{1.07 \times 10^5} = 9.32 \times 10^{-6}$$

or

$$\frac{1}{\beta_0} = 1.07 \times 10^5$$

$$S = \left(\frac{K_{sp}}{\beta_0}\right)^{1/2} = [K_{sp}(1/\beta_0)]^{1/2}$$

$$= [(5.25 \times 10^{-13})(1.07 \times 10^5)]^{1/2} = 2.37 \times 10^{-4}M$$

≡ 3.7 PROBLEMS ≡

(Use Tables 3.1 and 3.2 for K_{sp} and formation constant values.)

1. Calculate the solubility of AgI in a solution where $[NH_3] = 0.010M$.
   Ans: $S = 3.1 \times 10^{-7}M$

2. Calculate the solubility of AgI in a solution initially $0.01M$ KCN.
   Ans: $S = 0.005M$

3. Calculate the solubility of Ag_2CrO_4 in a solution where $[NH_3] = 0.010M$;
   Ans: $S = 7.3 \times 10^{-3}M$

≡ 3.8 DETAILED SOLUTIONS TO PROBLEMS ≡

1. Here, $S = \left(\dfrac{K_{sp}}{\beta_0}\right)^{1/2}$ (3.13)

   From Equation 3.14,

   $$\beta_0 = \frac{1}{1 + k_1[NH_3] + k_1k_2[NH_3]^2}$$

   $$= \frac{1}{1 + (1585)(1.0 \times 10^{-2}) + (1585)(6761)(1.0 \times 10^{-2})^2} = \frac{1}{1.1 \times 10^3}$$

   $$\frac{1}{\beta_0} = 1.1 \times 10^3$$

   $$S = [K_{sp}\,\frac{1}{\beta_0}]^{1/2} = [(9.0 \times 10^{-17})(1.1 \times 10^3)]^{1/2} = 3.1 \times 10^{-7}M$$

2. Here we find that CN^- does a better job than NH_3. Equation 3.13 applies

   $$S = \left(\frac{K_{sp}}{\beta_0}\right)^{1/2}$$ (3.13)

   From Equation 3.14,

   $$\beta_0 = \frac{1}{1 + k_1[CN^-] + k_1k_2[CN^-]^2}$$

   $$= \frac{1}{1 + (1)(0.01) + (1)(7.94 \times 10^{20})(0.01)^2} = \frac{1}{8 \times 10^{16}}$$

$$\frac{1}{\beta_0} = 8 \times 10^{16}$$

$$S = K_{sp}\left(\frac{1}{\beta_0}\right)^{1/2} = [(9 \times 10^{-17})(8 \times 10^{16})]^{1/2} = 2.7M$$

This is impossible, because the solubility is greater than the initial KCN concentration. Try a stoichiometric approach. The predominant reaction is

$$Ag^+ + 2CN^- \rightleftharpoons Ag(CN)_2^- \quad k_2 = 7.94 \times 10^{20}$$

With a k_2-value this high, *all* the CN$^-$ complexes silver. The stoichiometry is

$$1 \quad : \quad 2 \quad \theta \quad 1$$
$$0.005 \quad : \quad 0.01 \quad \theta \quad 0.005$$

The solubility is thus about 0.005M.

3. Here, $S \neq (K_{sp}/\beta_0)^{1/2}$. Why does Equation 3.13 not apply?

$$Ag_2CrO_4 \rightleftharpoons 2Ag^+ + CrO_4^{2-} \quad K_{sp} = [Ag^+]^2 \cdot [CrO_4^{2-}]$$

$$S = \text{solubility}$$
$$[CrO_4^{2-}] = S$$
$$C_{Ag} = 2S$$
$$\frac{[Ag^+]}{C_{Ag}} = \beta_0$$
$$[Ag^+] = \beta_0 \cdot C_{Ag}$$
$$[Ag^+] = 2 \cdot \beta_0 \cdot S$$
$$K_{sp} = [Ag^+]^2[CrO_4^{2-}] = (2\beta_0 S)^2 \cdot S = 4\beta_0^2 S^3$$

$$S = \left(\frac{K_{sp}}{4\beta_0^2}\right)^{1/3}$$

Where $[NH_3] = 0.010M$, $1/\beta_0 = 1.1 \times 10^3$ (from Problem 1).

$$S = \left[\frac{1.31 \times 10^{-12}}{4}(1.1 \times 10^3)^2\right]^{1/3} = 7.3 \times 10^{-3}M$$

Table 3.2
SUCCESSIVE FORMATION CONSTANTS FOR
METAL-LIGAND COMPLEXATION

Complexation	Constant
$Ag^+ + Cl^- \rightleftharpoons AgCl_{(solution)}$	$k_1 = 501$
$AgCl_{(solution)} + Cl^- \rightleftharpoons AgCl_2^-$	$k_2 = 63.9$
$AgCl_2^- + Cl^- \rightleftharpoons AgCl_3^{2-}$	$k_3 = 2.00$
$Ag^+ + S_2O_3^{2-} \rightleftharpoons AgS_2O_3^-$	$k_1 = 6.6 \times 10^8$
$AgS_2O_3^- + S_2O_3^{2-} \rightleftharpoons Ag(S_2O_3)_2^{3-}$	$k_2 = 4.4 \times 10^4$
$Ag(S_2O_3)_2^{3-} + S_2O_3^{2-} \rightleftharpoons Ag(S_2O_3)_3^{5-}$	$k_3 = 4.9$
$Ag^+ + CN^- \rightleftharpoons AgCN_{(solution)}$	$k_1 = 1$
$AgCN_{(solution)} + CN^- \rightleftharpoons Ag(CN)_2^-$	$k_2 = 7.94 \times 10^{20}$
$Ag^+ + NH_3 \rightleftharpoons AgNH_3^+$	$k_1 = 1585$

Table 3.2 continued

Complexation	Constant
$AgNH_3^+ + NH_3 \rightleftharpoons Ag(NH_3)_2^+$	$k_2 = 6761$
$Cd^{2+} + NH_3 \rightleftharpoons CdNH_3^{2+}$	$k_1 = 4.47 \times 10^2$
$CdNH_3^{2+} + NH_3 \rightleftharpoons Cd(NH_3)_2^{2+}$	$k_2 = 1.26 \times 10^2$
$Cd(NH_3)_2^{2+} + NH_3 \rightleftharpoons Cd(NH_3)_3^{2+}$	$k_3 = 2.75 \times 10^1$
$Cd(NH_3)_3^{2+} + NH_3 \rightleftharpoons Cd(NH_3)_4^{2+}$	$k_4 = 8.51$
$Cd(NH_3)_4^{2+} + NH_3 \rightleftharpoons Cd(NH_3)_5^{2+}$	$k_5 = 4.79 \times 10^{-1}$
$Cd(NH_3)_5^{2+} + NH_3 \rightleftharpoons Cd(NH_3)_6^{2+}$	$k_6 = 2.19 \times 10^{-2}$

≡ 3.9 FORMATION OF A WEAK ACID WITH THE ANION OF THE PRECIPITATE ≡

The means of solubility control discussed so far have included: (1) the simple common-ion effect, and (2) the complexation of the metal ion by a ligand. Can solubility be controlled through chemical reactions—other than precipitation—of the anion? The answer is yes if the anion is the anion of a weak acid. A weak acid is one that does not give up its proton, or protons, easily. Oxalic acid is such an acid, and the solubility of calcium oxalate, one of its salts, can be controlled through the manipulation of acidity.

The dissociation constants for oxalic acid immediately indicate that this is indeed a weak acid.

$$H_2C_2O_4 \rightleftharpoons H^+ + HC_2O_4^- \qquad K_1 = \frac{[H^+][HC_2O_4^-]}{[H_2C_2O_4]} = 5.6 \times 10^{-2}$$

$$HC_2O_4^- \rightleftharpoons H^+ + C_2O_4^{2-} \qquad K_2 = \frac{[H^+][C_2O_4^{2-}]}{[HC_2O_4^-]} = 6.2 \times 10^{-5}$$

In a system where solid CaC_2O_4 and hydrogen ions are present, the following competing equilibra would describe the dissolution of the precipitate.

The dissolution of the solid:

$$CaC_2O_4 \rightleftharpoons Ca^{2+} + C_2O_4^{2-} \qquad K_{sp} = 2.6 \times 10^{-9}$$

$$K_{sp} = [Ca^{2+}][C_2O_4^{2-}]$$

The protonation of $C_2O_4^{2-}$:

$$H^+ + C_2O_4^{2-} \rightleftharpoons HC_2O_4^- \qquad \frac{1}{K_2} = \frac{[HC_2O_4^-]}{[H^+][C_2O_4^{2-}]} = \frac{1}{6.2 \times 10^{-5}} = 1.6 \times 10^4$$

The protonation of $HC_2O_4^-$:

$$H^+ + HC_2O_4^- \rightleftharpoons H_2C_2O_4 \qquad \frac{1}{K_1} = \frac{[H_2C_2O_4]}{[H^+][HC_2O_4^-]} = \frac{1}{5.6 \times 10^{-2}} = 1.8 \times 10^1$$

By Le Chatelier's principle, as oxalate anions are protonated to form hydrogen oxalate (or binoxalate), $HC_2O_4^-$, and oxalic acid, $H_2C_2O_4$, more of the precipitate must dissolve. The greater the concentration of the hydrogen ions, the greater the solubility of the precipitate becomes.

This is a qualitative picture. How can solubility be calculated quantitatively? The calculation is best done with the aid of the α-functions studied in weak acid chemistry. Let us consider the solubility of CaC_2O_4 and examine the relationship it bears to the α-functions.

$$CaC_2O_4 \rightleftharpoons Ca^{2+} + C_2O_4^{2-} \qquad K_{sp} = [Ca^{2+}][C_2O_4^{2-}]$$

S = solubility

Because nothing is complexing the Ca^{2+},

$$[Ca^{2+}] = S \tag{3.20}$$

It is a mistake to say that

$$[C_2O_4^{2-}] = S$$

The truth is that

$$[C_2O_4^{2-}] \neq S$$

Because there are three oxalate-bearing species in solution, solubility is given by

$$S = [C_2O_4^{2-}] + [HC_2O_4^-] + [H_2C_2O_4] \tag{3.21}$$
$$C_{ox} = [C_2O_4^{2-}] + [HC_2O_4^-] + [H_2C_2O_4] \tag{3.22}$$
$$S = C_{ox} \tag{3.23}$$

The useful α-function here is α_2. For oxalic acid, from Eq. 2.33,

$$\alpha_2 = \frac{[C_2O_4^{2-}]}{[H_2C_2O_4] + [HC_2O_4^-] + [C_2O_4^{2-}]} = \frac{[C_2O_4^{2-}]}{C_{ox}} \tag{2.33}$$

or

$$\alpha_2 = \frac{[C_2O_4^{2-}]}{C_{ox}}$$

We rearrange Eq. 2.33

$$[C_2O_4^{2-}] = \alpha_2 \cdot C_{ox}$$

and substitute from Eq. 3.23.

$$[C_2O_4^{2-}] = \alpha_2 \cdot S \tag{3.24}$$

We write the expression for K_{sp}.

$$K_{sp} = [Ca^{2+}][C_2O_4^{2-}]$$

and then substitute from Eqs. 3.20 and 3.24

$$K_{sp} = (S)(\alpha_2 \cdot S) = \alpha_2 S^2$$

$$S = \left(\frac{K_{sp}}{\alpha_2}\right)^{1/2} \tag{3.25}$$

The only problem that remains is to express α_2 in terms of $[H^+]$, K_1, and K_2. This is easily done with Eq. 2.36.

$$\alpha_2 = \frac{K_1 K_2}{[H^+]^2 + [H^+]K_1 + K_1 K_2} \tag{2.36}$$

The equilibrium concentration of $[H^+]$ is often known. The pH is measured and defined as a logarithmic function of $[H^+]$.

$$pH = -\log [H^+] \tag{2.2}$$

Thus, if pH is known, and K_1, K_2, and K_{sp} can be found in a table somewhere, it is easy to calculate the solubility of a precipitate such as CaC_2O_4.

Example: Calculate the solubility of CaC_2O_4 in solutions where pH = 7.00, pH = 5.00, and pH = 3.00.

$$K_{sp} = 2.6 \times 10^{-9}$$

at pH 7, $[H^+] = 1.00 \times 10^{-7} M$

$$S = \left(\frac{K_{sp}}{\alpha_2}\right)^{1/2} \tag{3.25}$$

You can either look up α_2 from Table 2.5 or calculate it using K_1 and K_2 values from Table 2.7.

From Table 2.5 at pH 7.00,

$$\alpha_2 = 1.0$$

we substitute into Eq. 3.25

$$S = \left(\frac{2.6 \times 10^{-9}}{1.0}\right)^{1/2} = 5.1 \times 10^{-5} M$$

At pH 5.00, Eq. 3.25 still applies.

$$S = \left(\frac{K_{sp}}{\alpha_2}\right)^{1/2} \tag{3.25}$$

From Table 2.5 at pH 5.00,

$$\alpha_2 = 0.86$$

We substitute into Eq. 3.25

$$S = \left(\frac{2.6 \times 10^{-9}}{0.86}\right)^{1/2} = 5.5 \times 10^{-5} M$$

At pH 3.00, Eq. 3.25 still applies.

$$S = \left(\frac{K_{sp}}{\alpha_2}\right)^{1/2} \tag{3.25}$$

From Table 2.5, at pH 3.00,

$$\alpha_2 = 0.057$$

$$S = \left(\frac{2.6 \times 10^{-9}}{5.7 \times 10^{-2}}\right)^{1/2} = 2.1 \times 10^{-4} M$$

Thus there is a considerable change in solubility with varying pH.
There are, to be sure, cases in which Eq. 3.25 does not work.

$$S \neq \left(\frac{K_{sp}}{\alpha_2}\right)^{1/2}$$

Let us consider three such cases.

1. Ag₂S.

$$Ag_2S \rightleftharpoons 2Ag^+ + S^{2-} \qquad K_{sp} = [Ag^+]^2 \cdot [S^{2-}]$$

$$H_2S \rightleftharpoons H^+ + HS^- \qquad K_1 = \frac{[H^+][HS^-]}{[H_2S]}$$

$$HS^- \rightleftharpoons H^+ + S^{2-} \qquad K_2 = \frac{[H^+][S^{2-}]}{[HS^-]}$$

S = solubility

$$[Ag^+] = 2S$$

$$S \neq [S^{2-}]$$

$$S = [S^{2-}] + [HS^-] + [H_2S]$$

$$C_S = [S^{2-}] + [HS^-] + [H_2S]$$

$$S = C_S$$

$$\alpha_2 = \frac{[S^{2-}]}{C_S}$$

$$\alpha_2 = \frac{[S^{2-}]}{S}$$

$$[S^{2-}] = \alpha_2 \cdot S$$

$$K_{sp} = [Ag^+]^2 \cdot [S^{2-}] = (2S)^2 \cdot (\alpha_2 S) = 4\,\alpha_2 S^3$$

$$S = \left(\frac{K_{sp}}{4\,\alpha_2}\right)^{1/3}$$

As before,

$$\alpha_2 = \frac{K_1 K_2}{[H^+]^2 + [H^+]K_1 + K_1 K_2}$$

2. CaF₂.

$$CaF_2 \rightleftharpoons Ca^{2+} + 2F^- \qquad K_{sp} = [Ca^{2+}][F^-]^2$$

$$HF \rightleftharpoons H^+ + F^- \qquad K_a = \frac{[H^+][F^-]}{[HF]}$$

Because HF is a monoprotic acid, the α-function used will differ from the α_2 functions previously used.

$$S = \text{solubility}$$

$$S = [Ca^{2+}]$$

$$[F^-] \neq 2S$$

$$2S = [HF] + [F^-]$$

$$C_F = [HF] + [F^-]$$

$$2S = C_F$$

After Eq. 2.28,

$$\alpha_1 = \frac{[F^-]}{C_F}$$

Then

$$[F^-] = \alpha_1 C_F$$

$$[F^-] = \alpha_1 \cdot 2 \cdot S$$

$$K_{sp} = [Ca^{2+}][F^-]^2 = (S)(2\alpha_1 S)^2 = 4\alpha_1^2 S^3$$

$$S = \left(\frac{K_{sp}}{4\alpha_1^2}\right)^{1/3}$$

Equation 2.30 gives α_1 as a function of $[H^+]$ and K_a. Thus α_1 can be easily evaluated if K_a and pH are known.

3. Ag_3PO_4.

$$Ag_3PO_4 \rightleftharpoons 3Ag^+ + PO_4^{3-} \qquad K_{sp} = [Ag^+]^3 \cdot [PO_4^{3-}]$$

$$H_3PO_4 \rightleftharpoons H^+ + H_2PO_4^- \qquad K_1 = \frac{[H^+][H_2PO_4^-]}{[H_3PO_4]}$$

$$H_2PO_4^- \rightleftharpoons H^+ + HPO_4^{2-} \qquad K_2 = \frac{[H^+][HPO_4^{2-}]}{[H_2PO_4^-]}$$

$$HPO_4^{2-} \rightleftharpoons H^+ + PO_4^{3-} \qquad K_3 = \frac{[H^+][PO_4^{3-}]}{[HPO_4^{2-}]}$$

Because H_3PO_4 is a triprotic acid, the α-functions used will differ from the α_2 and α_1 functions used for the salts of di- and monoprotic acids.

$$S = \text{solubility}$$

$$3S = [Ag^+]$$

$$[PO_4^{3-}] \neq S$$

$$S = [H_3PO_4] + [H_2PO_4^-] + [HPO_4^{2-}] + [PO_4^{3-}]$$

$$C_{PO_4} = [H_3PO_4] + [H_2PO_4^-] + [HPO_4^{2-}] + [PO_4^{3-}]$$

$$S = C_{PO_4}$$

After Eq. 2.40,

$$\alpha_3 = \frac{[PO_4^{3-}]}{C_{PO_4}}$$

Then

$$[PO_4^{3-}] = \alpha_3 \cdot C_{PO_4} = \alpha_3 \cdot S$$

$$K_{sp} = [Ag^+]^3 \cdot [PO_4^{3-}] = (3S)^3 \cdot (\alpha_3 \cdot S) = 27\alpha_3 S^4$$

$$S = \left(\frac{K_{sp}}{27\alpha_3}\right)^{1/4}$$

Then α_3 is evaluated numerically by Eq. 2.40.

$$\alpha_3 = \frac{K_1 K_2 K_3}{[H^+]^3 + [H^+]^2 K_1 + [H^+] K_1 K_2 + K_1 K_2 K_3} \tag{2.40}$$

≡ 3.10 PROBLEMS ≡

(Use Tables 3.1 and 3.2 for K_{sp} and formation constant values. Tables 2.1, 2.2, 2.5, 2.7, 2.8, and 2.9 are used for K_a, K_b, and α-values.)

1. Calculate the solubility of strontium carbonate, $SrCO_3$, in a solution where pH is found to be 5.00. Remember that H_2CO_3 is a weak acid.
   Ans: $2.3 \times 10^{-2} M$

2. Calculate the solubility of silver oxalate, $Ag_2C_2O_4$, in a solution where pH is found to be 5.00.
   Ans: $2.2 \times 10^{-4} M$

3. Calculate the solubility of strontium fluoride, SrF_2, in a solution where pH is found to be 4.00.
   Ans: $9.7 \times 10^{-4} M$

4. Calculate the solubility of bismuth phosphate, $BiPO_4$, in a solution whose pH is found to be 12.00. Assume that no $Bi(OH)_3$ forms.
   Ans: $6.5 \times 10^{-12} M$

5. Calculate the solubility of calcium phosphate, $Ca_3(PO_4)_2$, in a solution where pH is found to be 12.00.
   Ans: $1.1 \times 10^{-6} M$

6. Calculate the solubility of cadmium sulfide, CdS, in a solution where $[NH_3] = 0.100M$, and where pH is observed to be 11.10.
   Ans: $4.3 \times 10^{-10} M$

7. Calculate the solubility of silver sulfide, Ag_2S, in a solution where $[NH_3] = 0.100M$, and where pH is observed to be 11.10.
   Ans: $3.4 \times 10^{-12} M$

≡ 3.11 DETAILED SOLUTIONS TO PROBLEMS ≡

1. $SrCO_3 \rightleftharpoons Sr^{2+} + CO_3^{2-}$ $K_{sp} = [Sr^{2+}] \cdot [CO_3^{2-}]$
   S = solubility

Because nothing is complexing the Sr^{2+},

$$[Sr^{2+}] = S$$

$$S = [CO_3^{2-}] + [HCO_3^-] + [H_2CO_3]$$

$$C_{carb} = [CO_3^{2-}] + [HCO_3^-] + [H_2CO_3]$$

$$S = C_{carb}$$

$$\alpha_2 = \frac{[CO_3^{2-}]}{[H_2CO_3] + [HCO_3^-] + [CO_3^{2-}]}$$

$$\alpha_2 = \frac{[CO_3^{2-}]}{C_{carb}}$$

$$[CO_3^{2-}] = \alpha_2 \cdot C_{carb} = \alpha_2 \cdot S$$

$$K_{sp} = [Sr^{2+}] \cdot [CO_3^{2-}] = (S) \cdot (\alpha_2 S) = \alpha_2 S^2$$

$$S = \left(\frac{K_{sp}}{\alpha_2}\right)^{1/2} \qquad (3.25)$$

From Table 2.9, at pH 5.00,

$$\alpha_2 = 2.0 \times 10^{-7}$$

we substitute into Equation 3.25,

$$S = \left(\frac{1.1 \times 10^{-10}}{2.0 \times 10^{-7}}\right)^{1/2} = 2.3 \times 10^{-2} M$$

2. $Ag_2C_2O_4 \rightleftharpoons 2Ag^+ + C_2O_4^{2-}$ $K_{sp} = [Ag^+]^2 \cdot [C_2O_4^{2-}]$

S = solubility

Equation 3.25 will not give the right answer. Because nothing is complexing the Ag^+, and because 2 moles of Ag^+ are formed when 1 mole of $Ag_2C_2O_4$ dissolves,

$$[Ag^+] = 2S$$

$$S = [C_2O_4^{2-}] + [HC_2O_4^-] + [H_2C_2O_4]$$

$$C_{ox} = [C_2O_4^{2-}] + [HC_2O_4^-] + [H_2C_2O_4]$$

$$S = C_{ox}$$

$$\alpha_2 = \frac{[C_2O_4^{2-}]}{C_{ox}}$$

$$[C_2O_4^{2-}] = \alpha_2 C_{ox} = \alpha_2 S$$

$$K_{sp} = [Ag^+]^2 \cdot [C_2O_4^{2-}] = (2S)^2 \cdot (\alpha_2 S) = 4\alpha_2 S^3$$

$$S = \left(\frac{K_{sp}}{4\alpha_2}\right)^{1/3}$$

From Table 2.5, at pH 5.00,

$$\alpha_2 = 0.86$$

$$S = \left(\frac{K_{sp}}{4\alpha_2}\right)^{1/3} = \left(\frac{3.5 \times 10^{-11}}{(4)(0.86)}\right)^{1/3} = 2.2 \times 10^{-4} M$$

3. $SrF_2 \rightleftharpoons Sr^{2+} + 2F^-$ $K_{sp} = [Sr^{2+}] \cdot [F^-]^2$

S = solubility

Equation 3.25 will not give the right answer. Because nothing is complexing the Sr^{2+},

$$[Sr^{2+}] = S$$

Because there are 2 moles of some fluoride-containing species for every mole of SrF_2 that is dissolved,

$$2S = [F^-] + [HF]$$

$$C_F = [F^-] + [HF]$$

$$2S = C_F$$

$$\alpha_1 = \frac{[F^-]}{C_F}$$

$$[F^-] = \alpha_1 C_F = 2\alpha_1 S$$

$$K_{sp} = [Sr^{2+}] \cdot [F^-]^2 = (S) \cdot (2\alpha_1 S)^2 = 4\alpha_1^2 S^3$$

$$S = \left(\frac{K_{sp}}{4\alpha_1^2}\right)^{1/3}$$

We evaluate α_1: $\alpha_1 = \dfrac{K_a}{K_a + [H^+]}$ $\qquad\qquad\qquad$ (2.30)

at pH 4.00, $[H^+] = 1.00 \times 10^{-4} M$.

$$\alpha_1 = \frac{6.7 \times 10^{-4}}{6.7 \times 10^{-4} + 1.00 \times 10^{-4}} = 0.87$$

Substituting into the expression for solubility, we obtain

$$S = \left(\frac{2.8 \times 10^{-9}}{(4) \cdot (.870)^2}\right)^{1/3} = 9.7 \times 10^{-4} M$$

4. $BiPO_4 \rightleftharpoons Bi^{3+} + PO_4^{3-}$ $K_{sp} = [Bi^{3+}] \cdot [PO_4^{3-}]$

S = solubility

Because nothing is complexing the Bi^{3+},

$$[Bi^{3+}] = S$$

$$S = [PO_4^{3-}] + [HPO_4^{2-}] + [H_2PO_4^-] + [H_3PO_4]$$

$$C_{PO_4} = [PO_4^{3-}] + [HPO_4^{2-}] + [H_2PO_4^-] + [H_3PO_4]$$

$$S = C_{PO_4}$$

$$\alpha_3 = \frac{[PO_4^{3-}]}{C_{PO_4}} \qquad (2.40)$$

$$[PO_4^{3-}] = \alpha_3 \cdot C_{PO_4}$$

$$[PO_4^{3-}] = \alpha_3 \cdot S$$

$$K_{sp} = [Bi^{3+}] \cdot [PO_4^{3-}]$$

$$K_{sp} = (S) \cdot (\alpha_3 \cdot S)$$

$$K_{sp} = \alpha_3 S^2$$

$$S = \left(\frac{K_{sp}}{\alpha_3}\right)^{1/2}$$

At pH 12.00, from Table 2.1,

$$\alpha_3 = 0.31$$

$$S = \left(\frac{1.3 \times 10^{-23}}{0.31}\right)^{1/2} = 6.5 \times 10^{-12}M$$

5. $Ca_3(PO_4)_2 \rightleftharpoons 3Ca^{2+} + 2PO_4^{3-}$ $\qquad K_{sp} = [Ca^{2+}]^3 \cdot [PO_4^{3-}]^2$

S = solubility

Because nothing is complexing the Ca^{2+}, and because 3 moles of Ca^{2+} are formed when 1 mole of $Ca_3(PO_4)_2$ dissolves,

$$[Ca^{2+}] = 3S$$

Because there are 2 moles of phosphate-containing species for every mole of $Ca_3(PO_4)_2$ that is dissolved,

$$2S = [PO_4^{3-}] + [HPO_4^{2-}] + [H_2PO_4^-] + [H_3PO_4]$$

$$C_{PO_4} = [PO_4^{3-}] + [HPO_4^{2-}] + [H_2PO_4^-] + [H_3PO_4]$$

$$2S = C_{PO_4}$$

$$\alpha_3 = \frac{[PO_4^{3-}]}{C_{PO_4}} \qquad (2.40)$$

$$[PO_4^{3-}] = \alpha_3 C_{PO_4} = 2\alpha_3 S$$

$$K_{sp} = [Ca^{2+}]^3 \cdot [PO_4^{3-}]^2 = (3S)^3 \cdot (2\alpha_3 S)^2 = 108\alpha_3^2 S^5$$

$$S = \left(\frac{K_{sp}}{108\alpha_3^2}\right)^{1/5}$$

From Problem 4, at pH 12.00,

$$\alpha_3 = 0.31$$

Then we substitute into the expression for solubility,

$$S = \left(\frac{2.0 \times 10^{-29}}{(108)(0.31)^2}\right)^{1/5} = 1.1 \times 10^{-6}M$$

6. Let S = solubility. First, we express $[Cd^{2+}]$ in terms of solubility.

$$S = [Cd^{2+}] + [CdNH_3^{2+}] + [Cd(NH_3)_2^{2+}] + [Cd(NH_3)_3^{2+}] + [Cd(NH_3)_4^{2+}] + [Cd(NH_3)_5^{2+}] + [Cd(NH_3)_6^{2+}]$$

$$C_{Cd} = [Cd^{2+}] + [CdNH_3^{2+}] + [Cd(NH_3)_2^{2+}] + [Cd(NH_3)_3^{2+}] + [Cd(NH_3)_4^{2+}] + [Cd(NH_3)_5^{2+}] + [Cd(NH_3)_6^{2+}]$$

$$S = C_{Cd}$$

$$\frac{[Cd^{2+}]}{C_{Cd}} = \beta_0$$

$$[Cd^{2+}] = \beta_0 C_{Cd} = \beta_0 S$$

Then we express $[S^{2-}]$ in terms of solubility:

$$S = [S^{2-}] + [HS^-] + [H_2S]$$

$$C_s = [S^{2-}] + [HS^-] + [H_2S]$$

$$S = C_s$$

$$\frac{[S^{2-}]}{C_s} = \alpha_2$$

$$[S^{2-}] = \alpha_2 C_s = \alpha_2 S$$

We recall that K_{sp}:

$$K_{sp} = [Cd^{2+}] \cdot [S^{2-}] = (\beta_0 S) \cdot (\alpha_2 S) = \alpha_2 \cdot \beta_0 S^2$$

$$S = \left(\frac{K_{sp}}{\alpha_2 \beta_0}\right)^{1/2}$$

Now we evaluate α_2.

$$pH = 11.10$$

$$[H^+] = 7.9 \times 10^{-12}M$$

$$\alpha_2 = \frac{(1.02 \times 10^{-7}) \cdot (1.2 \times 10^{-15})}{(7.9 \times 10^{-12})^2 + (7.9 \times 10^{-12})(1.02 \times 10^{-7}) + (1.02 \times 10^{-7})(1.2 \times 10^{-15})}$$

$$\alpha_2 = 1.5 \times 10^{-4}$$

Next, we write down and evaluate β_0.

$$\beta_0 = \frac{1}{1 + k_1[NH_3] + k_1k_2[NH_3]^2 + \ldots + k_1k_2k_3k_4k_5k_6[NH_3]^6}$$

There are so many terms that a table might be useful:

First term

$$1 = 1 \qquad\qquad\qquad = 1$$

Second term

$$k_1[NH_3] = (4.47 \times 10^{+2}) \cdot (10^{-1}) = 4.47 \times 10^1 \qquad = 44.7$$

Third term

$$k_1k_2[NH_3]^2 = (k_1[NH_3])k_2[NH_3]$$

$$k_1k_2[NH_3]^2 = (4.47 \times 10^1) \cdot (1.26 \times 10^2) \cdot (10^{-1}) = 5.63 \times 10^2 \qquad = 563$$

Fourth term

$$k_1k_2k_3[NH_3]^3 = (k_1k_2[NH_3]^2)k_3[NH_3] = (5.63 \times 10^2) \cdot (2.75 \times 10^1) \cdot (10^{-1})$$

$$k_1k_2k_3[NH_3]^3 = 1.548 \times 10^3 \qquad\qquad\qquad = 1548$$

Fifth term

$$k_1k_2k_3k_4[NH_3]^4 = (1.548 \times 10^3) \cdot (8.51) \cdot (10^{-1}) = 1.317 \times 10^3 \qquad = 1317$$

Sixth term

$$k_1k_2k_3k_4k_5[NH_3]^5 = (1.317 \times 10^3) \cdot (4.79 \times 10^{-1}) \cdot (10^{-1})$$

$$= 6.3 \times 10^1 \qquad\qquad\qquad = 63$$

Seventh term

$$k_1k_2k_3k_4k_5k_6[NH_3]^6 = (6.3 \times 10^1) \cdot (2.19 \times 10^{-2}) \cdot (10^{-1})$$

$$= 1.38 \times 10^{-1} \qquad\qquad\qquad = 0.138$$

$$\Sigma = 3538 \approx 3.54 \times 10^3$$

$$\beta_0 = \frac{1}{\Sigma} = \frac{1}{3.54 \times 10^3}$$

$$\frac{1}{\beta_0} = 3.54 \times 10^3$$

$$S = \left(\frac{K_{sp}}{\alpha_2}\left(\frac{1}{\beta_0}\right)\right)^{1/2} = \left(\frac{7.8 \times 10^{-27}}{1.5 \times 10^{-4}} \cdot 3540\right)^{1/2}$$

$$S = 4.3 \times 10^{-10} M$$

(This ammonia solution will not dissolve much CdS.)

7. Let S = solubility. First, express $[Ag^+]$ in terms of solubility.

$$C_{Ag} = [Ag^+] + [AgNH_3^+] + [Ag(NH_3)_2^+]$$

Because there are 2 moles of some silver-bearing species for every mole of Ag_2S that dissolves

$$2S = [Ag^+] + [AgNH_3^+] + [Ag(NH_3)_2^+]$$

$$C_{Ag} = 2S$$

$$\frac{[Ag^+]}{C_{Ag}} = \beta_0$$

$$[Ag^+] = \beta_0 C_{Ag} = 2\beta_0 S$$

Then we express $[S^{2-}]$ in terms of solubility,

$$S = [S^{2-}] + [HS^-] + [H_2S]$$
$$C_s = [S^{2-}] + [HS^-] + [H_2S]$$
$$S = C_s$$
$$\frac{[S^{2-}]}{C_s} = \alpha_2$$
$$[S^{2-}] = \alpha_2 C_s = \alpha_2 S$$

We recall that

$$K_{sp} = [Ag^+]^2 \cdot [S^{2-}] = (2\beta_0 S)^2 \cdot (\alpha_2 S) = 4\beta_0^2 \alpha_2 S^3$$
$$S = \left(\frac{K_{sp}}{4\beta_0^2 \alpha_2}\right)^{1/3}$$

or

$$S = \left(\frac{K_{sp}}{4\alpha_2}\left(\frac{1}{\beta_0}\right)^2\right)^{1/3}$$

Now we evaluate α_2, as we did for Problem 6. At pH 11.10, $\alpha_2 = 1.5 \times 10^{-5}$. Next we write down and evaluate β_0. From the AgBr example in Section 3.6, when $[NH_3] = 0.100M$,

$$\beta_0 = \frac{1}{1.07 \times 10^5}$$
$$\frac{1}{\beta_0} = 1.07 \times 10^5$$
$$S = \left[\frac{2 \times 10^{-49}}{(4)(1.5 \times 10^{-5})}(1.07 \times 10^5)^2\right]^{1/3} = 3.4 \times 10^{-12}M$$

(still not very soluble.)

chapter 4
Absorption Spectrophotometry

≡ 4.1 BASIC RELATIONSHIPS AND INSTRUMENTATION ≡

For quite a long time, light absorption has been used in chemical analysis, although the instruments used have been much simpler than the sleek spectrophotometers found in most laboratories today. The intuitive basis of spectrophotometry may be shown by the following example.

Solutions of $Co(NO_3)_2 \cdot 6H_2O$ are pink. The higher the concentration of the salt, the pinker the solution. This may be shown by making up solutions $0.15M$, $0.075M$, and $0.0375M$ in $Co(NO_3)_2 \cdot 6H_2O$. The pinkest solution is the $0.15M$ solution, and the least pink the $0.0375M$ solution. The degree of pinkness, or depth of pinkness, varies with concentration in a regular manner. With a good eye and many standards, we can readily determine the cobalt concentration of an unknown by a matching process.

The example just given is a rather fair representation of visual colorimetry. Visual colorimetry may be used in analyses of varied substances; hemoglobin, ammonia in water, and phosphates in water come to mind at once. More sophisticated techniques than visual ones are available today. These sophisticated techniques require terms less vague than "pink," "degree of pinkness," or "depth of pinkness."

254

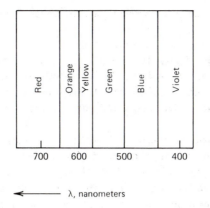

Figure 4.1 A rough representation of the visible spectrum.

First, consider the matter of color. The colors visible to human eyes range from red through orange, yellow, green, blue, and violet in that order. The wavelength scale corresponding to these colors runs from about 700 to about 400 nm. Figure 4.1 gives some idea of the visible spectrum, and Table 4.1 lists some of the units popularly used to describe wavelength.

Table 4.1
WAVELENGTH UNITS

Preferred: $1 \text{ nm} = 1 \text{ m}\mu = 10^{-9}\text{m} = 10^{-7} \text{ cm} = 10 \text{ Å}$
(nanometer)
Still used: $1 \text{ m}\mu = 1 \text{ nm} = 10^{-9}\text{m} = 10^{-7} \text{ cm} = 10 \text{ Å}$
(millimicron)
Often used
in emission
work: $1 \text{ Å} = 10^{-1} \text{ nm} = 10^{-1} \text{ m}\mu = 10^{-1}\text{m} = 10^{-8} \text{ cm}$
(Ångstrom)

| | λ-Values in Various Units | |
In nm (nanometers)	In mμ (millimicrons)	In Å (Ångstroms)
700	700	7000
600	600	6000
500	500	5000
400	400	4000

From Table 4.1, it is apparent that nanometers (nm) and millimicrons (mμ) are quite interchangeable. Nanometers, being part of the S.I. system of units, are the units most often used nowadays, but millimicrons are still encountered in the literature. The

Ångstrom is a favorite unit of emission spectroscopists, X-ray spectroscopists, and crystallographers.

What determines the color of a solution? Color is determined by the light of some wavelengths being absorbed and the light of other wavelengths being transmitted. In the case of the pink solutions of Co^{2+}, the light of longer wavelengths (red) is transmitted and the light of shorter wavelengths is absorbed. A blue solution, by contrast, absorbs the light of longer wavelengths (red) and transmits the light of shorter wavelengths (blue).

The chemical nature of the species in solution determines the wavelength of light that is absorbed. The concentration of the species in solution determines the amount of light of that wavelength that is absorbed. Thus, if monochromatic light (light of a single color or of a single wavelength) is passed through a solution, and if the solution absorbs light of that wavelength, the amount of that light absorbed ought somehow to be proportional to the concentration of the absorbing species. In the sections following, relationships among light transmitted, light absorbed, and the concentration of absorbing species will be derived, and a simple instrument will be described. The mathematical relationship is known as Beer's law, and the instrument is called a spectrophotometer.

4.2 A Derivation of Beer's Law

Suppose that a parallel beam of monochromatic light shines upon a solution. Suppose that the only absorbing species in the solution is a solute, M. Solute M absorbs the light of the wavelength passing through the solution. The absorption process is

$$M + photon \rightarrow M^*$$

M^* is species M which has absorbed a photon and has become excited. Most excited molecules or ions, M^*, do not re-emit the absorbed photons but dissipate their energy as heat rather quickly.

$$M^* \rightarrow M + heat$$

Thus most absorbed photons do not re-appear.

The derivation that follows is a simple derivation, making no direct use of the calculus.[*] In Figure 4.2, the starting assumptions for the derivation are shown.

How may P_0 and P_{T_1} be related to the concentration of absorbing species, C, and the thickness of the layer, Δb?

Suppose that there are Δn absorbing molecules, uniformly distributed, in the layer. Let the layer have a volume of ΔV cm^3, and remember that the area of its surface exposed to the light is W cm^2.

$$\Delta V \text{ cm}^3 = (W \text{ cm}^2)(\Delta b \text{ cm})$$

Remember that the photon capture cross section for each molecule is a. Then the photon capture cross section for all the molecules is $a\Delta n$. The fraction of the total

---

[*] *Courtesy of Professor J. L. Schrag, Department of Chemistry, University of Wisconsin.*

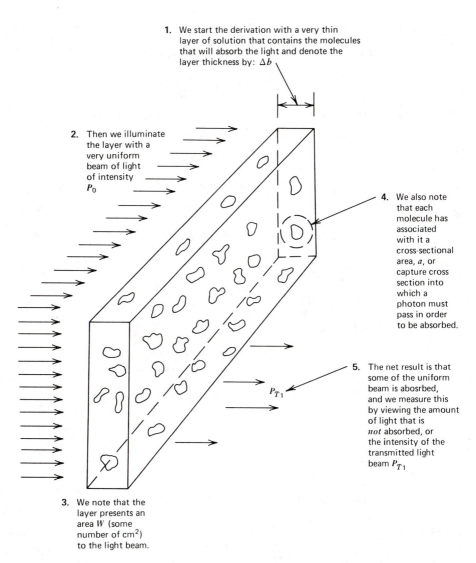

1. We start the derivation with a very thin layer of solution that contains the molecules that will absorb the light and denote the layer thickness by: Δb

2. Then we illuminate the layer with a very uniform beam of light of intensity P_0

3. We note that the layer presents an area W (some number of cm^2) to the light beam.

4. We also note that each molecule has associated with it a cross-sectional area, a, or capture cross section into which a photon must pass in order to be absorbed.

5. The net result is that some of the uniform beam is abosrbed, and we measure this by viewing the amount of light that is *not* absorbed, or the intensity of the transmitted light beam P_{T_1}

Figure 4.2 *Starting the derivation. Courtesy of Profs. John P. Walters and John L. Schrag, Dep't. of Chemistry, University of Wisconsin.*

exposed surface area, W, occupied by the capture cross section for all molecules, $a\Delta n$, is $a\Delta n/W$.

The probability that a photon will be absorbed in the layer is proportional to two quantities: P_0, the initial intensity of the beam; and $a\,\Delta n/W$, the fraction of the total area W occupied by the capture cross section for all molecules.

$$\text{(number of photons absorbed)} \;=\; P_0\,\frac{a\,\Delta n}{W}$$

The intensity of the transmitted beam is P_{T_1}.

$$P_{T_1} = P_0 - \text{(number of photons absorbed)}$$

$$= P_0 - P_0 \frac{a \, \Delta n}{W} = P_0 \left(1 - \frac{a \, \Delta n}{W} \right)$$

Now the equation above can be redefined in terms of useful quantities such as

C = concentration of absorbing species, in moles per liter

b = the sum of the thicknesses of all the thin layers

$$\frac{\Delta n}{6.023 \times 10^{23}} = \Delta m = \text{(number of moles in thin layers)}$$

Remember that

$$\Delta V = W(\Delta b)$$

$$W = \frac{\Delta V}{\Delta b}$$

$$P_{T_1} = P_0 \left(1 - \frac{a(\Delta b)(\Delta n)}{\Delta V} \right)$$

$$\Delta n = 6.023 \times 10^{23} \, \Delta m \text{ moles}$$

$$P_{T_1} = P_0 \left(1 - (a)(6.023 \times 10^{23}) \left(\frac{\Delta m}{\Delta V} \right) \Delta b \right)$$

ΔV is in cubic centimeters (cm^3) or milliliters; $\Delta m / \Delta V$ is in $\dfrac{\text{moles}}{\text{milliliters}}$

$$\frac{\Delta m}{\Delta V} (1000) = C \frac{\text{moles}}{\text{liter}}$$

$$\frac{\Delta m}{\Delta V} = \frac{C}{1000}$$

$$P_{T_1} = P_0 \left[1 - \left(\frac{6.023 \times 10^{23}}{1000} \right) (a)(C)(\Delta b) \right]$$

The quantity

$$\left[\frac{6.023 \times 10^{23}}{1000} a \right]$$

depends on what the absorbing species is and on the wavelength of the beam of light. It is constant for a given species at a given wavelength.

$$k = \frac{6.023 \times 10^{23}}{1000} a = \text{constant}$$

$$P_{T_1} = P_0 \left[1 - kC\Delta b \right]$$

$$\frac{P_{T_1}}{P_0} = 1 - kC\Delta b$$

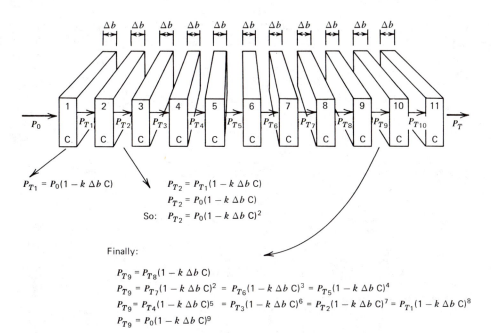

$P_{T_1} = P_0(1 - k \, \Delta b \, C)$

$P_{T_2} = P_{T_1}(1 - k \, \Delta b \, C)$
$P_{T_2} = P_0(1 - k \, \Delta b \, C)$
So: $P_{T_2} = P_0(1 - k \, \Delta b \, C)^2$

Finally:

$P_{T_9} = P_{T_8}(1 - k \, \Delta b \, C)$
$P_{T_9} = P_{T_7}(1 - k \, \Delta b \, C)^2 = P_{T_6}(1 - k \, \Delta b \, C)^3 = P_{T_5}(1 - k \, \Delta b \, C)^4$
$P_{T_9} = P_{T_4}(1 - k \, \Delta b \, C)^5 = P_{T_3}(1 - k \, \Delta b \, C)^6 = P_{T_2}(1 - k \, \Delta b \, C)^7 = P_{T_1}(1 - k \, \Delta b \, C)^8$
$P_{T_9} = P_0(1 - k \, \Delta b \, C)^9$

Figure 4.3 *Passage of light through many layers of solution. Courtesy of Profs. John P. Walters and John L. Schrag, Dep't. of Chemistry, University of Wisconsin.*

Can a relationship among P_0, k, C, Δb, and the number of layers be derived for light passing through the entire sample of solution, and not just one thin layer? The answer is yes. Figure 4.3 shows the derivation graphically.

P_{T_1} = intensity of light after passage through layer 1

P_{T_2} = intensity of light after passage through layer 2

P_{T_3} = intensity of light after passage through layer 3

$\vdots \qquad \vdots$

P_{T_N} = P = intensity of light after passage through all N layers of solution

N = total number of layers of solution, each of width Δb

$N\Delta b$ = b = total width of solution through which light passes

P_{T_1} = $P_0 (1 - k \, \Delta b \, C)$

P_{T_2} = $P_{T_1} (1 - k \, \Delta b \, C)$

By substituting,

$$P_{T_2} = P_0 (1 - k \, \Delta b \, C)(1 - k \, \Delta b \, C)$$

$$P_{T_2} = P_0 (1 - k \, \Delta b \, C)^2$$

$$P_{T_3} = P_{T_2} (1 - k \Delta b\ C)$$

and

$$P_{T_3} = P_0 (1 - k \Delta b\ C)^2 \cdot (1 - k \Delta b\ C)$$

$$P_{T_3} = P_0 (1 - k \Delta b\ C)^3$$

$$\begin{array}{cc} \bullet & \bullet \\ \bullet & \bullet \\ \bullet & \bullet \end{array}$$

Finally, we obtain

$$P_{T_N} = P = P_0 (1 - k \Delta b\ C)^N$$

$$\frac{P}{P_0} = (1 - k \Delta b\ C)^N$$

$$\log \frac{P}{P_0} = \log (1 - k \Delta b\ C)^N = N \log (1 - k \Delta b\ C)$$

It is a fact noted by mathematicians that, when any variable $x \ll 1$,

$$\log (1 - x) = \frac{-x}{2.303}$$

If $x = k \Delta b\ C$ and $k \Delta b\ C \ll 1$ (a very thin layer), then

$$\log(1 - k \Delta b\ C) = \frac{-k \Delta b\ C}{2.303}$$

$$\log \frac{P}{P_0} = N \frac{-k \Delta b\ C}{2.303} = \frac{-k}{2.303} (N \Delta b)\ C$$

But

$$N \Delta b = b$$

$$\log \frac{P}{P_0} = \frac{-k}{2.303}\ bC$$

The quantity $k/2.303$ is a constant dependent upon the absorbing species and the wavelength of light absorbed. It can be condensed.

$$\epsilon = \frac{k}{2.303} = \text{molar absorptivity}$$

When b is in cm and C is in moles/liter, ϵ, the molar absorptivity, has units of liter/ (mole $\cdot$ centimeter).

$$\log \frac{P}{P_0} = -\epsilon bC \qquad \text{This equation can be called Beer's law}$$

Thus the ratio of the intensity of transmitted light to the initial intensity of light falling on the solution bears a logarithmic relationship to the cell length (or the length of solution through which light passes) and the concentration of absorbing species. It is useful to define some new terms to make the equation a little easier to manipulate. We define:

$$\frac{P}{P_o} = T$$

$$T = \text{transmittance}$$

$$\log\frac{P}{P_o} = \log T$$

$$\log T = -\epsilon bC$$

or

$$-\log T = \epsilon bC$$

and

$$A = \text{absorbance}$$

$$A = -\log T$$

or

$$A = \log \frac{1}{T}$$

and

$$\%T = (T)(100)$$

$$A = -\log T = 2 - \log T - 2 = 2 - \log T - \log 100$$

$$= 2 - \log [(T)(100)] = 2 - \log \% T$$

The form of Beer's law that is written most often is

$$A = 2 - \log \% T = \epsilon bC \qquad (4.1)$$

The absorbance, A, varies linearly with the concentration of a given species (ϵ is constant), when the cell length, or b, is constant. Absorbance is a term defined for convenience. Note how a plot of absorbance versus concentration (Figure 4.4) is linear, while that of $\% T$ versus concentration is curved (Figure 4.5).

It is worth noting that there are many spectrophotometric terms which have gone out of use. These are given in Table 4.2. Students will often find the older terms in pre-1970 and biological literature.

Table 4.2
SPECTROPHOTOMETRIC TERMS

Accepted Terms		
Absorbance, A	Transmittance, T	Molar absorptivity, ϵ
Old Terms:		
Absorbancy	Transmittancy	Molar absorbancy index
Extinction	Transmission	Molar extinction coefficient
Optical density		Molar absorption coefficient

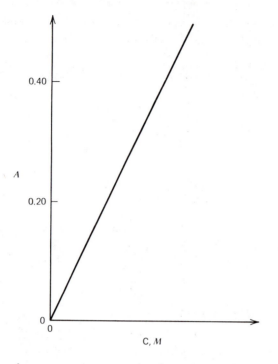

Figure 4.4 *Absorbance versus concentration for a system obeying Beer's law.*

4.3 A Simple Spectrophotometric Instrument

A simple spectrophotometer for use in the visible region of the spectrum is easy to construct for any firm having access to mechanical, optical, and electrical components. It need not be terribly expensive; quite a decent instrument may be had for $500.* Figure 4.6 is a simple block diagram of a spectrophotometer. Because only one beam of light goes through the device, it is called a single-beam spectrophotometer.

The light source is often an automobile light bulb. This source is regulated, so that it emits light of constant intensity. The light emitted is "white light," a mixture of all the visible wavelengths.

The monochromator allows the isolation of one wavelength or a band of wavelengths. Typically, a band of wavelengths 10 to 20 nanometers wide is passed. Turning a knob allows wavelength selection. The wavelength dispersing element in the monochromator is most often a replica diffraction grating, which may be bought very cheaply. Crude monochromation may be achieved with colored glass or gelatine filters, but this practice is dying out.

*This is a 1978 price.

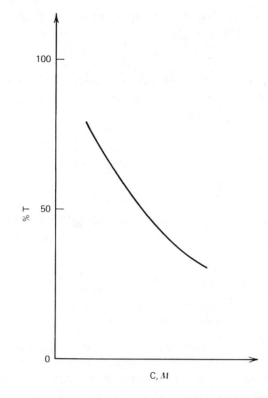

Figure 4.5 % T *versus Concentration for a system obeying Beer's law.*

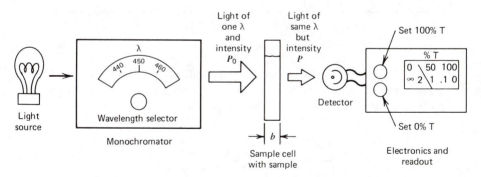

Figure 4.6 *Simple single-beam spectrophotometer.*

Monochromatic light of intensity P_0 impinges upon the sample, which is held in a cell of width b. The cell may be flat-sided, and of high-quality glass. Often, however, the cell is a modified test tube. Light of intensity P emerges from the cell.

The detector is a photoelectric device, which emits a small current proportional to the intensity of light falling upon it.

In the electronics and readout block, the current is amplified for readout on a meter. The meter is calibrated in % T (linear scale) and absorbance (nonlinear scale). For calibration purposes, both 0% T ($A = \infty$) and 100% T ($A = 0$) may be set with manual control knobs.

Because it is not easy to get absolute intensity measurements, relative measurements are used. Light is blocked completely from the detector, and the needle of the readout dial is set to 0% T ($A = \infty$). Here

$$P = 0 \qquad\qquad T = \frac{P}{P_0} = 0$$

$$P_0 = \text{finite} \qquad \% \ T = (T)(100) = 0$$

$$A = -\log\frac{P}{P_0} = \infty$$

With only solvent, and *no* absorbing species in the cell, the instrument dial is set to 100% T.

$$\text{Let } P = P_0 \qquad T = \frac{P}{P_0} = \frac{P_0}{P_0} = 1$$

$$\% \ T = (T)(100) = 100$$

$$A = -\log\frac{P}{P_0} = -\log 1 = 0$$

Then the % T values of solutions containing the absorbing species will fall between 0 and 100, and absorbances will fall between infinity and zero.

A great many analyses may be run on just such a spectrophotometer as the one described. Information on physical constants may be gained. Applications to chemical systems will be discussed at length in the next section.

$$\equiv \ 4.4 \ \text{PROBLEMS} \ \equiv$$

1. A laser bought for \$200 has an output of 2.0 milliwatts of monochromatic light. A neutral density filter (same % T at all wavelengths) allows 70% of impinging light to pass. How many milliwatts of power are received at point B in the set–up of Figure 4.7?

Laser   Filter

Figure 4.7 *Absorption by one filter.*

Ans: *1.4 milliwatts*

2. If two of the filters of Problem 1 are placed one behind the other, how many milliwatts of power are received at point C in the set-up of Figure 4.8?

Laser

Two filters

Figure 4.8 *Absorption by two filters.*

Ans: *1.0 milliwatts*

3. Suppose that a solution of $Co(NO_3)_2$ has a molar absorptivity of 5.1 liters/(mole· centimeter) at 505 nanometer. Plot a graph of A versus C (moles/liter) for solutions 0.020, 0.040, 0.060, 0.080, and 0.10 M in $Co(NO_3)_2$, each in a 1.0-cm cell. Plot on the same graph, the % T of each solution versus concentration. (λ = 505 nanometers)

Ans: *See detailed solutions.*

4. (a) Plot the absorbances of 0.040M $Co(NO_3)_2$ solutions in 0.50-, 2.00-, and 2.50-cm cells at 505 nanometers versus the cell length.
   (b) On the same graph, plot % T versus cell length b.

Ans: *See detailed solutions.*

$\equiv$ 4.5   DETAILED SOLUTIONS TO PROBLEMS $\equiv$

1. Let

$$P_1 = \text{power after passage through filter}$$

$$P_0 = \text{power } before \text{ passage through filter}$$

$$\% T = 70\% = (T)(100)$$

$$T = \frac{P_1}{P_0}$$

$$\% T = \frac{P_1}{P_0} (100)$$

$$70 = \frac{P_1}{P_0} (100)$$

But

$$P_0 = 2.0 \text{ milliwatts}$$

$$70 = \frac{P_1}{2.0} (100)$$

$$P_1 = (2.0)\frac{(70)}{(100)} = 1.4 \text{ milliwatts}$$

2. This problem can be solved by the use of the absorbance function.

$$A = \epsilon bC = 2 - \log \% T$$

The molar absorptivity, ϵ, is the same for either filter, or for both filters. It is independent of how many filters are stacked up. The concentration of the absorber, C, is the same for either filter, or for both filters. It is also independent of how many filters are stacked up.

Let

$\% \, T_1$ = % T after passage of light through one filter = 70%

A_1 = absorbance after passage of light through one filter = $2 - \log \% \, T_1$

Also, let

b_1 = path length through one filter = 1.0 unit

Then

$$A_1 = 2 - \log \% \, T_1 = \epsilon b_1 C$$

Let

$\% \, T_2$ = % T after passage of light through both filters = unknown

A_2 = absorbance after passage of light through two filters = $2 - \log \% \, T_2$

Also, let

b_2 = path length through two filters = 2.0 units

Then

$$A_2 = 2 - \log \% \, T_2 = \epsilon b_2 C$$

Now we put in some numbers.

$$A_1 = 2 - \log \% \, T_1 = 2 - \log (70.) = 0.15$$
$$b_1 = 1.0$$
$$b_2 = 2.0$$
$$A_2 = \epsilon b_2 C$$
$$A_1 = \epsilon b_1 C$$
$$\frac{A_2}{A_1} = \frac{\epsilon b_2 \cancel{C}}{\cancel{\epsilon} b_1 \cancel{C}} = \frac{b_2}{b_1}$$
$$A_2 = \frac{b_2}{b_1} A_1$$
$$A_2 = \left(\frac{2.0}{1.0}\right)(0.15) = 0.30 = 2 - \log \% \, T_2$$
$$0.30 = 2.00 - \log \% \, T_2$$
$$\log \% \, T_2 = 1.69$$
$$\% \, T_2 = 50.\%$$

Let

P_2 = power after passage through two filters

$$\left(\frac{P_2}{P_o}\right)(100) = \% \, T_2 = 50.$$

$$P_o = 2.0 \text{ milliwatts}$$

$$\left(\frac{P_2}{2.0}\right)(100) = 50.$$

$$P_2 = \frac{(2.0)(50)}{(100)}$$

$$P_2 = 1.0 \text{ milliwatts}$$

Alternatively, the problem could be solved thus:

Let

$$P_o = \text{initial power} = 2.0 \text{ milliwatts}$$
$$P_1 = \text{power after passage through one filter}$$
$$P_2 = \text{power after passage through two filters}$$
$$P_1 = 0.70\, P_o = (0.70)\,(2.0) = 1.4$$
$$P_2 = (0.7)\, P_1 = (0.70)\,(1.4) = 1.0 \text{ milliwatts}$$

This method of solution is more awkward than the one using absorbance, particularly if three or more filters are involved.

3. Beer's law is applicable (Equation 4.1).

$$A = 2 - \log \% T = \epsilon b C$$
$$\epsilon = 5.1 \text{ liters/(mole} \cdot \text{centimeters)}$$
$$b = 1.0 \text{ centimeter}$$

In the case where

$$c = 0.020M$$
$$A = \epsilon b C = (5.1)\,(1.0)\,(0.020) = 0.10$$

By Beer's law

$$A = 2.00 - \log \% T$$
$$0.10 = 2.00 - \log \% T$$
$$\log \% T = 1.90$$
$$\% T = 79.$$

Other values of A and $\% T$ can be calculated in a similar fashion.

We arrange these data:

C, M	A	$\% T$
0.020	0.10	79.
0.040	0.20	63.
0.060	0.31	49.
0.080	0.41	39.
0.10	0.51	31.

Now we plot the data (Figure 4.9). Note how the absorbance increases linearly with concentration, while the percent transmittance decreases nonlinearly with concentration.

4. Beer's law is applicable.

$$A = 2 - \log \% T = \epsilon b c$$
$$\epsilon = 5.1 \text{ liters/(mole} \cdot \text{centimeters)}$$
$$C = 4.0 \times 10^{-2} M$$

where

$$b = 0.50 \text{ cm}$$
$$A = \epsilon b c = (5.1)\,(0.50)\,(0.040) = 0.10 = 2.00 - \log \% T$$
$$0.10 = 2.00 - \log \% T$$
$$\log \% T = 1.90$$

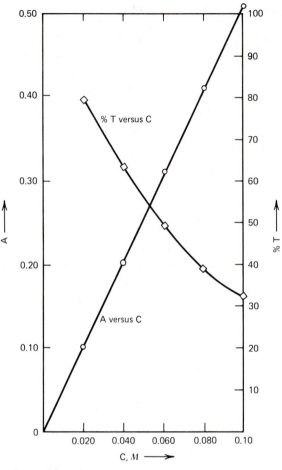

Figure 4.9 *Plot for Problem 4.3.*

$$\% \, T \; = \; 79\%$$

Other values of A and $\% \, T$ can be calculated in a similar fashion. We list these data:

b, cm	A	% T
0.50	0.10	79.
1.00	0.20	63.
2.00	0.41	39.
2.50	0.51	31.

Now we plot the data (Figure 4.10). Note how the absorbance increases linearly with the cell length, while the percent transmittance decreases nonlinearly with the cell length.

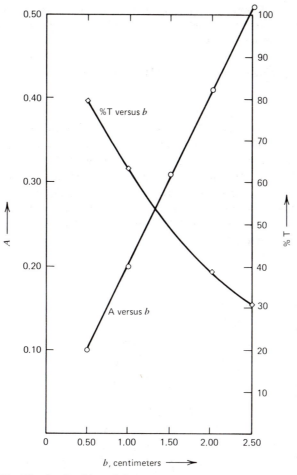

Figure 4.10 *Plot for Problem 4.4.*

≡ 4.6 SOME APPLICATIONS OF ABSORPTION SPECTROPHOTOMETRY ≡

Beer's law is applicable to a number of chemical systems. Here, four systems will be considered: (1) simple applications to one-component analyses, (2) simultaneous analysis of two or more components, (3) spectrophotometric titration, and (4) determination of physical constants of complexes and indicators.

4.7 Simple Applications

In the simple application of Beer's law the concentration of a single colored component in solution is sought. The steps are simple: (1) finding the wavelength of maximum absorption (popularly known as λ_{max}), (2) making a Beer's law calibration plot, (3) trying to determine the concentration of the colored species in an unknown sample, and (4) trying to minimize the error.

4.8 Finding the Wavelength of Maximum Absorption

We make up a solution of the species of interest and measure the absorbance as a function of the wavelength. Usually, in published work, molar absorptivity, ϵ, or its logarithm, log ϵ, is plotted versus wavelength. Plotting ϵ or log ϵ insures a spectrum that is independent of concentration. For much laboratory work, however, a plot of A versus λ will serve. Such an absorption spectrum is shown in Figure 4.11. The data from which the spectrum was drawn are shown in Table 4.3. The spectrum is typical of pinkish solutions. It was taken on a $0.0750M$ solution of $Co(NO_3)_2$. The absorbance maximum is somewhere between 500 and 520 nanometers. (Closer resolution would show it to be at 505 nanometers.)

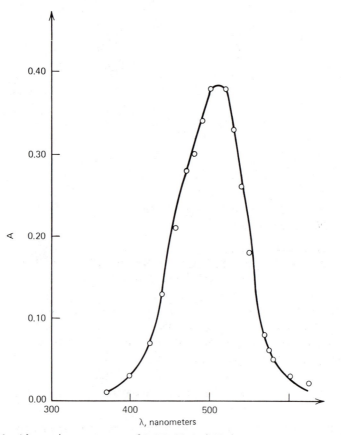

Figure 4.11 *Absorption spectrum of 0.075 M $Co(NO_3)_2$.*

Table 4.3
ABSORBANCE DATA FOR A 0.075M Co(NO₃)₂ SOLUTION[a]

λ, nm	A
375	0.01
400	0.03
425	0.07
440	0.13
455	0.21
470	0.28
480	0.30
490	0.34
500	0.38
520	0.38
530	0.33
540	0.26
550	0.18
570	0.08
575	0.06
580	0.05
600	0.03
625	0.02

[a] *Courtesy of V. McGuffin.*

In visible and ultraviolet solution spectrophotometry, smooth curves like the one in Figure 4.11 are often observed. In the case of cobaltous nitrate just illustrated, the wavelength of maximum absorption (λ_{max}) is found to be 505 nanometers, and it is this value that is chosen as the wavelength for analysis. By using λ_{max} as the wavelength for analysis, we get the maximum possible sensitivity. If the peak is a smooth, nearly flat one, the error due to poor monochromation or faulty wavelength setting is minimized.

4.9 Plotting the Beer's Law Graph
≡

A calibration plot is easily made. A series of solutions of varying concentrations is prepared and a plot of absorbance versus concentration drawn. This is the calibration plot. The data for calibration are given in Table 4.4. Figure 4.12 is the calibration curve made for Co(NO₃)₂ solutions from these data. Because $A = \epsilon bc$, any system obeying Beer's law ought to exhibit a straight line, passing through the origin, for the graph of absorbance versus concentration. The cobalt system illustrated here does indeed live up to these expectations.

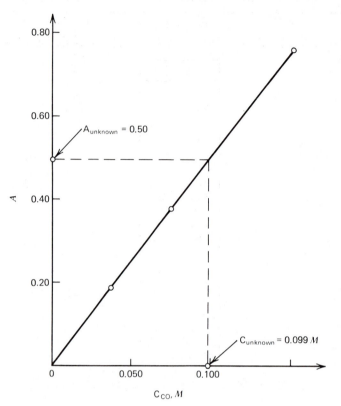

Figure 4.12 *Det'n of C_{Co} in an unknown from a calibration plot (505 nm).*

Table 4.4
ABSORBANCE DATA FOR $Co(NO_3)_2$ SOLUTIONS, TAKEN AT 505 NANOMETERS

C_{Co}, M	A
0.0375	0.19
0.0750	0.38
0.1500	0.76

4.10 Finding the Concentration of an Unknown
≡

First, of course, one makes sure that the unknown does contain the species sought. Then, the absorbance of the unknown is measured under the same conditions used for the absorbances of the standard solutions (same spectrophotometer, same λ, and same cell length). The concentration corresponding to this absorbance is read from the abscissa of the graph, and the task is complete.

Example:

What is the cobalt concentration of a solution which, treated the same way as the standards used to get Figure 4.12, gives an absorbance of 0.50?

Answer:

$$C_{unknown} = 0.099M. \text{ See Figure 4.12.}$$

If it is known that a system obeys Beer's law, you can sometimes get away with measuring two absorbances only—those of *one* standard and the unknown. The procedure is fraught with peril, however, for errors that would quickly be revealed in a calibration plot are hidden in the two-measurement method and are reflected in the concentration figures.

Example:

A 0.0750M sample of $Co(NO_3)_2$ gives an absorbance of 0.38 at 505 nanometers in a 1-centimeter cell. What is the cobalt concentration of a solution giving an absorbance of 0.26 in the same cell, in the same spectrophotometer?

Answer:

$$\text{Let} \begin{cases} A_1 = 0.38 & \epsilon = ? \\ c_1 = 0.0750 & b = ? \\ A_1 = \epsilon b c_1 \end{cases}$$

$$\text{Let} \begin{cases} A_2 = 0.26 & \epsilon = ? \\ c_2 = ? & b = ? \\ A_2 = \epsilon b c_2 \end{cases}$$

$$\frac{A_2}{A_1} = \frac{\epsilon b c_2}{\epsilon b c_1}$$

$$c_2 = \frac{A_2}{A_1} c_1 = \left(\frac{0.26}{0.38}\right)(0.0750) = 0.051M$$

4.11 Error

≡

The possibility of error is worth serious consideration. If one can eliminate errors of dilution—the sort that make the two-measurement method so very risky—there are still plenty of other errors that can occur. These can be categorized as: (1) chemical errors, (2) errors due to poor monochromation, and (3) errors arising from uncertainty in the readout of common instruments. Sometimes such errors cannot be eliminated, but it is important that they be understood.

≫ 4.12 Chemical Errors / Chemical errors can arise when a number of species, all absorbing significantly at a given wavelength, but each with a different molar absorptivity, are present in a solution whose absorbance is being measured. An example is not hard to find; the determination of chromium as Cr(VI) by spectrophotometry is often cited. Depending on the pH of the solution in which it finds itself, Cr(VI) can exist either as $Cr_2O_7^{2-}$ (orange) or CrO_4^{2-} (yellow).

$$Cr_2O_7^{2-} + H_2O \rightleftharpoons 2HCrO_4^- \rightleftharpoons 2H^+ + CrO_4^{2-}$$

The dichromate and chromate ions have different absorptivities at almost all wavelengths, so that considerable error may result from using solutions not buffered well enough to force all of the Cr(VI) into either the dichromate or chromate form.

≫ 4.13 Errors of Monochromation / Deviations from Beer's law may also result from poor monochromation. Nowadays, instruments with reasonable monochromation may be purchased for about $500, but once upon a time, good monochromation was far too expensive for the average laboratory. Good monochromation is a result of *narrow bandpass*. What is bandpass? When we set a wavelength dial on an instrument to, say, 508 nanometers, we do not get light exclusively of 508.0000 . . . and so forth, nanometers passing through the sample. Rather, we get a *band* of wavelengths or a range of wavelengths. In one popular instrument, this band of wavelengths extends from 498 to 518 nanometers.

bandpass = 518 nanometers – 498 nanometers = 20 nanometers

In another popular instrument, the same band of wavelengths around a setting of 508 nanometers would range from 503 to 513 nanometers.

bandpass = 513 nanometers – 503 nanometers = 10 nanometers

An example of error resulting from a combination of wide bandpass and an incorrect wavelength setting can be illustrated by Figure 4.13. The wide bandpass ranges over the sloping portion of the curve. The absorbance read for a solution will not be the average of A_1 and A_2, the absorbances at the extreme ends of the bandpass. The observed transmittance, with which instrument response varies linearly, *may* be a mean value of T_1 and T_2. Because of the logarithmic relationship between T and A, the observed absorbance will not be the average of A_1 and A_2. A negative deviation from Beer's law, illustrated in Figure 4.14 will result.

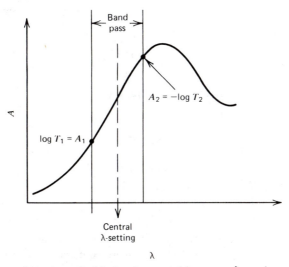

Figure 4.13 *A combination of wide bandpass and incorrect λ-setting.*

≫ 4.14 Error Resulting from Uncertainty in Readout / Error can result from the uncertainty in the readout of a great many common instruments. In most instruments, the uncertainty in % T may be taken as ±0.5% T. In 1933, Twyman and Lothian* published a paper in which they derived the relationship of relative error in concentration to observed value of % T. The derivation is of interest, but the result is important to use here. It is valid for instruments like the Bausch and Lomb Spectronic 20.

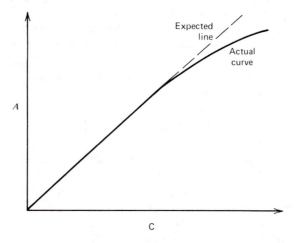

Figure 4.14 *Result of errors of Figure 4.13.*

* F. Twyman and G. F. Lothian, Proc. Phys. Soc. (London), 45, 643 (1933).

The Twyman-Lothian relationship:

$$\frac{\Delta C}{C} = \frac{0.434}{T \log T} \Delta T$$

ΔC = absolute error in concentration

C = concentration

$\dfrac{\Delta C}{C}$ = relative error in concentration   (4.2)

T = transmittance value observed

ΔT = uncertainty in transmittance value (usually taken as ± 0.005, corresponding to $\pm 0.5\%$ T)

A graph of $\Delta C/C$ versus T will tell more than the bare formula above. The graph is shown in Figure 4.15. Note that this *un*symmetrical curve goes through a minimum between T-values of 0.36 and 0.37. Moreover, the values of $\Delta C/C$ are close to the minimum over a range from $T = 0.15$ to $T = 0.70$ (15 to 70% T). The conclusion is that, for a low relative error in concentration, we ought to strive to make measurements between 15 and 70% T. For the very best accuracy, measurements ought to be made between 36 and 37% T. It is not hard to make measurements on concentrated solutions of low % T; such solutions can be diluted to bring their %'s T within the favorable range. Very dilute solutions, having high values of % T, are not easy to concentrate, however.

The foregoing arguments, table, and graph simply confirm what the reader probably suspected all along: (1) If a solution is almost too dark for a spectrophotometer to "see" through it, accurate measurements are hard to make; and (2) if a solution is so light that it looks like a blank, accurate measurements are also hard to make. There are, however, ways to improve the relative accuracy of spectrophotometric measurements. The first is to reduce the value of ΔT or the uncertainty of the transmittance measurement. Reducing ΔT is easy *on paper*, but costs a great deal of money in the laboratory, because it means buying or building a new instrument.

4.15 Simultaneous Analysis of Two or More Components

Very often, a chemist would like to determine more than one component of a system spectrophotometrically *without* going through separation procedures that are tiresome and capable of introducing contamination. Examples that come to mind are chromium and manganese in steels; chromium and cobalt in other alloys; and chlorophylls a, b, and c in extracts from plants. Simultaneous spectrophotometric determinations of two or more components are possible if (1) the chemical species to be determined do not react with each other, and (2) each species follows Beer's law over the wavelength range of interest. If these criteria are met, then, at any wavelength the absorbance of the solution should be the sum of the absorbances of its components.

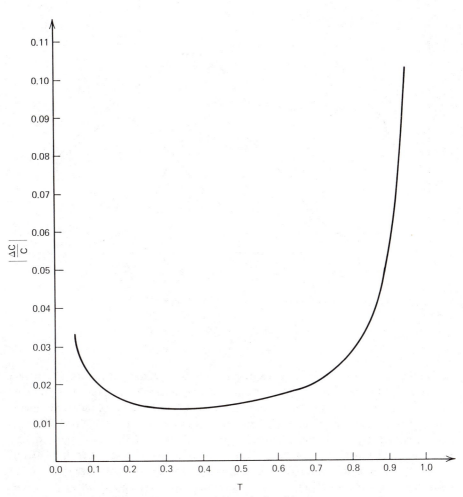

Figure 4.15 *Graph of the Twyman-Lothian relationship.*

Suppose that a solution is made up of chromium and cobalt nitrates, $Cr(NO_3)_3$ and $Co(NO_3)_2$, and that, in the visible region of the spectrum Cr^{3+} and Co^{2+} are the only absorbing species in that solution. The total absorbance of the solution at any wavelength is just the sum of the absorbances of chromium and cobalt at that wavelength.

(4.3)

$$A_{T\lambda} = A_{Cr\lambda} + A_{Co\lambda}$$

$A_{T\lambda}$ = total absorbance of the solution at wavelength λ

$A_{Cr\lambda}$ = absorbance of Cr^{3+} at wavelength λ

$A_{Co\lambda}$ = absorbance of Co^{2+} at wavelength λ

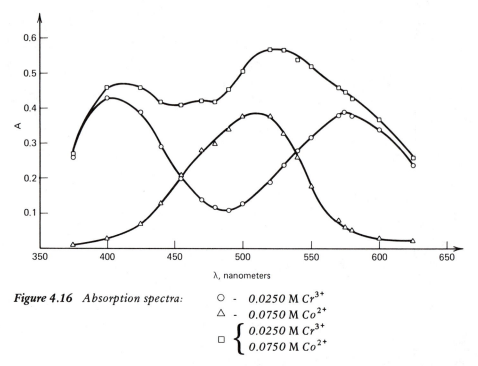

Figure 4.16 *Absorption spectra:*   ◯ - 0.0250 M Cr^{3+}
△ - 0.0750 M Co^{2+}
▢ $\begin{cases} 0.0250 \text{ M } Cr^{3+} \\ 0.0750 \text{ M } Co^{2+} \end{cases}$

Naturally, each ion alone will have its own absorbance spectrum. The total absorbance spectrum will be the sum of the spectra of the individual ions. To prove this point, V. McGuffin in 1974 accumulated the data in Table 4.5. The data are plotted in Figure 4.16, which contains three curves: (1) the absorbance spectrum of $0.0750M$ $Co(NO_3)_2$ alone, designated by △; (2) the absorbance spectrum of 0.0250 $Cr(NO_3)_3$ alone, designated by ◯; and (3) the absorbance spectrum of a solution at once $0.0750M$ in $Co(NO_3)_2$ and $0.0250M$ in $Cr(NO_3)_3$, designated by ▢. Each absorbance value on the third spectrum, the spectrum, that is, of both $Cr(NO_3)_3$ and $Co(NO_3)_2$ together (▢), is the sum of the absorbance values for $Cr(NO_3)_3$ (◯) and $Co(NO_3)_2$ (△) at that wavelength. The data tabulated in Table 4.5 and graphed in Figure 4.16 confirm that the absorbances of $Cr(NO_3)_3$ and $Co(NO_3)_2$ in solution are additive. With this confirmation of additivity, it may be possible to determine the concentrations of chromium and cobalt in a solution simultaneously, that is, without going through a throublesome separation procedure. How?

First, an unknown solution must be procured. Then its absorbances at two wavelengths must be measured. What two wavelengths are to be chosen? Chromium has an absorbance peak at 400 nanometers; cobalt has an absorbance peak at 505 nanometers. Thus 400 and 505 nanometers are chosen as wavelengths for analysis—an analysis at once chemical and mathematical.

At 400 nanometers, the absorbance of the combined solution, A_{400}, is the sum of the absorbances of cobalt and chromium at 400 nanometers. It is found experimentally to equal 0.400.

Table 4.5
ABSORBANCE DATA FOR 0.0250M Cr(NO$_3$)$_3$,
0.0750M Co(NO$_3$)$_2$, AND A SOLUTION AT ONCE
0.0250M IN Cr(NO$_3$)$_3$ AND 0.0750M IN Co(NO$_3$)$_2$.
ABSORBANCES IN THIS TABLE ARE TAKEN TO
TWO SIGNIFICANT FIGURES ONLY.

λ, nm	A, 0.0250M Cr^{3+}	A, 0.0750M Co^{2+}	A, 0.0250 M Cr^{3+} & 0.0750M Co^{2+}
375	0.26	0.01	0.27
400	0.43	0.03	0.46
425	0.39	0.07	0.46
440	0.29	0.13	0.42
455	0.20	0.21	0.41
470	0.14	0.28	0.42
480	0.12	0.30	0.42
490	0.11	0.34	0.45
500	0.13	0.38	0.51
520	0.19	0.38	0.57
530	0.24	0.33	0.57
540	0.28	0.26	0.54
550	0.32	0.18	0.50
570	0.38	0.08	0.46
575	0.39	0.06	0.45
580	0.38	0.05	0.43
600	0.34	0.03	0.37
625	0.24	0.02	0.26

Data taken by V. McGuffin, 1974.

(4.4)

$$A_{400} = A_{Co\,400} + A_{Cr400}$$

$$= 0.400$$

= absorbance of combined solution at 400 nanometers

$A_{Co\,400}$ = absorbance of Co^{2+} at 400 nanometers

A_{Cr400} = absorbance of Cr^{3+} at 400 nanometers

This is a valid equation, but it says nothing whatever about the concentrations of either cobalt or chromium. Beer's law must be used (Eq. 4.1).

$$A_{Co400} = \epsilon_{Co400} \cdot b \cdot C_{Co}$$

$$A_{Cr400} = \epsilon_{Cr400} \cdot b \cdot C_{Cr}$$

ϵ_{Co400} = molar absorptivity of Co at 400 nanometers

b = cell length, centimeters

C_{Co} = concentration of Co in moles per liter

ϵ_{Cr400} = molar absorptivity of Cr at 400 nanometers

b = cell length, centimeters

C_{Cr} = concentration of Cr in moles per liter

$$A_{400} = A_{Co400} + A_{Cr400} \tag{4.4}$$

$$A_{400} = 0.400$$

We substitute into Eq. 4.4

$$\Rightarrow \quad 0.400 = \epsilon_{Co400} \cdot b \cdot C_{Co} + \epsilon_{Cr400} \cdot b \cdot C_{Cr} \tag{4.5}$$

At 505 nanometers, the absorbance of the combined solution, A_{505}, is the sum of the absorbances of cobalt and chromium at 505 nanometers. It is found experimentally to equal 0.530.

$$A_{505} = A_{Co505} + A_{Cr505} \tag{4.6}$$

$$= 0.530$$

= absorbance of combined solution at 505 nanometers

A_{Co505} = absorbance of Co at 505 nanometers

A_{Cr505} = absorbance of Cr at 505 nanometers

Once more, this is a valid equation, but it says nothing at all about the concentrations of either chromium or cobalt. Beer's law must be invoked once more.

280 / Absorption Spectrophotometry

$$A_{Co505} = \epsilon_{Co505} \cdot b \cdot C_{Co}$$

$$A_{Cr505} = \epsilon_{Cr505} \cdot b \cdot C_{Cr}$$

ϵ_{Co505} = molar absorptivity of Co at 505 nanometers

ϵ_{Cr505} = molar absorptivity of Cr at 505 nanometers

$$A_{505} = A_{Co505} + A_{Cr505}$$

$$A_{505} = 0.530$$

We substitute into Eq. 4.6.

$$\Rightarrow 0.530 = \epsilon_{Co505} \cdot b \cdot C_{Co} + \epsilon_{Cr505} \cdot b \cdot C_{Cr} \qquad (4.7)$$

Now, at last, there are two useful equations, Eqs. 4.5 and 4.7, in two unknowns, both designated by an arrow.

$$\Rightarrow 0.400 = \underbrace{\epsilon_{Co400}}_{} \cdot \underbrace{b}_{} \cdot \underbrace{C_{Co}}_{} + \underbrace{\epsilon_{Cr400}}_{} \cdot \underbrace{b}_{} \cdot \underbrace{C_{Cr}}_{} \qquad (4.5)$$

$$\qquad\qquad\quad \text{measured} \quad \text{known} \quad \text{unknown} \quad \text{known} \quad \text{unknown}$$

$$\Rightarrow 0.530 = \overbrace{\epsilon_{Co505}}^{} \cdot \overbrace{b}^{} \cdot \overbrace{C_{Co}}^{} + \overbrace{\epsilon_{Cr505}}^{} \cdot \overbrace{b}^{} \cdot \overbrace{C_{Cr}}^{} \qquad (4.7)$$

There is only one problem, namely, that of the allegedly known values of $\epsilon_{Co400} \cdot b$, $\epsilon_{Cr400} \cdot b$, $\epsilon_{Co505} \cdot b$, $\epsilon_{Cr505} \cdot b$. The values must be determined some way. One *could* use the absorbance measurements, for example, for a $0.0250M$ $Cr(NO_3)_3$ solution at 400 and 505 nanometers to get $\epsilon_{Cr400} \cdot b$ and $\epsilon_{Cr505} \cdot b$. With this approach, however, one must be tolerant of error, for an error in measured absorbance will show

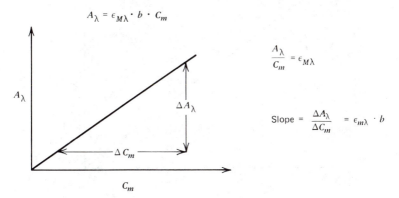

Figure 4.17 Finding $(\epsilon_{m\lambda} \cdot b)$.

282

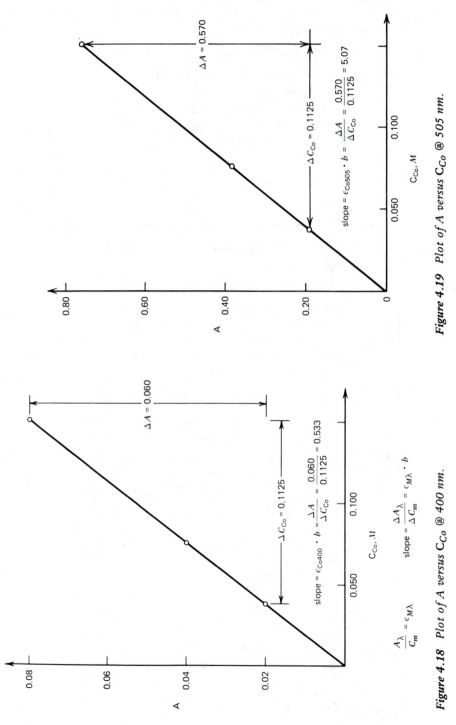

$$\frac{A_\lambda}{C_m} = \epsilon_{M\lambda}$$

$$\text{slope} = \epsilon_{Co400} \cdot b = \frac{\Delta A}{\Delta C_{Co}} = \frac{0.060}{0.1125} = 0.533 \qquad \text{slope} = \frac{\Delta A_\lambda}{\Delta C_m} = \epsilon_{M\lambda} \cdot b$$

Figure 4.18 Plot of A versus C_{Co} @ 400 nm.

$$\text{slope} = \epsilon_{Co505} \cdot b = \frac{\Delta A}{\Delta C_{Co}} = \frac{0.570}{0.1125} = 5.07$$

Figure 4.19 Plot of A versus C_{Co} @ 505 nm.

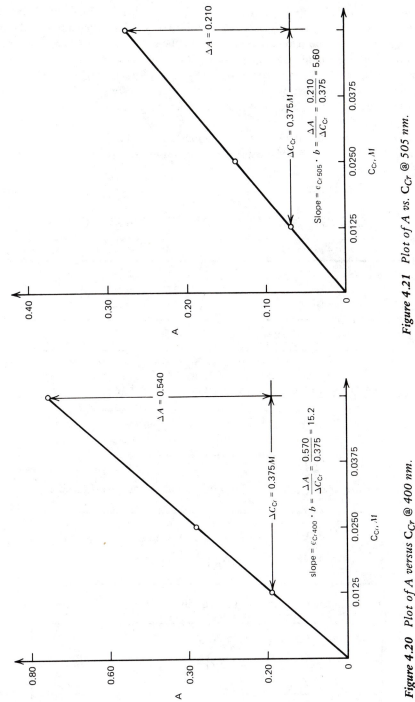

Figure 4.21 Plot of A vs. C_{Cr} @ 505 nm.

Figure 4.20 Plot of A versus C_{Cr} @ 400 nm.

Slope = $\epsilon_{Cr,505} \cdot b = \dfrac{\Delta A}{\Delta C_{Cr}} = \dfrac{0.210}{0.375} = 5.60$

slope = $\epsilon_{Cr,400} \cdot b = \dfrac{\Delta A}{\Delta C_{Cr}} = \dfrac{0.570}{0.375} = 15.2$

283

up in the calculated values of $\epsilon_{Cr400} \cdot b$ and $\epsilon_{Cr505} \cdot b$. It is much safer, and almost as easy, to make up a series of $Cr(NO_3)_3$ solutions at varying concentrations and plot absorbance versus chromium concentration for the series at both 400 and 505 nanometers. Then, one could make up a series of $Co(NO_3)_2$ at varying concentrations and plot absorbance versus cobalt concentration for this series at both 400 and 505 nanometers. The *slope* of each plot is equal to the product ϵb for the species at that wavelength. Why? For any species, call it M, that follows Beer's law, at a given wavelength, λ, Figure 4.17 applies.

Table 4.6 lists some data for both chromium and cobalt solutions. Figures 4.18 through 4.21 are absorbance versus concentration plots. The results are compiled in Table 4.7.

Table 4.6
A VERSUS C DATA FOR Cr AND Co AT 400 AND 505 nm

C_{Co}, M	A at 400 nm	A at 505 nm	C_{Cr}, M	A at 400 nm	A at 505 nm
0.0375	0.020	0.190	0.0125	0.190	0.070
0.0750	0.040	0.380	0.0250	0.380	0.140
0.1500	0.080	0.760	0.0500	0.760	0.280

Data taken by V. McGuffin, 1974.

Table 4.7
ϵb-VALUES FOR Co-Cr SYSTEMS (UNITS OF LITERS/MOLE)

$$\epsilon_{Co400} \cdot b = 0.533 \qquad \epsilon_{Cr400} \cdot b = 15.2$$
$$\epsilon_{Co505} \cdot b = 5.07 \qquad \epsilon_{Cr505} \cdot b = 5.60$$

Now, we must recall the two equations in two unknowns, Eqs. 4.5 and 4.7.

$$0.400 = \epsilon_{Co400} \cdot b \cdot C_{Co} + \epsilon_{Cr400} \cdot b \cdot C_{Cr} \qquad (4.5)$$
$$0.530 = \epsilon_{Co505} \cdot b \cdot C_{Co} + \epsilon_{Cr505} \cdot b \cdot C_{Cr} \qquad (4.7)$$

We substitute numbers for symbols and then solve.

$$0.400 = 0.533 C_{Co} + 15.2 C_{Cr}$$
$$0.530 = 5.07 C_{Co} + 5.60 C_{Cr}$$
$$C_{Cr} = \frac{0.400 - 0.533 C_{Co}}{15.2}$$

$$0.530 = 5.07 C_{Co} + 5.60 \frac{0.400 - 0.533 C_{Co}}{15.2}$$

$$0.530 = 5.07 C_{Co} + \frac{(5.60)(0.400)}{15.2} - \frac{(5.60)(0.533)}{15.2} C_{Co}$$

$$0.383 = 4.87 C_{Co}$$

Ans:

$\Rightarrow C_{Co} = 0.0780 M$

$$C_{Cr} = \frac{0.400 - (0.533)(0.0786)}{15.2}$$

Ans:

$$= 0.0236 M$$

With a little arithmetic and some careful measurements, simultaneous determinations of two, or sometimes even more, components are possible.

4.16   Spectrophotometric Titrations

Spectrophotometric techniques may also be applied to the determination of titration endpoints. Many titration endpoints are marked by a sharp color change, and these are detectable by eye. Others are marked by more gradual color changes. It is to the latter changes that spectrophotometric techniques may be most profitably applied. A spectrophotometric titration rig can be very simple, like the one shown in Figure 4.22.

Devices are available, for under $20, which will allow common spectrophotometric instruments, such as the Bausch and Lomb Spectronic 20, to be used for titration.* In a spectrophotometric titration, we plot the absorbance versus the volume of titrant

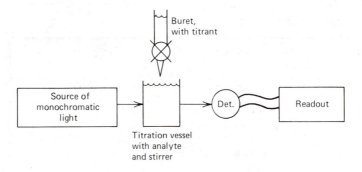

Figure 4.22   Simple spectrophotometric titration apparatus.

* Kontes Glass Co., of Vineland, N.J., is one supplier.

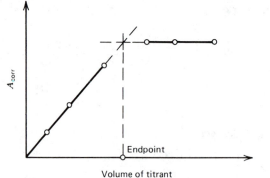

Figure 4.23 *A spectrophotometric titration curve.*

added, and often see titration curves where the breakpoints, or endpoints, are very sharp, such as the one shown in Figure 4.23.

The symbol A_{corr} just stands for absorbance corrected for dilution. A_{corr} is obtained as follows:

$$A_{obs} = \text{observed absorbance}$$

$$V = \text{orig. volume of analyte}$$

$$x = \text{volume of titrant added}$$

$$A_{corr} = \text{corrected absorbance}$$

$$A_{corr} = A_{obs}\left(\frac{V + x}{V}\right) \qquad (4.8)$$

An example of spectrophotometric titration is that of Cu(II) by EDTA (ethyl-enediaminetetraacetic acid); EDTA is represented as H_4Y. The ethylenediaminetetra-acetate ion, abbreviated as Y^{4-}, actually complexes the copper. The complex, CuY^{2-}, absorbs strongly at 625 nanometers.

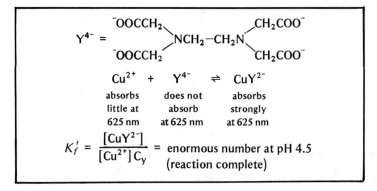

286 / Absorption Spectrophotometry

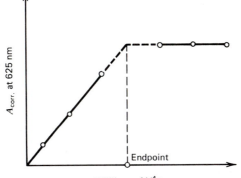

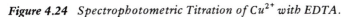

Figure 4.24 *Spectrophotometric Titration of Cu^{2+} with EDTA.*

If the reaction is followed spectrophotometrically, and the absorbance at 625 nanometers plotted versus the milliliters of Y^{4-} solution added, a titration curve like the one in Figure 4.24 results. Up to the endpoint, absorbance increases with added Y^{4-}, because more and more of the absorbing complex, CuY^{2-} is being formed. After the endpoint, that is, after all the Cu^{2+} has been complexed as CuY^{2-}, the absorbance stays constant because only nonabsorbing Y^{4-} solution is being added.

Other photometric titration curve shapes may be observed. Suppose that this general reaction takes place, and goes very nearly to completion:

M +	**X** ⇌	**MX**
analyte in	titrant	product in
vessel in	in buret	vessel in
light path		light path

In Figure 4.25, some possible curve shapes are shown.

There are advantages to photometric titration: (1) it may be conducted in the ultraviolet region; and (2) in the cases we have considered, it is not vital to specify absorbance readings close to the endpoint because the intersection of extrapolated straight lines *is* the endpoint. The second advantage will seem most attractive to those who have done acid-base titrations with a pH meter, taking readings every 0.10 milliliter or so in the region of the endpoint.

4.17 Spectrophotometric Determination of Physical Constants

The discussion of spectrophotometric titrations can lead directly to a discussion of spectrophotometric methods of determining physical constants. First, ways of determining stoichiometry and formation constant values for metal-ligand complexes will be considered. Then, K_a-values for acid-base indicators will be taken up.

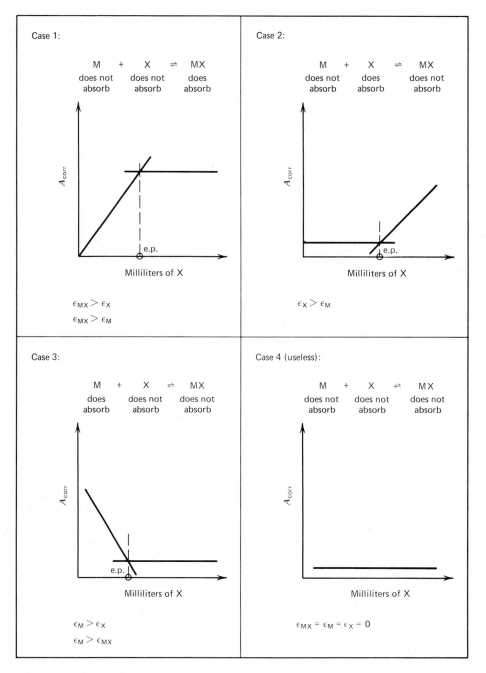

Figure 4.25 *Possible spectrophotometric titration curve shapes.*

There are three generally used methods for finding the stoichiometry of complexes: (1) the mole-ratio method (also known as the Yoe-Jones method), (2) the method of continuous variations (also known as Job's method), and (3) the slope-ratio method. The mole-ratio method and the method of continuous variations may be used to find formation constants, as well as stoichiometry, of metal-ligand complexes. The mole-ratio method is the most closely allied to spectrophotometric titrations and will be considered first.

≫ **4.19 The Mole-Ratio Method** / Ferrous iron (Fe^{2+}) and its complexes are especially well suited to study by the mole-ratio method. When ferrous iron (colorless) reacts with the ligand *ortho*-phenanthroline (also colorless), a bright orange product with a molar absorptivity of about 11,100 liter/(mole · centimeter) at 510 nm is formed. The structure of *ortho*-phenanthroline follows.

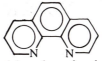

For reasons that will not be labored, *ortho*-phenanthroline will be represented as "Oph." The reaction is then:

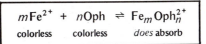

$$m Fe^{2+} + n Oph \rightleftharpoons Fe_m Oph_n^{2+}$$

colorless   colorless   *does* absorb

The ligand-to-metal ratio, n/m, can be called the stoichiometry of the complex. How is it found?

Example: Five (5.00) milliliters of a solution 100.0 milligram per liter in Fe^{2+} are added to each of a series of 100.0-milliliter volumetric flasks. A different amount of $5.05 \times 10^{-3}M$ Oph is added to each flask. Each solution is diluted to the mark with water. From the data in Table 4.8 compute the value of n/m for the complex.

Table 4.8
MOLE-RATIO DATA FOR THE
Fe^{2+}-Oph COMPLEX

Flask Number	milliters of $5.05 \times 10^{-3}M$ Oph added	A at 510 nm
1	2.00	0.374
2	3.00	0.560
3	4.00	0.747
4	6.00	0.993
5	8.00	0.993
6	10.00	0.993
7	12.00	0.993

Solution: We plot the abosrbance versus the volume of Oph added; this plot is shown in Figure 4.26. At the point where the extrapolated lines meet, like the endpoint of a photometric titration, all of the iron has been complexed by the added Oph. The number of milliliters of Oph needed to complex all the iron may be read off on the abscissa of the graph. The molarity of the Oph is known ($5.05 \times 10^{-3} M$), and the amount of iron present in each flask (5.00 milliliters of 100.0 milligrams/liter solution) is also known. From these known quantities, n/m can be calculated.

$$\frac{n}{m} = \left(\frac{\text{number of moles of Oph}}{\text{number of moles of Fe}^{2+}} \right)$$

(number of moles of Fe^{2+})

$$= \left(\frac{100.00 \times 10^{-3} \text{ grams of Fe}}{\text{liters}} \right) \left(\frac{5.00 \times 10^{-3} \text{ liters}}{55.85 \frac{\text{grams of Fe}}{\text{moles of Fe}}} \right) = 8.95 \times 10^{-6} \text{ moles of Fe}$$

From Figure 4.26

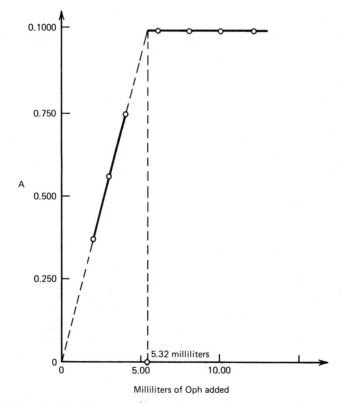

Figure 4.26 *Determination of Fe^{2+}-ortho-phenanthroline complex stoichiometry by mole-ratio method.*

(number of moles of Oph)

$$= (5.32 \times 10^{-3} \text{ liter}) \left(5.05 \times 10^{-3} \frac{\text{moles}}{\text{liters}}\right) = 2.69 \times 10^{-5} \text{ moles Oph}$$

$$\frac{n}{m} = \frac{2.69 \times 10^{-5} \text{ moles of Oph}}{8.95 \times 10^{-6} \text{ moles of Fe}} \cong \frac{3 \text{ moles of Oph}}{1 \text{ mole of Fe}}$$

The reaction between ferrous iron and *ortho*-phenanthroline is *probably*

$$Fe^{2+} + 3 \text{ Oph} \rightleftharpoons FeOph_3^{2+}$$

Because the ratio n/m, and *not* individual values of n and m are known, it is possible that the reaction is

$$2 Fe^{2+} + 6 \text{ Oph} \rightleftharpoons Fe_2Oph_6^{4+}$$

Structural considerations just about rule out the latter possibility, however, and it may be assumed that the complex composition is $FeOph_3^{2+}$.

There are about as many ways of determining iron spectrophotometrically as a St. Bernard has fleas. With this in mind the student may allow his imagination to roam and to suppose that at an obscure midwestern college, which is always nearly bankrupt, a new complexing agent for Fe^{2+} is found. The ligand is an organic molecule of enormously complicated structure, and is called versatilene, abbreviated Vers. The Fe^{2+}-Vers complex absorbs at 515 nanometers, where neither Fe^{2+} nor Vers alone absorbs. In the hope of finding the stoichiometry *and* formation constant for the reaction

$$m Fe^{2+} + n Vers \rightleftharpoons Fe_m Vers_n^{2+}$$

$$K_f = \frac{[Fe_m Vers_n^{2+}]}{[Fe^{2+}]^m \cdot [Vers]^n}$$

a mole-ratio experiment is run.

The Experiment: A series of solutions is made up in 50.0-milliliter volumetric flasks. The same amount of Fe^{2+} is added to each flask. This amount is 2.00 milliliters of $1.00 \times 10^{-3}M$ Fe^{2+}. Varying amounts of $1.00 \times 10^{-3}M$ Vers are added to each flask. Each flask is diluted to the mark. The absorbance, at 515 nm, of each solution is taken in a 1.00 centimeter cell. The results appear in Table 4.9. What is n/m? What is K_f?

Solution: First, n/m is not too hard to find. A plot of A versus added milliliters of versatilene will serve. Figure 4.27 is just such a plot.

$$\frac{n}{m} = \frac{(\text{number of moles of Vers at break point})}{(\text{number of moles of Fe})}$$

(number of moles of Fe^{2+})

$$= (1.00 \times 10^{-3}M)(2.00 \times 10^{-3} \text{ liters}) = 2.00 \times 10^{-6} \text{ moles of Fe}$$

(number moles Vers at break point)

$$= (1.00 \times 10^{-3} M)(6.00 \times 10^{-3} \text{ liters}) = 6.00 \times 10^{-6} \text{ moles Vers}$$

From Figure 4.27

$$\frac{n}{m} = \frac{6.00 \times 10^{-6} \text{ moles Vers at breakpoint}}{2.00 \times 10^{-6} \text{ moles Fe}} = \frac{3}{1}$$

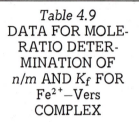

Table 4.9
DATA FOR MOLE-
RATIO DETER-
MINATION OF
n/m AND K_f FOR
Fe^{2+}–Vers
COMPLEX

Added Milliliters of Vers	A
2.00	0.240
3.00	0.360
4.00	0.480
5.00	0.593
6.00	0.700
8.00	0.720
10.00	0.720
12.00	0.720

The reaction is probably

$$Fe^{2+} + 3 \text{ Vers} = Fe \text{ Vers}_3^{2+}$$

$$K_f = \frac{[Fe \text{ Vers}_3^{2+}]}{[Fe^{2+}][\text{Vers}]^3}$$

How may a numerical value for K_f be found? Three quantities are needed: $[Fe \text{ Vers}_3^{2+}]$, $[Fe^{2+}]$, and [Vers]. They are most apt to have appreciable values around the breakpoint, where some dissociation can occur. A point on the curved portion of the graph is selected, where $A = 0.700$.

The calculation is

$$A = \epsilon b [Fe \text{ Vers}_3^{2+}]$$

$$[Fe \text{ Vers}_3^{2+}] = \frac{A}{\epsilon b}$$

The chemical bookkeeping follows:

$$C_{Fe} = [Fe^{2+}] + [Fe \text{ Vers}_3^{2+}]$$

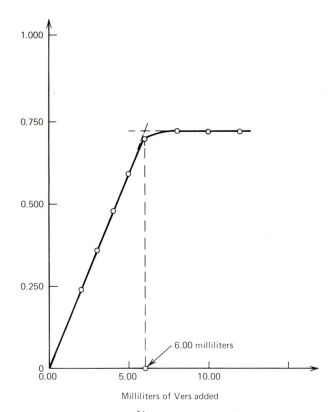

Figure 4.27 *Determination of* Fe^{2+}*-versatilene complex stoichiometry and* K_f *by mole-ratio method.*

$$[Fe^{2+}] = C_{Fe} - [Fe\,Vers_3^{2+}] = C_{Fe} - \frac{A}{\epsilon b}$$

$$C_{Vers} = [Vers] + 3[Fe\,Vers_3^{2+}]$$

$$[Vers] = C_{Vers} - 3[Fe\,Vers_3^{2+}] = C_{Vers} - 3\frac{A}{\epsilon b}$$

$$K_f = \frac{\dfrac{A}{\epsilon b}}{\left(C_{Fe} - \dfrac{A}{\epsilon b}\right) \cdot \left(C_{Vers} - \dfrac{3A}{\epsilon b}\right)^3} \tag{4.9}$$

It only remains to get numerical values for all the quantities in the last equation. The absorbance value chosen is that corresponding to 6.00 milliliters of Vers added, $A = 0.700$, because it is on a curved portion of the graph. The value of molar absorptivity, ϵ, is easily found. *Past* the breakpoint, virtually *all* of the Fe^{2+} is complexed.

Past the breakpoint:

$$A = 0.720$$

$$C_{Fe} = [\text{Fe Vers}_3^{2+}] = \left(\frac{2.00 \text{ milliliters}}{50.00 \text{ milliliters}}\right)(1.00 \times 10^{-3} \text{ moles/liter})$$

$$= 4.00 \times 10^{-5} M$$

$$[\text{Fe Vers}_3^{2+}] = 4.00 \times 10^{-5} M$$

$$A = \epsilon b [\text{Fe Vers}_3^{2+}]$$

$$\epsilon = \frac{A}{(b)[\text{Fe Vers}_3^{2+}]} = \frac{0.720}{(1.00)(4.00 \times 10^{-5})}$$

$$= 1.80 \times 10^4 \frac{\text{liter}}{(\text{mole})(\text{centimeter})}$$

Just at the breakpoint, $A = 0.700$.

$$\frac{A}{\epsilon b} = \frac{0.700}{(1.80 \times 10^4)(1.00)}$$

$$\frac{A}{\epsilon b} = 3.89 \times 10^{-5}$$

$$C_{Fe} = 4.00 \times 10^{-5} M$$

$$C_{Vers} = \left(\frac{6.00 \text{ milliliters}}{50.00 \text{ milliliters}}\right)\left(1.00 \times 10^{-3} \frac{\text{moles}}{\text{liter}}\right)$$

$$= 12.0 \times 10^{-5} M$$

We recall Eq. 4.9:

$$K_f = \frac{\dfrac{A}{\epsilon b}}{\left(C_{Fe} - \dfrac{A}{\epsilon b}\right)\left(C_{Vers} - \dfrac{3A}{\epsilon b}\right)^3}$$

$$= \frac{3.89 \times 10^{-5}}{(4.00 \times 10^{-5} - 3.89 \times 10^{-5})[12.0 \times 10^{-5} - (3)(3.89 \times 10^{-5})]^3}$$

$$= 9.84 \times 10^{17}$$

Where complexes are strong, the mole-ratio method may be used to great advantage to find both stoichiometry and formation constants for complexes.

≫ 4.20 The Method of Continuous Variations / There is another method that can be used to find the stoichiometry and formation constants of complexes. It is known as the method of continuous variations, or Job's method, after its originator. Suppose that a reaction takes place between a colorless metal M and a colorless ligand L to give a brightly colored complex, $M_m L_n$.

$$mM + nL \rightleftharpoons M_mL_n \qquad K_f = \frac{[M_mL_n]}{[M]^m \cdot [L]^n}$$

If just one complex is formed, and if K_f is high, Job's method is useful in determining n/m and K_f.

The method works like this: We prepare a series of flasks in which the sum of the number of moles of ligand plus the sum of the number of moles of metal is kept constant.

(number of moles of M) + (number of moles of L) = constant

So long as the *sum* remains *constant*, the addends that comprise it may be varied. We prepare a series of solutions of varying mole fraction of the metal and plot their absorbances versus the mole fraction of metal or the mole fraction of ligand.

mole fraction of M = X_M

$$= \frac{\text{(number of moles of M)}}{\text{(number of moles of M)} + \text{(number of moles of L)}}$$

mole fraction of L = X_L

$$= \frac{\text{(number of moles of L)}}{\text{(number of moles of M)} + \text{(number of moles of L)}}$$

$$X_M + X_L = 1$$

The resultant plot is a mound-shaped curve like that shown in Figure 4.28. The straight-line portions of the curve are extrapolated. The point on the abscissa corresponding to their intersection is called X_{MINT}, which is the mole fraction of M at which the maximum amount of complex that can be formed is formed and the maximum possible absorbance is observed. Dissociation accounts for the curvature.

$$\frac{n}{m} = \frac{X_{LINT}}{X_{MINT}}$$

$$X_{LINT} = 1 - X_{MINT}$$

$$\frac{n}{m} = \frac{1 - X_{MINT}}{X_{MINT}}$$

How does one find K_f? This will be explained in the following example.

A good example of the application of Job's method to a chemical system is given by

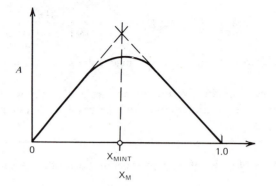

Figure 4.28 Example of Job's plot.

Carmody[*]. Around 500 nanometers, neither Fe^{3+} in a nitric acid solution nor the thiocyanate ion, SCN^-, absorbs very much. When the two are combined under these conditions, they react to form a brilliant red complex that has a reasonable molar absorptivity at 500 nanometers.

$$m Fe^{3+} + n SCN^- \rightleftharpoons Fe_m SCN_n^{(3m-n)}$$

$$K_f = \frac{[Fe_m SCN_n]}{[Fe^{3+}]^m \cdot [SCN^-]^n}$$

Carmody prepared the solutions listed in Table 4.10 and measured their absorbances in a 2.50 centimeter cell at around 500 nanometers. From these data may be determined n/m and K_f for the complex. How?

The data do not include one single mole fraction, but they do not have to. The number of milliliters of either Fe^{3+} solution or SCN^- solution is proportional to either the mole fraction of Fe^{3+} or of SCN^-, as the case may be. Why? The concentrations of the two solutions are equal, and the total volumes of the mixtures whose absorbances are measured are always 24.00 milliliters. Therefore, absorbance may be plotted against either the milliliters of Fe^{3+} solution or the milliliters of SCN^- solution. The plot is shown in Figure 4.29. The extrapolated straight lines intersect at 12.00 milliliters of KSCN solution.

$$\frac{n}{m} = \frac{\text{(number of moles of KSCN)}}{\text{(number of moles of } Fe^{3+})}$$

$$= \frac{(12.00 \times 10^{-3} \text{ liters})(1.00 \times 10^{-3} \text{ mole/liter})}{(12.00 \times 10^{-3} \text{ liters})(1.00 \times 10^{-3} \text{ mole/liter})}$$

$$= 1.00$$

* W. R. Carmody, J Chem. Educ., *41, 615 (1964)*.

Table 4.10
DATA ON Fe^{3+} - SCN^- COMPLEX FOR JOB'S METHOD.[a]

Solution Number	Milliliters of $1.00 \times 10^{-3}M$ $Fe(NO_3)_3$	Milliliters of $1.00 \times 10^{-3}M$ KSCN	A
1	24.00	0.00	0.000
2	22.00	2.00	0.125
3	20.00	4.00	0.237
4	18.00	6.00	0.339
5	16.00	8.00	0.415
6	14.00	10.00	0.460
7	12.00	12.00	0.473
8	10.00	14.00	0.460
9	8.00	16.00	0.424
10	6.00	18.00	0.344
11	4.00	20.00	0.254
12	2.00	22.00	0.152
13	0.00	24.00	0.000

[a] *From W. R. Carmody,* J. Chem. *Educ.,* **41,** *615 (1964), by permission.*

$$Fe^{3+} + SCN^- \rightleftharpoons FeSCN^{2+}$$

$$K_f = \frac{[FeSCN^{2+}]}{[Fe^{3+}][SCN^-]}$$

How can K_f be found? Three quantities, $[FeSCN^{2+}]$, $[Fe^{3+}]$, and $[SCN^-]$, need to be known. They are all best calculated from a point where all are large and finite. At the maximum absorbance observed in the Job's plot, all are large and finite. Thus, for our calculations, we shall deal with that maximum, where $A = 0.473$.

$$A = 0.473$$

$$\text{milliliters of } Fe^{3+} = 12.00 \text{ milliliters}$$

$$\text{milliliters of } SCN^- = 12.00 \text{ milliliters}$$

$$C_{Fe} = [Fe^{3+}] + [Fe\,SCN^{2+}]$$

$$[Fe^{3+}] = C_{Fe} - [Fe\,SCN^{2+}]$$

$$C_{SCN} = [SCN^-] + [Fe\,SCN^{2+}]$$

$$[SCN^-] = C_{SCN} - [Fe\,SCN^{2+}]$$

$$C_{Fe} = \frac{(12.00 \text{ milliliters})(1.00 \times 10^{-3}M)}{(24.00 \text{ milliliters})} = 5.00 \times 10^{-4}M$$

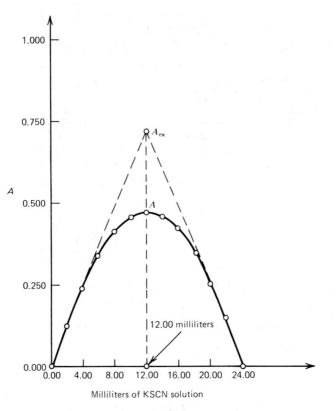

Figure 4.29 Determination of Fe^{3+}-SCN^- complex stoichiometry by Job's method.

$$C_{SCN} = \frac{(12.00 \text{ milliliters})(1.00 \times 10^{-3}M)}{(24.00 \text{ milliliters})} = 5.00 \times 10^{-4}M$$

What is $[Fe\ SCN^{2+}]$?

To find $[Fe\ SCN^{2+}]$, we define a new term, A_{ex}, the extrapolated absorbance. A_{ex} is that absorbance which would be observed if the complex did not dissociate at all. If the complex did not dissociate at all, the equilibrium concentration of the complex $[Fe\ SCN^{2+}]$ would be equal to the analytical concentration of the iron, C_{Fe}.

$$\begin{cases} [Fe\ SCN^{2+}]_{ex} = C_{Fe} \\ A_{ex} = \epsilon b[Fe\ SCN^{2+}]_{ex} = \epsilon b C_{Fe} \end{cases}$$

$$\begin{cases} A = observed \text{ absorbance} \\ A = \epsilon b[Fe\ SCN^{2+}] \end{cases}$$

$$\frac{A}{A_{ex}} = \frac{\cancel{\epsilon b}[Fe\ SCN^{2+}]}{\cancel{\epsilon b}C_{Fe}}$$

$$[Fe\,SCN^{2+}] = \frac{A}{A_{ex}} C_{Fe}$$

From Figure 4.29

$$A = 0.473$$

$$A_{ex} = 0.718$$

$$K_f = \frac{[Fe\,SCN^{2+}]}{[Fe^{3+}][SCN^-]} = \frac{[Fe\,SCN^{2+}]}{(C_{Fe} - [Fe\,SCN^{2+}])(C_{SCN} - [Fe\,SCN^{2+}])}$$

$$= \frac{\dfrac{A}{A_{ex}} C_{Fe}}{\left(C_{Fe} - \dfrac{A}{A_{ex}} C_{Fe}\right)\left(C_{SCN} - \dfrac{A}{A_{ex}} C_{Fe}\right)}$$

$$= \frac{\dfrac{A}{A_{ex}} \cancel{C_{Fe}}}{\cancel{C_{Fe}}\left(1 - \dfrac{A}{A_{ex}}\right)\left(C_{SCN} - \dfrac{A}{A_{ex}} C_{Fe}\right)}$$

$$= \frac{\dfrac{0.473}{0.718}}{\left(1 - \dfrac{0.473}{0.718}\right)\left[5.00 \times 10^{-4} - \left(\dfrac{0.473}{0.718}\right)(5.00 \times 10^{-4})\right]}$$

$$= 1.13 \times 10^4$$

(This does not disagree too badly with some other K_f - values found in the literature.)

≫ 4.21 The Slope-Ratio Method / There is one more method for determining the stoichiometry of a complex spectrophotometrically. It is called the slope-ratio method and is especially applicable to determining the stoichiometry of weak complexes. (The mole-ratio method, applied to weak complexes, often gives curves without any straight-line portions to extrapolate.) Unfortunately, the slope-ratio method cannot be used for finding the value of the formation constant for a complex.

Harvey and Manning, in 1950*, proposed and worked out the details of the slope-ratio method. Much of the discussion that follows is based on their treatment. Suppose that metal M and ligand L react to a form a complex $M_m L_n$.

$$mM + nL \rightleftharpoons M_m L_n$$

Suppose further that M and L, uncombined, are colorless, but that $M_m L_n$ absorbs. The determination of n/m is a two-step process.

First, we acquire, by some means, a series of volumetric flasks, and put into each enough L so that the L concentration will be constant from flask to flask and far in excess of the concentration of M to be added. The excess L insures complexation of all the M in each flask. The equilibrium concentration value of the complex, $[M_m L_n]$ in each flask will be determined by the value of the analytical concentration of M, C_M, added to each flask.

* A. E. Harvey and D. L. Manning, J. Am. Chem. Soc., 72, 4488 (1950).

$$\boxed{[M_mL_n] = \frac{C_M}{m}}$$

We measure the absorbance of each solution at the λ_{max} of the complex.

$$\boxed{\begin{aligned} A &= \epsilon b [M_mL_n] \\ A &= \epsilon b \frac{C_M}{m} \end{aligned}}$$

We then make a plot of the absorbance versus C_M, as shown in Figure 4.30. The slope of the curve is measured, and related to ϵ, b, and m.

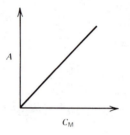

Figure 4.30 *Plot for slope-ratio method.*

$$\boxed{\begin{aligned} A &= \frac{\epsilon b}{m} \cdot C_M \\[4pt] \frac{A}{C_M} &= \frac{\epsilon b}{m} \\[4pt] slope_1 &= \frac{\Delta A}{\Delta C_M} = \frac{\epsilon b}{m} \end{aligned}}$$

(4.10)

This is half of the experiment.

Next, we get another set of volumetric flasks (or wash out the set just used) and put into each flask enough M so that the M concentration will be constant and far in excess of the concentration of L to be added to each flask. The excess M insures complexation of all the L in each flask. The equilibrium concentration of the complex, $[M_mL_n]$, in each flask will be determined by the value of the analytical concentration of L, C_L, added to each flask.

$$\boxed{[M_mL_n] = \frac{C_L}{n}}$$

We measure the absorbance of each solution at the λ_{max} of the complex.

$$A = \epsilon b [M_m L_n]$$

$$A = \epsilon b \frac{C_L}{n}$$

Then we make a plot of absorbance versus C_L, as shown in Figure 4.31. The slope of the curve is measured, and related to ϵ, b, and n.

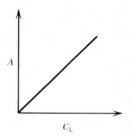

Figure 4.31 *Another plot for slope-ratio method.*

$$A = \frac{\epsilon b}{n} C_L$$

$$\frac{A}{C_L} = \frac{\epsilon b}{n}$$

$$\text{slope}_2 = \frac{\Delta A}{\Delta C_L} = \frac{\epsilon b}{n} \qquad (4.11)$$

This is the other half of the experiment. One slope is then divided by the other.

$$\frac{\text{slope}_1}{\text{slope}_2} = \frac{\dfrac{\cancel{\epsilon b}}{m}}{\dfrac{\cancel{\epsilon b}}{n}}_1$$

$$\frac{\text{slope}_1}{\text{slope}_2} = \frac{n}{m} \qquad (4.12)$$

Harvey and Manning give numerous examples of stoichiometry determination. Among their examples is that of a blue ($\lambda_{\max} = 620$ nanometers) complex of Fe^{3+} and 1,2-dihydroxybenzene-3,5-disulfonate, the latter being called Tiron, and abbreviated Trn. Their data are shown in Table 4.11. The data are plotted in Figures 4.32 and 4.33. From the graphs, slope_1 and slope_2, as defined above, are easily obtained.

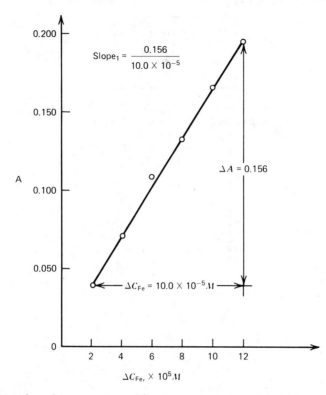

Slope$_1$ = $\dfrac{0.156}{10.0 \times 10^{-5}}$

$\Delta A = 0.156$

$\Delta C_{Fe} = 10.0 \times 10^{-5} M$

$\Delta C_{Fe}, \times 10^5 M$

Figure 4.32 *Plot of A versus* C_{Fe} *with* C_{Tm} *constant. Reprinted with permission from A. E. Harvey and D. L. Manning, J. Amer. Chem. Soc., 72, 4488 (1950). Copyright by the American Chemical Society.*

Table 4.11
DATA FOR SLOPE-
RATIO DETERMINA-
TION FROM HARVEY
AND MANNING[a]
FOR SLOPE$_1$:
C_{Tm} = CONSTANT
= $30.0 \times 10^{-5} M$

C_{Fe}, M	A
2.00×10^{-5}	0.038
4.00×10^{-5}	0.072
6.00×10^{-5}	0.109
8.00×10^{-5}	0.133
10.00×10^{-5}	0.166
12.00×10^{-5}	0.195

Table 4.11 continued

FOR SLOPE$_2$:
C$_{Fe}$ = CONSTANT
= 30 × 10^{-5}M

C_{Trn}, M	A
2.00 × 10^{-5}	0.017
4.00 × 10^{-5}	0.048
6.00 × 10^{-5}	0.082
8.00 × 10^{-5}	0.114
10.00 × 10^{-5}	0.147
12.00 × 10^{-5}	0.179

[a]Reprinted with permission from A. E. Harvey and D. L. Manning, J. Am. Chem. Soc., 72, 4488 (1950). Copyright by the American Chemical Society.

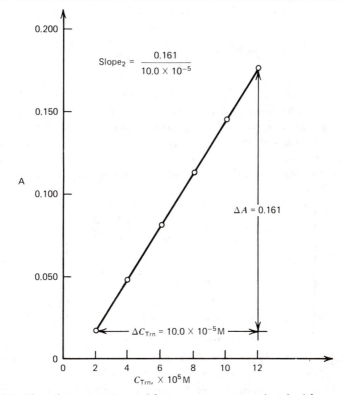

Figure 4.33 Plot of A versus C$_{Trn}$ with C$_{Fe}$ constant. Reprinted with permission from A. E. Harvey and D. L. Manning, J. Amer. Chem. Soc., 72, 4488 (1950). Copyright by the American Chemical Society.

$$\frac{n}{m} = \frac{\text{slope}_1}{\text{slope}_2} \qquad (4.12)$$

$$\frac{n}{m} = \frac{\dfrac{0.156}{10.0 \times 10^{-5}}}{\dfrac{0.161}{10.0 \times 10^{-5}}}$$

$$\frac{n}{m} = 0.969$$

$$\frac{n}{m} \cong 1$$

The reaction between Fe^{3+} and Tiron to form the blue complex is probably one-to-one.

$$Fe^{3+} + Trn \rightleftharpoons Fe\,Trn^{3+}$$

The equilibrium constant for the reaction would have to be found by some other means.

4.22 K_a-Values for Indicator Dyes

Complex ions are not the only brightly colored species of interest to the analytical chemist. Indicator dyes also excite the analyst's imagination if only for the reason that it is far easier to run a titration whose endpoint is signaled by a color change than to follow a titration with a pH meter, carefully taking readings near the endpoint. Many indicator dyes are weak acids. They can be represented quite simply as HIn, - called the acid or protonated form. H^+ is a proton, and In^- is some organic ion, called the *un*-protonated, or basic form.

$$HIn \rightleftharpoons H^+ + In^- \qquad K_a = \frac{[H^+][In^-]}{[HIn]}$$

one color   another
acid form   color
       basic form

$$\frac{[In^-]}{[HIn]} = \frac{K_a}{[H^+]}$$

The ratio of $[In^-]/[HIn]$ is determined by the value of $[H^+]$. At low values of $[H^+]$, $[In^-] > [HIn]$, and the basic form predominates. At high values of $[H^+]$, $[In^-] < [HIn]$, and the acid form predominates. If, as we have said, there are two forms of the indicator, HIn and In^-, each of a different color, and if the ratio of their concentrations is determined by the value of $[H^+]$, the absorbance spectrum of an indicator dye ought to show some variation with pH. The point is illustrated by Figure 4.34, taken

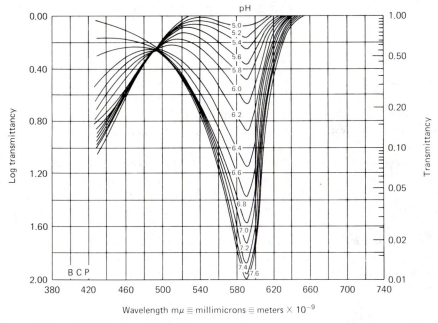

Figure 4.34 *pH-Dependent absorption spectra.**

from a 1924 publication by Brode*.(Note that the left-hand ordinate, called "log transmittancy" runs down from 0.00 to 2.00.)

$$\text{log transmittancy} = \log T = -A$$

High absorbance is found at the bottom of the graph. A number of curves are plotted on the graph. Each curve is an absorption spectrum for the dye bromcresol purple in a buffer of the pH indicated on the curve.The basic form of the dye has an absorbance maximum around 590 nanometers. As the pH increases, the absorbance at 590 nanometers increases—just as it should, for more of the In^- is being formed as the pH increases. The acid form of the indicator has an absorbance maximum somewhere below 440 nanometers. Unfortunately, Brode's instrument (this was done in 1924, remember) did not operate very far below 440 nanometers. As the pH increases, the absorbance of the acid form at 440 nanometers decreases just as it should, for more of the In^- is formed at the expense of HIn as pH increases. All of the curves seem to intersect at one point, about 495 nanometers. This point of intersection is known as the *isosbestic point.*

---

* *W. R. Brode*, J. Am. Chem. Soc., *46, 581 (1924).*

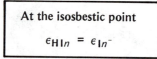

At the isosbestic point

$$\epsilon_{HIn} = \epsilon_{In^-}$$

The isosbestic point provides a handy wavelength for determining the analytical concentration of the dye, C_L (C_L = [HIn] + [In$^-$]), without resort to a pH adjustment. It also indicates that there are *probably* only two chemical species containing any form of the indicator present at the pH-values investigated. These two chemical species are HIn and In$^-$.

How should we determine the value of K_a for the indicator? First, we select a wavelength at which either the acid or base form absorbs most strongly. In the case of bromcresol purple, 590 nanometers is the wavelength chosen. At this wavelength, the unprotonated, or basic form of the indicator, In$^-$, has its absorbance maximum. Then, we vary the pH and observe the value of the absorbance as a function of pH. Brode's data are shown in Table 4.12. Figure 4.35 shows the variation of absorbance with pH. Above pH 7.6, it is safe to assume that all of the indicator is in the basic form, and that absorbance is 2.00. When the absorbance drops to half this value, half of the In$^-$ has been protonated.

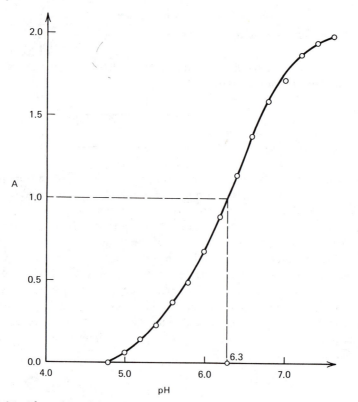

Figure 4.35 *Plot of Brode's data for bromcresol purple (**A versus pH** @ λ = 590 nm).*

Table 4.12
BRODE'S DATA FOR BROMCRESOL PURPLE[a]

pH	A
4.8	0.000
5.0	0.057
5.2	0.143
5.4	0.229
5.6	0.363
5.8	0.486
6.0	0.672
6.2	0.887
6.4	1.144
6.6	1.373
6.8	1.587
7.0	1.716
7.2	1.873
7.4	1.945
7.6	1.973

[a]From W. R. Brode, J. Am. Chem. Soc., 46, 581 (1924).

At

$$A = 1.00$$

$$[HIn] = [In^-]$$

$$K_a = \frac{[H^+][In^-]}{[HIn]}$$

$$K_a = [H^+]$$

$$pK_a = pH$$

From Figure 4.35, the value of pK_a for bromcresol purple is 6.3.

Another way of plotting the data yields a straight line, and a value of pK_a for the indicator. It offers the advantage of not requiring the experimenter to guess at the best curve to draw with a French curve. The mathematics involved is simple.

$$K_a = \frac{[H^+][In^-]}{[HIn]}$$

$$C_{In} = [HIn] + [In^-]$$

$$[\text{HI}n] = C_{\text{I}n} - [\text{I}n^-]$$

$$\frac{[\text{I}n^-]}{(C_{\text{I}n} - [\text{I}n^-])} = \frac{[\text{I}n^-]}{[\text{HI}n]}$$

$$K_a = [\text{H}^+] \frac{[\text{I}n^-]}{(C_{\text{I}n} - [\text{I}n^-])}$$

At pH 7.6 and above;

$$C_{\text{I}n} = [\text{I}n^-]$$

Let A at pH 7.6 = A_{basic},

$$A_{\text{basic}} = \epsilon b C_{\text{I}n}$$

$$C_{\text{I}n} = \frac{A_{\text{basic}}}{\epsilon b}$$

At *any* point;

$$A = \epsilon b [\text{I}n^-]$$

$$[\text{I}n^-] = \frac{A}{\epsilon b}$$

$$[\text{HI}n] = C_{\text{I}n} - [\text{I}n^-]$$

$$[\text{HI}n] = \frac{A_{\text{basic}}}{\epsilon b} - \frac{A}{\epsilon b}$$

$$K_a = \frac{[\text{H}^+][\text{I}n^-]}{(C_{\text{I}n} - [\text{I}n])}$$

$$K_a = [\text{H}^+] \left\{ \frac{\dfrac{A}{\epsilon b}}{\dfrac{A_{\text{basic}}}{\epsilon b} - \dfrac{A}{\epsilon b}} \right\}$$

$$K_a = [\text{H}^+] \left\{ \frac{A}{A_{\text{basic}} - A} \right\}$$

$$\log K_a = \log [\text{H}^+] + \log \left(\frac{A}{A_{\text{basic}} - A} \right)$$

$$\log [\text{H}^+] = \log K_a - \log \left(\frac{A}{A_{\text{basic}} - A} \right)$$

$$\text{pH} = pK_a + \log \left(\frac{A}{A_{\text{basic}} - A} \right)$$

$$\log \left(\frac{A_{\text{basic}} - A}{A} \right) = pK_a - \text{pH} \qquad (4.13)$$

From Eq. 4.13,

If $\log \left(\dfrac{A_{\text{basic}} - A}{A} \right) = 0$

Then $\qquad \text{pH} = pK_a$

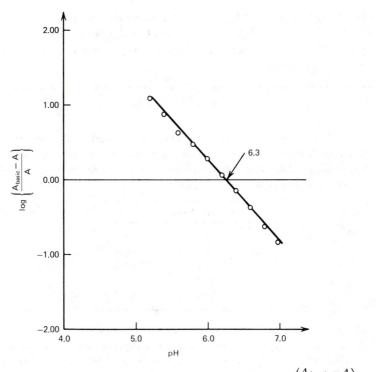

Figure 4.36 *Linear plot of Brode's data for bromcresol purple* $\dfrac{(A_{\text{basic}} - A)}{A}$ *versus pH @ λ = 590 nm].*

Table 4.13
BRODE'S DATA FOR
BROMCRESOL PURPLE(a)

pH	$A_{\text{basic}} - A$	$\dfrac{A_{\text{basic}} - A}{A}$	$\log\left(\dfrac{A_{\text{basic}} - A}{A}\right)$
5.2	1.830	12.80	1.107
5.4	1.744	7.62	0.882
5.6	1.610	4.44	0.647
5.8	1.487	3.06	0.486
6.0	1.301	1.94	0.287
6.2	1.086	1.24	0.088
6.4	0.829	0.724	−0.140
6.6	0.600	0.437	−0.360
6.8	0.386	0.243	−0.613
7.0	0.257	0.150	−0.824

[a]*From W. R. Brode,* J. Am. Chem. Soc., *46, 581 (1924).*

A plot of $\log((A_{basic} - A)/A)$ versus pH should yield a straight line. The pH value where $\log((A_{basic} - A)/A) = 0$ should be the pK_a value for the indicator. Table 4.13 and Figure 4.36 contain Brode's data. The value of pK_a is 6.3, in agreement with the value of 6.3 found earlier.

≡ 4.23 PROBLEMS ≡

1. Copper(II) is complexed by bis-cyclohexanone dioxaldihydrazone, popularly known as "cuprizone." The complex shows an absorption maximum at 606 nm. A chemist in Columbus, Ohio, gathered the following data on the complex at 606 nanometers in 1.00 centimeter cells*

μgCu per 25.00ml	A
44.0	0.553
56.0	0.699
68.0	0.860
85.0	0.998
95.0	1.160

(a) Another chemist, on the same day, determined the copper concentration of a solution, using the same cell, spectrophotometer, and wavelength. He found the absorbance of his solution to be 0.750. What was the concentration of copper in his solution, in moles per liter?
   Ans: $3.90 \times 10^{-5} M$

(b) What is the molar absorptivity of the copper-cuprizone complex at 606 nanometers? Assume 1 mole of copper to every mole of complex.
   Ans: 1.95×10^4 liter per (mole · centimeter)

2. Cobalt(II) is readily complexed by the thiocyanate ion to give a blue species. The procedure is simple: A solution of three parts acetone to one part of 50 percent aqueous solution of NH_4SCN is made. This solution is added to a flask containing Co^{2+}, and the resultant complex is a bright blue. Vicki McGuffin in 1974 prepared the following standard solutions and read their absorbances at 620 nanometers on a Bausch and Lomb Spectronic 20.

Co, ppm (mg / liter)	A
3.0	0.10
6.0	0.21
9.0	0.32
12.0	0.42

---

* Reprinted by permission from G. F. Smith, The Trace Element Determination of Copper and Mercury in Pulp and Paper. The G. Frederick Smith Chemical Company, Columbus, Ohio, 1954.

Then she took 4.0 milliliters of the effluent from a separation column, and diluted this to 100 milliliters in a volumetric flask with the solution of acetone, water, and NH_4SCN. The absorbance of the resultant solution, measured at 620 nanometers in the same cell used for the standards, turned out to be 0.25. Data Courtesy of V. McGuffin.

(a) What was the concentration of cobalt, in ppm, in the 100-milliliter flask used to contain the unknown?

Ans: 7.1 ppm

(b) What was the concentration of cobalt, in ppm, in the 4.0 milliliter sample?

Ans: 1.8×10^2 ppm

(c) How many milligrams of cobalt were in the 4.0-milliliter sample?

Ans: 0.72 milligrams

3. Manganese is often determined spectrophotometrically as the permanganate ion (MnO_4^-), whose aqueous solutions are a deep purple color ($\lambda_{max} = 525$ nm). A 1.00×10^{-4} M solution of $KMnO_4$ gives an absorbance of 0.585 when a 1.00 centimeter cell is used at 525 nm. A 0.500 gram sample of a maganese-containing alloy is dissolved in acid, and all the manganese is converted to MnO_4^- by periodate oxidation. The sample is then diluted to 500 milliliters in a volumetric flask, and its absorbance, taken at 525 nanometers in a 1.00 centimeter cell, is found to be 0.400. Assume that the permanganate system follows Beer's law and calculate the weight percent of manganese in the unknown.

Ans: *0.376%*

4. Copper(I) and neo-cuproine (abbreviated as "Nec") form a strong complex.

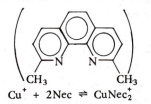

$$Cu^+ + 2Nec \rightleftharpoons CuNec_2^+$$

The complex is yellow and has a molar absorptivity of 7950 liters/ (mole · centimeter) at 450 nanometers in chloroform. The complex obeys Beer's law at 450 nanometers. A solution of the $CuNec_2^+$ complex in chloroform, in which $C_{Cu} = 6.30 \times 10^{-5} M$, gives an absorbance of 0.500 at 450 nanometers in the 1.00 centimeter cell of a spectrophotometer. An unknown solution is prepared in this way: A penny, weighing 3.0011 grams is soaked for a week in 25.00 milliliters of slightly acidic distilled H_2O. The distilled H_2O with some copper from the penny dissolved in it is transferred to a special vessel. The Cu^{2+} is reduced to Cu^+. The Cu^+ is complexed with neo-cuproine. The complex is completely transferred to 20.00 milliliters of chloroform. Some of this chloroform solution is placed in a 1-centimeter cell, and its absorbance at 450 nanometers is found to be 0.125. How many grams of copper were lost from the penny?

Ans: *2.01×10^{-5} grams*

5. Ferrous ion, Fe^{2+}, and bipyridyl (structure shown: two pyridine rings each with N, abbreviated "Bipy") form a pinkish-red complex.

$$Fe^{2+} + 3Bipy \rightleftharpoons FeBipy_3^{2+}$$

The complex obeys Beer's law over a wide concentration range. There is a device that once enjoyed great propularity in colorimetric determinations, called a Duboscq colorimeter. It employs visual color (absorbance) matching of two solutions, a known and an unknown. The cell lengths of known and unknown solutions are varied until a color match is achieved (until the absorbances of the two are equal). A color match between a standard of 4.0 ppm Fe and an unknown sample, the Fe^{2+} in both being complexed by bipyridyl, is achieved when the light path through the standard is 2.00 centimeters and that through the unknown is 1.25 centimeters. What is the concentration of iron in the unknown, in ppm?

Ans: *6.4 ppm*

6. Manganese may be determined as permanganate by visual colorimetry with a Duboscq colorimeter as well as by spectrophotometry with more sophisticated instruments. The deep purple of permanganate solutions is quite visible. A 0.450 gram sample of a manganese-containing alloy is dissolved in acid, and all the manganese oxidixed to MnO_4^-. The sample is diluted to 200 milliliters in a volumetric flask. A color match between a $1.00 \times 10^{-4} M$ standard solution of $KMnO_4$ and the unknown is reached when the light path through the standard is 4.00 centimeters and that through the unknown is 5.50 centimeters. What is the weight % of manganese in the sample?

Ans: *0.177%*

7. The absorbances of $Co(NO_3)_2$ and $Cr(NO_3)_3$ solutions are additive over the visible spectrum. A student decides to analyze a solution simultaneously spectrophotometrically for both Co^{2+} and Cr^{3+}. He chooses wavelengths of 400 and 505 nanometers for his analysis and uses a 1.00 centimeter cell. His results follow, together with some other data.

$$A_{400} = 1.167$$

$$A_{505} = 0.674$$

Absorptivities, liters/(mole · centimeter)	
$\epsilon_{Co400} = 0.530$	$\epsilon_{Cr400} = 15.2$
$\epsilon_{Co505} = 5.07$	$\epsilon_{Cr505} = 5.60$

Calculate the concentrations of chromium and cobalt, in moles/liter, in the unknown mixture.

Ans: $C_{Co} = 0.500M$; $C_{Cr} = 0.0750M$

8. It is sometimes desired to determine chromium and manganese in steels simultaneously. This analysis can be done spectrophotometrically, if one has the time. The manganese is oxidized to MnO_4^-, and the chromium to $Cr_2O_7^{2-}$. The absorbances of the resulting solutions, according to Lingane and Collat*, are best measured at 440 and 545 nanometers (see Table 4.14). The absorbances of dichromate and permanganate are additive. A 1.000-gram sample of a steel is dissolved, and the chromium and manganese are oxidized to $Cr_2O_7^{2-}$ and MnO_4^-. The solution is diluted to 50 milliliters in a volumetric flask. Absorbance readings made on this solution at 440 and 545 nanometers are:

$$A_{440} = 0.204$$
$$A_{545} = 0.860$$

In order to estimate absorptivities for both MnO_4^- and $Cr_2O_7^{2-}$ at both 440 and 545 nanometers, two sets of solutions are made up. Their absorbances are measured at 440 and 545 nanometers. Define

$$C_{Cr} = \text{concentration of } Cr_2O_7^{2-}, M$$
$$C_{Mn} = \text{concentration of } MnO_4^-, M$$

Table 4.14[a]

C_{Cr}, M	A at 440 nm	A at 545 nm	C_{Mn}, M	A at 440 nm	A at 545 nm
2.00×10^{-4}	0.074	—	1.00×10^{-4}	—	0.235
4.00×10^{-4}	0.148	—	2.00×10^{-4}	—	0.470
8.00×10^{-4}	0.295	—	3.00×10^{-4}	—	0.705
1.20×10^{-3}	0.443	—	4.00×10^{-4}	—	0.940
1.00×10^{-2}	—	0.110	1.00×10^{-3}	0.095	—
2.00×10^{-2}	—	0.220	2.00×10^{-3}	0.190	—
3.00×10^{-2}	—	0.330	3.00×10^{-3}	0.285	—
4.00×10^{-2}	—	0.440	4.00×10^{-3}	0.380	—

[a] Reprinted with permission from J. J. Lingane and J. W. Collat, Anal. Chem., 22, 166 (1950). Copyright by the American Chemical Society.

What are the weight percents of chromium and manganese in the alloy?
Ans: *0.239% Cr; 0.100% Mn*

9. Copper may be determined by spectrophotometric titration with a sodium salt EDTA (Na_2H_2Y). The reaction is one-to-one.

$$Cu^{2+} + Y^{4-} \rightleftharpoons CuY^{2-}$$

does	does	absorbs
not	not	strongly
absorb	absorb	at 625 nm
much		

* J. J. Lingane and J. W. Collat, Anal. Chem., 22, 166 (1950).

Fifty (50.00) milliliters of a copper solution are taken from a 250-milliliter volumetric flask and titrated with $0.1000M$ Na_2H_2Y solution. The data are as follows:

Milliliters of Y^{4-} Solution	Corrected A at 625 nm
2.00	0.080
4.00	0.160
6.00	0.240
8.00	0.320
10.00	0.400
12.00	0.420
14.00	0.420
16.00	0.420

How many milligrams of copper are present in the 250-milliliter volumetric flask?
Ans: *333.6 milligrams Cu*

10. Bismuth and copper may be determined simultaneously by spectrophotometric titration with EDTA.

$$Cu^{2+} + Y^{4-} \rightleftharpoons CuY^{2-} \quad K_f \cong 10^{+18}$$

does	does	absorbs
not	not	strongly
absorb	absorb	at 625 nanometers
much		

$$Bi^{3+} + Y^{4-} \rightleftharpoons BiY^{-} \quad K_f \cong 10^{+40}$$

does	does	does
not	not	not
absorb	absorb	absorb

A sample containing both Cu^{2+} and Bi^{3+} is spectrophotometrically titrated at 625 nanometers with EDTA $(0.1000M)$. The data follow.

Milliliters of Y^{4-}	A_{Corr}	Milliliters of Y^{4-}	A_{Corr}
2.00	0.050	18.00	0.250
4.00	0.050	20.00	0.300
6.00	0.050	22.00	0.350
8.00	0.050	26.00	0.360
12.00	0.100	28.00	0.360
14.00	0.150	30.00	0.360
16.00	0.200		

How many milligrams of copper and how many milligrams of bismuth are present in the sample?
Ans: *79.4 milligrams Cu; 209 milligrams Bi*

11. Ferrous iron, Fe^{2+}, forms a brightly colored complex with bathophenanthroline (abbreviated "Baph").

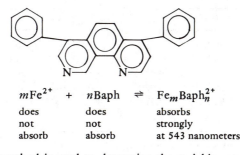

$$m\text{Fe}^{2+} \;+\; n\text{Baph} \;\rightleftharpoons\; \text{Fe}_m\text{Baph}_n^{2+}$$

does	does	absorbs
not	not	strongly
absorb	absorb	at 543 nanometers

The mole-ratio method is used to determine the stoichiometry of the complex.* To each of a series of 100 milliliter volumetric flasks is added 4.50×10^{-6} moles of Fe^{2+}. Varying amounts of $2.50 \times 10^{-3}M$ Baph are added to each flask, and the flasks are filled to the mark. The absorbance of each solution in a 1.00 centimeter cell at 543 nanometers is read. The results are as follows.

Milliliters of Baph	A	Milliliters of Baph	A
0.00	0.000	7.00	1.000
1.00	0.185	8.00	1.000
2.00	0.370	9.00	1.000
3.00	0.555	10.00	1.000
4.00	0.740		
6.00	1.000		

What is n/m for the complex?
Ans: *3*

12. Let us allow our imaginations to roam. Suppose that a new organic ligand, Gelbine (Gel), which complexes Cu(I) very strongly to form a yellow product, is discovered. The stoichiometry and formation constant for the complex are desired.

$$m\text{Cu}^+ \;+\; n\text{Gel} \;\rightleftharpoons\; \text{Cu}_m\text{Gel}_n^+$$

colorless	colorless	yellow

$$K_f = \frac{[\text{Cu}_m\text{Gel}_n^+]}{[\text{Cu}^+]^m \cdot [\text{Gel}]^n}$$

The mole-ratio method is applied to the problem. In a series of volumetric flasks, The analytical concentration of Cu^+ is kept constant at $6.90 \times 10^{-5}M$. The analytical concentration of Gel is varied from flask to flask. The data collected for these solutions at 450 nanometers where the complex absorbs most strongly are

---

* *G. F. Smith, W. H. McCurdy, and H. Diehl,* Analyst, *77, 418 (1952).*

C_{Gel}, M	A	C_{Gel}, M	A
1.50×10^{-5}	0.070	2.50×10^{-4}	0.609
3.50×10^{-5}	0.163	3.00×10^{-4}	0.626
6.00×10^{-5}	0.275	3.50×10^{-4}	0.630
1.00×10^{-4}	0.420	4.00×10^{-4}	0.630
1.50×10^{-4}	0.515	4.50×10^{-4}	0.630
2.00×10^{-4}	0.575		

Evaluate n/m and K_f for the complex.

Ans: **$n/m = 2$; $K_f = 3.23 \times 10^{+9}$**

13. Chloride is not *always* determined by gravimetry. Some years ago, Gerlach and Frazier* developed a method for the rapid determination of chloride in sweat and serum. The method involves a complex between the mercuric ion, Hg(II) and diphenyl carbazone (abbreviated Dip).

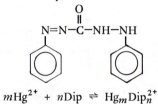

$$m\text{Hg}^{2+} + n\text{Dip} \rightleftharpoons \text{Hg}_m\text{Dip}_n^{2+}$$

The complex, $\text{Hg}_m\text{Dip}_n^{2+}$, absorbs most strongly at 520 nanometers, at a pH of 3.2. In an attempt to determine n/m for the complex, the authors used the method of continuous variations. Their data are as follows.

X_{Hg}	A	X_{Hg}	A
0.10	0.416	0.60	1.531
0.20	0.804	0.70	1.116
0.30	1.220	0.80	0.727
0.40	1.609	0.90	0.363
0.50	1.920		

From these data, find n/m for the complex.

Ans: **$n/m = 1$**

14. Ferric iron, Fe^{3+}, and 8-hydroxyquinoline (abbreviated HQ)

form a soluble complex that absorbs at 645 nanometers, a spectral region in which neither Fe^{3+} nor HQ absorbs. Sandell and Spindler[†] determined the stoi-

* *J. L. Gerlach and R. G. Frazier,* Anal. Chem., *30, 1142 (1958).*
† *E. B. Sandell and D. C. Spindler,* J. Am. Chem. Soc., *71, 3806 (1949).*

chiometry of the complex by Job's method. Their data (taken at pH 1.92) follow.

$$C_{Fe} + C_Q = 1.19 \times 10^{-3}M$$

X_{Fe}	A	X_{Fe}	A
0.100	0.058		
0.200	0.111	0.600	0.163
0.300	0.143	0.700	0.141
0.400	0.163	0.800	0.106
0.500	0.169	0.900	0.058

(a) From these data, determine the stoichiometry of the complex.

Ans: **1:1**

(b) Determine a *conditional* formation constant, K'_f, for the complex. What does this mean? *If* the complex were 1:1,

$$K'_f = \frac{[FeQ^{2+}]}{(y)[Fe^{3+}]}$$

$$y = C_Q - [FeQ^{2+}]$$

$$C_Q = [H_2Q^+] + [HQ] + [Q^-] + [FeQ^{2+}]$$

$$y = [H_2Q^+] + [HQ] + [Q^-]$$

Ans: $K'_f = 7.06 \times 10^{+3}$

15. A complex, with its λ_{max} at 380 nanometers, is formed between Ti(IV) and 1,2-dihydroxybenzene-3,5-disulfonate, called Tiron (abbreviated Trn). Harvey and Manning[*] investigated the composition of the complex by the slope-ratio method. Their data follow; find the composition of the complex.

Slope₁ C_{Trn} = constant = 24 × 10⁻⁴M		Slope₂ C_{Ti} = constant = 4.0 × 10⁻⁴M	
C_{Ti}, M	A	C_{Trn}, M	A
1.0×10^{-5}	0.150	2.0×10^{-5}	0.077
2.0×10^{-5}	0.291	4.0×10^{-5}	0.131
3.0×10^{-5}	0.384	6.0×10^{-5}	0.165
4.0×10^{-5}	0.495	8.0×10^{-5}	0.209
5.0×10^{-5}	0.616	10.0×10^{-5}	0.257
6.0×10^{-5}	0.752	12.0×10^{-5}	0.297

[*] *A. E. Harvey and D. L. Manning, J. Am. Chem. Soc., 74, 4744 (1952).*

Ans: *Probably 5 or 6 tirons to 1 Ti*

16. The author, as a graduate student during the Early Bronze Age, determined with a partner the pK_a of the indicator dye bromthymol blue. The results of this instrumental analysis experiment were encouraging. Bromthymol blue may be represented as a weak acid, HI*n*. At 432 nm, the acid form of the indicator (HI*n*) absorbs strongly, but the basic form, In⁻, absorbs only a little. The spectrophotometer was set at 432 nm and absorbances of a series of solutions in which the sum ([HI*n*] + [In⁻]) was held constant, but in which pH was varied, were read. The data follow.

pH	A	pH	A
1.60	0.80	6.90	0.48
4.20	0.77	7.20	0.37
5.95	0.74	7.50	0.28
6.40	0.70	8.50	0.16
6.55	0.58	11.20	0.15

Determine the pK_a of the indicator by a simple plot of A versus pH.
Ans: *6.9 to 7.0*

17. Bromcresol green is an indicator which may be represented as HI*n*⁻

$$HIn^- \rightleftharpoons H^+ + In^{2-} \qquad K_a = \frac{[H^+][In^{2-}]}{[HIn^-]}$$

yellow   blue

At 615 nanometers, only the In²⁻, or basic form absorbs. A student in a quantitative analysis lab made a series of solutions in which the sum ([HI*n*⁻] + [In²⁻]) was kept constant, but the pH was varied. One solution was of *very* high pH. Its absorbance at 615 nanometers was called A_{basic}

$$A_{basic} = 0.481$$

The pH values and absorbances for the other solutions are

pH	A
4.18	0.115
4.48	0.187
4.62	0.234
4.79	0.287
5.10	0.360

Use a straight-line plot to determine the pK_a of bromcresol green.
Ans: $pK_a = 4.65$

≡ 4.24 DETAILED SOLUTIONS TO PROBLEMS ≡

1. (a) First, determine the copper concentration in moles/liter
   Example:

$$\frac{\text{gram of Cu}}{\text{liter}} = \frac{44.0 \times 10^{-6} \text{ grams}}{25.00 \text{ milliliters}} \times \frac{1000 \text{ milliliter}}{1 \text{ liter}} = 1.76 \times 10^{-3} \frac{\text{gram}}{\text{liter}}$$

$$C_{Cu}, \frac{\text{moles of Cu}}{\text{liter}} = \left(1.76 \times 10^{-3} \frac{\text{gram}}{\text{liter}}\right)\left(\frac{1 \text{ mole}}{63.54 \text{ gram}}\right)$$

$$C_{Cu} = 2.77 \times 10^{-5} \text{ moles/liter}$$

Then, prepare a table.

Micrograms of Cu/25 ml	C_{Cu}, M	A
44.0	2.77×10^{-5}	0.553
56.0	3.52×10^{-5}	0.699
68.0	4.28×10^{-5}	0.860
85.0	5.35×10^{-5}	0.998
95.0	5.98×10^{-5}	1.160
Unknown		0.750

Next, plot A versus C_{Cu} (Figure 4.37) and determine the concentration of copper in the unknown. It is found to be $3.90 \times 10^{-5} M$.

(b) First, find the slope of the graph plotted in part a.

$$\text{slope} = \frac{\Delta A}{\Delta C_{Cu}} = \frac{0.390}{2.00 \times 10^{-5}} = 1.95 \times 10^4$$

If

$$A = \epsilon b C_{Cu}$$

$$\frac{A}{C_{Cu}} = \epsilon b$$

$$\text{slope} = \frac{\Delta A}{\Delta C_{Cu}} = \epsilon b$$

$$\epsilon = \frac{\text{slope}}{b} = \frac{1.95 \times 10^4}{1.00 \text{ centimeters}} \text{liters/mole}$$

$$= 1.95 \times 10^4 \text{ liters/(mole} \cdot \text{centimeter)}$$

2. (a) First, plot A versus C_{Co}. From Figure 4.38;

$$A_{unknown} = 0.25$$
$$C_{unknown} = 7.1 \text{ ppm Co}$$
$$C_{Co} \text{ in 100-milliliter flask} = 7.1 \text{ ppm Co}$$

(b) Remember the dilution:

$$C_1 v_1 = C_2 v_2 \qquad\qquad C_1 = ?$$

$$C_1 = \frac{C_2 v_2}{v_1} \qquad\qquad v_1 = 4.0 \text{ milliliters}$$

320

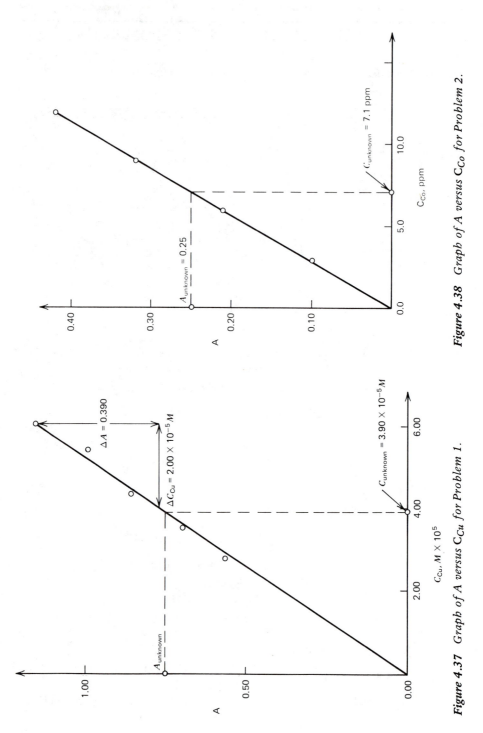

Figure 4.38 *Graph of A versus C_{Co} for Problem 2.*

Figure 4.37 *Graph of A versus C_{Cu} for Problem 1.*

$$= \frac{(7.1)(100)}{(4.0)} \qquad\qquad C_2 = 7.1 \text{ ppm}$$

$$= 1.8 \times 10^2 \text{ ppm} = 1.8 \times 10^2 \text{ milligrams/liter} \qquad v_2 = 100 \text{ milliliters}$$

(c) (number of milligrams of Co) = $(1.8 \times 10^2 \text{ milligram/liter})(4.0 \times 10^{-3} \text{ liter})$
$$= 0.72 \text{ milligrams Co}$$

3. Beer's law: $\qquad A = \epsilon bc$

$\qquad\qquad\qquad\qquad\qquad\qquad\qquad\qquad\qquad\quad \begin{cases} A_u = 0.400 \\[4pt] C_u = ? \\[4pt] A_k = 0.585 \\[4pt] C_k = 1.00 \times 10^{-4} \end{cases}$

For unknown: $\qquad A_u = \epsilon b C_u$

For known: $\qquad\;\; A_k = \epsilon b C_k$

$$\frac{A_u}{A_k} = \frac{\cancel{\epsilon b}C_u}{\cancel{\epsilon b}C_k}$$

$$C_u = \frac{A_u}{A_k} C_k$$

$$C_u = \left(\frac{0.400}{0.585}\right)(1.00 \times 10^{-4})$$

$$C_u = 6.84 \times 10^{-5} M \text{ in } MnO_4^-$$

Then find the weight % manganese in the alloy.

$$\text{Weight \% Mn} = \frac{(\text{number of grams of Mn})}{(\text{number of grams of alloy})} \times 100\%$$

To find (number of grams of Mn):

(number of moles of Mn) = (number of moles of MnO_4^-)

(number of moles of MnO_4^-) = $(C_u)(5 \times 10^{-1} \text{ liters})$

$$= \left(6.84 \times 10^{-5}\, \frac{\text{mole}}{\cancel{\text{liter}}}\right)(5.00 \times 10^{-1}\, \cancel{\text{liter}})$$

$$= 3.42 \times 10^{-5} \text{ moles}$$

(number of moles of Mn) = 3.42×10^{-5} moles of Mn

(number of grams of Mn) = (number of moles of Mn) $\left(54.94\, \dfrac{\text{grams of Mn}}{\text{moles of Mn}}\right)$

$$= (3.42 \times 10^{-5}\, \cancel{\text{moles of Mn}})\left(54.94\, \frac{\text{grams of Mn}}{\cancel{\text{moles of Mn}}}\right)$$

$$= 1.88 \times 10^{-3} \text{ grams of Mn}$$

$$\text{wt \% Mn} = \left(\frac{1.88 \times 10^{-3} \text{ grams}}{5.00 \times 10^{-1} \text{ grams}}\right)(100\%) = 0.376\%$$

4. Beer's law: $\qquad A = \epsilon bc$

$\qquad\qquad\qquad\qquad\qquad\qquad\qquad\qquad\qquad\quad \begin{cases} A_u = 0.125 \\[4pt] C_u = ? \\[4pt] A_k = 0.500 \\[4pt] C_k = 6.30 \times 10^{-5} M \end{cases}$

For unknown: $A_u = \epsilon b C_u$

For known: $\quad\; A_k = \epsilon b C_k$

$$\frac{A_u}{A_k} = \frac{\epsilon b C_u}{\epsilon b C_k}$$

$$C_u = \frac{A_u}{A_k} C_k = \left(\frac{0.125}{0.500}\right)(6.30 \times 10^{-5}M) = 1.58 \times 10^{-5}M$$

Now, find the number of grams of copper lost.

(number of moles of Cu) $= (C_u)(20.00 \times 10^{-3}$ liters)

$$= (1.58 \times 10^{-5})(20.00 \times 10^{-3}) = 3.16 \times 10^{-7} \text{ moles}$$

(number of grams of Cu) $= (3.16 \times 10^{-7} \text{ moles}) \left(63.54 \frac{\text{grams of Cu}}{\text{mole}}\right)$

$$= 2.01 \times 10^{-5} \text{ gram}$$

This could *not* have been weighed on an ordinary analytical balance.

5. Beer's law: $\quad A = \epsilon b c$ $\qquad \left\{ \begin{array}{l} A_u = A_k \\[4pt] b_u = 1.25 \text{ centimeters} \\[4pt] C_u = ? \\[4pt] b_k = 2.00 \text{ centimeters} \\[4pt] C_k = 4.0 \text{ ppm} \end{array} \right.$

For unknown: $\quad A_u = \epsilon \cdot b_u \cdot C_u$

For standard: $\quad A = \epsilon \cdot b_k \cdot C_k$

$$A_u = A_k$$
$$\cancel{\epsilon} \cdot b_u \cdot C_u = \cancel{\epsilon} \cdot b_k \cdot C_k$$

$$C_u = \frac{b_k}{b_u} C_k = \left(\frac{2.00}{1.25}\right)(4.0) = 6.4 \text{ ppm Fe}$$

6. Beer's law: $\quad A = \epsilon b c$ $\qquad \left\{ \begin{array}{l} A_u = A_k \\[4pt] b_u = 5.50 \text{ centimeters} \\[4pt] C_u = ? \\[4pt] b_k = 4.00 \text{ centimeters} \\[4pt] C_k = 1.00 \times 10^{-5}M \end{array} \right.$
(which we
hope is
obeyed)

For unknown: $\quad A_u = \epsilon b_u C_u$

For standard: $\quad A_k = \epsilon \cdot b_k \cdot C_k$

$$A_u = A_k$$
$$\cancel{\epsilon} \cdot b_u \cdot C_u = \cancel{\epsilon} \cdot b_k \cdot C_k$$

$$C_u = \frac{b_k}{b_u} C_k = \left(\frac{4.00}{5.50}\right)(1.00 \times 10^{-4}) = 7.27 \times 10^{-5}M$$

This looks familiar. Next we find the weight % Mn in the alloy.

$$\text{weight \% Mn} = \frac{\text{(number of grams of Mn)}}{\text{(number of grams of alloy)}} \times 100\%$$

Next find (number of grams of Mn):

(number of moles of Mn) $=$ (number of moles of MnO_4^-)

(number of moles of MnO_4^-) $= (C_u)(2.00 \times 10^{-1}$ liter)

$$= \left(7.27 \times 10^{-5} \frac{\text{moles}}{\cancel{\text{liter}}}\right)(2.00 \times 10^{-1} \cancel{\text{liter}}) = 1.45 \times 10^{-5} \text{ moles of } MnO_4^-$$

(number of moles of Mn) = 1.45 $\times$ 10^{-5} moles of Mn

(number of grams of Mn) = (1.45 $\times$ 10^{-5} ~~moles of Mn~~) $\left(54.94 \dfrac{\text{grams of Mn}}{\text{moles of Mn}}\right)$

$\qquad\qquad\qquad\qquad$ = 7.97 $\times$ 10^{-4} grams of Mn

weight % Mn = $\dfrac{\text{(number of grams of Mn)}}{\text{(number of grams of alloy)}}$ $\times$ 100% = $\dfrac{7.97 \times 10^{-4}}{0.450}$ $\times$ 100%

$\qquad\qquad\quad$ = 0.177% Mn

7. This question may be answered by the solution of two equations in two unknowns. The example given in the preceding section will serve as a guide here.

$$A_{400} = \epsilon_{Co400} \cdot b \cdot C_{Co} + \epsilon_{Cr400} \cdot b \cdot C_{Cr}$$
$$A_{505} = \epsilon_{Co505} \cdot b \cdot C_{Co} + \epsilon_{Cr505} \cdot b \cdot C_{Cr}$$

Numbers can be inserted into the equations.

$$1.167 = (0.530)(1.00)C_{Co} + (15.2)(1.00)C_{Cr}$$
$$0.674 = (5.07)(1.00)C_{Co} + (5.60)(1.00)C_{Cr}$$

$$C_{Cr} = \frac{1.167 - 0.530C_{Co}}{15.2}$$

$$0.674 = 5.07C_{Co} + 5.60\left\{\frac{1.167 - 0.530C_{Co}}{15.2}\right\}$$

$$0.674 = 5.07C_{Co} + \frac{(5.60)(1.167)}{15.2} - \frac{(5.60)(0.530)}{15.2}C_{Co}$$

$$0.674 = 5.07C_{Co} + 0.430 - 0.195\,C_{Co}$$

$$0.244 = 4.88C_{Co}$$

$$C_{Co} = \frac{0.244}{4.88} = 0.0500M$$

$$C_{Cr} = \frac{1.167 - (0.530)(0.050)}{15.2} = 0.0750M$$

8. First, find the molar absorptivities for $Cr_2O_7^{2-}$ and MnO_4^- at 440 and 545 nanometers. Four plots are required, as shown in Figures 4.39 to 4.42.

$\qquad\qquad\qquad\qquad$ ϵb-values (liters per mole)

$(\epsilon_{Cr440} \cdot b) = 369.0$ $\qquad$ $(\epsilon_{Mn440} \cdot b) = 95.0$

$(\epsilon_{Cr545} \cdot b) = 11.0$ $\qquad$ $(\epsilon_{Mn545} \cdot b) = 2.35 \times 10^3$

Two equations in two unknowns are then set up.

$$A_{440} = \epsilon_{Cr440} \cdot b \cdot C_{Cr} + \epsilon_{Mn440} \cdot b \cdot C_{Mn}$$
$$A_{545} = \epsilon_{Cr545} \cdot b \cdot C_{Cr} + \epsilon_{Mn545} \cdot b \cdot C_{Mn}$$

Numbers are added next.

$$0.204 = 3.69C_{Cr} + 95.0C_{Mn}$$
$$0.860 = 11.0C_{Cr} + 2350C_{Mn}$$

$$C_{Mn} = \frac{0.204 - 369C_{Cr}}{95}$$

$$0.860 = 11.0C_{Cr} + (2350)\left\{\frac{0.204 - 369\,C_{Cr}}{95}\right\}$$

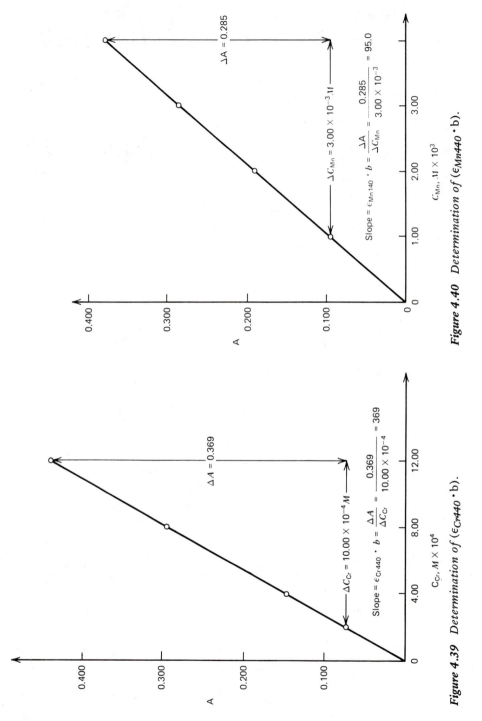

324

Figure 4.39 Determination of $(\epsilon_{Cr440} \cdot b)$.

Figure 4.40 Determination of $(\epsilon_{Mn440} \cdot b)$.

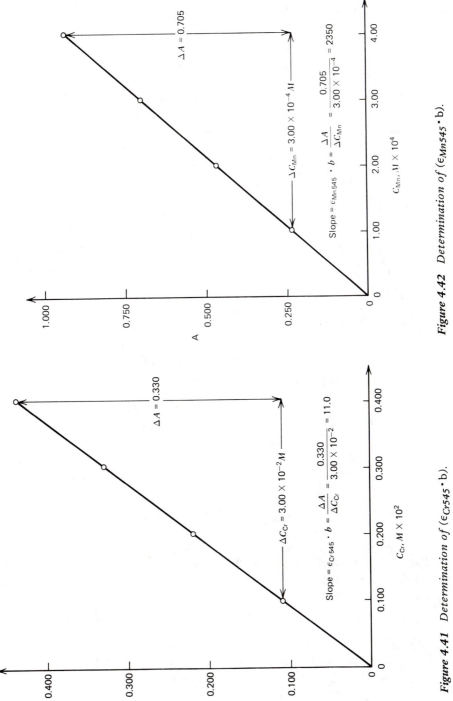

Figure 4.42 Determination of $(\epsilon_{Mn545} \cdot b)$.

Figure 4.41 Determination of $(\epsilon_{Cr545} \cdot b)$.

$$0.860 = 11.0C_{Cr} + \frac{(2350)(0.204)}{95} - \frac{(2350)(369)}{95}C_{Cr}$$

$$0.860 = 11.0C_{Cr} + 5.046 - 9128C_{Cr}$$

$$-4.186 = -9117C_{Cr}$$

$$C_{Cr} = 4.59 \times 10^{-4}M$$

$$C_{Mn} = \frac{0.204 - 369C_{Cr}}{95} = \frac{0.204 - (369)(4.59 \times 10^{-4})}{95} = 3.64 \times 10^{-4}M$$

The problem is not completely done yet. The percents of chromium and manganese in the steel must still be determined.

$$\% \text{ Cr} = \frac{(\text{number of grams of Cr})}{(\text{number of grams of sample})} \times 100\%$$

$$(\text{number of moles of Cr}) = (2)(\text{number of moles of } Cr_2O_7^{2-})$$

$$(\text{number of moles of } Cr_2O_7^{2-}) = \left(4.59 \times 10^{-4} \frac{\text{moles}}{\text{liter}}\right)(5.00 \times 10^{-2} \text{ liter})$$

$$= 2.30 \times 10^{-5} \text{ moles}$$

$$(\text{number of moles of Cr}) = (2)(2.30 \times 10^{-5}) = 4.60 \times 10^{-5} \text{ moles}$$

$$(\text{number of grams of Cr}) = (4.60 \times 10^{-5} \text{ moles})\left(52.01 \frac{\text{grams of Cr}}{\text{mole}}\right)$$

$$= 2.39 \times 10^{-3} \text{ grams of Cr}$$

$$\% \text{ Cr} = \frac{2.39 \times 10^{-3}}{1.0000} \times 100\% = 0.239\%$$

$$\% \text{ Mn} = \frac{(\text{number of grams of Mn})}{(\text{number of grams of sample})} \times 100\%$$

$$(\text{number of moles of Mn}) = (\text{number of moles of } MnO_4^-)$$

$$(\text{number of moles of } MnO_4^-) = \left(3.64 \times 10^{-4} \frac{\text{moles}}{\text{liter}}\right)(5.00 \times 10^{-2} \text{ liter})$$

$$= 1.82 \times 10^{-5} \text{ moles}$$

$$(\text{number of moles of Mn}) = 1.82 \times 10^{-5} \text{ mole}$$

$$(\text{number of grams of Mn}) = (1.82 \times 10^{-5} \text{ mole})\left(54.94 \frac{\text{grams of Mn}}{\text{moles}}\right)$$

$$= 1.00 \times 10^{-3} \text{ grams of Mn}$$

$$\% \text{ Mn} = \frac{1.00 \times 10^{-3}}{1.0000} \times 100\% = 0.100\%$$

9. First graph the data, A versus milliliters of Na_2H_2Y as in Figure 4.43. The endpoint of the titration is seen to be 10.50 milliliters of Y^{4-} added. At the endpoint,

(moles of Cu in 50 milliliters)

$$= (\text{moles of } Y^{4-})$$

$$= (10.50 \times 10^{-3} \text{ liter})\left(1.000 \times 10^{-1} \frac{\text{mole}}{\text{liter}}\right) = 10.50 \times 10^{-4} \text{ mole}$$

(moles of Cu in 250 milliliters)

$$= (5)(\text{moles of Cu in 50 milliliters}) = (5)(10.50 \times 10^{-4} \text{ mole})$$

$$= 5.250 \times 10^{-3} \text{ mole}$$

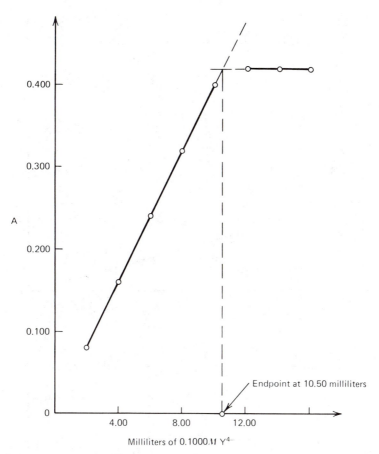

Figure 4.43 *Titration of* Cu^{2+}.

(milligrams of Cu)

$$= \text{(moles of Cu in 250 milliliters)(atomic weight of Cu)} \left(\frac{1000 \text{ milligrams}}{\text{gram}} \right)$$

$$= (5.250 \times 10^{-3} \text{ mole}) \left(63.54 \frac{\text{grams}}{\text{mole}} \right) \left(1000 \frac{\text{milligrams}}{\text{grams}} \right)$$

$$= 333.6 \text{ milligrams of Cu}$$

10. The BiY^- complex is formed first because its K_f is greater than the K_f for the CuY^{2-} complex. Thus the absorbance remains constant until all the bismuth is complexed. Absorbance rises while the copper is being complexed and levels off after all the copper has been complexed. Figure 4.44 shows these facts.

(moles of Bi) = (moles of Y^{4-} to first endpoint)

$$= (10.00 \times 10^{-3} \text{ liter})(0.1000 \text{ mole per liter}) = 1.000 \times 10^{-3} \text{ mole}$$

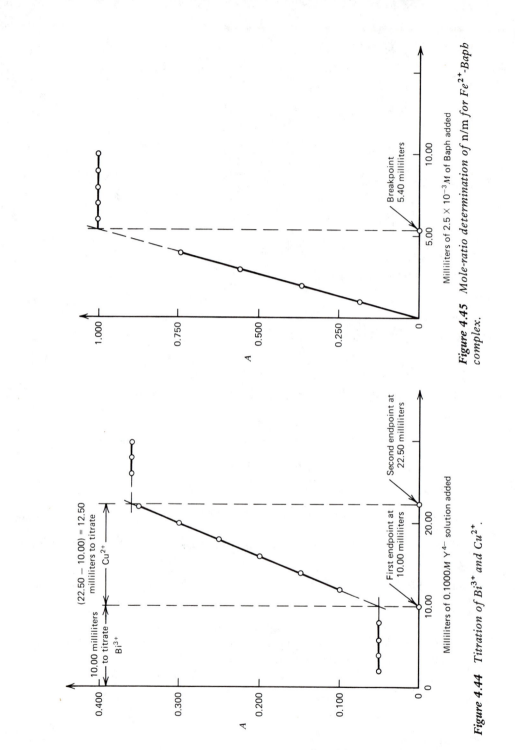

Figure 4.45 *Mole-ratio determination of n/m for Fe²⁺-Baph complex.*

Figure 4.44 *Titration of Bi³⁺ and Cu²⁺.*

$$\text{(milligrams of Bi)} = \text{(moles of Bi)(atomic weight of Bi)(1000 milligrams per gram)}$$
$$= (1.000 \times 10^{-3})(209.00)(1000) = 209 \text{ milligrams of Bi}$$
$$\text{(moles of Cu)} = \text{(moles of } Y^{4-} \text{ between first and second endpoints)}$$
$$= (12.50 \times 10^{-3} \text{ liter})(0.1000 \text{ mole per liter})$$
$$= 1.250 \times 10^{-3} \text{ mole}$$
$$\text{(milligrams of Cu)} = \text{(moles of Cu)(atomic weight of Cu)(1000 milligrams per gram)}$$
$$= (1.250 \times 10^{-3})(63.54)(1000) = 79.4 \text{ milligrams of Cu}$$

11. Another plot is needed. Plot A versus milliliters of Baph added as in Figure 4.45. This will yield the typical mole-ratio plot. At the breakpoint, all the Fe^{2+} has been complexed.

$$\frac{n}{m} = \frac{\text{(moles of Baph at breakpoint)}}{\text{(moles of Fe)}}$$

$$\text{(moles of Baph at breakpoint)} = (5.40 \times 10^{-3} \text{ liter})\left(2.50 \times 10^{-3} \frac{\text{mole}}{\text{liter}}\right)$$

$$= 1.35 \times 10^{-5}$$

$$\frac{n}{m} = \frac{1.35 \times 10^{-5}}{4.50 \times 10^{-6}} = 3.00$$

Thus the reaction is *probably*
$$Fe^{2+} + 3Baph \rightleftharpoons FeBaph_3^{2+}$$

12. A plot, just like the one in Problem 11, is needed. Plot A versus C_{Gel} (Figure 4.46). At the breakpoint:

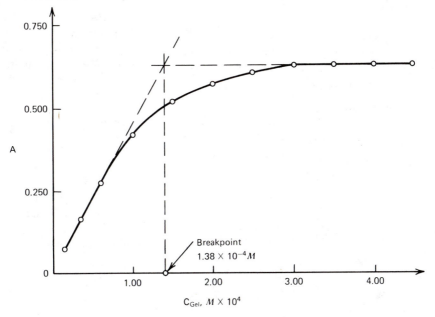

Figure 4.46 Mole-ratio plot for Cu^+-Gel complex.

$$\frac{n}{m} = \frac{C_{Gel}}{C_{Cu}} = \frac{1.38 \times 10^{-4}}{6.90 \times 10^{-5}} = \frac{2}{1}$$

The reaction is probably

$$Cu^+ + 2Gel \rightleftharpoons CuGel_2^+$$

$$K_f = \frac{[CuGel_2^+]}{[Cu^+][Gel]^2}$$

Now, numbers must be found for $[CuGel_2^+]$, $[Cu^+]$, and $[Gel]$.

$$A = \epsilon b [CuGel_2^+]$$

$$[CuGel_2^+] = \frac{A}{\epsilon b}$$

The chemical bookkeeping follows:

$$C_{Cu} = [Cu^+] + [CuGel_2^+]$$

$$[Cu^+] = C_{Cu} - [CuGel_2^+]$$

$$[Cu^+] = C_{Cu} - \frac{A}{\epsilon b}$$

$$C_{Gel} = [Gel] + 2[CuGel_2^+]$$

$$[Gel] = C_{Gel} - 2[CuGel_2^+]$$

$$[Gel] = C_{Gel} - 2\frac{A}{\epsilon b}$$

$$K_f = \frac{\dfrac{A}{\epsilon b}}{\left(C_{Cu} - \dfrac{A}{\epsilon b}\right) \cdot \left(C_{Gel} - 2\dfrac{A}{\epsilon b}\right)^2}$$

Numerical values must be found for all terms in the last equation. The absorbance value chosen corresponds to $C_{Gel} = 1.50 \times 10^{-4}M$. At this point, $A = 0.515$. Here, dissociation is apt to be largest, and an accurate value of K_f may be found. The value of ϵ, or molar absorptivity, is easy to find. *Past* the breakpoint, on the flat line, *all* of the Cu^+ is complexed.

Past the breakpoint;

$$A = 0.630$$

$$C_{Cu} = [CuGel_2^+]$$

$$C_{Cu} = 6.90 \times 10^{-5}M$$

$$[CuGel_2^+] = 6.90 \times 10^{-5}M$$

$$A = \epsilon b [CuGel_2^+]$$

$$\epsilon = \frac{A}{(b)[CuGel_2^+]} = \frac{0.630}{(1.00)(6.90 \times 10^{-5})}$$

$$= 9.13 \times 10^3 \text{ liter/(mole} \cdot \text{centimeter)}$$

On the curved portion;

$$A = 0.515$$

$$\frac{A}{\epsilon b} = \frac{0.515}{(9.13 \times 10^3)(1.00)} = 5.64 \times 10^{-5}$$

$$C_{Cu} = 6.90 \times 10^{-5}M$$

$$C_{Gel} = 15.0 \times 10^{-5} M$$

Remember that

$$K_f = \frac{\dfrac{A}{\epsilon b}}{\left(C_{Cu} - \dfrac{A}{\epsilon b}\right) \cdot \left(C_{Gel} - \dfrac{2A}{\epsilon b}\right)^2}$$

$$= \frac{5.64 \times 10^{-5}}{(6.90 \times 10^{-5} - 5.64 \times 10^{-5})[15.0 \times 10^{-5} - (2)(5.64 \times 10^{-5})]^2}$$

$$= 3.23 \times 10^{+9}$$

13. Make a plot of A versus X_{Hg}, dropping a perpendicular from the intersection to the abscissa and calling the point X_{Hgint} (Figure 4.47).

$$X_{Hgint} = 0.50$$
$$X_{Dipint} = 1.0 - X_{Hgint} = 1.0 - 0.50 = 0.50$$
$$\frac{n}{m} = \frac{X_{Dipint}}{X_{Hgint}} = \frac{0.50}{0.50} = 1.0$$

Thus the reaction most likely is

$$Hg^{2+} + Dip \rightleftharpoons HgDip^{2+}$$

14. (a) Plot A versus X_{Fe}, dropping a perpendicular from the intersection to the abscissa and calling the point X_{Feint} (Figure 4.48).

$$X_{Feint} = 0.500$$
$$X_{HQint} = 1.000 - X_{Feint} = 1.000 - 0.500 = 0.500$$
$$\frac{n}{m} = \frac{X_{Feint}}{X_{HQint}} = \frac{0.500}{0.500} = 1.00$$
$$m Fe^{3+} + n Q^- \rightleftharpoons Fe_m Q_n$$

is

$$Fe^{3+} + Q^- \rightleftharpoons FeQ^{2+}$$

(b) This is a little harder. Why find a conditional formation constant? We look at the true formation constant for

$$Fe^{3+} + Q^- \rightleftharpoons FeQ^{2+}$$
$$K_f = \frac{[FeQ^{2+}]}{[Fe^{3+}][Q^-]}$$

$[Q^-]$ is needed. At pH 1.92, there is *no way* that we can assume that all of the uncomplexed 8-hydroxyquinoline is present as Q^-. Let

$$C_Q = [H_2Q^+] + [HQ] + [Q^-] + [FeQ^{2+}]$$

(The total concentrations of all uncomplexed Q-species) $= y$.

$$y = [H_2Q^+] + [HQ] + [Q^-]$$
$$C_{HQ} = y + [FeQ^{2+}]$$

Define K_f', the conditional formation constant;

$$K_f' = \frac{[FeQ^{2+}]}{[y][Fe^{3+}]}$$

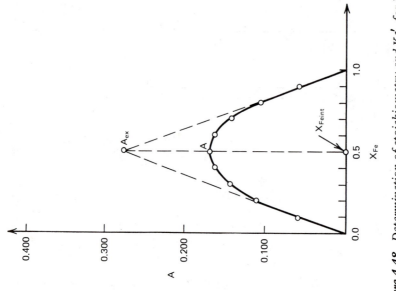

Figure 4.48 Determination of stoichiometry and K_f' for Fe^{3+}-HQ complex. Reprinted with permission from E. B. Sandell and D. C. Spindler, J. Amer. Chem. Soc., 71, 3806 (1949). Copyright by the American Chemical Society.

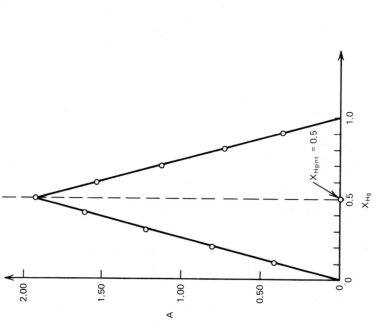

Figure 4.47 Determination of stoichiometry of an Hg^{2+} complex. Reprinted with permission from J. L. Gerlach and R. G. Frazier, Anal. Chem., 30, 1142 (1958). Copyright by the American Chemical Society.

Here K_f' is the conditional formation constant at pH 1.92.

Now, recall the $FeSCN^{2+}$ example. The steps we follow will closely parallel those taken in that example. Refer to the graph for values for A_{ex} and A.

$$A = 0.169$$
$$C_{Fe} = [Fe^{3+}] + [FeQ^{2+}]$$
$$[Fe^{3+}] = C_{Fe} - [FeQ^{2+}]$$
$$C_Q = y + [FeQ^{2+}]$$
$$y = C_Q - [FeQ^{2+}]$$
$$C_{Fe} = (0.500)(1.19 \times 10^{-3}M) = 5.95 \times 10^{-4}M$$
$$C_Q = (0.500)(1.19 \times 10^{-3}M) = 5.95 \times 10^{-4}M$$

What is $[FeQ^{2+}]$?

Recall our arguments for $FeSCN^{2+}$.

$$A_{ex} = \epsilon b[FeQ^{2+}]_{ex}$$
$$[FeQ^{2+}]_{ex} = C_{Fe}$$
$$A_{ex} = \epsilon b C_{Fe}$$
$$A = observed \text{ absorbance}$$
$$A = \epsilon b [FeQ^{2+}]$$
$$\frac{A}{A_{ex}} = \frac{\cancel{\epsilon b}[FeQ^{2+}]}{\cancel{\epsilon b}C_{Fe}}$$
$$[FeQ^{2+}] = \frac{A}{A_{ex}} C_{Fe}$$

From the graph;

$A = 0.169$

$A_{ex} = 0.274$

$$K_f' = \frac{[FeQ^{2+}]}{[y][Fe^{2+}]} = \frac{\dfrac{A}{A_{ex}}C_{Fe}}{\left(C_Q - \dfrac{A}{A_{ex}}C_{Fe}\right)\left(C_{Fe} - \dfrac{A}{A_{ex}}C_{Fe}\right)}$$

$$= \frac{\dfrac{A}{A_{ex}}\cancel{C_{Fe}}}{\left(C_Q - \dfrac{A}{A_{ex}}C_{Fe}\right)\left(1 - \dfrac{A}{A_{ex}}\right)\cancel{C_{Fe}}}$$

$$= \frac{\dfrac{0.169}{0.274}}{\left\{5.95 \times 10^{-4} - \left(\dfrac{0.169}{0.274}\right)(5.95 \times 10^{-4})\right\}\left(1 - \dfrac{0.169}{0.274}\right)} = 7.06 \times 10^3$$

15. First, draw two graphs both of A versus C. In Figure 4.49, A versus C_{Ti}, (C_{Tm} = constant), slope$_1$, is measured.

$$\text{slope}_1 = \frac{\Delta A}{\Delta C_{Ti}} \tag{4.10}$$

$$mTi + nTrn \rightleftharpoons Ti_mTrn_n$$
$$A = \epsilon b [Ti_mTrn_n]$$

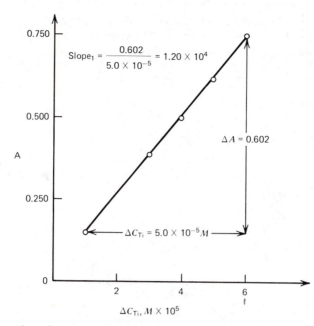

Figure 4.49 *Plot of A versus* C_{Ti} *(*C_{Trn} = *const.). Reprinted with permission from A. E. Harvey and D. L. Manning,* J. Amer. Chem. Soc., **74**, 4744 *(1952). Copyright by the American Chemical Society.*

$$[Ti_m Trn_n] = \frac{C_{Ti}}{m}$$

$$A = \frac{\epsilon b}{m} C_{Ti}$$

$$\frac{A}{C_{Ti}} = \frac{\epsilon b}{m}$$

$$\frac{\Delta A}{\Delta C_{Ti}} = \frac{\epsilon b}{m}$$

$$slope_1 = \frac{\Delta A}{\Delta C_{Ti}} = \frac{\epsilon b}{m} = 1.20 \times 10^4$$

Figure 4.50 is a plot of A versus C_{Trn} (C_{Ti} = constant); slope$_2$ is measured.

$$slope_2 = \frac{\Delta A}{\Delta C_{Trn}} \tag{4.11}$$

$$mTi + nTrn \rightleftharpoons Ti_m Trn_n$$

$$A = \epsilon b [Ti_m Trn_n]$$

$$[Ti_m Trn_n] = \frac{C_{Trn}}{n}$$

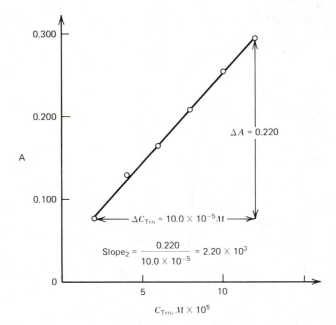

Figure 4.50 Plot of A versus C_{Trn} (C_{Ti} = const.). Reprinted with permission from A. E. Harvey and D. L. Manning, J. Amer. Chem. Soc., **74**, 4744 (1952). Copyright by the American Chemical Society.

$$A = \frac{\epsilon b}{n} C_{Trn}$$

$$\frac{A}{C_{Trn}} = \frac{\epsilon b}{n}$$

$$\frac{\Delta A}{\Delta C_{Trn}} = \frac{\epsilon b}{n}$$

$$slope_2 = \frac{\Delta A}{\Delta C_{Trn}} = \frac{\epsilon b}{n} = 2.20 \times 10^3$$

$$\frac{slope_1}{slope_2} = \frac{\frac{\epsilon b}{m}}{\frac{\epsilon b}{n}} = \frac{n}{m} = \frac{1.20 \times 10^4}{2.20 \times 10^3} = 5.45 \qquad (4.12)$$

$$\frac{n}{m} = 5.45$$

This tells us that the complex is either $TiTrn_6$ or $TiTrn_5$. On the basis of this *one* experiment, either is a possibility.

16. Refer to Figure 4.51. The indicator is almost all in the acid form at pH 1.6 (A = 0.80). It is almost all in the base form at pH 11.2 (A = 0.15). Midway between

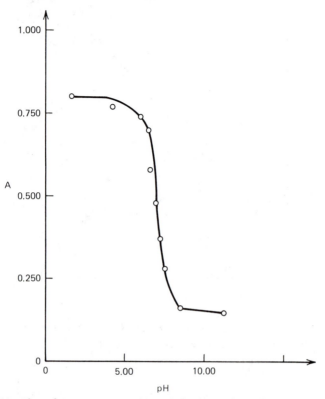

Figure 4.51 *Plot of A versus pH @ 432 nm for bromthymol blue.*

$A = 0.80$ and $A = 0.15$, $[\text{HIn}] = [\text{In}^-]$. This midway point is $(A = 0.47, \text{pH} = 6.9$ to 7.0). At this point, $\text{pH} = \text{p}K_a$.

$$\text{p}K_a = 6.9 \text{ to } 7.0 \quad \text{Why?}$$

$$K_a = \frac{[\text{H}^+][\text{In}^-]}{[\text{HIn}]}$$

If

$$[\text{In}^-] = [\text{HIn}]$$
$$K_a = [\text{H}^+]$$
$$\text{p}K_a = \text{pH}$$

The numerical answer is somewhat uncertain because the curve could be drawn differently.

17. Recall the derivation of Eq. 4.13.

$$\log\left(\frac{A_{\text{basic}} - A}{A}\right) = \text{p}K_a - \text{pH} \tag{4.13}$$

If

$$\log\left(\frac{A_{\text{basic}} - A}{A}\right) = 0$$

then

$$pH = pK_a$$

A_{basic} = absorbance of indicator at very high pH

A = absorbances of sample solutions

A_{basic} = 0.481

pH	A	$\left(\dfrac{A_{\text{basic}} - A}{A}\right)$	$\log\left\{\dfrac{A_{\text{basic}} - A}{A}\right\}$
4.18	0.115	3.18	0.503
4.48	0.187	1.57	0.196
4.62	0.234	1.06	0.023
4.79	0.287	0.676	−0.170
5.10	0.360	0.336	−0.474

Now plot $\log\left\{\dfrac{A_{\text{basic}} - A}{A}\right\}$ versus the pH. From Figure 4.52 we find that pK_a = 4.65.

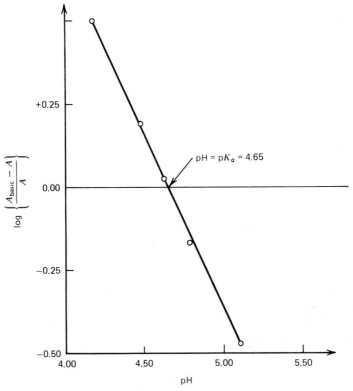

Figure 4.52 Plot to find pK_a of bromcresol green.

chapter 5
Potentiometry

≡ 5.1 INTRODUCTION ≡

Potentiometry is a name that conjures up for many a vision of wire-wound resistors, dry cells, flickering galvanometers, and wires running everywhere in a laboratory with high windows and a creaky floor. Yet potentiometry is a study of great chemical importance. It is as up-to-date as this morning's measurement of the pH of a city water supply. It is as close to life processes as the direct potentiometric monitoring of ionic activities in blood. It is an accessible science; the student who has mastered logarithmic reckoning can master its calculations.

Our discussion of potentiometry will be divided into four parts. The first part will set forth basic ideas, including the Nernst equation. The second will deal with potentiometric titrations, a subject already touched on in the study of pH titration curves. In the third part, the relationships of potentiometric values to equilibrium constants will be explored, and a quantity called *activity*, closely related to concentrations, will be defined. Direct concentration sensing with specific ion electrodes will be treated in the fourth part.

≡ 5.2 BASIC CONSIDERATIONS ≡

Suppose that you are asked to calculate the potential of the cell illustrated in Figure 5.1.

You would first have to know that V represents a potentiometer, a device that measures electrical potential but through which no current flows. In potentiometric measurements, no current, or only the tiniest current, is allowed to flow. Next, you would divide the cell into right- and left-hand parts. The right-hand part would be the Cu-CuSO$_4$ *half-cell*. The left-hand part would be the Zn-ZnSO$_4$ *half-cell*. The division into half-cells is shown by the dashed line running up the middle of Figure 5.1. It is a generally accepted convention that the potential of any cell is just the algebraic difference between the potential of the right-hand half-cell and the left-hand half-cell.

Let

$$E_{cell} = \text{potential of cell}$$

Define

$$E_{cell} = E_{right} - E_{left}$$

In Figure 5.1

$$E_{right} = E_{cu} = \text{the potential of the Cu-CuSO}_4 \text{ half-cell}$$

$$E_{left} = E_{Zn} = \text{the potential of the Zn-ZnSO}_4 \text{ half-cell}$$

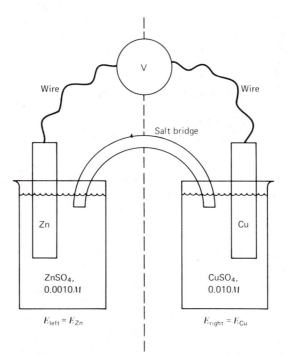

Figure 5.1 An electrochemical cell.

$$E_{cell} = E_{Cu} - E_{Zn}$$

The cell potential, E_{cell}, may be either positive or negative. To find E_{cell}, values of E_{Cu} and E_{Zn} are needed. These may be determined from standard reduction potentials and the Nernst equation. The values of E_{Cu} and E_{Zn} cannot be measured separately, although E_{cell} can be measured. Standard reduction potentials (E^0-values) for half cells are given in Table 5.1, together with formal potentials ($E^{0'}$-values). Formal potentials are those that are determined under a specific set of solution conditions.

Table 5.1
STANDARD REDUCTION POTENTIALS
AND FORMAL POTENTIALS (E°-VALUES)
AT 25°C[a]

Half-Reaction	E°, Volts	$E^{0'}$ Formal Potential, Volts
$F_2 + 2H^+ + 2e^- \rightleftharpoons 2HF$	3.06	
$O_3 + 2H^+ + 2e^- \rightleftharpoons O_2 + H_2O$	2.07	
$S_2O_8^{2-} + 2e^- \rightleftharpoons 2SO_4^{2-}$	2.01	
$Co^{3+} + e^- \rightleftharpoons Co^{2+}$	1.842	
$H_2O_2 + 2H^+ + 2e^- \rightleftharpoons 2H_2O$	1.77	
$MnO_4^- + 4H^+ + 3e^- \rightleftharpoons MnO_2 + 2H_2O$	1.695	
$Ce^{4+} + e^- \rightleftharpoons Ce^{3+}$		1.70 ($1M$ $HClO_4$); 1.61 ($1M$ HNO_3); 1.44 ($1M$ H_2SO_4)
$HClO + H^+ + e^- \rightleftharpoons \frac{1}{2}Cl_2 + H_2O$	1.63	
$H_5IO_6 + H^+ + 2e^- \rightleftharpoons IO_3^- + 3H_2O$	1.6	
$BrO_3^- + 6H^+ + 5e^- \rightleftharpoons \frac{1}{2}Br_2 + 3H_2O$	1.52	
$MnO_4^- + 8H^+ + 5e^- \rightleftharpoons Mn^{2+} + 4H_2O$	1.51	
$Mn^{3+} + e^- \rightleftharpoons Mn^{2+}$		1.51 ($8M$ H_2SO_4)
$ClO_3^- + 6H^+ + 5e^- \rightleftharpoons \frac{1}{2}Cl_2 + 3H_2O$	1.47	
$PbO_2 + 4H^+ + 2e^- \rightleftharpoons Pb^{2+} + 2H_2O$	1.455	
$Cl_2 + 2e^- \rightleftharpoons 2Cl^-$	1.359	
$Cr_2O_7^{2-} + 14H^+ + 6e^- \rightleftharpoons 2Cr^{3+} + 7H_2O$	1.33	
$Tl^{3+} + 2e^- \rightleftharpoons Tl^+$	1.25	0.77 ($1M$ HCl)
$IO_3^- + 2Cl^- + 6H^+ + 4e^- \rightleftharpoons ICl_2^- + 3H_2O$	1.24	
$MnO_2 + 4H^+ + 2e^- \rightleftharpoons Mn^{2+} + 2H_2O$	1.23	
$O_2 + 4H^+ + 4e^- \rightleftharpoons 2H_2O$	1.229	
$2IO_3^- + 12H^+ + 10e^- \rightleftharpoons I_2 + 6H_2O$	1.20	
$SeO_4^{2-} + 4H^+ + 2e^- \rightleftharpoons H_2SeO_3 + H_2O$	1.15	
$Br_2(aq) + 2e^- \rightleftharpoons 2Br^-$	1.087*	
$Br_2(1) + 2e^- \rightleftharpoons 2Br^-$	1.065*	
$ICl_2^- + e^- \rightleftharpoons \frac{1}{2}I_2 + 2Cl^-$	1.06	
$VO_2^+ + 2H^+ + e^- \rightleftharpoons VO^{2+} + H_2O$	1.000	
$HNO_2 + H^+ + e^- \rightleftharpoons NO + H_2O$	1.00	
$Pd^{2+} + 2e^- \rightleftharpoons Pd$	0.987	
$NO_3^- + 3H^+ + 2e^- \rightleftharpoons HNO_2 + H_2O$	0.94	

Table 5.1 continued

Half-Reaction	E°, Volts	$E^{0'}$ Formal Potential, Volts
$2Hg^{2+} + 2e^- \rightleftharpoons Hg_2^{2+}$	0.920	
$H_2O_2 + 2e^- \rightleftharpoons 2OH^-$	0.88	
$Cu^{2+} + I^- + e^- \rightleftharpoons CuI$	0.86	
$Hg^{2+} + 2e^- \rightleftharpoons Hg$	0.854	
$Ag^+ + e^- \rightleftharpoons Ag$	0.799	0.228 (1M HCl); 0.792 (1M HClO$_4$)
$Hg_2^{2+} + 2e^- \rightleftharpoons 2Hg$	0.789	0.274 (1M HCl)
$Fe^{3+} + e^- \rightleftharpoons Fe^{2+}$	0.771	
$H_2SeO_3 + 4H^+ + 4e^- \rightleftharpoons Se + 3H_2O$	0.740	
$PtCl_4^{2-} + 2e^- \rightleftharpoons Pt + 4Cl^-$	0.73	
$C_6H_4O_2(quinone) + 2H^+ + 2e^- \rightleftharpoons C_6H_4(OH)_2$	0.699	0.696 (1M HCl, H$_2$SO$_4$, HClO$_4$)
$O_2 + 2H^+ + 2e^- \rightleftharpoons H_2O_2$	0.682	
$PtCl_6^{2-} + 2e^- \rightleftharpoons PtCl_4^{2-} + 2Cl^-$	0.68	
$I_2(aq) + 2e^- \rightleftharpoons 2I^-$	0.619[†]	
$Hg_2SO_4 + 2e^- \rightleftharpoons 2Hg + SO_4^{2-}$	0.615	
$Sb_2O_5 + 6H^+ + 4e^- \rightleftharpoons 2SbO^+ + 3H_2O$	0.581	
$MnO_4^- + e^- \rightleftharpoons MnO_4^{2-}$	0.564	
$H_3AsO_4 + 2H^+ + 2e^- \rightleftharpoons H_3AsO_3 + H_2O$	0.559	0.577 (1M HCl, HClO$_4$)
$I_3^- + 2e^- \rightleftharpoons 3I^-$	0.5355	
$I_2(s) + 2e^- \rightleftharpoons 2I^-$	0.5345[†]	
$Mo^{6+} + e^- \rightleftharpoons Mo^{5+}$		0.53 (2M HCl)
$Cu^+ + e^- \rightleftharpoons Cu$	0.521	
$H_2SO_3 + 4H^+ + 4e^- \rightleftharpoons S + 3H_2O$	0.45	
$Ag_2CrO_4 + 2e^- \rightleftharpoons 2Ag + CrO_4^{2-}$	0.446	
$VO^{2+} + 2H^+ + e^- \rightleftharpoons V^{3+} + H_2O$	0.361	
$Fe(CN)_6^{3-} + e^- \rightleftharpoons Fe(CN)_6^{4-}$	0.36	0.72 (1M HClO$_4$, H$_2$SO$_4$)
$Cu^{2+} + 2e^- \rightleftharpoons Cu$	0.337	
$UO_2^{2+} + 4H^+ + 2e^- \rightleftharpoons U^{4+} + 2H_2O$	0.334	
$BiO^+ + 2H^+ + 3e^- \rightleftharpoons Bi + H_2O$	0.32	
$Hg_2Cl_2(s) + 2e^- \rightleftharpoons 2Hg + 2Cl^-$	0.268	0.246 (saturated KCl-SCE); 0.282 (1M KCl)
$AgCl + e^- \rightleftharpoons Ag + Cl^-$	0.222	0.228 (1M KCl)
$SO_4^{2-} + 4H^+ + 2e^- \rightleftharpoons H_2SO_3 + H_2O$	0.17	
$BiCl_4^- + 3e^- \rightleftharpoons Bi + 4Cl^-$	0.16	
$Sn^{4+} + 2e^- \rightleftharpoons Sn^{2+}$	0.154	0.14 (1M HCl)
$Cu^{2+} + e^- \rightleftharpoons Cu^+$	0.153	
$S + 2H^+ + 2e^- \rightleftharpoons H_2S$	0.141	
$TiO^{2+} + 2H^+ + e^- \rightleftharpoons Ti^{3+} + H_2O$	0.1	
$Mo^{4+} + e^- \rightleftharpoons Mo^{3+}$		0.1 (4M H$_2$SO$_4$)
$AgBr + e^- \rightleftharpoons Ag + Br^-$	0.095	
$S_4O_6^{2-} + 2e^- \rightleftharpoons 2S_2O_3^{2-}$	0.08	
$Ag(S_2O_3)_2^{3-} + e^- \rightleftharpoons Ag + 2S_2O_3^{2-}$	0.01	
$2H^+ + 2e^- \rightleftharpoons H_2$	0.000	

Table 5.1 continued

Half-Reaction	E°, Volts	$E^{\circ\,\prime}$ Formal Potential, Volts
$Pb^{2+} + 2e^- \rightleftharpoons Pb$	-0.126	
$CrO_4^{2-} + 4H_2O + 3e^- \rightleftharpoons Cr(OH)_3 + 5OH^-$	-0.13	
$Sn^{2+} + 2e^- \rightleftharpoons Sn$	-0.136	
$AgI + e^- \rightleftharpoons Ag + I^-$	-0.151	
$CuI + e^- \rightleftharpoons Cu + I^-$	-0.185	
$N_2 + 5H^+ + 4e^- \rightleftharpoons N_2H_5^+$	-0.23	
$Ni^{2+} + 2e^- \rightleftharpoons Ni$	-0.250	
$V^{3+} + e^- \rightleftharpoons V^{2+}$	-0.255	
$Co^{2+} + 2e^- \rightleftharpoons Co$	-0.277	
$Ag(CN)_2^- + e^- \rightleftharpoons Ag + 2CN^-$	-0.31	
$Tl^+ + e^- \rightleftharpoons Tl$	-0.336	-0.551 (1M HCl)
$PbSO_4 + 2e^- \rightleftharpoons Pb + SO_4^{2-}$	-0.356	
$Ti^{3+} + e^- \rightleftharpoons Ti^{2+}$	-0.37	
$Cd^{2+} + 2e^- \rightleftharpoons Cd$	-0.403	
$Cr^{3+} + e^- \rightleftharpoons Cr^{2+}$	-0.41	
$Fe^{2+} + 2e^- \rightleftharpoons Fe$	-0.440	
$2CO_2(g) + 2H^+ + 2e^- \rightleftharpoons H_2C_2O_4$	-0.49	
$Cr^{3+} + 3e^- \rightleftharpoons Cr$	-0.74	
$Zn^{2+} + 2e^- \rightleftharpoons Zn$	-0.763	
$2H_2O + 2e^- \rightleftharpoons H_2 + 2OH^-$	-0.828	
$Mn^{2+} + 2e^- \rightleftharpoons Mn$	-1.18	
$Al^{3+} + 3e^- \rightleftharpoons Al$	-1.66	
$Mg^{2+} + 2e^- \rightleftharpoons Mg$	-2.37	
$Na^+ + e^- \rightleftharpoons Na$	-2.714	
$Ca^{2+} + 2e^- \rightleftharpoons Ca$	-2.87	
$Ba^{2+} + 2e^- \rightleftharpoons Ba$	-2.90	
$K^+ + e^- \rightleftharpoons K$	-2.925	
$Li^+ + e^- \rightleftharpoons Li$	-3.045	

[a] *Reprinted by permission from G. D. Christian,* Analaytical Chemistry, 2nd ed., *John Wiley & Sons, New York, 1977.*

Note: *Certain quantities, called* Formal Potentials, *are also listed in the table. They represent the half-cell potentials under the conditions noted.*

[*] E° *for* Br_2 *(l) is used for saturated solutions of* Br_2, *while* E° *for* Br_2 *(aq) is used for unsaturated solutions.*

[†] E° *for* I_2 *(s) is used for saturated solutions of* I_2, *while* E° *for* I_2 *(aq) is used for unsaturated solutions.*

Of course, the absolute value for a standard reduction potential cannot be measured. All of the standard reduction potentials listed in Table 5.1 are the potential differences between the half-cells in question and the hydrogen electrode. The potential of the hydrogen electrode is arbitrarily taken as 0.000 volt. Luckily, the assignment of 0.000 volt as the E°-value for the hydrogen half-cell is accepted by all chemists.

The E°-values are those values of potentials which would be observed under standard conditions. These standard conditions are generally taken to be unit activity of each species. Activity may be regarded as a sort of effective concentration, and, in the limiting case of extreme dilution, activities and molar concentrations are taken as equal. Under nonstandard conditions, the potentials of half-cells are calculated by the *Nernst equation*.

The *Nernst equation* states that the potential, E, of any half-cell whose half-reaction is given by

$$a\text{Ox} + ne^- \rightleftharpoons b\text{ Red}$$

(we always write this as a reduction) where a moles of Ox, an oxidized species, acquire n electrons to form b moles of Red, a reduced species, is just

$$E = E^\circ - \frac{RT}{nF} \ln \frac{[\text{Red}]^b}{[\text{Ox}]^a}$$

R is the ideal gas constant, T the absolute temperature in degrees Kelvin, and F the Faraday. At 25°C, the Nernst equation becomes

$$E = E^\circ - \frac{0.059}{n} \log \frac{[\text{Red}]^b}{[\text{Ox}]^a} \tag{5.1}$$

The concentrations of dissolved species are expressed in units of moles per liter. The partial pressures of gases, in units of atmospheres, are used when gases are involved. Pure solids are considered to have effective concentrations of unity.

Now it is possible to calculate the potential of the cell shown in Figure 5.1.

$$E_{\text{cell}} = E_{\text{Cu}} - E_{\text{Zn}}$$

The cell contains $CuSO_4$ which dissociates to give Cu^{2+}.

For copper

$$Cu^{2+} + 2e^- \rightleftharpoons Cu$$

The effective concentration of solid Cu is unity.

$$E_{\text{Cu}} = E^\circ_{\text{Cu}} - \frac{0.059}{2} \log \frac{1}{[Cu^{2+}]}$$

number of electrons in half-reaction

A number for E°_{Cu} is found in Table 5.1. $[Cu^{2+}] = 0.010M$ from Figure 5.1.

$$E_{\text{Cu}} = 0.337 - \frac{0.059}{2} \log \frac{1}{1.0 \times 10^{-2}} = +0.278 \text{ volt}$$

For zinc,

$$Zn^{2+} + 2e^- \rightleftharpoons Zn$$

The effective concentration of solid Zn is unity.

$$E_{\text{Zn}} = E^\circ_{\text{Zn}} - \frac{0.059}{2} \log \frac{1}{[Zn^{2+}]}$$

number of electrons in half-reaction

A number for E°_{Zn} is found in Table 5.1; $[Zn^{2+}] = 0.0010M$ from Figure 5.1

$$E_{\text{Zn}} = -0.763 - \frac{0.059}{2} \log \frac{1}{1.0 \times 10^{-3}} = -0.852 \text{ volt}$$

$$E_{cell} = E_{Cu} - E_{Zn} = 0.278 - (-0.852) = +1.130 \text{ volts}$$

This is the voltage that would be seen on the potentiometer V in Figure 5.1.

It is worth the few words needed to describe a sort of shorthand notation for the cell in Figure 5.1. Your hands can become very cramped drawing beakers and salt bridges. In conventional cell notation, a single vertical line ($|$) represents a boundary between two phases, and two parallel vertical lines ($||$) a salt bridge. The cell of Figure 5.1 can be written:

$$Zn|ZnSO_4 \ (0.0010M)||CuSO_4 \ (0.010M)|Cu$$

The rules for determining cell potentials follow.

    ◻ **1.** $E_{cell} = E_{right} - E_{left}$. ($E_{cell}$ may be positive or negative.)

    ◻ **2.** Calculate E_{right}.

        (a) Find the half-reaction for the right-hand half-cell.
        (b) Write this reaction as a reduction (electrons on the left).
        (c) Find $E°$ for the right-hand half-cell.
        (d) Calculate E_{right} from the Nernst equation (Eq. 5.1).

    ◻ **3.** Calculate E_{left}.

        (a) Find the half-reaction for the left-hand half-cell.
        (b) Write this reaction as a reduction (electrons on the left).
        (c) Find $E°$ for the left-hand half-cell.
        (d) Calculate E_{left} from the Nernst equation (Eq. 5.1).

    ◻ **4.** Now that you know E_{right} and E_{left}, calculate E_{cell}.

$$E_{cell} = E_{right} - E_{left}$$

The voltage of the cell of Fig. 5.1 is measured with a potentiometer that allows the passage of no current. When no current flows, there is no motion of electrons from electrode to electrode. Unless electrons flow, neither reduction half-reactions

$$Cu^{2+} + 2e^- \rightleftharpoons Cu$$

$$Zn^{2+} + 2e^- \rightleftharpoons Zn$$

nor oxidation half-reactions

$$Cu \rightleftharpoons Cu^{2+} + 2e^-$$

$$Zn \rightleftharpoons Zn^{2+} + 2e^-$$

can occur. If no half-reactions can happen, no overall reactions can happen. Only if current flows can reactions take place in the two halves of the cell.

Current can flow and reaction can take place if the cell is short-circuited, or shorted. A cell can be shorted by connecting the right electrode to the left with a wire. Metal wires conduct electrons. A shorted cell is shown in Figure 5.2.

What reaction occurs in a shorted cell? What does the cell potential become after the reaction occurs? The answer to the first question requires a few lines, while the answer to the second is *very* brief, with a somewhat longer explanation. There is a formal way to predict the reaction that will occur in a shorted cell.

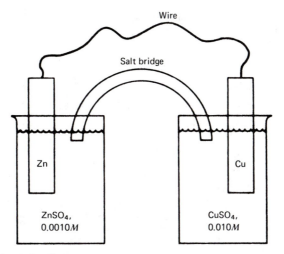

Figure 5.2 *A shorted cell.*

¤ **1.** Subtract the left-hand half-reaction from the right-hand half-reaction, with appropriate multiplications of each half-reaction so that no electrons are left in the final equation. In other words, balance a redox equation.

¤ **2.** Calculate the potential of the cell before the shorting wire was connected.

¤ **3.** If the cell potential is positive, the overall reaction of Rule 1 will proceed from left to right as written.

¤ **4.** If the cell potential is negative, the overall reaction of Rule 1 will *not* proceed from left to right as written. The *reverse* of the reaction of Rule 1 will proceed.

Take as an example the cell of Figure 5.2.

$$\text{(right-hand half-reaction)} = Cu^{2+} + 2e^- \rightleftharpoons Cu$$

$$-\text{ (left-hand half-reaction)} = -(Zn^{2+} + 2e^- \rightleftharpoons Zn)$$

$$\text{(overall reaction)} = Cu^{2+} - Zn^{2+} \rightleftharpoons Cu - Zn$$

$$\text{(overall reaction)} = Cu^{2+} + Zn \rightleftharpoons Cu + Zn^{2+}$$

The potential of the cell before shorting is +1.130 volts. The overall reaction

$$Cu^{2+} + Zn \rightleftharpoons Cu + Zn^{2+}$$

proceeds as written, from left to right, because the cell potential is positive.

Take another example:

$$Pt, H_2 \ (1.0 \ atm)|HCl \ (0.10M)||ZnSO_4 \ (0.0010M)|Zn$$

What is the voltage of the unshorted cell? What reaction will occur if the cell is shorted?

$$E_{cell} = E_{right} - E_{left}$$

$$E_{right} = E_{Zn}$$

$$E_{left} = E_{H}$$

From the previous example,

$$E_{Zn} = -0.852 \text{ volt}$$

For the hydrogen half-cell (Pt undergoes no reaction; it is simply an indicator electrode):

$$2H^+ + 2e^- \rightleftharpoons H_2 \qquad E^\circ_H = 0.000 \text{ volt}$$

$$E_H = E^\circ_H - \frac{0.059}{2} \log \frac{p_{H_2}}{[H^+]^2} = 0.000 - \frac{0.059}{2} \log \frac{1.0}{(0.10)^2} = -0.059 \text{ volt}$$

$$E_{cell} = E_{right} - E_{left} = E_{Zn} - E_H = -0.852 - (-0.059) = -0.793 \text{ volt}$$

Now write the overall reaction

$$\text{(right-hand half-reaction)} = (Zn^{2+} + 2e^- \rightleftharpoons Zn)$$

$$- \text{ (left-hand half-reaction)} = -(2H^+ + 2e^- \rightleftharpoons H_2)$$

$$\text{(overall reaction)} = Zn^{2+} - 2H^+ \rightleftharpoons Zn - H_2$$

$$\text{(overall reaction)} = Zn^{2+} + H_2 \rightleftharpoons 2H^+ + Zn$$

The potential of the cell before shorting is

$$-0.793 \text{ volt}$$

The overall reaction

$$Zn^{2+} + H_2 \rightleftharpoons 2H^+ + Zn$$

does not proceed as written. If the cell were shorted, the reverse reaction

$$2H^+ + Zn \rightleftharpoons Zn^{2+} + H_2$$

would proceed, because the potential is positive.

Half-cells are also known as electrodes. In potentiometry, the more positive electrode is known as the *cathode*. The less positive electrode is known as the *anode*. If a cell is shorted, reduction takes place at the cathode. Oxidation takes place at the anode.

Which electrode in the cell of Figure 5.1 is the cathode? What reaction would occur there if the cell were shorted? Which electrode is the anode? What reaction would occur there if the cell were shorted? From prior work,

$$E_{right} = E_{Cu} = -0.278 \text{ volt}$$

$$E_{left} = E_{Zn} = -0.852 \text{ volt}$$

The copper electrode is the more positive; it is the cathode. The zinc electrode is the less positive; it is the anode. Reduction would take place at the cathode (copper). This means that the Cu^{2+}-Cu half-cell reaction would proceed as a reduction:

$$Cu^{2+} + 2e^- \rightleftharpoons Cu$$

Oxidation would take place at the anode. This means that the Zn^{2+}-Zn half-cell reaction would be written as an oxidation:

$$Zn \rightleftharpoons Zn^{2+} + 2e^-$$

This system of designating anode and cathode is consistent with the overall reaction that would occur if the cell were shorted.

$$Cu^{2+} + Zn \rightleftharpoons Cu + Zn^{2+}$$

Note that in the overall reaction Cu^{2+} is reduced to Cu and Zn is oxidized to Zn^{2+}, as predicted by the anode-cathode rule just stated.

Electrons flow from anode to cathode in the shorted cell. This is consistent with the fact that zinc metal (the anode) loses electrons, which flow through the shorting wire to the copper cathode, which provides electrons to Cu^{2+} ions, which are then reduced to copper metal. Figure 5.3 shows the electron path in the shorted cell.

After a cell is shorted for a while, its potential is no longer what it was before shorting. This is illustrated by what happens to people who leave their cars parked with the headlights on. Leaving the headlights on for a long time has the same effect

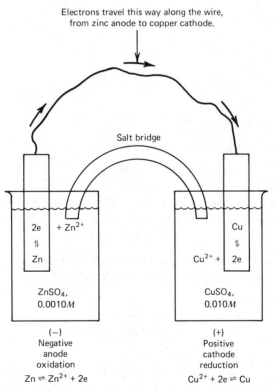

Figure 5.3 *A shorted cell, in detail.*

as shorting each of the cells in the car battery. When these unhappy people turn their keys to the start position, they hear only a most disheartening click, or at best a faint "rowr." The voltage, or potential, of each cell has dropped to some point near zero volts. *The potential of a cell that has been shorted for a long time is zero volts. When the potential of a shorted cell is zero volts, all the species in the cell are at equilibrium with each other.*

How does the potential of a shorted cell reach zero volts? Consider the cell of Figure 5.3.

$$E_{cell} = E_{right} - E_{left}$$

$$E_{right} = E_{Cu}$$

$$E_{left} = E_{Zn}$$

$$E_{cell} = E_{Cu} - E_{Zn}$$

$$E_{Cu} = E°_{Cu} - \frac{0.059}{2} \log \frac{1}{[Cu^{2+}]}$$

$$E_{Zn} = E°_{Zn} - \frac{0.059}{2} \log \frac{1}{[Zn^{2+}]}$$

$$E_{Cu} = 0.337 - \frac{0.059}{2} \log \frac{1}{[Cu^{2+}]}$$

or

$$E_{Cu} = 0.337 + \frac{0.059}{2} \log [Cu^{2+}]$$

$$E_{Zn} = -0.763 - \frac{0.059}{2} \log \frac{1}{[Zn^{2+}]}$$

$$E_{Zn} = -0.763 + \frac{0.059}{2} \log [Zn^{2+}]$$

In the copper half-cell, this reaction proceeds when the cell is shorted:

$$Cu^{2+} + 2e^- \rightarrow Cu$$

Cu^{2+} is used up. The copper ion concentration, $[Cu^{2+}]$, decreases. The term $0.059/2$ $\log [Cu^{2+}]$ becomes more and more negative. The value of E_{Cu} falls.

In the zinc half-cell, this reaction proceeds when the cell is shorted:

$$Zn \rightarrow Zn^{2+} + 2e^-$$

Zn^{2+} is generated. The zinc ion concentration, $[Zn^{2+}]$, increases. The term $0.059/2$ $\log [Zn^{2+}]$ becomes more and more positive. The value of E_{Zn} rises.

After a while,

$$E_{Zn} = E_{Cu}$$

$$E_{cell} = E_{Cu} - E_{Zn}$$

$$E_{cell} = 0$$

(The same result would be achieved more quickly by mixing all the reagents together, but it would be difficult to tap the electron current if there were no half-cells with a wire carrying current running between them.)

≡ 5.3 PROBLEMS ≡

Assume that all cells are at $25°C$ and use Table 5.1 for half-reactions and $E°$-values.

1. Using the data of Table 1, balance the following redox reactions:
   (a) $Cr_2O_7^{2-} + Fe^{2+} \rightarrow Cr^{3+} + Fe^{3+}$ (acid).
   (b) $Fe^{3+} + Sn^{2+} \rightarrow Fe^{2+} + Sn^{4+}$.
   (c) $Ce^{4+} + H_3AsO_3 \rightarrow Ce^{3+} + H_3AsO_4$ (acid)
   (d) $Fe^{2+} + V(OH)_4^+ \rightarrow VO^{2+} + Fe^{3+}$ (acid).
   (e) $MnO_4^- + H_2C_2O_4 \rightarrow Mn^{2+} + CO_2 + H_2O$ (acid).
   (f) $Cu^{2+} + Pb \rightarrow Pb^{2+} + Cu$.
   Ans: *See detailed solutions to problems.*

2. Calculate the potential of the half-cell

$$\|Pb(NO_3)_2 \ (0.0010M)|Pb$$

   Ans: *-0.214 volt*

3. Calculate the potential of the half-cell

$$\|CdCl_2 \ (0.010M)|Cd$$

   Ans: *-0.462 volt*

4. Calculate the potential of the half-cell

$$\|FeCl_3 \ (0.001M), FeCl_2 \ (0.010M)|Pt$$

   (Platinum does not undergo reaction; it is simply an indicator electrode.)
   Ans: *0.712 volt*

5. Calculate the potential of the half-cell

$$\|HClO_4 \ (1M), Ce^{3+} \ (0.10M), Ce^{4+} \ (0.050M)|Pt$$

   Ans: *1.68 volt*

6. Calculate the potential of the half-cell

$$\|HCl \ (0.010M)|H_2 \ (1.0 \ atm), Pt$$

   Ans: *0.12 volt*

7. Calculate the potential of the half-cell

$$\|HCl \ (0.010M)|H_2 \ (0.90 \ atm), Pt$$

   Ans: *0.012 volt*

8. Calculate the potential of the cell

$$Pb|Pb(NO_3)_2 \ (0.10M)\|CuSO_4 \ (0.10M)|Cu$$

   Identify the anode and cathode. Write the reaction for the shorted cell. Will the reaction proceed as written?
   Ans: *0.464 volt*

9. Calculate the potential of the cell

$$Zn|ZnSO_4 \ (1.0M)\|K_2Cr_2O_7 \ (0.10M), \ [H^+] , \ (1.0M) \ Cr^{3+} \ (0.10M)|Pt$$

Identify the anode and cathode. Write the reaction for the shorted cell. Will the reaction proceed as written?

Ans: *See detailed solutions.*

10. Calculate the potential of the cell

$$Cu|CuSO_4 \ (0.10M)\|HCl \ (0.010M)|H_2 \ (1.0 \ atm), \ Pt$$

Identify the anode and cathode. Write the reaction for the shorted cell. Will the reaction proceed as written?

Ans: *See detailed solutions.*

11. Calculate the potential of the cell

$$Pt, \ H_2 \ (1.0 \ atm)|HCl \ (1.00 \times 10^{-4}M)\|HCl \ (1.00 \times 10^{-1}M)|H_2 \ (1.0 \ atm), \ Pt$$

Identify the anode and cathode.

Ans: *0.18 volt; the right-hand half-cell is the cathode.*

≡ 5.4 DETAILED SOLUTIONS TO PROBLEMS ≡

1. Balancing redox reactions with the aid of Table 5.1 is not difficult.

   (a) Write the appropriate half-reactions, one as a reduction,

   $$Cr_2O_7^{2-} + 14H^+ + 6e^- \rightarrow 2Cr^{3+} + 7H_2O,$$

   the other as an oxidation,

   $$Fe^{2+} \rightarrow Fe^{3+} + e^-$$

   Now multiply each through by numbers to make the numbers of electrons in each equal, and add the two.

   $$+\begin{cases} Cr_2O_7^{2-} + 14H^+ + \quad 6e^- \rightarrow 2Cr^{3+} + 7H_2O \quad\quad (\times 1) \\ \underline{\qquad\qquad\quad 6Fe^{2+} \rightarrow 6Fe^{3+} + 6e^- \quad\quad (\times 6)} \end{cases}$$
   $$Cr_2O_7^{2-} + 14H^+ + 6Fe^{2+} \rightarrow 2Cr^{3+} + 6Fe^{3+} + 7H_2O$$

   (b) $Fe^{3+} + e^- \rightarrow Fe^{2+}$

   $Sn^{2+} \rightarrow Sn^{4+} + 2e^-$

   $$+\begin{cases} 2Fe^{3+} + 2e^- \rightarrow 2Fe^{2+} \quad\quad\quad\quad (\times 2) \\ \underline{\qquad\quad Sn^{2+} \rightarrow Sn^{4+} + 2e^- \quad\quad (\times 1)} \end{cases}$$
   $$2Fe^{3+} + Sn^{2+} \rightarrow Sn^{4+} + 2Fe^{2+}$$

   (c) $Ce^{4+} + e^- \rightarrow Ce^{3+}$

   $H_3AsO_3 + H_2O \rightarrow H_3AsO_4 + 2H^+ + 2e^-$

   $$+\begin{cases} 2Ce^{4+} + 2e^- \qquad\qquad\qquad \rightarrow 2Ce^{3+} \qquad\qquad\qquad\qquad\qquad (\times 2) \\ \underline{\qquad\quad H_3AsO_3 + H_2O \rightarrow \qquad\qquad H_3AsO_4 + 2H^+ + 2e^- \quad (\times 1)} \end{cases}$$
   $$2Ce^{4+} + H_3AsO_3 + H_2O \rightarrow 2Ce^{3+} + H_3AsO_4 + 2H^+$$

(d)
$$
+ \begin{cases}
Fe^{2+} \rightarrow & Fe^{3+} + e^- \\
V(OH)_4^+ + 2H^+ + e^- \rightarrow VO^{2+} & + 3H_2O
\end{cases}
$$
$$
\overline{V(OH)_4^+ + 2H^+ + Fe^{2+} \rightarrow VO^{2+} + Fe^{3+} + 3H_2O}
$$

(e) $MnO_4^- + 8H^+ + 5e^- \rightarrow Mn^{2+} + 4H_2O$

$H_2C_2O_4 \rightarrow 2CO_2 + 2H^+ + 2e^-$

$$
+ \begin{cases}
2MnO_4^- + 16H^+ + 10e^- \rightarrow 2Mn^{2+} + 8H_2O & (\times 2) \\
5H_2C_2O_4 \rightarrow 10CO_2 + 10H^+ + 10e^- & (\times 5)
\end{cases}
$$
$$
\overline{2MnO_4^- + 5H_2C_2O_4 + 16H^+ \rightarrow 2Mn^{2+} + 8H_2O + 10CO_2 + 10H^+}
$$

or

$2MnO_4^- + 5H_2C_2O_4 + 6H^+ \rightarrow 2Mn^{2+} + 8H_2O + 10CO_2$

(f)
$$
+ \begin{cases}
Cu^{2+} + 2e^- \rightarrow Cu \\
Pb \rightarrow Pb^{2+} + 2e^-
\end{cases}
$$
$$
\overline{Cu^{2+} + Pb \rightarrow Cu + Pb^{2+}}
$$

2. First, write the half-cell reaction and associated E°.

$$Pb^{2+} + 2e^- \rightleftharpoons Pb \qquad E^\circ = -0.126 \text{ volt}$$

Then, write the Nernst equation for the reaction.

$$E = E^\circ - \frac{0.059}{2} \log \frac{1}{[Pb^{2+}]}$$

Substitute numbers and calculate

$$E = -0.126 - \frac{0.059}{2} \log \frac{1}{1.0 \times 10^{-3}} = -0.214 \text{ volt}$$

3. $Cd^{2+} + 2e^- \rightleftharpoons Cd \qquad E^\circ = -0.403 \text{ volt}$

$$E = E^\circ - \frac{0.059}{2} \log \frac{1}{[Cd^{2+}]} = -0.403 - \frac{0.059}{2} \log \frac{1}{(1.0 \times 10^{-2})} = -0.462 \text{ volt}$$

4. $Fe^{3+} + e^- \rightleftharpoons Fe^{2+} \qquad E^\circ = +0.771 \text{ volt}$

$$E = E^\circ - \frac{0.059}{1} \log \frac{[Fe^{2+}]}{[Fe^{3+}]} = 0.771 - 0.059 \log \left(\frac{1.0 \times 10^{-2}}{1.0 \times 10^{-3}}\right) = 0.712 \text{ volt}$$

5. $Ce^{4+} + e^- \rightarrow Ce^{3+} \qquad E^{\circ\prime} = 1.70 \text{ volts}$

(Use the formal potential here.)

$$E = E^{\circ\prime} - \frac{0.059}{1} \log \frac{[Ce^{3+}]}{[Ce^{4+}]} = 1.70 - 0.059 \log \frac{(1.0 \times 10^{-1})}{(5.0 \times 10^{-2})} = 1.68 \text{ volts}$$

6. $2H^+ + 2e^- \rightleftharpoons H_2 \qquad E^\circ = 0.000 \text{ volt}$

$$E = E^\circ - \frac{0.059}{2} \log \frac{p_{H_2}}{[H^+]^2} = 0.000 - \frac{0.059}{2} \log \frac{1.0}{(1.0 \times 10^{-2})^2} = -0.12 \text{ volt}$$

7. $2H^+ + 2e^- \rightleftharpoons H_2$ $E^\circ = 0.000$ volt

$$E = E^\circ - \frac{0.059}{2} \log \frac{p_{H_2}}{[H^+]^2} = 0.000 - \frac{0.059}{1} \log \frac{0.90}{(1.0 \times 10^{-2})^2} = -0.12 \text{ volt}$$

8. Proceed according to the rules given in the text. (The numbers and letters correspond to those of rules given in the text.)

(1) $E_{cell} = E_{right} - E_{left}$

$E_{right} = E_{Cu}$

$E_{left} = E_{Pb}$

(2) Calculate E_{right} ($= E_{Cu}$)

(a)
(b) } $Cu^{2+} + 2e^- \rightleftharpoons Cu$ $E^\circ = +0.337$ volt
(c)

(d) $E_{Cu} = E^\circ - \dfrac{0.059}{2} \log \dfrac{1}{[Cu^{2+}]} = 0.337 - \dfrac{0.059}{2} \log \left(\dfrac{1}{0.10}\right)$

$\qquad = 0.308$ volt

(3) Calculate E_{left} ($= E_{Pb}$)

(a)
(b) } $Pb^{2+} + 2e^- \rightleftharpoons Pb$ $E^\circ = -0.126$ volt
(c)

(d) $E_{Pb} = E^\circ - \dfrac{0.059}{2} \log \dfrac{1}{[Pb^{2+}]} = -0.126 - \dfrac{0.059}{2} \log \dfrac{1}{(1.0 \times 10^{-1})}$

$\qquad = -0.156$ volt

(4) Now that you know E_{right} and E_{left}, calculate E_{cell}.

$E_{cell} = E_{right} - E_{left} = E_{Cu} - E_{Pb} = +0.308 - (-0.156) = +0.464$ volt
The copper half-cell is the more positive ($E_{Cu} = +0.308$ volt). Hence it is the cathode. The lead half-cell is the less positive ($E_{Pb} = -0.156$ volt). Hence it is the anode.

To find the reaction for the shorted cell, subtract the left-hand half-reaction from the right-hand half-reaction.

(right-hand half-reaction) $= Cu^{2+} + 2e^- \rightleftharpoons Cu$

$-$ (left-hand half-reaction) $= -(Pb^{2+} + 2e^- \rightleftharpoons Pb)$

(overall reaction) $= Cu^{2+} - Pb^{2+} \rightleftharpoons Cu - Pb$

(overall reaction) $= Cu^{2+} + Pb \rightleftharpoons Pb^{2+} + Cu$

The potential of the cell before shorting is $+0.464$ volt. The overall reaction proceeds as written above, from left to right, because the cell potential is positive.

9. Again, proceed according to the rules given in the text.

(1) $E_{cell} = E_{right} - E_{left}$

$E_{right} = E_{Cr}$

$E_{left} = E_{Zn}$

(2) Calculate $E_{right} (= E_{Cr})$

$\left.\begin{array}{l}\text{(a)}\\\text{(b)}\\\text{(c)}\end{array}\right\}$ $Cr_2O_7^{2-} + 14H^+ + 6e^- \rightleftharpoons 2Cr^{3+} + 7H_2O \qquad E^\circ = 1.33$ volts

(d) $E_{Cr} = E^\circ - \dfrac{0.059}{6} \log \dfrac{[Cr^{3+}]^2}{[Cr_2O_7^{2-}][H^+]^{14}}$

$\qquad = 1.33 - \dfrac{0.059}{6} \log \dfrac{(1.0 \times 10^{-1})^2}{(1.0 \times 10^{-1})(1.0)^{14}} = 1.34$ volts

(3) Calculate $E_{left} (= E_{Zn})$

$\left.\begin{array}{l}\text{(a)}\\\text{(b)}\\\text{(c)}\end{array}\right\}$ $Zn^{2+} + 2e^- \rightleftharpoons Zn \qquad E^\circ = -0.763$ volt

(d) $E_{Zn} = E^\circ - \dfrac{0.059}{2} \log \dfrac{1}{[Zn^{2+}]} = -0.763 - \dfrac{0.059}{2} \log (1.0)$

$\qquad = -0.763$ volt

(4) Now that you have E_{right} and E_{left}, calculate E_{cell}.

$E_{cell} = E_{right} - E_{left} = E_{Cr} - E_{Zn} = 1.34 - (-0.763) = 2.10$ volts

The chromium half-cell is the more positive ($E_{Cr} = +1.34$ volts). Hence it is the cathode. The zinc half-cell is the less positive ($E_{Zn} = -0.763$ volt). Hence it is the anode.

To find the reaction for the shorted cell, subtract the left-hand half-reaction from the right-hand half-reaction.

$\qquad$ (right-hand half-reaction) $= Cr_2O_7^{2-} + 6e^- + 14H^+ \rightleftharpoons 2Cr^{3+} + 7H_2O$

$- (3)$(left-hand half-reaction) $= -(3Zn^{2+} + 6e^- \rightleftharpoons 3Zn)$

$\qquad\qquad$ (overall reaction) $= Cr_2O_7^{2-} + 14H^+ - 3Zn^{2+} \rightleftharpoons 2Cr^{3+} + 7H_2O - 3Zn$

$\qquad\qquad$ (overall reaction) $= Cr_2O_7^{2-} + 3Zn + 14H^+ \rightleftharpoons 2Cr^{3+} + 3Zn^{2+} + 7H_2O$

The potential of the cell before shorting is 2.10 volts. (Multiplying a half-reaction by an integer does *not* affect the value of E, E°, or $E^{\circ\prime}$.) The overall reaction

$$Cr_2O_7^{2-} + 3Zn + 14H^+ \rightleftharpoons 2Cr^{3+} + 3Zn^{2+} + 7H_2O$$

proceeds as written, from left to right, because the cell potential is positive.

10. Again proceed according to the rules given in the text.

(1) $E_{cell} = E_{right} - E_{left}$

$E_{right} = E_H$

$E_{left} = E_{Cu}$

(2) Calculate E_{right} $(= E_H)$

(a)
(b) } $2H^+ + 2e^- \rightleftharpoons H_2 \qquad E° = 0.000$
(c)

(d) $E_H = E° - \dfrac{0.059}{2}\log\dfrac{p_{H_2}}{[H^+]^2} = 0.000 - \dfrac{0.059}{2}\log\dfrac{1.0}{(1.0 \times 10^{-2})^2}$

$= -0.118$ volt

(3) Calculate E_{left} $(= E_{Cu})$

(a)
(b) } $Cu^{2+} + 2e^- \rightleftharpoons Cu \qquad E° = +0.337$ volt
(c)

(d) $E_{Cu} = E° - \dfrac{0.059}{2}\log\dfrac{1}{[Cu^{2+}]} = 0.337 - \dfrac{0.059}{2}\log\dfrac{1}{(1.0 \times 10^{-1})}$

$= 0.308$ volt

(4) Now that you have E_{right} and E_{left}, calculate E_{cell}

$E_{cell} = E_{right} - E_{left} = E_H - E_{Cu} = -0.118 - 0.308$ volt $= -0.426$ volt

The copper half-cell is the more positive ($E_{Cu} = +0.308$ volt). Hence it is the cathode. The hydrogen half-cell is the less positive ($E_H = -0.118$ volt). Hence it is the anode. To find the reaction for the shorted cell, subtract the left-hand half-reaction from the right-hand half-reaction.

(right-hand half-reaction) $= 2H^+ + 2e^- \rightleftharpoons H_2$

$-$ (left-hand half-reaction) $= -(Cu^{2+} + 2e^- \rightleftharpoons Cu)$

(overall reaction) $= 2H^+ - Cu^{2+} \rightleftharpoons H_2 - Cu$

(overall reaction) $= 2H^+ + Cu \rightleftharpoons Cu^{2+} + H_2$

The potential of the cell before shorting is -0.426 volt.

The overall reaction

$$2H^+ + Cu \rightleftharpoons Cu^{2+} + H_2$$

does not proceed as written, because the cell potential is negative. The reverse reaction

$$Cu^{2+} + H_2 \rightleftharpoons 2H^+ + Cu$$

would occur if the cell were shorted.

11. This is also called a concentration cell. Its potential is found by the same rules as the potentials of other cells.

(1) $E_{cell} = E_{right} - E_{left}$.

(2) Calculate E_{right}.

(a)
(b) } $2H^+ + 2e^- \rightleftharpoons H_2 \qquad E° = 0.000$ volt
(c)

(d) $E_{right} = E° - \dfrac{0.059}{2}\log\dfrac{p_{H_2}}{[H^+]^2} = 0.000 - \dfrac{0.059}{2}\log\dfrac{1}{(1.00 \times 10^{-1})^2}$

$$= -0.059$$

(3) Calculate E_{left}.

$$\left.\begin{matrix}(a) \\ (b) \\ (c)\end{matrix}\right\}\ 2H^{+} + 2e^{-} \rightleftharpoons H_2 \qquad E^{\circ} = 0.000\ volt$$

(d) $E_{left} = E^{\circ} - \dfrac{0.059}{2} \log \dfrac{p_{H_2}}{[H^+]^2} = 0.000 - \dfrac{0.059}{2} \log \dfrac{1.0}{(1.00 \times 10^{-4})^2}$

$$= -0.24\ volt$$

(4) Now that you have E_{right} and E_{left}, calculate E_{cell}.

$$E_{cell} = E_{right} - E_{left} = -0.059 - (-0.24) = 0.18\ volt$$

The right-hand half-cell is the more positive (-0.059 volt). Hence it is the cathode. The left-hand half-cell is the less positive (-0.236 volt). Hence it is the anode.

≡ 5.5 POTENTIOMETRIC TITRATIONS ≡

The Nernst equation provides a link between measured potentials and the logarithms of species concentrations. Indeed, with the proper electrode arrangements, direct potentiometric measurement of concentration is not only possible but is done every hour of every day. Direct potentiometric measurement of concentrations with conventional electrodes and with ion-selective electrodes is quick and allows great sensitivity in a number of applications.

5.6 Prediction of Curve Shapes

The direct potentiometric method has yet to supplant the titration for work requiring high accuracy. A single measurement error in direct potentiometry can invalidate a result. Sources for such measurement errors are many: (1) drift and instability in the measuring instrument; (2) substances in the analyzed solution that can poison the measuring electrode into giving seriously unstable potentials; and (3) changes, ever so slight, in the reference electrode voltage.

In titration methods in the 0.001 to 0.1M range, however, the absolute value of a single potential difference (voltage) between two electrodes is not too important, so long as the reading does not drift too violently. The quantity of interest is the change, or jump, in potential difference in the endpoint region. In a typical potentiometric titration, a plot of potential difference, E, versus the volume of titrant added looks like the curve shown in Figure 5.4. Note the swift change of potential around the endpoint. The change in potential with the volume of titrant added, not absolute potential values, is important, so small errors in potentials are easily tolerated. Even a \$100 pH meter can register such a change accurately enough so that the accuracy of the endpoint reading depends on the accuracy of the buret used to deliver the titrant.

The apparatus used in potentiometric titration is shown in Figure 5.5. It bears a resemblance to that often used in acid base titration. The potential of the solution in

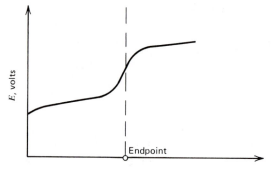

Figure 5.4 *Potentiometric titration curve.*

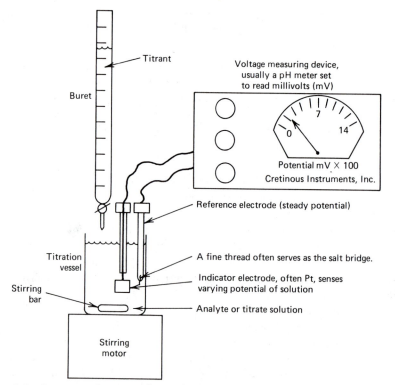

Figure 5.5 *Apparatus for potentiometric titration.*

the titration vessel is sensed by the indicator electrode. The solution potential varies as the titrant is added from the buret. The indicator electrode is often platinum. The reference electrode is often a saturated calomel (calomel is Hg_2Cl_2) electrode. Its voltage, at $25°C$, is a steady +0.246 volt. The calomel half-reaction is

$$Hg_2Cl_2 + 2e^- \rightleftharpoons 2Hg + 2Cl^- \quad E^{\circ\prime} = +0.246 \text{ volt}$$

It may be represented as

$$Hg|Hg_2Cl_2(\text{saturated}), KCl (\text{saturated})\|$$

The entire titration rig could be represented as a cell. Suppose that the indicator electrode is platinum and the reference electrode a saturated calomel electrode (abbreviated SCE). The cell can be written:

Reference   Indicator

$$Hg|Hg_2Cl_2(\text{saturated}), KCl(\text{saturated})\|\text{solution in titration vessel}|Pt$$

This can be shortened:

Reference   Indicator

$$\text{SCE}\|\text{solution in titration vessel}|Pt$$

The potential, E, read on the meter is the difference between the potential of the indicator electrode, E_{solution}, and the potential of the reference electrode, $E_{\text{reference}}$.

$$E = E_{\text{solution}} - E_{\text{reference}}$$

$E_{\text{reference}}$ is constant because concentrations of reacting species within the reference electrode do not change. Because $E_{\text{reference}}$ is constant, any changes in E_{solution} result from the addition of titrant and are greatest near the endpoint.

It is possible to predict the shapes of potentiometric titration curves. The accurate prediction of titration curve shapes will allow intelligent judgment of the feasibility of titrations and of the suitability of indicators. An example of a potentiometric titration is that of Fe^{2+} by Ce^{4+}.

Example:

$$Ce^{4+} + Fe^{2+} \rightleftharpoons \underline{Ce^{3+} + Fe^{3+}}$$

titrant   analyte   These species
in titration   are found in the
titration vessel
as soon as some
Ce^{4+} is added.

Calculate the potentials read on the potentiometer dial if 100.00 milliliters of $0.1000M$ Fe^{2+} solution are titrated with $0.1000M$ Ce^{4+} solution. Calculate the potentials when 25.00, 50.00, 75.00, 99.00, 100.00, 101.00, 150.00, 200.00, and 400.00 milliliters of Ce^{4+} solution are added. The electrochemical cell is

$E^{\circ\prime}_{\text{SCE}}$ E_{solution}

$$\text{SCE}\|\text{solution in titration vessel}|Pt$$

$$Ce^{4+} + e^- \rightleftharpoons Ce^{3+} \qquad E^{\circ}_{Ce} = 1.61 \text{ volts}$$

$$Fe^{3+} + e^- \rightleftharpoons Fe^{2+} \qquad E^{\circ}_{Fe} = 0.771 \text{ volt}$$

$$Hg_2Cl_2 + 2e^- \rightleftharpoons 2Hg + 2Cl^- \quad E^{\circ\prime}_{SCE} = 0.246 \text{ volt}$$

At all points in the titration, the potential, E, read on the potentiometer is the difference between the potential, E_{solution} of the solution in the titration vessel, and the potential, $E^{\circ\prime}_{SCE}$, of the saturated calomel electrode (SCE).

$$E = E_{\text{solution}} - E^{\circ\prime}{}_{\text{SCE}}$$
$$E^{\circ\prime}{}_{\text{SCE}} = 0.246 \text{ volt} \quad (\text{does not change})$$
$$E = E_{\text{solution}} - 0.246 \text{ volt}$$

What is E_{solution}? At all times after the titration has begun, Fe^{2+}, Fe^{3+}, Ce^{3+}, and Ce^{4+} are in equilibrium with each other. When these are in equilibrium with each other, it is as if an Fe—Ce cell had been shorted. Remember, mixing is the same as shorting a cell. The potential, E_{Fe}, of the iron couple is equal to the potential, E_{Ce}, of the cerium couple. Both the potential, E_{Fe}, of the iron couple and the potential, E_{Ce}, of the cerium couple are equal to the solution potential, E_{solution}.

$$E_{\text{solution}} = E_{Fe} = E_{Ce}$$

The iron potential, E_{Fe}, and cerium potential, E_{Ce}, are further defined by Nernst equations.

$$Fe^{3+} + e^- \rightleftharpoons Fe^{2+}$$
$$E_{\text{solution}} = E_{Fe} = E^{\circ}{}_{Fe} - 0.059 \log \frac{[Fe^{2+}]}{[Fe^{3+}]}$$
$$Ce^{4+} + e^- \rightleftharpoons Ce^{3+}$$
$$E_{\text{solution}} = E_{Ce} = E^{\circ}{}_{Ce} - 0.059 \log \frac{[Ce^{3+}]}{[Ce^{4+}]}$$

Either the Nernst equation for iron or the Nernst equation for cerium may be used to calculate E_{solution} at any point during the titration. It is most convenient, however, to divide the titration into three regions. Region 1 is the region between zero milliliter of titrant added and the endpoint. Region 2, a very small region, is the endpoint. Region 3 is the region past the endpoint.

In region 1, it is most convenient to use the Nernst equation for the iron couple for calculations.

$$E_{\text{solution}} = E_{Fe} = E^{\circ}{}_{Fe} - 0.059 \log \frac{[Fe^{2+}]}{[Fe^{3+}]}$$

After a little Ce^{4+} has been added, the concentrations of Fe^{2+} and Fe^{3+} are easily calculated from reaction stoichiometry. The Nernst equation for the cerium couple is also valid in region 1

$$E_{\text{solution}} = E_{Ce} = E^{\circ}{}_{Ce} - 0.059 \log \frac{[Ce^{3+}]}{[Ce^{4+}]}$$

but $[Ce^{4+}]$ is a very small quantity. To calculate it, we need to know the equilibrium constant for the reaction, $[Fe^{2+}]$, $[Fe^{3+}]$, and $[Ce^{3+}]$. These quantities can be found, but it is much easier to compute $[Fe^{2+}]$ and $[Fe^{3+}]$. In this system, in region 1

$$C_{Fe^{2+}} = [Fe^{2+}]$$

and

$$C_{Fe^{3+}} = [Fe^{3+}]$$

nearly up to the endpoint.

Now, let us calculate potentials at 0.00, 25.00, 50.00, 75.00, and 99.00 milliliters of Ce^{4+} added (or at 0.00, 25.00, 75.00, and 99.00% titrated, respectively). At 0.00 milliliter of Ce^{4+} added, the potential of the solution is indeterminate. Some Fe^{3+} is present, but its concentration is both small and unknown. The best thing to do is to proceed to the next point. At 25.00 milliliters of Ce^{4+} added (25.00% titrated), remember that every millimole of Ce^{4+} added before the endpoint is changed to a millimole of Ce^{3+}, and changes a millimole of Fe^{2+} to a millimole of Fe^{3+}.

$$Ce^{4+} + Fe^{2+} \rightleftharpoons Ce^{3+} + Fe^{3+}$$

(number of millimoles of Fe^{3+} formed) = (number of millimoles of Ce^{4+} added)

(number of millimoles of Fe^{3+} formed) = (25.00)(0.1000)

(number of millimoles of Fe^{3+} formed) = 2.500 millimoles

$$[Fe^{3+}] = \frac{\text{(number of millimoles of } Fe^{3+} \text{ formed)}}{\text{(number of milliliters of solution)}} = \frac{2.500}{125.00} M$$

(number of millimoles of Fe^{2+} left) = (number of millimoles of Fe^{2+} to start) −
(number of millimoles of Fe^{3+} formed)

(number of millimoles of Fe^{2+} left) = (100.00)(0.1000) − (25.00)(0.1000)

(number of millimoles of Fe^{2+} left) = 7.500 millimoles

$$[Fe^{2+}] = \frac{\text{(number of millimoles of } Fe^{2+} \text{ left)}}{\text{(number of milliliters of solution)}} = \frac{7.500}{125.00} M$$

Now we write the Nernst equation:

$$Fe^{3+} + e^- \rightleftharpoons Fe^{2+}$$

$$E_{\text{solution}} = E_{Fe} = E^{\circ}_{Fe} - 0.059 \log \frac{[Fe^{2+}]}{[Fe^{3+}]}$$

$$E_{\text{solution}} = 0.771 - 0.059 \log \frac{(7.500/125.00)}{(2.500/125.00)}$$

$$E_{\text{solution}} = 0.743 \text{ volt}$$

We calculate the potential, E, observed on the meter:

$$E = E_{\text{solution}} - E^{\circ\prime}_{SCE} = 0.743 - 0.246 = 0.497 \text{ volt}$$

We return now to the term $[Fe^{2+}]/[Fe^{3+}]$. Note that, at 25.00 titrated

$$\frac{[Fe^{2+}]}{[Fe^{3+}]} = \frac{\dfrac{75.00}{125.00}}{\dfrac{25.00}{125.00}} = \frac{100.00 - 25.00}{25.00} = 3.000$$

For this titration, before the endpoint, if x = % titrated,

$$\frac{[Fe^{2+}]}{[Fe^{3+}]} = \frac{100.00 - x}{x}$$

The relationship works for this simple titration with one-to-one stiochiometry. It is not necessarily valid for other titrations. At 50.00 milliliters of Ce^{4+} added, the solution is 50.00% titrated; $x = 50.00$.

$$E_{\text{solution}} = E_{Fe} = E^{\circ}_{Fe} - 0.059 \log \frac{[Fe^{2+}]}{[Fe^{3+}]} = 0.771 - 0.059 \log \left(\frac{100.00 - x}{x} \right)$$

$$= 0.771 - 0.059 \log \left(\frac{100.00 - 50.00}{50.00} \right) = 0.771 \text{ volt}$$

Now we calculate the potential, E, observed on the meter:

$$E = E_{\text{solution}} - E^{\circ}_{SCE} = 0.771 - 0.246 = 0.525 \text{ volt}$$

(Some writers would call this potential, E, "0.525 volt versus SCE.") At 75.00 milliliters of Ce^{4+} added, the solution is 75.00 titrated; $x = 75.00$.

$$E_{solution} = E_{Fe} = E^{\circ}_{Fe} - 0.059 \log \frac{[Fe^{2+}]}{[Fe^{3+}]} = 0.771 - 0.059 \log \left(\frac{100.00 - x}{x}\right)$$

$$= 0.771 - 0.059 \log \left(\frac{100.00 - 75.00}{75.00}\right) = 0.799 \text{ volt}$$

Now calculate the potential, E, observed on the meter:

$$E = E_{solution} - E^{\circ\prime}_{SCE} = 0.799 - 0.246 = 0.553 \text{ volt}$$

At 99.00 milliliters of Ce^{4+} added, the solution is 99.00 titrated; $x = 99.00$.

$$E_{solution} = E_{Fe} = E^{\circ}_{Fe} - 0.059 \log \frac{[Fe^{2+}]}{[Fe^{3+}]} = 0.771 - 0.059 \log \left(\frac{100.00 - x}{x}\right)$$

$$= 0.771 - 0.059 \log \left(\frac{100.00 - 99.00}{99.00}\right) = 0.889 \text{ volt}$$

Now calculate the potential, E, observed on the meter:

$$E = E_{solution} - E^{\circ\prime}_{SCE} = 0.889 - 0.246 = 0.643 \text{ volt}$$

In region 2, which is the endpoint, the number of moles of Ce^{4+} added is just exactly equal to the number of moles of Fe^{2+} initially in the titration vessel; Ce^{4+}, Ce^{3+}, Fe^{3+}, and Fe^{2+} are in equilibrium.

$$Ce^{4+} + Fe^{2+} \rightleftharpoons Ce^{3+} + Fe^{3+}$$

At the endpoint,

$$[Ce^{3+}] = [Fe^{3+}] \quad \text{(both large)}$$
$$[Ce^{4+}] = [Fe^{2+}] \quad \text{(both small)}$$
$$\frac{[Ce^{3+}]}{[Ce^{4+}]} = \frac{[Fe^{3+}]}{[Fe^{2+}]}$$
$$[Ce^{3+}][Fe^{2+}] = [Ce^{4+}][Fe^{3+}]$$

$$+ \begin{cases} E_{solution} = E^{\circ}_{Ce} - 0.059 \log \dfrac{[Ce^{3+}]}{[Ce^{4+}]} \\[2mm] E_{solution} = E^{\circ}_{Fe} - 0.059 \log \dfrac{[Fe^{2+}]}{[Fe^{3+}]} \end{cases}$$

$$2E_{solution} = E^{\circ}_{Ce} + E^{\circ}_{Fe} - 0.059 \log \frac{[Ce^{3+}]}{[Ce^{4+}]} - 0.059 \log \frac{[Fe^{2+}]}{[Fe^{3+}]}$$

$$= E^{\circ}_{Ce} + E^{\circ}_{Fe} - 0.059 \left\{ \log \frac{[Ce^{3+}]}{[Ce^{4+}]} + \log \frac{[Fe^{2+}]}{[Fe^{3+}]} \right\}$$

$$= E^{\circ}_{Ce} + E^{\circ}_{Fe} - 0.059 \log \left\{ \frac{([Ce^{3+}] \cdot [Fe^{2+}])}{([Ce^{4+}] \cdot [Fe^{3+}])} \right\}$$

We recall that

$$[Ce^{3+}][Fe^{2+}] = [Ce^{4+}][Fe^{3+}]$$

$$2E_{solution} = E^{\circ}_{Ce} + E_{Fe} - 0.059 \log \left\{ \frac{(\cancel{[Ce^{3+}]} \cdot \cancel{[Fe^{2+}]})}{(\cancel{[Ce^{4+}]} \cdot \cancel{[Fe^{3+}]})} \right\}$$

$$2E_{solution} = E^{\circ}_{Ce} + E^{\circ}_{Fe}$$

$$E_{solution} = \frac{E^{\circ}_{Ce} + E^{\circ}_{Fe}}{2} = \frac{(1.61 + .771)}{2} = 1.19 \text{ volts}$$

Now we calculate the potential, E, observed on the meter:

$$E = E_{solution} - E^{\circ'}_{SCE} = 1.19 - 0.246 = 0.94 \text{ volt}$$

In region 3, the region past the endpoint, it is most convenient to use the Nernst equation for the cerium redox couple for calculations.

$$E_{solution} = E_{Ce} = E^{\circ}_{Ce} - 0.059 \log \frac{[Ce^{3+}]}{[Ce^{4+}]}$$

Both $[Ce^{4+}]$ and $[Ce^{3+}]$ are large and easily calculated from the reaction stoichiometry. The Nernst equation for the iron couple is also valid in region 3.

$$E_{solution} = E_{Fe} = E^{\circ}_{Fe} - 0.059 \log \frac{[Fe^{2+}]}{[Fe^{3+}]}$$

but $[Fe^{2+}]$ is a very small quantity. To calculate it, we need to know the equilibrium constant for the overall reaction, $[Ce^{4+}]$, $[Ce^{3+}]$, and $[Fe^{3+}]$. These quantities can be found, but it is much easier just to compute $[Ce^{3+}]$ and $[Ce^{4+}]$. In this system, in region 3,

$$C_{Ce^{3+}} = [Ce^{3+}]$$

and

$$C_{Ce^{4+}} = [Ce^{4+}]$$

Now, let us calculate potentials at 101.00, 150.00, 200.00, and 400.00 milliliters of Ce^{4+}. For 101.00 milliliters of Ce^{4+} added (101.00% titrated) we recall that every millimole of Ce^{4+} added before the endpoint is changed to a millimole of Ce^{3+}. Every millimole of Ce^{4+} added after the endpoint remains a millimole of Ce^{4+}.

$$Ce^{4+} + Fe^{2+} \rightleftharpoons Ce^{3+} + Fe^{3+}$$

(number of millimoles of Ce^{4+} excess) = (number of millimoles of Ce^{4+} added) – (number of millimoles of Fe^{2+} to start)

(number of millimoles of Ce^{4+} excess) = (101.00)(0.1000) – (100.00)(0.1000)

(number of millimoles of Ce^{4+} excess) = 0.1000 millimoles

$$[Ce^{4+}] = \frac{\text{(number of millimoles of } Ce^{4+} \text{ excess)}}{\text{(number of milliliters of solution)}}$$

$$[Ce^{4+}] = \frac{0.1000}{201.00} M$$

(number of millimoles of Ce^{3+}) = (number of millimoles of Fe^{2+} to start)

(number of millimoles of Ce^{3+}) = (100.00)(0.1000)

(number of millimoles of Ce^{3+}) = 10.00 millimoles

$$[Ce^{3+}] = \frac{\text{(number of millimoles of } Ce^{3+}\text{)}}{\text{(number of milliliters of solution)}}$$

$$[Ce^{3+}] = \frac{10.00}{201.00} M$$

Now we write the Nernst equation:

$$Ce^{4+} + e^{-} \rightleftharpoons Ce^{3+}$$

$$E_{\text{solution}} = E_{\text{Ce}} = E^\circ{}_{\text{Ce}} - 0.059 \log \frac{[\text{Ce}^{3+}]}{[\text{Ce}^{4+}]} = 1.61 - 0.059 \log \frac{(10.00/201.00)}{(0.1000/201.00)}$$

$$= 1.49 \text{ volts}$$

Then we calculate the potential, E, observed on the meter.

$$E = E_{\text{solution}} - E^{\circ\prime}{}_{\text{SCE}} = 1.49 - 0.246 = 1.24 \text{ volts}$$

We return now to the term $[\text{Ce}^{3+}]/[\text{Ce}^{4+}]$. Note that, at 101% titrated,

$$\frac{[\text{Ce}^{3+}]}{[\text{Ce}^{4+}]} = \frac{(100.00/201.00)}{(1.000/201.00)} = \frac{100.00}{101.00 - 100.00} = 100.0$$

For this titration, after the endpoint, if x = % titrated,

$$\frac{[\text{Ce}^{3+}]}{[\text{Ce}^{4+}]} = \frac{100.00}{x - 100.00}$$

This relationship is not necessarily valid for other titrations. At 150.00 milliliters of Ce^{4+} added, the solution is 150.00% titrated; $x = 150.00$

$$E_{\text{solution}} = E_{\text{Ce}} = E^\circ{}_{\text{Ce}} - 0.059 \log \frac{[\text{Ce}^{3+}]}{[\text{Ce}^{4+}]} = E^\circ{}_{\text{Ce}} - 0.059 \log \left(\frac{100.00}{x - 100.00} \right)$$

$$= 1.61 - 0.059 \log \left(\frac{100.00}{150.00 - 100.00} \right) = 1.59 \text{ volts}$$

Now we calculate the potential, E, observed on the meter:

$$E = E_{\text{solution}} - E^{\circ\prime}{}_{\text{SCE}} = 1.59 - 0.246 = 1.34 \text{ volts}$$

At 200.00 milliliters of Ce^{4+} added, the solution is 200.00% titrated; $x = 200.00$.

$$E_{\text{solution}} = E_{\text{Ce}} = E^\circ{}_{\text{Ce}} - 0.059 \log \frac{[\text{Ce}^{3+}]}{[\text{Ce}^{4+}]} = E^\circ{}_{\text{Ce}} - 0.059 \log \frac{100.00}{x - 100.00}$$

$$= 1.61 - 0.059 \log \frac{100.00}{200.00 - 100.00} = 1.61 \text{ volts}$$

Then we calculate the potential, E, observed on the meter:

$$E = E_{\text{solution}} - E^{\circ\prime}{}_{\text{SCE}} = 1.61 - 0.246 = 1.36 \text{ volts}$$

Table 5.2
TITRATION OF Fe^{2+}
BY Ce^{4+}

Milliliters of Ce^{4+} added	Potential, E, observed on meter, volts
25.00	0.497
50.00	0.525
75.00	0.553
99.00	0.643
100.00	0.94
101.00	1.24
150.00	1.34
200.00	1.36
400.00	1.39

At 400.00 milliliters of Ce^{4+} added, the solution is 400.00% titrated; $x = 400.00$.

$$E_{solution} = E_{Ce} = E^{\circ}_{Ce} - 0.059 \log \frac{[Ce^{3+}]}{[Ce^{4+}]} = E^{\circ}_{Ce} - 0.059 \log \frac{100.00}{x - 100.00}$$

$$= 1.61 - 0.059 \log \frac{100.00}{400.00 - 100.00} = 1.64 \text{ volts}$$

We calculate the potential, E, observed on the meter:

$$E = E_{solution} - E^{\circ\prime}_{SCE} = 1.64 - 0.246 = 1.39 \text{ volts}$$

The titration data are listed in Table 5.2.

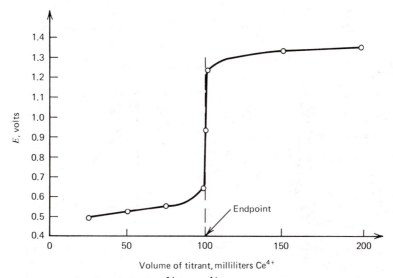

Figure 5.6 *Curve for titration of Fe^{2+} by Ce^{4+}.*

The data are plotted in Figure 5.6. The titration curve is called sigmoid (S-shaped), and is typical of potentiometric titrations. Because pH titrations are a subspecies of potentiometric titrations, pH titration curves are also sigmoid. The sharp rise in potential at the endpoint allows easy endpoint determination. The location and height of the rise govern the indicator choices.

5.7 Endpoint Detection

In a potentiometric titration such as that of Fe(II) by Ce(IV) just illustrated, how can the endpoint be found in the laboratory? There are a number of methods of endpoint detection.

The first method can involve a bit of guesswork. A plot of potential, E, versus milliliters of titrant added, V, is made, as shown in Figure 5.7a. The endpoint is taken as that volume of titrant corresponding to the place where the curve is steepest, or in other words, at the inflection point of the curve. In one-to-one titrations, such as that

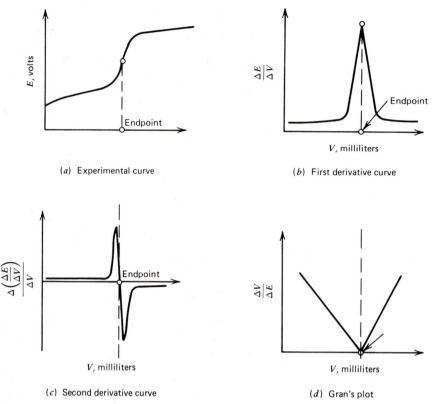

(a) Experimental curve

(b) First derivative curve

(c) Second derivative curve

(d) Gran's plot

Figure 5.7 *Endpoints in potentiometric titrations.*

of Fe(II) by Ce(IV), this point is halfway up the rising portion of the curve. It is hard, however, to estimate this point by eye, and if the whole curve is displayed, hard to read the volume, V, on the abscissa to the nearest 0.02 milliliter.

The application of a little mathematics—in practice just arithmetic—can help out. A plot of $\Delta E/\Delta V$, called the first derivative, versus V will yield a graph with a much more sharply defined endpoint, such as the one shown in Figure 5.7(b). The endpoint is just that volume corresponding to the highest value of $\Delta E/\Delta V$. *Because $\Delta E/\Delta V$ is just the slope of the E versus V curve, the V-value where $\Delta E/\Delta V$ is highest is the same as the V-value at the steepest point of the E versus V curve.*

In practice, another plot, called a *second derivative plot*, may be made. This is a plot [Figure 5.7(c)] of $\left[\Delta\left(\dfrac{\Delta E}{\Delta V}\right)\right]/\Delta V$ versus V. The endpoint is the V-value where the curve crosses the abscissa. Here, $\left[\Delta\left(\dfrac{\Delta E}{\Delta V}\right)\right]/\Delta V = 0$, because, just at this point, $\Delta E/\Delta V$ goes from being positive to being negative.

(Both the first and the second derivative curves can be plotted on an expanded abscissa. The only interesting part of the derivative curve appears around the endpoint so that nothing is lost by letting the milliliter or so around the endpoint take up the entire graphed length of the abscissa.)

An interesting way to use the *reciprocal* of a derivative to find the endpoint of a potentiometric titration was first described in 1950. A scientist in a Swedish forest products laboratory, Gunnar Gran,* demonstrated that a plot of $\Delta V/\Delta E$, the reciprocal of $\Delta E/\Delta V$, will yield, in the endpoint region, two intersecting straight lines [Figure 5.7(d)]. The endpoint will be the V-value corresponding to the point of intersection of the two lines. The drawback to the use of Gran's plots, as they have come to be called, is that there can be a great deal of point scatter, even in the endpoint region.

A numerical example of the use of graphical methods for endpoint detection will be helpful.

Example: 100.00 milliliters of 0.1000M Fe^{2+} are titrated with 0.1000M Ce^{4+}. The titration is followed potentiometrically with a platinum electrode and a saturated calomel electrode. Under the conditions of this titration, $E^{\circ\prime}_{Fe} = 0.770$ volt; $E^{\circ\prime}_{SCE} =$ constant $= 0.246$ volt. The data are shown in Table 5.3.

Table 5.3
TITRATION OF Fe^{2+} BY Ce^{4+}

V, ml Ce^{4+} added	E, volts	V, ml Ce^{4+} added	E, volts	V, ml Ce^{4+} added	E, volts
25.00	0.496	99.50	0.660	103.00	1.274
37.50	0.511	99.60	0.665	104.00	1.282
50.00	0.524	99.70	0.673	105.00	1.287
62.50	0.537	99.80	0.683	107.50	1.298
75.00	0.552	99.90	0.701	110.00	1.305
87.50	0.574	100.00	0.944	112.50	1.311
90.00	0.580	100.10	1.187	125.00	1.328
92.50	0.588	100.20	1.205	137.50	1.339
95.00	0.599	100.30	1.215	150.00	1.346
96.00	0.605	100.40	1.222	175.00	1.357
97.00	0.613	100.50	1.228	200.00	1.364
98.00	0.624	100.60	1.233		
99.00	0.641	100.70	1.237		
99.10	0.644	100.80	1.240		
99.20	0.648	100.90	1.243		
99.30	0.651	101.00	1.246		
99.40	0.655	102.00	1.263		

Note: The E-values past the endpoint have an extra significant figure, carried to facilitate derivative calculations.

Find the endpoint of the titration using a plot of E versus V, an expanded plot of E versus V, a first derivative plot, a second derivative plot, and a Gran's plot.

* G. Gran, Acta Chem. Scand., *4, 559 (1950).*

Table 5.4
DERIVATIVE DATA

V, ml Ce^{4+}	E, volts	ΔE	ΔV	$\dfrac{\Delta E}{\Delta V}$	$\dfrac{\Delta V}{\Delta E}$	V, ml	$\Delta\left(\dfrac{\Delta E}{\Delta V}\right)$	ΔV	$\Delta\left(\dfrac{\Delta E/\Delta V}{\Delta V}\right)$	V
25.00	0.496									
37.50	0.511	0.015	12.50	0.0012	833	31.25				
50.00	0.524	0.013	12.50	0.0010	962	43.75				
62.50	0.537	0.013	12.50	0.0010	962	56.25	+0.0002	12.50		62.50
75.00	0.552	0.015	12.50	0.0012	833	68.75	0.0006	12.50		75.00
87.50	0.574	0.022	12.50	0.0018	568	81.25	0.0006	7.50		85.00
90.00	0.580	0.006	2.50	0.0024	417	88.75	0.0008	2.50		90.00
92.50	0.588	0.008	2.50	0.0032	312	91.25	0.0012	2.50		92.50
95.00	0.599	0.011	2.50	0.0044	227	93.75	0.0016	1.75		96.38
96.00	0.605	0.006	1.00	0.0060	167	95.50	0.0020	1.00		96.00
97.00	0.613	0.008	1.00	0.0080	125	96.50	0.0030	1.00	0.003	97.00
98.00	0.624	0.011	1.00	0.011	90.9	97.50	0.0060	1.00	0.006	98.00
99.00	0.641	0.017	1.00	0.017	58.8	98.50	0.0130	0.55	0.024	98.78
99.10	0.644	0.003	0.10	0.030	33.3	99.05	0.0100	0.10	0.100	99.10
99.20	0.648	0.004	0.10	0.040	25.0	99.15	−0.0100	0.10	−0.100	99.20
99.30	0.651	0.003	0.10	0.030	33.3	99.25	0.0100	0.10	0.100	99.30
99.40	0.655	0.004	0.10	0.040	25.0	99.35	0.0500	0.10	+0.500	99.40
99.50	0.660	0.005	0.10	0.050	20.0	99.45	0.0000	0.10	0.000	99.50
99.60	0.665	0.005	0.10	0.050	20.0	99.55	0.0300	0.10	+0.30	99.60
99.70	0.673	0.008	0.10	0.080	12.5	99.65	0.0200	0.10	+0.20	99.70
99.80	0.683	0.010	0.10	0.100	10.0	99.75	0.0800	0.10	+0.80	99.80
99.90	0.701	0.018	0.10	0.180	5.56	99.85	2.25	0.10	22.5	99.90
100.00	0.944	0.243	0.10	2.43	0.412	99.95	0.00	0.10	00.0	100.00

100.10	1.18	0.243	0.10	2.43	0.412	100.05
100.20	1.205	0.018	0.10	0.180	5.56	100.15
100.30	1.215	0.010	0.10	0.100	10.0	100.25
100.40	1.222	0.007	0.10	0.070	14.2	100.35
100.50	1.228	0.006	0.10	0.060	16.7	100.45
100.60	1.233	0.005	0.10	0.050	20.0	100.55
100.70	1.237	0.004	0.10	0.040	25.0	100.65
100.80	1.240	0.003	0.10	0.030	33.3	100.75
100.90	1.243	0.003	0.10	0.030	33.3	100.85
101.00	1.246	0.003	0.10	0.030	33.3	100.95
102.00	1.263	0.017	1.00	0.017	58.8	101.50
103.00	1.274	0.011	1.00	0.011	90.9	102.50
104.00	1.282	0.008	1.00	0.008	125	103.50
105.00	1.287	0.005	1.00	0.005	200	104.50
107.50	1.298	0.011	2.50	0.004	227	106.25
110.00	1.305	0.007	2.50	0.003	357	108.75
112.50	1.311	0.006	2.50	0.0024	417	111.25
125.00	1.328	0.017	12.50	0.0014	735	118.75
137.50	1.339	0.011	12.50	0.001	1136	131.25
150.00	1.346	0.007	12.50	*	1785	143.75
175.00	1.357	0.011	25.00	*	2273	162.50
200.00	1.364	0.007	25.00	*	3571	187.50

-2.25	0.10	-22.5	100.10
-0.080	0.10	-0.80	100.20
-0.030	0.10	-0.30	100.30
-0.010	0.10	-0.10	100.40
-0.010	0.10	-0.10	100.50
-0.010	0.10	-0.10	100.60
-0.010	0.10	-0.10	100.70
0.00	0.10	0.00	100.80
0.00	0.10	0.00	100.90
-0.013	0.55	-0.024	101.22
-0.006	1.00	-0.006	102.00
-0.003	1.00	-0.003	103.00
-0.003	1.00	-0.003	104.00
-0.001	1.75		105.38
-0.001	2.50		107.50
-0.0006	2.50		110.00
-0.001	7.50		115.00
-0.0004	12.50		125.00
*	12.50		137.50
*	18.75		153.12
*	25.00		175.00

* Finite, but not distinguishable from zero on the graph.

Answer: We know that the endpoint occurs when 100.00 milliliters of Ce^{4+} solution have been added. This helps. A table can be made, like Table 5.4. Note the values of V in the seventh column of the table. These volumes correspond to to the values of $\Delta E/\Delta V$ and $\Delta V/\Delta E$ in the fifth and sixth columns. The first value in the seventh column is 31.25 milliliters. It is midway between the first and second values in the first column, 25.00 and 37.50 milliliters. The second value in the seventh column is 43.75 milliliters. This is midway between the second and third values in the first column, 37.50 and 50.00 milliliters. V-values to correspond to $\Delta E/\Delta V$ and $\Delta V/\Delta E$ values are taken in this way. Next, the five plots called for are made (Figures 5.8 through 5.12). Note the scatter in the Gran's plot (Figure 5.12).

Following reactions by potentiometric means is not the most pleasant task in the world, although there are many that are far worse. In your laboratory work, you may well titrate a solution of Fe(II) with one of dichromate, and look for a visible endpoint. The following discussion gives you some idea of how a redox indicator works, and how one may be selected for use in titrations.

At one time, no really good indicators for redox titrations had been found. Many titrations were done with solutions of potassium permanganate, $KMnO_4$, as oxidant. An example is the titration of ferrous iron by permanganate.

$$MnO_4^- + 8H^+ \quad + \quad 5Fe^{2+} \rightleftharpoons Mn^{2+} + 5Fe^{3+} + 4H_2O$$

deep purple		very	colorless in
$\epsilon = 2350$ liter/mole · cm	colorless	nearly	presence
@ 545 nm		colorless	of phosphate

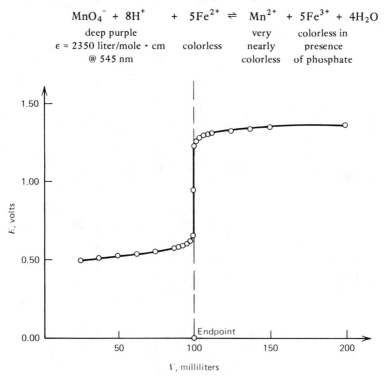

Figure 5.8 *Condensed plot of titration of Fe^{2+} by Ce^{4+}.*

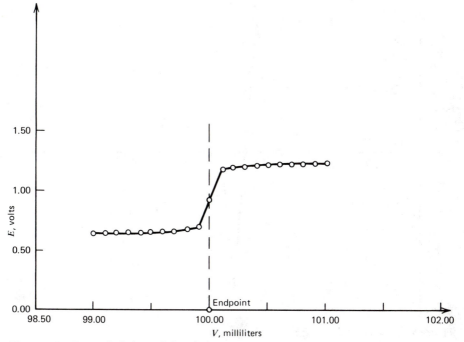

Figure 5.9 *Expanded view of titration curve.*

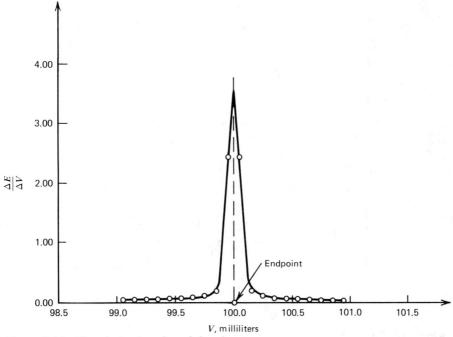

Figure 5.10 *First derivative plot of titration.*

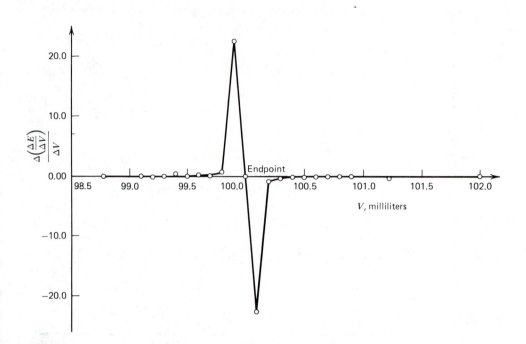

Figure 5.11 *Second derivative plot of titration.*

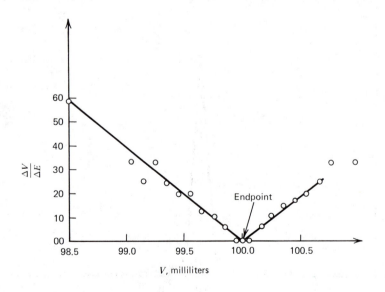

Figure 5.12 *Gran's plot of titrations.*

After all the ferrous iron has been titrated, any MnO_4^- added in excess will give the solution a slight coloration. Very little excess permanganate is needed. This is shown in the example below.

Example: Assume that a solution in a titration vessel need only have a %T of 95%, at 545 nm, for a permanganate excess to be visible. The light path through the titration vessel is 10. cm. The $KMnO_4$ in the buret is 0.0200M, and 40.00 milliliters of $KMnO_4$ have been added to the titration vessel when the first tinge of color appears. When that first color appears, the total volume of solution in the titration vessel—added from the buret, wash solution, and all the rest— is 250. milliliters.

From Chapter 4:

$$A = 2.00 - \log \%T = 2 - \log 95 = 0.022$$

$$A = \epsilon bc$$

$$A = \epsilon b \underset{\text{excess, unreacted titrant}}{[MnO_4^-]}$$

$$0.022 = (2350)(10.)[MnO_4^-]$$

$$[MnO_4^-] = \frac{0.022}{(2350)(10.)}$$

or

$$[MnO_4^-] = 9.4 \times 10^{-7}M$$

or

$$[MnO_4^-] \cong 10^{-6}M \text{ in excess}$$

(number of moles of MnO_4^- in excess)

$$= \left(10^{-6}\frac{\text{moles}}{\text{liter}} \right)(0.250 \text{ liter}) = 2.5 \times 10^{-7} \text{ moles}$$

(number of moles of MnO_4^- delivered)

$$= (40.00 \text{ ml})\left(\frac{1 \text{ liter}}{1000 \text{ milliliters}} \right)\left(0.0200 \frac{\text{moles}}{\text{liter}} \right) = 8.00 \times 10^{-4} \text{ moles}$$

$$\% \text{ error} = \frac{(\text{number of moles of } MnO_4^- \text{ in excess})}{(\text{number of moles of } MnO_4^- \text{ delivered})} (100\%)$$

$$= \frac{(2.5 \times 10^{-7})}{(8.00 \times 10^{-4})} (100\%) \cong 0.03\%$$

With so small a percent error, it would seem that permanganate is the ideal oxidant. It is *not*. Why? First, potassium permanganate is not a primary standard. Its solutions, once made up, must be standardized against a primary standard reductant, such as standard iron wire. Second, solutions of permanganate decompose on exposure to light, and over extended (about three or four days) periods of time, require frequent restandardization. Other oxidants, $K_2Cr_2O_7$, $Ce(NH_4)_2(NO_3)_6$ and $Ce(SO_4)_2 \cdot 2(NH_4)_2$-$SO_4 \cdot 2H_2O$ can be used as primary standards to make indefinitely stable solutions. Yet none of these other oxidants can serve as a self-indicator as does permanganate. Their use depends on potentiometric monitoring of the course of the titration, or some indicator added to the titration mixture that will change color just at the endpoint—or awfully near it. The color change will depend upon the potential of the solution. Let us take an example of such an indicator.

The *ortho*-phenanthroline complexes of ferrous and ferric iron were discussed in Chapter 4. *Ortho*-phenanthroline is represented as "Oph"; its structure is

The value of $E^°$ has been measured for reduction of ferric *ortho*-phenanthroline ($FeOph_3^{3+}$) to ferrous *ortho*-phenanthroline ($FeOph_3^{2+}$), also known as ferroin.[*]

$$FeOph_3^{3+} + e^- \rightleftharpoons FeOph_3^{2+} \quad E^° = +1.06 \text{ volts}$$

very light	bright
blue	orange
"Ferriin"	"Ferroin"

A Nernst equation may be written for the reaction.

$$E_{solution} = E^° - 0.059 \log \frac{[FeOph_3^{2+}]}{[FeOph_3^{3+}]}$$

In the titration of Fe(II) by Ce(IV) previously considered, ferroin may serve as a visible indicator. Initially, before the addition of much Ce(IV) from the buret, the potential of the solution is low, ferroin predominates, and the solution is bright orange. At what potential will the indicator appear blue to the eye? To answer this, let us suppose that, for the solution to appear blue, the ferriin concentration must be at least 10 times greater than that of ferroin.

For orange to disappear, and blue to appear:

$$\frac{[FeOph_3^{3+}]}{[FeOph_2^{2+}]} = 10$$

$$E_{solution} = E^° - 0.059 \log \frac{[FeOph_2^{2+}]}{[FeOph_3^{3+}]} = 1.06 - 0.059 \log \left(\frac{1}{10}\right) = 1.12 \text{ volts}$$

The theoretical endpoint of the titration, as we have seen earlier, is at a solution potential of 1.19 volts. Is a solution potential of 1.12 volts very far off? Is there a significant endpoint error when this indicator is used? To answer the questions, calculate the ratio of $[Fe^{3+}]/[Fe^{2+}]$ at a solution potential of 1.12 volts.

$$E_{solution} = 1.12 \text{ volts} = E^°_{Fe} - 0.059 \log \frac{[Fe^{2+}]}{[Fe^{3+}]} = E^°_{Fe} + 0.059 \log \frac{[Fe^{3+}]}{[Fe^{2+}]}$$

$$1.12 = 0.771 + 0.059 \log \frac{[Fe^{3+}]}{[Fe^{2+}]}$$

[*] *D. N. Hume and I. M. Kolthoff, J. Am. Chem. Soc., 65, 1895 (1943).*

$$1.12 - 0.771 = 0.059 \log \frac{[Fe^{3+}]}{[Fe^{2+}]}$$

$$0.35 = 0.059 \log \frac{[Fe^{3+}]}{[Fe^{2+}]}$$

$$\frac{0.35}{0.059} = \log \frac{[Fe^{3+}]}{[Fe^{2+}]}$$

$$5.9 = \log \frac{[Fe^{3+}]}{[Fe^{2+}]}$$

$$\frac{[Fe^{3+}]}{[Fe^{2+}]} \cong 8.6 \times 10^5$$

Therefore, when the indicator changes color, at a solution potential of 1.12 volts, only about 1 part in 860,000 parts of the iron in solution is in the ferrous form. The error is insignificant.

≡ 5.8 PROBLEMS ≡

(Assume 25°C)

1. Consider the titration of Fe^{2+} with dissolved Br_2. The reaction is

$$2Fe^{2+} + Br_2 \rightleftharpoons 2Fe^{3+} + 2Br^-$$

Prove that the endpoint solution potential, $E_{solution}$, is given by the equation

$$E_{solution} = \frac{E^\circ_{Fe} + 2E^\circ_{Br}}{3} - \frac{0.059}{3} \log \left\{ 2[Br^-] \right\}$$

Remember, $[Fe^{2+}] \neq [Fe^{3+}]$ at the endpoint.

Ans: *see detailed solutions.*

2. When dichromate is used to titrate Fe^{2+}, the titration reaction is

$$Cr_2O_7^{2-} + 6Fe^{2+} + 14H^+ \rightleftharpoons 2Cr^{3+} + 6Fe^{3+} + 7H_2O$$

Prove that the endpoint solution potential, $E_{solution}$, is

$$E_{solution} = \frac{E^\circ_{Fe} + 6E^\circ_{Cr}}{7} - \frac{0.059}{7} \log \left\{ \frac{2[Cr^{3+}]}{[H^+]^{14}} \right\}$$

Ans: *see detailed solutions.*

3. It is not very often that titrations are run in 8M perchloric acid ($HClO_4$).* However, ferrous iron and V(IV) may be titrated with Ce(IV) in 8M $HClO_4$. The following formal potentials are found for this system.†

---

* CAUTION: *Perchloric acid, $HClO_4$, presents a fire and explosion hazard.*

† G. F. Smith, *"Cerate Oxidimetry,"* 2nd ed., The G. Frederick Smith Chemical Company, Columbus, Ohio, 1964.

$$Fe^{3+} + e^- \rightleftharpoons Fe^{2+} \qquad\qquad E^{\circ\prime}{}_{Fe} = 0.825 \text{ volts}$$

$$VO_2{}^+ + 2H^+ + e^- \rightleftharpoons VO^{2+} + H_2O \qquad E^{\circ\prime}{}_V = 1.120 \text{ volts}$$

$$Ce^{4+} + e^- \rightleftharpoons Ce^{3+} \qquad\qquad E^{\circ\prime}{}_{Ce} = 1.790 \text{ volts}$$

Trace the titration of 20.00 ml of a solution $0.1000M$ in Fe^{2+} and $0.1000M$ in VO^{2+} with $0.1000M$ Ce^{4+}. Assume $8M$ $HClO_4$, and that all species follow the Nernst equation throughout. Draw graphs of $E_{solution}$ versus V, $\Delta E/\Delta V$ versus V, $\Delta(\Delta E/\Delta V)/\Delta V$ versus V, and $\Delta V/\Delta E$ versus V.

Ans: *see detailed solutions.*

4. There has been much discussion of the titration of Fe^{2+} with Ce^{4+}. Under the conditions of this problem, the E° values for the half reactions of Fe and Ce are as given as follows

$$Fe^{3+} + e^- \rightleftharpoons Fe^{2+} \qquad E^{\circ}{}_{Fe} = 0.77 \text{ volts}$$

$$Ce^{4+} + e^- \rightleftharpoons Ce^{3+} \qquad E^{\circ}{}_{Ce} = 1.61 \text{ volts}$$

The endpoint solution potential for a titration of Fe^{2+} by Ce^{4+} is found to be 1.19 volts. It would be nice to detect the endpoint with an indicator. Two new indicators are

$$\text{di-Bolane} \quad Inox + 2e^- \rightleftharpoons Inred \quad E^{\circ}{}_{dip} = 0.76 \text{ volt}$$
$$\text{("Dip")} \qquad \text{violet} \qquad\qquad \text{colorless}$$

$$p\text{-nitro-di-Bolane} \quad Inox + 2e^- \rightleftharpoons Inred \quad E^{\circ}{}_{pn} = 1.01 \text{ volts}$$
$$\text{("pn")} \qquad\qquad \text{violet} \qquad\qquad \text{colorless}$$

Assume that, for either indicator, the color change from colorless to violet is visible when $[Inox]/[Inred] = 10$. Would either, or both indicators be suitable for the Fe^{2+}-Ce^{4+} titration?

Ans: *only p-nitro-di-Bolane.*

≡ 5.9 DETAILED SOLUTION TO PROBLEMS ≡

1. $2Fe^{2+} + Br_2 \rightleftharpoons 2Fe^{3+} + 2Br^-$

Write the two half-reactions as reductions.

$$Fe^{3+} + e^- \rightleftharpoons Fe^{2+}$$

$$Br_2 + 2e^- \rightleftharpoons 2Br^-$$

Remember, all species are at equilibrium.

$$E_{solution} = E_{Br} = E_{Fe}$$

Write the Nernst equations.

$$E_{solution} = E_{Fe} = E^{\circ}{}_{Fe} - 0.059 \log \frac{[Fe^{2+}]}{[Fe^{3+}]}$$

$$E_{\text{solution}} = E_{\text{Br}} = E^\circ{}_{\text{Br}} - \frac{0.059}{2} \log \frac{[\text{Br}^-]^2}{[\text{Br}_2]}$$

Now, rearrange the Nernst equations and add:

$$+ \begin{cases} E_{\text{solution}} = E^\circ{}_{\text{Fe}} - 0.059 \log \dfrac{[\text{Fe}^{2+}]}{[\text{Fe}^{3+}]} \\[4mm] 2E_{\text{solution}} = 2E^\circ{}_{\text{Br}} - 0.059 \log \dfrac{[\text{Br}^-]^2}{[\text{Br}_2]} \end{cases}$$

$$3E_{\text{solution}} = E^\circ{}_{\text{Fe}} + 2E^\circ{}_{\text{Br}} - 0.059 \left\{ \log \frac{[\text{Fe}^{2+}]}{[\text{Fe}^{3+}]} + \log \frac{[\text{Br}^-]^2}{[\text{Br}_2]} \right\}$$

$$3E_{\text{solution}} = E^\circ{}_{\text{Fe}} + 2E^\circ{}_{\text{Br}} - 0.059 \log \left\{ \frac{[\text{Fe}^{2+}][\text{Br}^-]^2}{[\text{Fe}^{3+}][\text{Br}_2]} \right\}$$

At the endpoint

$$[\text{Fe}^{2+}] = 2[\text{Br}_2]$$
$$[\text{Fe}^{3+}] = [\text{Br}^-]$$

Substitute

$$3E_{\text{solution}} = E^\circ{}_{\text{Fe}} + 2E^\circ{}_{\text{Br}} - 0.059 \log \left\{ \frac{2\,[\cancel{\text{Br}_2}] \cdot [\text{Br}^-]^{\cancel{2}}}{[\cancel{\text{Br}^-}]\,[\cancel{\text{Br}_2}]} \right\}$$

$$3E_{\text{solution}} = E^\circ{}_{\text{Fe}} + 2E_{\text{Br}} - 0.059 \log \left\{ 2[\text{Br}^-] \right\}$$

$$E_{\text{solution}} = \frac{E^\circ{}_{\text{Fe}} + 2E^\circ{}_{\text{Br}}}{3} - \frac{0.059}{3} \log \left\{ 2[\text{Br}^-] \right\}$$

2. $Cr_2O_7^{2-} + 6Fe^{2+} + 14H^+ \rightleftharpoons 2Cr^{3+} + 6Fe^{3+} + 7H_2O$

Write the two half-reactions as reductions

$$Fe^{2+} + e^- \rightleftharpoons Fe^{3+}$$
$$Cr_2O_7^{2-} + 14H^+ + 6e^- \rightleftharpoons 2Cr^{3+} + 7H_2O$$

Remember, all species are at equilibrium.

$$E_{\text{solution}} = E_{\text{Cr}} = E_{\text{Fe}}$$

Write the Nernst equations.

$$E_{\text{solution}} = E_{\text{Fe}} = E^\circ{}_{\text{Fe}} - 0.059 \log \frac{[\text{Fe}^{2+}]}{[\text{Fe}^{3+}]}$$

$$E_{\text{solution}} = E_{\text{Cr}} = E^\circ{}_{\text{Cr}} - \frac{0.059}{6} \log \frac{[\text{Cr}^{3+}]^2}{[\text{Cr}_2O_7^{2-}][\text{H}^+]^{14}}$$

Now, rearrange the Nernst equations and add:

$$+ \begin{cases} E_{\text{solution}} = E^\circ{}_{\text{Fe}} - 0.059 \log \dfrac{[\text{Fe}^{2+}]}{[\text{Fe}^{3+}]} \\[4mm] 6E_{\text{solution}} = 6E^\circ{}_{\text{Cr}} - 0.059 \log \dfrac{[\text{Cr}^{3+}]^2}{[\text{Cr}_2O_7^{2-}][\text{H}^+]^{14}} \end{cases}$$

$$7E_{solution} = E^{\circ}{}_{Fe} + 6E^{\circ}{}_{Cr} - 0.059 \left\{ \log \frac{[Fe^{2+}]}{[Fe^{3+}]} + \log \frac{[Cr^{3+}]^2}{[Cr_2O_7^{2-}][H^+]^{14}} \right\}$$

$$7E_{solution} = E^{\circ}{}_{Fe} + 6E^{\circ}_{Cr} - 0.059 \log \left\{ \frac{[Fe^{2+}][Cr^{3+}]^2}{[Fe^{3+}][Cr_2O_7^{2-}][H^+]^{14}} \right\}$$

At the endpoint

$$[Fe^{2+}] = 6[Cr_2O_7^{2-}]$$

$$[Fe^{3+}] = 3[Cr^{3+}]$$

Substitute

$$7E_{solution} = E^{\circ}{}_{Fe} + 6E^{\circ}{}_{Cr} - 0.059 \log \left\{ \frac{6[\cancel{Cr_2O_7^{2-}}][Cr^{3+}]^2}{3[\cancel{Cr^{3+}}][\cancel{Cr_2O_7^{2-}}][H^+]^{14}} \right\}$$

$$7E_{solution} = E^{\circ}{}_{Fe} + 6E^{\circ}{}_{Cr} - 0.059 \log \left\{ \frac{2[Cr^{3+}]}{[H^+]^{14}} \right\}$$

$$E_{solution} = \frac{E^{\circ}{}_{Fe} + 6E^{\circ}{}_{Cr}}{7} - \frac{0.059}{7} \log \left\{ \frac{2[Cr^{3+}]}{[H^+]^{14}} \right\}$$

3. First, there are two reactions that take place:

$$Ce^{4+} + Fe^{2+} \rightleftharpoons Ce^{3+} + Fe^{3+} \quad (E^{\circ\prime}{}_{Ce} - E^{\circ\prime}{}_{Fe}) = +0.97 \text{ volts}$$

$$Ce^{4+} + VO^{2+} + H_2O \rightleftharpoons Ce^{3+} + VO_2^+ + 2H^+ (E^{\circ\prime}{}_{Ce} - E^{\circ\prime}{}_{V}) = +0.67 \text{ volts}$$

The Fe-Ce reaction takes place first, because the Fe-Ce potential difference, +0.97 volts, is greater than the V-Ce potential difference, +0.67 volts. There will be two endpoints, the first one that of the Fe-Ce reaction, the second that of the V-Ce reaction.

There are five (5) different regions of the curve.

(1) Large amounts of *both* Fe^{2+} and Fe^{3+}.
Use the Fe^{2+}-Fe^{3+} couple to determine $E_{solution}$.

$$E_{solution} = E^{\circ\prime}{}_{Fe} - 0.059 \log \frac{[Fe^{2+}]}{[Fe^{3+}]}$$

(2) First e.p. Some special calculations are necessary.

(3) Large amounts of *both* VO^{2+} and VO_2^+. Use the VO^{2+}-VO_2^+ couple to determine $E_{solution}$.

$$E_{solution} = E^{\circ\prime}{}_{V} - 0.059 \cdot \log \frac{[VO^{2+}]}{[VO_2^+][H^+]^2}$$

(4) Second e.p. Some special calculations are necessary.

(5) Excess Ce^{4+} is present, as is a lot of Ce^{3+}. Use the Ce^{3+}-Ce^{4+} couple.

$$E_{solution} = E^{\circ\prime}{}_{Ce} - 0.059 \log \frac{[Ce^{3+}]}{[Ce^{4+}]}$$

Points should be selected for determining $E_{solution}$ as a function of volume (milliliters) of Ce^{4+} added. Note that many points, 20 or more, should be taken within

±1.00 milliliter of the endpoints. This is an arithmetical pain, but does allow good first derivative, second derivative, and Gran's plots.

Example of calculation in region 1.

At 5.00 milliliters of Ce^{4+} added:

$$Because\ Ce^{4+} + Fe^{2+} \rightleftharpoons Fe^{3+} + Ce^{3+}$$

(millimoles of Fe^{3+} formed) = (millimoles of Ce^{4+} added) = (5.00 milliliters)(0.1000M)

$$[Fe^{3+}]\ =\ C_{Fe^{3+}} = \frac{(\text{millimoles of } Fe^{3+} \text{ formed})}{(\text{milliliters of solution})} = \frac{(5.00)(0.1000)}{25.00}$$

(millimoles of Fe^{2+} left) = (millimoles of Fe^{2+} to start) −
$$(\text{millimoles of } Fe^{3-} \text{ formed})$$

$$= (20.00 \text{ milliliters})(0.100M) - (5.00 \text{ milliliters})(0.1000M)$$

$$[Fe^{2+}]\ =\ C_{Fe^{2+}} = \frac{(\text{millimoles of } Fe^{2+} \text{ left})}{(\text{milliliters of solution})} = \frac{(20.00)(0.1000) - (5.00)(0.1000)}{(25.00)}$$

$$E_{\text{solution}}\ =\ E^{\circ\prime}_{Fe}\ -\ 0.059 \log \frac{[Fe^{2+}]}{[Fe^{3+}]}$$

$$= 0.825 - 0.059 \log \frac{\dfrac{(20.00)(0.1000) - (5.00)(0.1000)}{25.00}}{\dfrac{(5.00)(0.1000)}{25.00}} = 0.797 \text{ volts}$$

In region 2, the first e.p., at 20.00 milliliters Ce^{4+} added, Fe^{3+}, VO^{2+}, and Ce^{3+} are the species present in large concentrations.

$$E_{\text{solution}}\ =\ E_{Fe}\ =\ E_V$$

From $Fe^{3+} + e^- \rightleftharpoons Fe^{2+}$

$$E_{\text{solution}}\ =\ E^{\circ\prime}_{Fe}\ -\ 0.059 \log \frac{[Fe^{2+}]}{[Fe^{3+}]}$$

From $VO_2^+ + 2H^+ + e^- \rightleftharpoons VO^{2+} + H_2O$

$$E_{\text{solution}}\ =\ E^{\circ\prime}_V\ -\ 0.059 \log \frac{[VO^{2+}]}{[VO_2^+][H^+]^2}$$

$$+ \begin{cases} E_{\text{solution}}\ =\ E^{\circ\prime}_{Fe}\ -\ 0.059 \log \frac{[Fe^{2+}]}{[Fe^{3+}]} \\ \\ E_{\text{solution}}\ =\ E^{\circ\prime}_V\ -\ 0.059 \log \frac{[VO^{2+}]}{[VO_2^+][H^+]^2} \end{cases}$$

$$2E_{\text{solution}}\ =\ E^{\circ\prime}_{Fe}\ +\ E^{\circ\prime}_V\ -\ 0.059 \log \frac{[Fe^{2+}]}{[Fe^{3+}]}\ -\ 0.059 \log \frac{[VO^{2+}]}{[VO_2^+][H^+]^2}$$

$$2E_{\text{solution}}\ =\ E^{\circ\prime}_{Fe}\ +\ E^{\circ\prime}_V\ -\ 0.059 \log \frac{[Fe^{2+}][VO^{2+}]}{[Fe^{3+}][VO_2^+][H^+]^2}$$

At the endpoint, because initial amounts of Fe^{2+} and VO^{2+} were equal:

$$[Fe^{3+}] = [VO^{2+}] = \frac{(20.00)(0.1000) \text{ millimoles}}{30.00 \text{ milliliters}}$$

From a reaction that does not go far:

$$Fe^{3+} + VO^{2+} + H_2O \rightleftharpoons Fe^{2+} + VO_2^+ + 2H^+ \quad (E^{o\prime} = E^{o\prime}_{Fe} - E^{o\prime}_V = -0.295 \text{ volt})$$

$[Fe^{2+}] = [VO_2^+] = ?$, but we don't have to know.

$$\frac{[Fe^{3+}]}{[Fe^{2+}]} = \frac{[VO^{2+}]}{[VO_2^+]}$$

$$[Fe^{3+}][VO_2^+] = [Fe^{2+}][VO^{2+}]$$

$$2E_{solution} = E^{o\prime}_{Fe} + E^{o\prime}_V - 0.059 \log \frac{1}{[H^+]^2}$$

$$E_{solution} = (E^{o\prime}_{Fe} + E^{o\prime}_V + 0.059 \log [H^+]^2)/2$$

$$= [0.825 + 1.120 + 0.059 \log (8)^2]/2 = 1.026 \text{ volts}$$

In region 3, between the first and second endpoints, remember that there are large amounts of both VO^{2+} and VO_2^+. $[H^+] = 8M$. Example of calculation in region 3. At 25.00 milliliters of Ce^{4+} added

$$\text{Because } Ce^{4+} + VO^{2+} + H_2O \rightleftharpoons Ce^{3+} + VO_2^+ + 2H^+$$

(millimoles of VO_2^+ formed)
$$= \text{(millimoles of } Ce^{4+} \text{ added)} - \text{(millimoles of } Ce^{4+} \text{ reacted with } Fe^{2+})$$
$$= (25.00)(0.1000) - (20.00)(0.1000)$$

$$[VO_2^+] = C_{VO_2^+} = \frac{(25.00)(0.1000) - (20.00)(0.1000)}{45.00}$$

(millimoles of VO^{2+} left)
$$= \text{(millimoles } VO^{2+} \text{ to start)} - \text{(millimoles of } VO_2^+ \text{ formed)}$$
$$= (20.00)(0.1000) - [(25.00)(0.1000) - (20.00)(0.1000)]$$

$$[VO^{2+}] = C_{VO^{2+}} = \frac{(20.00)(0.1000) - [(25.00)(0.1000) - (20.00)(0.1000)]}{45.00}$$

$$E_{solution} = E^{o\prime}_V - 0.059 \log \frac{[VO^{2+}]}{[VO_2^+][H^+]^2}$$

$$= 1.120 - 0.059 \log \frac{[VO^{2+}]}{[VO_2^+]} - 0.059 \log \frac{1}{[H^+]^2}$$

$$= 1.120 - 0.059 \log \frac{[VO^{2+}]}{[VO_2^+]} + 0.059 \log (8)^2 = 1.226 - 0.059 \log \frac{[VO^{2+}]}{[VO_2^+]}$$

$$= 1.226 - 0.059 \log \frac{\dfrac{(20.00)(0.1000) - [(25.00)(0.1000) - (20.00)(0.1000)]}{45.00}}{\dfrac{(25.00)(0.1000) - (20.00)(0.1000)}{45.00}}$$

$$= 1.198 \text{ volts}$$

In region 4, the second e.p., @ 40.00 milliliters Ce^{4+} added, Fe^{3+}, VO_2^+, and Ce^{3+} are the species present in large concentrations. The reaction

$$Ce^{3+} + VO_2^+ + 2H^+ \rightleftharpoons Ce^{4+} + VO^{2+} + H_2O \quad E^{o\prime} = -0.670 \text{ volts}$$

takes place to a *very* small extent.

From $VO_2^+ + 2H^+ + e^- \rightleftharpoons VO^{2+} + H_2O$

$$E_{\text{solution}} = E^{o\prime}{}_V - 0.059 \log \frac{[VO^{2+}]}{[VO_2^+][H^+]^2}$$

$$= 1.120 - 0.059 \log \frac{[VO^{2+}]}{[VO_2^+]} + 0.059 \log [8]^2 = 1.226 - 0.059 \log \frac{[VO^{2+}]}{[VO_2^+]}$$

From $Ce^{4+} + e^- \rightleftharpoons Ce^{3+}$

$$E_{\text{solution}} = E^{o\prime}{}_{Ce} - 0.059 \log \frac{[Ce^{3+}]}{[Ce^{4+}]} = 1.790 - 0.059 \log \frac{[Ce^{3+}]}{[Ce^{4+}]}$$

$$+ \begin{cases} E_{\text{solution}} = 1.226 - 0.059 \log \dfrac{[VO^{2+}]}{[VO_2^+]} \\[2em] E_{\text{solution}} = 1.790 - 0.059 \log \dfrac{[Ce^{3+}]}{[Ce^{4+}]} \end{cases}$$

$$2E_{\text{solution}} = 1.226 + 1.790 - 0.059 \log \frac{[VO^{2+}]}{[VO_2^+]} - 0.059 \log \frac{[Ce^{3+}]}{[Ce^{4+}]}$$

$$2E_{\text{solution}} = 3.016 - 0.059 \log \frac{[VO^{2+}][Ce^{3+}]}{[VO_2^+][Ce^{4+}]}$$

But, from the reaction above (between Ce^{3+} and VO_2^+), and from the Ce^{3+} produced by titration of Fe^{2+}:

$$[Ce^{3+}] = 2[VO_2^+] = (20.00)(0.1000)/(60.00) + (20.00)(0.1000)/(60.00)$$

$$[Ce^{4+}] = [VO^{2+}] = ? \text{ but we don't have to know.}$$

$$\frac{[Ce^{3+}]}{[Ce^{4+}]} = \frac{2[VO_2^+]}{[VO^{2+}]}$$

$$[VO^{2+}][Ce^{3+}] = 2[VO_2^+][Ce^{4+}]$$

$$2E_{\text{solution}} = 3.016 - 0.059 \log (2) = 2.998$$

$$E_{\text{solution}} = 1.499 \text{ volts}$$

Beyond the second endpoint, in region 5, excess Ce^{4+} is being added. There are present large amounts of both Ce^{4+} and Ce^{3+}. At 45.00 ml added

(millimoles of Ce^{3+} formed) $= (40.00)(0.1000)$

$$[Ce^{3+}] = C_{Ce^{3+}} = \frac{(40.00)(0.1000)}{(65.00)}$$

(millimoles of Ce^{4+}) = (millimoles Ce^{4+} added) - (millimoles of Ce^{3+} formed)

$$= (45.00)(0.1000) - (40.00)(0.1000)$$

$$[Ce^{4+}] = C_{Ce^{4+}} = \frac{(45.00)(0.1000) - (40.00)(0.1000)}{65.00}$$

$$E_{\text{solution}} = E^{\circ\prime}_{Ce} - 0.059 \log \frac{[Ce^{3+}]}{[Ce^{4+}]}$$

$$= 1.790 - 0.059 \log \frac{\dfrac{(40.00)(0.1000)}{(65.00)}}{\dfrac{(45.00)(0.1000) - (40.00)(0.1000)}{65.00}} = 1.737 \text{ volts}$$

By analogy to this example:

At 50.00 milliliters of Ce^{4+} added:

$$E_{\text{solution}} = 1.790 - 0.059 \log \left(\frac{40.00}{50.00 - 40.00} \right) = 1.754 \text{ volts}$$

All the titration data can be tabulated and graphed as were the Fe-Ce data in the text.

4. To answer the question, take the approach of the discussion of ferroin indicators. For di-Bolane,

$$E_{\text{solution}} = E^{\circ}_{\text{dip}} + \frac{0.059}{2} \log \frac{[\text{Inox}]}{[\text{Inred}]}$$

When $[\text{Inox}]/[\text{Inred}] = 10$

$$E_{\text{solution}} = 0.76 + \frac{0.059}{2} \log 10 = 0.76 + \frac{0.059}{2} = 0.79$$

At 0.79 volt, calculate $[Fe^{3+}]/[Fe^{2+}]$

$$E_{\text{solution}} = E^{\circ}_{Fe} - \frac{0.059}{1} \log \frac{[Fe^{2+}]}{[Fe^{3+}]}$$

$$E_{\text{solution}} = 0.77 + 0.059 \log \frac{[Fe^{3+}]}{[Fe^{2+}]}$$

$$0.79 = 0.77 + 0.059 \log \frac{[Fe^{3+}]}{[Fe^{2+}]}$$

$$\frac{0.02}{0.059} = \log \frac{[Fe^{3+}]}{[Fe^{2+}]}$$

$$0.34 = \log \frac{[Fe^{3+}]}{[Fe^{2+}]}$$

$$\frac{[Fe^{3+}]}{[Fe^{2+}]} = 2.2$$

No—dip just won't work.

For p-nitro-di-Bolane

$$E_{\text{solution}} = E^{\circ}_{\text{pn}} + \frac{0.059}{2} \log \frac{[\text{Inox}]}{[\text{Inred}]}$$

When $[\text{Inox}]/[\text{Inred}] = 10$

$$E_{\text{solution}} = 1.01 + \frac{0.059}{2} \log 10 = 1.01 + \frac{(0.059)(1)}{(2)} = 1.04 \text{ volts}$$

At 1.04 volts, calculate $[\text{Fe}^{3+}]/[\text{Fe}^{2+}]$

$$E_{\text{solution}} = E^{\circ}{}_{\text{Fe}} - \frac{0.059}{1} \log \frac{[\text{Fe}^{2+}]}{[\text{Fe}^{3+}]}$$

$$E_{\text{solution}} = 0.77 + 0.059 \log \frac{[\text{Fe}^{3+}]}{[\text{Fe}^{2+}]}$$

$$1.04 = 0.77 + 0.059 \log \frac{[\text{Fe}^{3+}]}{[\text{Fe}^{2+}]}$$

$$0.27 = 0.059 \log \frac{[\text{Fe}^{3+}]}{[\text{Fe}^{2+}]}$$

$$\frac{0.27}{0.059} = \log \frac{[\text{Fe}^{3+}]}{[\text{Fe}^{2+}]}$$

$$4.58 = \log \frac{[\text{Fe}^{3+}]}{[\text{Fe}^{2+}]}$$

$$3.80 \times 10^4 = \frac{[\text{Fe}^{3+}]}{[\text{Fe}^{2+}]}$$

Thus, the second of the two indicators would serve better than the first.

≡ 5.10 PHYSICAL CONSTANTS AND ACTIVITIES ≡

5.11 Physical Constants

Equilibrium has been discussed in terms of titration curves and solution potentials. Potentiometric data may be used to determine equilibrium constants. The mathematics is only slightly subtler than that involved in finding equilibrium constants from spectrophotometric data. To introduce the topic, consider the hydrogen electrode, whose potential serves as a reference for all potentiometric work. This electrode may be pressed into service as a pH indicator electrode. In past years, the hydrogen electrode did valuable service in this application. The physical arrangement is shown in Figure 5.13.

$$\text{Pt, H}_2 \text{ (1.00 atm)} | [\text{H}^+] = \text{?} \| \text{SCE}$$

Hydrogen gas ($p_{\text{H}_2} = 1.00$ atm) is bubbled over a black Pt surface, which dips into a solution of unknown pH. A salt bridge connects the saturated calomel electrode (SCE) to the solution of unknown pH. Both the Pt electrode and the SCE are connected to the terminals of a potentiometer, marked V. The potential, E, of the cell can be written as a function of pH.

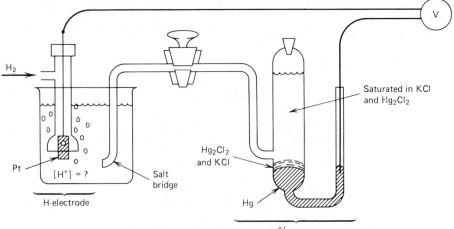

Figure 5.13 *pH Measurement with hydrogen electrode.*

$$E = E^{o'}_{SCE} - E_H$$

E = observed potential of cell

E_H = potential of hydrogen half-cell

E^{o}_{H} = 0.000 volts

$E^{o'}_{SCE}$ = 0.246 volts at 25°C

From:

$$2H^+ + 2e^- \rightleftharpoons H_2(g)$$

$$E_H = E^{o}_{H} - \frac{0.059}{2} \log \frac{p_{H_2}}{[H^+]^2}$$

$$p_{H_2} = 1.00 \text{ atm}$$

$$[H^+] = ?$$

$$E = E^{o'}_{SCE} - \left\{ E^{o}_{H} - \frac{0.059}{2} \log \frac{p_{H_2}}{[H^+]^2} \right\}$$

$$= 0.246 - 0.000 + \frac{0.059}{2} \log \left(\frac{1.00}{[H^+]^2} \right) = 0.246 - \frac{0.059}{2} \log [H^+]^2$$

$$= 0.246 - \frac{(0.059)(2)}{2} \log [H^+] = 0.246 + (0.059)(-\log [H^+])$$

$$E = 0.246 + 0.059 \text{ (pH)} \tag{5.2}$$

Potential of the cell is a linear function of pH. This linear dependence of potential on the logarithm of hydrogen ion concentration is one of the reasons Sørensen's choice of definitions of his pH scale is such a convenient one. It is easy to see how the scale of an electrical instrument could be calibrated linearly in pH units.

The hydrogen electrode has its disadvantages. The first is that hydrogen gas presents a fire and explosion hazard. The second is that 5 or 10 minutes are needed for the H_2

and H^+ to reach equilibrium at the Pt surface. Finally, many interferences—oxidants such as Fe^{3+}, $Cr_2O_7^{2-}$, HNO_3, Cl_2, and H_2O_2 or reductants such as SO_2 and H_2S, aniline dyes, and even phenol—will cause drift and render readings inaccurate. Thus, when the glass electrode, combined with the SCE and special voltage measuring devices, became widely available, it was eagerly adopted in many quarters.

Before the hydrogen electrode is completely forgotten, it might be useful to look at how it may be used to determine acid dissociation constants. Consider the following cell.

Example: The voltage of the cell shown below

$$Pt, H_2 \text{ (1.00 atm)} | HAc \text{ (0.0100}M\text{)} \| SCE$$

is just +0.445 volts. From this figure, and using any reasonable approximaations, calculate K_a for acetic acid.

Solution: First, find pH and $[H^+]$, using Equation 5.2.

$$E_{cell} = 0.246 + 0.059 \text{ (pH)} \qquad (5.2)$$

$$0.445 = 0.246 + 0.059 \text{ (pH)}$$

$$0.199 = 0.059 \text{ (pH)}$$

$$pH = 3.37$$

Remember

$$pH = -\log [H^+]$$

or

$$[H^+] = 10^{-pH} = 10^{-3.37} = 4.3 \times 10^{-4}M$$

Remember

$$[H^+]^2 = K_a(C_{HAc} - [H^+])$$

$$K_a = \frac{[H^+]^2}{C_{HAc} - [H^+]} = \frac{(4.3 \times 10^{-4})^2}{(1.00 \times 10^{-2} - 4.3 \times 10^{-4})} = 1.8 \times 10^{-5}$$

Other sorts of equilibrium constants may be found from potentiometric data. The example we will take is that of a rather nasty metal, thallium, whose salts are frequently used to poison coyotes and other carrion fanciers in the West.

The following data have been taken on the thallous ion and one of its sparingly soluble salts, thallous chloride.

$$Tl^+ + e^- \rightleftharpoons Tl \qquad\qquad E^\circ_{Tl} = -0.336 \text{ volts}$$

$$TlCl + e^- \rightleftharpoons Tl^\circ + Cl^- \qquad E^\circ_{TlCl} = -0.557 \text{ volts}$$

From these data, calculate the value of K_{sp} for the dissolution of thallous chloride.

$$TlCl \rightleftharpoons Tl^+ + Cl^- \qquad K_{sp} = [Tl^+][Cl^-]$$

To do this, call the potential of any half-cell involving Tl^+ and Tl E. If Cl^- and/or TlCl are introduced, the potential is still E.

$$E = E^\circ_{Tl} - 0.059 \log \frac{1}{[Tl^+]}$$

or

$$E = E^\circ_{TlCl} - 0.059 \log [Cl^-]$$

$$E^{\circ}{}_{Tl} - 0.059 \log \frac{1}{[Tl^+]} = E^{\circ}{}_{TlCl} - 0.059 \log [Cl^-]$$

$$E^{\circ}{}_{Tl} - E^{\circ}{}_{TlCl} = 0.059 \log \frac{1}{[Tl^+]} - 0.059 \log [Cl^-]$$

$$E^{\circ}{}_{Tl} - E^{\circ}{}_{TlCl} = -0.059 \log [Tl^+] - 0.059 \log [Cl^-]$$

$$E^{\circ}{}_{Tl} - E^{\circ}{}_{TlCl} = -0.059 [\log [Tl^+] + \log [Cl^-]]$$

$$E^{\circ}{}_{Tl} - E^{\circ}{}_{TlCl} = -0.059 \log ([Tl^+][Cl^-])$$

$$E^{\circ}{}_{Tl} - E^{\circ}{}_{TlCl} = -0.059 \log K_{sp}$$

$$-0.336 + 0.557 = -0.059 \log K_{sp}$$

$$+0.221 = -0.059 \log K_{sp}$$

$$\frac{0.221}{-0.059} = \log K_{sp}$$

$$\log K_{sp} = -3.74$$

$$K_{sp} = 1.8 \times 10^{-4}$$

Values of K_{sp} can be found if the right half-cell potentials are known. It is also possible to find values of half-cell potentials if the right combination of K_{sp} and another half-cell potential is known.

Example: Calculate the voltage of the cell below, assuming no complexation of the silver ion by oxalate to form soluble species. Assume no hydrolysis of $C_2O_4^{2-}$.

$$SCE \| Na_2C_2O_4 (1.00 \times 10^{-3}M), Ag_2C_2O_4 \text{ (saturated)} | Ag$$

$$Ag^+ + e^- \rightleftharpoons Ag \quad E^{\circ}{}_{Ag} = 0.799 \text{ volts}$$

$$Ag_2C_2O_4 \rightleftharpoons 2Ag^+ + C_2O_4^{2-} \quad K_{sp} = 1.1 \times 10^{-11}$$

$$E_{cell} = E_{Ag} - E^{\circ\prime}{}_{SCE}$$

$$E_{Ag} = ?$$

$$Ag^+ + e^- \rightleftharpoons Ag$$

$$E_{Ag} = E^{\circ}{}_{Ag} - 0.059 \log \frac{1}{[Ag^+]}$$

$$E_{Ag} = E^{\circ}{}_{Ag} + 0.059 \log [Ag^+]$$

But:

$$K_{sp} = [Ag^+]^2 \cdot [C_2O_4^{2-}]$$

$$[Ag^+] = \sqrt{\frac{K_{sp}}{[C_2O_4^{2-}]}}$$

Substitute:

$$E_{Ag} = E^{\circ}{}_{Ag} + 0.059 \log \sqrt{\frac{K_{sp}}{[C_2O_4^{2-}]}}$$

$$= 0.799 + 0.059 \log \sqrt{\frac{1.1 \times 10^{-11}}{1.00 \times 10^{-3}}}$$

$$= 0.564 \text{ volts}$$

$$E_{cell} = E_{Ag} - E^{\circ\prime}{}_{SCE} = 0.564 - 0.246 = 0.318 \text{ volts}$$

5.12 Activities and Activity Coefficients

So far, we have dealt with concentrations of species in solution. We have examined precipitation equilibria, acid-base equilibria, redox equilibria, and situations in which equilibria compete. Equilibrium concentrations have been used in writing equilibrium constants. The treatment would appear to have been pretty thorough. Yet something *has* been neglected. To define that something more closely, let us examine the dissolution of two common precipitates.

$$AgCl \rightleftharpoons Ag^+ + Cl^- \qquad K_{sp} = [Ag^+][Cl^-] = 1.8 \times 10^{-10}$$
$$BaSO_4 \rightleftharpoons Ba^{2+} + SO_4^{2-} \qquad K_{sp} = [Ba^{2+}][SO_4^{2-}] = 1.08 \times 10^{-10}$$

If one were to add solid KNO_3 to saturated solutions of these precipitates, one would expect neither enhancement nor depression of their solubilities. Most of the salts of K^+ and NO_3^- are quite soluble. Potassium and nitrate ions do not complex Ba^{2+}, SO_4^{2-}, Ag^+, and Cl^-. The acid-base behavior of K^+ and Cl^- is not important. Precipitate solubility ought to be independent of the concentration of KNO_3. As it turns out, the solubilities of both precipitates increase with increasing KNO_3 concentration.

Precipitates aren't the only species affected by the presence of soluble salts. Weak acids show enhanced dissociation in the presence of such salts. In short, there is a real "salt effect," if we should want to name it that.

What is the cause of this "salt effect"? Electrostatic attraction is most likely the chief cause. Positive ions attract negative ions, and vice versa. Thus, the addition of KNO_3 to a saturated solution of $BaSO_4$ will enhance the solubility of the precipitate simply because the K^+ ions attract SO_4^{2-} ions into solution and the NO_3^- ions attract Ba^{2+} ions into solution. As the concentration of KNO_3 increases, so do the numbers of Ba^{2+} and SO_4^{2-} ions drawn into solution increase. The higher the *ionic strength* of the solution, at least up to *ionic strength* values of 0.1, the more soluble the precipitate.

Ionic strength is a term that could be utterly vague, but it is quite exactly defined. Chemists define ionic strength, μ, like this:

$$\mu = 1/2 \sum_{i=1}^{n} m_i Z_i^2$$

μ = ionic strength
m_i = concentration of any ion i, in moles/liter
Z_i = charge on any ion i

Example: What is the ionic strength of a solution $0.01M$ in KCl?

$$m_K = 0.01 \qquad\qquad Z_K = 1$$
$$m_{Cl} = 0.01 \qquad\qquad Z_{Cl} = -1$$
$$\mu = 1/2 \, [(0.01)(1)^2 + (0.01)(-1)^2] = 0.01$$

Example: What is the ionic strength of a solution $0.01M$ in K_2SO_4?

$$m_K = 0.02 \qquad\qquad Z_K = 1$$
$$m_{SO_4} = 0.01 \qquad\qquad Z_{SO_4} = -2$$

$$\mu = 1/2\ [(0.02)(1)^2 + (0.01)(-2)^2] = 0.03$$

Example: What is the ionic strength of a solution 0.01M in $MgCl_2$?

$$m_{Mg} = 0.01 \qquad\qquad Z_{Mg} = 2$$
$$m_{Cl} = 0.02 \qquad\qquad Z_{Cl} = -1$$
$$\mu = 1/2\ [(0.01)(2)^2 + (0.02)(-1)^2] = 0.03$$

Example: What is the ionic strength of a solution 0.01M in $MgSO_4$?

$$m_{Mg} = 0.01 \qquad\qquad Z_{Mg} = 2$$
$$m_{SO_4} = 0.01 \qquad\qquad Z_{SO_4} = -2$$
$$\mu = 1/2\ [(0.01)(2)^2 + (0.01)(-2)^2] = 0.04$$

Note how, for any given molarity of a salt, ionic strength increases with the charges on the ions.

The ionic strength of a solution and enhanced solubility of a precipitate are not related to each other in any simple fashion. A new term, *activity*, must be defined and introduced to clarify the relationship. *Activity* may be thought of as a sort of effective concentration. The platinum electrodes so often used in potentiometry, the glass electrodes used in pH measurement, and the newer ion sensitive electrodes all measure activities of solution species.

$$a_i = \text{activity of species } i$$
$$a_i = f_i\ [C_i]$$
$$f_i = \text{activity coefficient of species } i$$
$$[C_i] = \text{equilibrium concentration of species } i$$

Note: For a pure solid s, $a_s = 1$, by definition.

The activity coefficient, f_i, is related to ionic strength. In 1923, Debye and Hückel derived a theoretical equation for estimating activity coefficients of ions. The full expression is:

$$-\log f_i = \frac{0.5085\ Z_i^2\ \sqrt{\mu}}{1 + 0.3281\alpha_i\sqrt{\mu}} \qquad \text{(at 25°C in water)} \qquad (5.3)$$

$$f_i = \text{activity coefficient for species } i$$
$$Z_i = \text{charge on species } i$$
$$\mu = \text{ionic strength of solution in which } i$$
$$\text{is found}$$
$$\alpha_i = \text{effective diameter of hydrated ion in Å}$$
$$(1\ \text{Å} = 10^{-8}\ \text{cm})$$

Plainly, Equation 5.3 isn't the easiest equation to apply. The least clearly defined quantity in it is α_i, the effective ionic diameter. A table of these ionic diameters was compiled in 1937 by Kielland.* Kielland also calculated activity coefficients for various ions at varying ionic strength values. Selected values from his table are given in Table 5.5. Naturally, if the table isn't handy, it would be necessary to calculate f_i for any given value of μ. Attempts have been made to simplify the Debye-Hückel equa-

---

* J. Kielland, J. Amer. Chem. Soc., *59*, 1675 (1937).

Table 5.5
CALCULATED ACTIVITY COEFFICIENTS, f_i, FOR INDIVIDUAL IONS IN WATER*

Ions	Diameter a_i, Å	Ionic Strength, μ							
		0.0005	0.001	0.0025	0.005	0.01	0.025	0.05	0.1
H^+	9	0.975	0.967	0.950	0.933	0.914	0.88	0.86	0.83
Li^+	6	.975	.965	.948	.929	.907	.87	.835	.80
Rb^+, Cs^+, NH_4^+, Tl^+, Ag^+	2.5	.975	.964	.945	.924	.898	.85	.80	.75
K^+, Cl^-, Br^-, I^-, CN^-, NO_2^-, NO_3^-	3	.975	.964	.945	.925	.899	.85	.805	.755
Na^+, HCO_3^-, $H_2PO_4^-$	4-4.5	.975	.964	.947	.928	.902	.86	.82	.775
Hg_2^{2+}, SO_4^{2-}, $S_2O_3^{2-}$, CrO_4^{2-}, HPO_4^{2-}	4	.903	.867	.803	.740	.660	.545	.445	.355
Pb^{2+}, CO_3^{2-}	4.5	.903	.868	.805	.742	.665	.55	.455	.37
Sr^{2+}, Ba^{2+}, Cd^{2+}, Hg^{2+}, S^{2-}	5	.903	.868	.805	.744	.67	.555	.465	.38
Ca^{2+}, Cu^{2+}, Zn^{2+}, Mn^{2+}, Fe^{2+}, Ni^{2+}, Co^{2+}	6	.905	.870	.809	.749	.675	.57	.485	.405

tion. The first simplification is particularly useful for calculating the activity coefficients of ions having diameters of 3 Å or so (K^+, Cl^-, Br^-, I^-, CN^-, NO_2^-, NO_3^-).

Because

$$(3)(0.3281) \cong 1$$

$$-\log f_i \cong \frac{0.5085\, Z_i^2\, \sqrt{\mu}}{1 + \sqrt{\mu}} \qquad (5.4)$$

Equation 5.4 is called the extended Debye-Hückel equation (EHDE).

If μ is quite small, $\sqrt{\mu}$ may also be small in comparison to 1. Thus, a new relationship, the Debye-Hückel limiting law (DHLL), may be written as:

$$-\log f_i \cong 0.5\, Z_i^2\, \sqrt{\mu} \qquad (5.5)$$

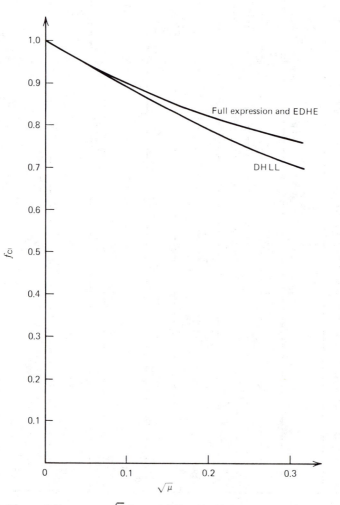

Figure 5.14 Plots of f_{Cl} versus $\sqrt{\mu}$ from Kielland's (full expression) data, EDHE, and DHLL.

Figure 5.14 shows f_{Cl}, the activity coefficient for chloride ion, computed by each of the three expressions given above, plotted as a function of $\sqrt{\mu}$. Kielland's data, calculated from the full expression, and points plotted from the extended Debye-Hückel equation (EHDE) fall on the same curve; this is not surprising, because Cl^- has an ionic diameter of about 3 Å. The points plotted from the Debye-Hückel limiting law begin to diverge from the Kielland and EHDE curve very early, where $\sqrt{\mu} \cong 0.05$ ($\mu = 0.0025$). This is not surprising; $\sqrt{\mu}$ was considered to be much, much less than 1 in DHLL calculations. The DHLL is of limited utility above $\mu = 0.0025$ in this case.

Do the points calculated from the EDHE always coincide with those calculated from the full expression? No. Not all ions are of 3 Å diameter. Take the case of Li^+ ($\alpha_{Li} \cong$ 6 Å). Figure 5.15 shows f_{Li}, the activity coefficient for lithium ion, computed by each

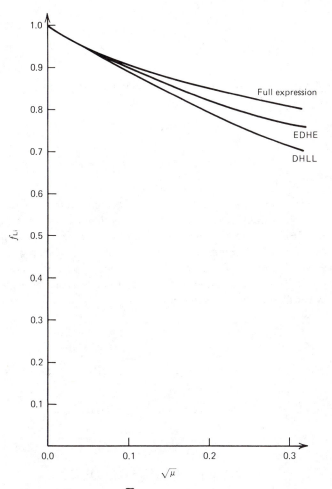

Figure 5.15 *Plots of f_{Li} versus $\sqrt{\mu}$ from Kielland's (full expression) data, EDHE, and DHLL.*

of the three expressions, again plotted as a function of $\sqrt{\mu}$. The plots of f_{Li} versus $\sqrt{\mu}$ for values of f_{Li} calculated from the full expression, the EDHE, and the DHLL begin to diverge at $\sqrt{\mu} = 0.05$ ($\mu = 0.0025$). The EDHE divergence can be explained simply by the fact that the EDHE assumes an ionic diameter of about 3 Å, while that of Li^+ is known to be about 6 Å. The DHLL divergence is greater than that of the EDHE curve. The DHLL does not take account of ionic size, nor of the fact that $\sqrt{\mu}$ may be a sizable fraction of 1. After examining these graphs, a great many would contend that a table of Kielland's data could be far more valuable than either the EDHE or the DHLL. None of the three is needed, however, in the limit of zero ionic strength, where f values approach 1. Why? In each of the three expressions for $-\log f_i$, the right-hand side is multiplied through by $\sqrt{\mu}$.

In each expression, when $\mu = 0$

$$\sqrt{\mu} = 0$$
$$\log f_i = 0$$

and

$$f_i = 1$$

Thus, in the limit of *very* low ionic strength, activity coefficients may be taken as equal to unity.

If

$$\mu = 0$$

then

$$f_i = 1$$

and

$$a_i = [C_i]$$

In the limit of low, or zero, ionic strength, activities and concentrations become equal.

Unfortunately, above ionic strengths in the neighborhood of 0.1 or so, the Debye-Hückel equation begins to require modification in order to fit it to observed facts. The brines and other very salty solutions, the Debye-Hückel equation simply doesn't work. It is. however, applicable to many analytical calculations involving moderately dilute solutions. To give the reader a notion of the uses of activity coefficients in analytical calculations, we will start by writing a number of equilibrium constants in terms of activities.

First, a K_{sp}:

$$BaSO_4 \rightleftharpoons Ba^{2+} + SO_4^{2-}$$

$$K_{sp} = a_{Ba} \cdot a_{SO_4} = f_{Ba} \cdot [Ba^{2+}] \cdot f_{SO_4} \cdot [SO_4^{2-}] = [Ba^{2+}] \cdot [SO_4^{2-}] \cdot (f_{Ba} \cdot f_{SO_4})$$

Then, a more complicated K_{sp}:

$$Ag_2CrO_4 \rightleftharpoons 2Ag^+ + CrO_4^{2-}$$

$$K_{sp} = a_{Ag}^2 \cdot a_{CrO_4}$$
$$f_{Ag}^2 \cdot [Ag^+]^2 \cdot f_{CrO_4} \cdot [CrO_4^{2-}]$$
$$= [Ag^+]^2 \cdot [CrO_4^{2-}] \cdot (f_{Ag}^2 \cdot f_{CrO_4})$$

Next, a K_a for a weak acid, HNO_2:

$$HNO_2 \rightleftharpoons H^+ + NO_2^-$$

$$K_a = \frac{a_H \cdot a_{NO_2}}{a_{HNO_2}} = \frac{f_H \cdot [H^+] \cdot f_{NO_2}[NO_2^-]}{f_{HNO_2} \cdot [HNO_2]} = \frac{[H^+][NO_2^-]}{[HNO_2]} \cdot \frac{f_H \cdot f_{NO_2}}{f_{HNO_2}}$$

(The activity coefficient of an uncharged species such as HNO_2 is often taken as unity.)

Electrode potentials are also defined most properly in terms of activities.

$$Ag^+ + e^- \rightleftharpoons Ag(s)$$

$$E_{Ag} = E^\circ_{Ag} - 0.059 \log \frac{1}{a_{Ag}}$$

$$E_{Ag} = E^\circ_{Ag} - 0.059 \log \frac{1}{f_{Ag} \cdot [Ag^+]}$$

(The activity of a pure solid, such as metallic Ag, is defined as unity.)

It would be instructive to work a few problems involving activity coefficients.

Problem: Calculate the solubility of $BaSO_4$ in (a) pure H_2O; (b) 0.0010M KNO_3; (c) 0.010M KNO_3; and (d) 0.10M KNO_3.

$$BaSO_4 \rightleftharpoons Ba^{2+} + SO_4^{2-}$$
$$K_{sp} = 1.08 \times 10^{-10}$$

for all cases:

$$K_{sp} = a_{Ba} \cdot a_{SO_4} = f_{Ba} \cdot [Ba^{2+}] \cdot f_{SO_4}[SO_4^{2-}] = [Ba^{2+}] \cdot [SO_4^{2-}] \cdot (f_{Ba} \cdot f_{SO_4})$$

$$\text{Solubility} \quad = S = [Ba^{2+}] = [SO_4^{2-}]$$

$$K_{sp} = (S)(S) \cdot (f_{Ba} \cdot f_{SO_4}) = S^2(f_{Ba} \cdot f_{SO_4})$$

$$\frac{K_{sp}}{f_{Ba} \cdot f_{SO_4}} = S^2$$

$$S = \sqrt{\frac{K_{sp}}{f_{Ba} \cdot f_{SO_4}}}$$

(a) Pure H_2O: Here, the only ions are from $BaSO_4$, and there aren't many of them. As $\mu \rightarrow 0, f \rightarrow 1$.

$$f_{Ba} = f_{SO_4} = 1$$

$$S = \sqrt{\frac{K_{sp}}{f_{Ba} \cdot f_{SO_4}}} = \sqrt{\frac{1.08 \times 10^{-10}}{(1)(1)}} = 1.04 \times 10^{-5} M$$

(b) 0.0010M KNO_3: Here, Kielland's values are helpful.

$$\mu = 1/2 \, [(0.0010)(1)^2 + (0.0010)(1)^2] = 0.0010$$
$$f_{Ba} = 0.868$$
$$f_{SO_4} = 0.867$$

$$S = \sqrt{\frac{K_{sp}}{f_{Ba} \cdot f_{SO_4}}} = \sqrt{\frac{1.08 \times 10^{-10}}{(0.868)(0.867)}} = 1.2 \times 10^{-5} M$$

(c) 0.010M KNO_3:

$$\mu = 1/2 \, [(0.010)(1)^2 + (0.010)(1)^2] = 0.010$$
$$f_{Ba} = 0.67$$

$$f_{SO_4} = 0.660$$

$$S = \sqrt{\frac{K_{sp}}{f_{Ba} \cdot f_{SO_4}}} = \sqrt{\frac{1.08 \times 10^{-10}}{(0.67)(0.660)}} = 1.6 \times 10^{-5} M$$

(d) $0.10M$ KNO_3:

$$\mu = 1/2 \, [(0.10)(1)^2 + (0.10)(1)^2] = 0.10$$
$$f_{Ba} = 0.38$$
$$f_{SO_4} = 0.355$$

$$S = \sqrt{\frac{K_{sp}}{f_{Ba} \cdot f_{SO_4}}} = \sqrt{\frac{1.08 \times 10^{-10}}{(0.38)(0.355)}} = 2.8 \times 10^{-5} M$$

Note how solubility is enhanced with increasing ionic strength.

Problem: Calculate the potentials of these three half-cells:

(a) $Ag|AgNO_3$ $(0.010M)\|$
(b) $Ag|AgNO_3$ $(0.010M)$, KNO_3 $(0.015M)\|$
(c) $Ag|AgNO_3$ $(0.010M)$, KNO_3 $(0.040M)\|$

$$Ag^+ + e^- \rightleftharpoons Ag_{(s)} \qquad E^\circ_{Ag} = +0.799 \text{ volt}$$

In each case:

$$E = E^\circ_{Ag} - 0.059 \log \frac{1}{a_{Ag}} = 0.799 - 0.059 \log \frac{1}{f_{Ag} \cdot [Ag^+]}$$

$$= 0.799 + 0.059 \log [f_{Ag} \cdot [Ag^+]]$$

$$= 0.799 + 0.059 \log f_{Ag} + 0.059 \log [Ag^+]$$

In each case, $[Ag^+] = 0.010M$

$$E = 0.799 + 0.059 \log (1.0 \times 10^{-2}) + 0.059 f_{Ag} = 0.681 + 0.059 \log f_{Ag}$$

(a) $0.10M$ $AgNO_3$:

$$\mu = 1/2 \, [(0.010)(1)^2 + (0.010)(1)^2] = 0.010$$
$$f_{Ag} = 0.898$$
$$E = 0.681 + 0.059 \log (0.898) = 0.678 \text{ volt}$$

(b) $0.010M$ $AgNO_3$, $0.015M$ KNO_3:

$$\mu = 1/2 \, [(0.010)(1)^2 + (0.010)(1)^2 + (0.015)(1)^2 + (0.015)(1)^2]$$
$$= 0.025$$
$$f_{Ag} = 0.85$$
$$E = 0.681 + 0.059 \log (0.85)$$
$$= 0.677 \text{ volt}$$

(c) $0.010M$ $AgNO_3$, $0.040M$ KNO_3:

$$\mu = 1/2 \, [(0.010)(1)^2 + (0.010)(1)^2 + (0.040)(1)^2 + (0.040)(1)^2]$$
$$= 0.050$$
$$f_{Ag} = 0.80$$
$$E = 0.681 + 0.059 \log (0.80)$$
$$= 0.675 \text{ volt}$$

≡ 5.13 PROBLEMS ≡

(Assume 25°C and that activity effects are negligible.)

1. Before the glass electrode and pH meter came into general use, there existed other ways to determine pH potentiometrically. One way involved the hydrogen electrode. The other involved a peculiar substance called quinhydrone. Quinhydrone is made up of quinone and hydroquinone, in a one-to-one mole ratio.

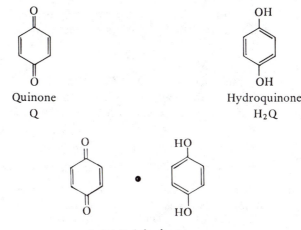

Quinone
Q

Hydroquinone
H_2Q

Solid Quinhydrone
$Q \cdot H_2Q$

When put into water, the solid dissolves only slightly. What does dissolve can dissociate to give a one-to-one mixture of quinone to hydroquinone.

$$Q \cdot H_2Q \rightleftharpoons Q + H_2Q$$

The reduction potential of quinone is known.

$$Q + 2H^+ + 2e^- \rightleftharpoons H_2Q \qquad E^\circ_Q = 0.6994 \text{ volts}$$

Given the following cell

$$SCE \| Q(\text{saturated}), H_2Q(\text{saturated}), [H^+] \text{ varying} | Au$$

plot the voltage of the cell as a function of pH for pH values of 0.00, 1.00, 2.00, 3.00, 4.00, 5.00, 6.00, and 7.00. (Above pH 7.9, the electrode does not work.) The Au electrode serves as an indicator electrode just like Pt. It doesn't react.
Ans: *see detailed answers.*

2. Given that

$$\text{and} \begin{cases} Ag^+ + e^- \rightleftharpoons Ag^\circ & E^\circ_{Ag} = 0.799 \text{ volts} \\ Ag_2CrO_4 + 2e^- \rightleftharpoons 2Ag^\circ + CrO_4^{2-} & E^\circ_{Ag_2CrO_4} = 0.446 \text{ volts} \end{cases}$$

calculate the value of K_{sp} for Ag_2CrO_4. Assume that no chromate complexes of Ag_2CrO_4 exist.
Ans: *1.1×10^{-12}*

3. Calculate the potential of the following cell:

$$Pt,H_2(1.00 \text{ atm})|HPro(0.10M); NaPro(0.050M)\|SCE$$

HPro is propionic acid.

$$HPro \rightleftharpoons H^+ + Pro^- \quad K_a = 1.3 \times 10^{-5}$$

Ans: *0.52 volt*

4. The potential of the following cell is found to be 0.323 volts:

$$SCE\|Q(\text{saturated}), H_2Q(\text{saturated}), H_2PO_4(0.100M), NaH_2PO_4(0.100M)|Au$$

What is the value of K_1 of phosphoric acid? Remember Problem 1 of this set. The right-hand side of the cell is a quinhydrone system. Phosphates react with neither quinhydrone nor solid gold.
Ans: *7.1 × 10⁻³*

5. Suppose that 100.00 milliliters of 0.1000M Ce^{4+} solution were added to 100.00 milliliters of 0.1000M Fe^{2+} solution. What would be the concentrations of Ce^{4+}, Ce^{3+}, Fe^{2+}, and Fe^{3+} after equilibrium was reached?

$$Ce^{4+} + e^- \rightleftharpoons Ce^{3+} \quad E^°_{Ce} = 1.61 \text{ volts}$$
$$Fe^{3+} + e^- \rightleftharpoons Fe^{2+} \quad E^°_{Fe} = 0.771 \text{ volt}$$

Ans: *[Fe²⁺] = [Ce⁴⁺] = 3.8 × 10⁻⁹M; [Fe³⁺] = [Ce⁴⁺] = 5.000 × 10⁻²M*

6. Suppose that 50.00 milliliters of 0.01667M $K_2Cr_2O_7$ solution were added to 50.00 milliliters of a solution 0.1000M in Fe^{2+} What would be the concentrations of Cr^{3+}, $Cr_2O_7^{2-}$, Fe^{2+}, and Fe^{3+} after equilibrium was reached?

$$Cr_2O_7^{2-} + 6e^- + 14H^+ \rightleftharpoons 2Cr^{3+} + 7H_2O \quad E^°_{Cr} = 1.33 \text{ volts}$$
$$Fe^{3+} + e^- \rightleftharpoons Fe^{2+} \quad E^°_{Fe} = 0.771 \text{ volt}$$
$$[H^+] = 1.00M \text{ throughout}$$

Ans: *[Fe²⁺] = 2.2 × 10⁻¹⁰M; [Cr₂O₇²⁻] = 3.7 × 10⁻¹¹M; [Fe³⁺] = 5.000 × 10⁻²M; [Cr³⁺] = 1.667 × 10⁻²M*

≡ 5.14 DETAILED SOLUTIONS TO PROBLEMS ≡

1. First, calculate the cell potential, E.

$$E = E_Q - E^{°'}_{SCE}$$
$$E_Q = ?$$
$$Q + 2H^+ + 2e^- \rightleftharpoons H_2Q$$

$$E_Q = E^°_Q - \frac{0.059}{2} \log \frac{[H_2Q]}{[Q][H^+]^2} = E^°_Q - \frac{0.059}{2} \log \frac{[H_2Q]}{[Q]} - \frac{0.059}{2} \log \frac{1}{[H^+]^2}$$

The stoichiometry of quinhydrone, $Q \cdot H_2Q$ dictates the ratio of hydroquinone, H_2Q, to quinone, Q.

$$[H_2Q] = [Q]$$
$$\frac{[H_2Q]}{[Q]} = 1$$

$$E_Q = E^\circ{}_Q - \frac{0.059}{2} \log 1 - \frac{0.059}{2} \log \frac{1}{[H^+]^2} = E^\circ{}_Q - \frac{0.059}{2} \log \frac{1}{[H^+]^2}$$

$$= 0.6994 - \frac{0.059}{2} \log \frac{1}{[H^+]^2} = 0.6994 + \frac{0.059}{2} \log [H^+]^2$$

$$= 0.6994 + \frac{(0.059)}{2} (2)\log [H^+] = 0.6994 - 0.059 (-\log [H^+])$$

$$= 0.6994 - 0.059 \text{ (pH)}$$

$$E = E_Q - E^{\circ\prime}{}_{SCE} = 0.6994 - 0.059 \text{ (pH)} - 0.246 = 0.453 - 0.059 \text{ (pH)}$$

Now, make a table as Table 5.6.

Table 5.6
QUINHYDRONE DATA

pH	E, v
0.00	0.453
1.00	0.394
2.00	0.335
3.00	0.276
4.00	0.217
5.00	0.158
6.00	0.099
7.00	0.040

A plot of the data is linear.

2. The potential of any half-cell involving Ag^+ and Ag can be written as E.

$$E = E^\circ{}_{Ag} - 0.059 \log \frac{1}{[Ag^+]}$$

$$E = E^\circ{}_{Ag_2CrO_4} - \frac{0.059}{2} \log [CrO_4^{2-}]$$

$$E^\circ{}_{Ag} - 0.059 \log \frac{1}{[Ag^+]} = E^\circ{}_{Ag_2CrO_4} - \frac{0.059}{2} \log [CrO_4^{2-}]$$

$$E^\circ{}_{Ag} - E^\circ{}_{Ag_2CrO_4} = 0.059 \log \frac{1}{[Ag^+]} - \frac{0.059}{2} \log [CrO_4^{2-}]$$

$$E^\circ{}_{Ag} - E^\circ{}_{Ag_2CrO_4} = -0.059 \log [Ag^+] - \frac{0.059}{2} \log [CrO_4^{2-}]$$

$$E^\circ{}_{Ag} - E^\circ{}_{Ag_2CrO_4} = \frac{-(0.059)\cdot(2)}{2} \log [Ag^+] - \frac{0.059}{2} \log [CrO_4^{2-}]$$

$$E^\circ{}_{Ag} - E^\circ{}_{Ag_2CrO_4} = \frac{-0.059}{2} \log [Ag^+]^2 - \frac{0.059}{2} \log [CrO_4^{2-}]$$

$$E^\circ{}_{Ag} - E^\circ{}_{Ag_2CrO_4} = \frac{-0.059}{2} [\log [Ag^+]^2 + \log [CrO_4^{2-}]]$$

$$E^\circ{}_{Ag} - E^\circ{}_{Ag_2CrO_4} = \frac{-0.059}{2} \log ([Ag^+]^2 \cdot [CrO_4^{2-}])$$

$$Ag_2CrO_4 \rightleftharpoons 2Ag^+ + CrO_4^{2-} \qquad K_{sp} = [Ag^+]^2 \cdot [CrO_4^{2-}]$$

$$E°_{Ag} - E°_{Ag_2CrO_4} = \frac{-0.059}{2} \log K_{sp}$$

$$(0.799 - 0.446) = \frac{-0.059}{2} \log K_{sp}$$

$$\log K_{sp} = \frac{-(0.353)(2)}{(0.059)} = -11.97$$

$$K_{sp} = 1.1 \times 10^{-12}$$

3.
$$E = E°'_{SCE} - E_H$$
$$E_H = ?$$

$$2H^+ + 2e^- \rightleftharpoons H_2$$

$$E_H = E°_H - \frac{0.059}{2}\log \frac{p_{H_2}}{[H^+]^2} = 0.000 - \frac{0.059}{2}\log \frac{p_{H_2}}{[H^+]^2}$$

$$= \frac{-0.059}{2} \log \frac{1.00}{[H^+]^2} = \frac{0.059}{2} \log [H^+]^2 = \frac{(0.059)}{2} (2) \log [H^+]$$

Now, find $[H^+]$. What we have is a *buffer*. $C_{HPro} = 0.10M$;

$$C_{Pro^-} = 0.050M$$
$$\uparrow$$
$$\text{From NaPro}$$

Remember:

$$[H^+]' = K_a \frac{C_{HPro}}{C_{Pro^-}} = (1.3 \times 10^{-5})\left(\frac{0.10}{0.050}\right) = 2.6 \times 10^{-5}M$$

This buffer is *acidic*.

$$[H^+]'' = K_a \frac{(C_{HPro} - [H^+]')}{(C_{Pro^-} + [H^+]')} = (1.3 \times 10^{-5})\left(\frac{0.10 - 1.3 \times 10^{-5}}{0.05 - 1.3 \times 10^{-5}}\right)$$

$$= 2.6 \times 10^{-5}M$$
$$[H^+]' = [H^+]''$$

The iterations have *converged*.

$$E_H = (0.059) \log (2.6 \times 10^{-5}) = -0.27$$
$$E = E°'_{SCE} - E_H = 0.246 - (-0.27) = 0.52 \text{ volt}$$

4. This is easier than it looks. Go back to Problem 1.

$$E = 0.453 - 0.059 \, (pH)$$

$$pH = -\left(\frac{E - 0.453}{0.059}\right) = -\left(\frac{0.323 - 0.453}{0.059}\right) = 2.20$$

$$[H^+] = 6.3 \times 10^{-3}M$$

The system is a phosphoric acid-dihydrogen phosphate buffer. From the work on weak acids in Chapter 2:

$$[H^+] = K_1 \frac{(C_{H_3PO_4} - [H^+])}{(C_{H \cdot PO_4^-} + [H^+])}$$

$$K_1 = [H^+] \frac{(C_{H_2PO_4^-} + [H^+])}{(C_{H_3PO_4} - [H^+])} = 6.3 \times 10^{-3} \frac{(0.100 + 6.3 \times 10^{-3})}{(0.100 - 6.3 \times 10^{-3})}$$

$$= 7.1 \times 10^{-3}$$

5. The overall reaction is

$$Ce^{4+} + Fe^{2+} \rightleftharpoons Ce^{3+} + Fe^{3+}.$$

It is a one-to-one reaction. In this solution, just enough Ce^{4+} has been added to react with all the Fe^{2+}. Still, tiny quantities of Fe^{2+} and Ce^{4+} will be generated from back reaction. How tiny? We need to determine an equilibrium constant first.

$$K = \frac{[Ce^{3+}][Fe^{3+}]}{[Ce^{4+}][Fe^{4+}]}$$

The solution is at equilibrium, so:

$$E_{solution} = E_{Fe} = E_{Ce}$$

$$E_{solution} = E_{Fe} = E°_{Fe} - 0.059 \log \frac{[Fe^{2+}]}{[Fe^{3+}]}$$

$$E_{solution} = E_{Ce} = E°_{Ce} - 0.059 \log \frac{[Ce^{3+}]}{[Ce^{4+}]}$$

$$E°_{Ce} - 0.059 \log \frac{[Ce^{3+}]}{[Ce^{4+}]} = E°_{Fe} - 0.059 \log \frac{[Fe^{2+}]}{[Fe^{3+}]}$$

$$E°_{Ce} - E°_{Fe} = 0.059 \log \frac{[Ce^{3+}]}{[Ce^{4+}]} - 0.059 \log \frac{[Fe^{2+}]}{[Fe^{3+}]}$$

$$E°_{Ce} - E°_{Fe} = 0.059 \log \frac{[Ce^{3+}]}{[Ce^{4+}]} + 0.059 \log \frac{[Fe^{3+}]}{[Fe^{2+}]}$$

$$E°_{Ce} - E°_{Fe} = 0.059 \left(\log \frac{[Ce^{3+}]}{[Ce^{4+}]} + \log \frac{[Fe^{3+}]}{[Fe^{2+}]} \right)$$

$$E°_{Ce} - E°_{Fe} = 0.059 \log \left(\frac{[Ce^{3+}][Fe^{3+}]}{[Ce^{4+}][Fe^{2+}]} \right)$$

$$E°_{Ce} - E°_{Fe} = 0.059 \log K$$

$$\log K = \frac{E°_{Ce} - E°_{Fe}}{0.059} = \frac{1.61 - 0.771}{0.059} = 14.24$$

$$K = 1.7 \times 10^{14}$$

Now, calculate concentrations.

$$C_{Fe} = [Fe^{3+}] + [Fe^{2+}] = \frac{(\text{total millimoles of Fe})}{(\text{total milliliters of solution})}$$

$$= \frac{(100.00)(0.1000)}{200.00} = 5.000 \times 10^{-2} M$$

$$5.000 \times 10^{-2} = [Fe^{3+}] + [Fe^{2+}]$$

$$[Fe^{3+}] = 5.000 \times 10^{-2} - [Fe^{2+}]$$

$$C_{Ce} = [Ce^{3+}] + [Ce^{4+}] = \frac{(\text{total millimoles of Ce})}{(\text{total milliliters of solution})}$$

$$= \frac{(100.00)(0.1000)}{200.00} = 5.000 \times 10^{-2} M$$

$$5.000 \times 10^{-2} = [Ce^{3+}] + [Ce^{4+}]$$

$$[Ce^{4+}] = [Fe^{2+}]$$

$$5.000 \times 10^{-2} = [Ce^{3+}] + [Fe^{2+}]$$

$$[Ce^{3+}] = 5.000 \times 10^{-2} - [Fe^{2+}]$$

Now:

$$K = \frac{[Ce^{3+}][Fe^{3+}]}{[Ce^{4+}][Fe^{2+}]}$$

Substitute:

$$K = \frac{(5.00 \times 10^{-2} - [Fe^{2+}])(5.000 \times 10^{-2} - [Fe^{2+}])}{[Fe^{2+}]^2}$$

$$[Fe^{2+}]^2 = \frac{(5.00 \times 10^{-2} - [Fe^{2+}])^2}{K}$$

$$[Fe^{2+}] = \left[\frac{(5.00 \times 10^{-2} - [Fe^{2+}])^2}{K} \right]^{1/2}$$

First approximation:

$$[Fe^{2+}]' = \left[\frac{(5.00 \times 10^{-2})^2}{K} \right]^{1/2}$$

Second approximation:

$$[Fe^{2+}]'' = \left[\frac{(5.00 \times 10^{-2} - [Fe^{2+}]')^2}{K} \right]^{1/2}$$

and so forth.

Will they converge? We'll see.

$$[Fe^{2+}]' = \left[\frac{(5.00 \times 10^{-2})^2}{1.7 \times 10^{14}} \right]^{1/2} = 3.8 \times 10^{-9} M$$

$$[Fe^{2+}]'' = \left[\frac{(5.00 \times 10^{-2} - 3.8 \times 10^{-9})^2}{1.7 \times 10^{14}} \right]^{1/2} = 3.8 \times 10^{-9} M$$

(As we see, they do converge.)

$$[Fe^{2+}] = [Fe^{2+}]' = [Fe^{2+}]'' = 3.8 \times 10^{-9} M$$

$$[Ce^{4+}] = [Fe^{2+}] = 3.8 \times 10^{-9} M$$

$$[Fe^{3+}] = 5.000 \times 10^{-2} - [Fe^{2+}]$$

$$= 5.000 \times 10^{-2} - 3.8 \times 10^{-9} = 5.000 \times 10^{-2} M$$

$$[Ce^{3+}] = 5.000 \times 10^{-2} - [Fe^{2+}]$$

$$= 5.000 \times 10^{-2} - 3.8 \times 10^{-9} = 5.000 \times 10^{-2} M$$

6. The overall reaction is

$$Cr_2O_7^{2-} + 6Fe^{2+} + 14H^+ \rightleftharpoons 6Fe^{3+} + 2Cr^{3+} + 7H_2O$$

It is *not* a one-to-one reaction:

(6)(number of millimoles of $Cr_2O_7^{2-}$ reduced)

$$= \text{(number of millimoles of } Fe^{2+} \text{ oxidized)}$$

$$(6)(50.00)(0.01667) = (50.00)(0.1000)$$

$$5.000 \text{ millimoles} = 5.000 \text{ millimoles}$$

In this solution, just enough $Cr_2O_7^{2-}$ has been added to react with *all* the Fe^{2+}. Still, tiny quantities of Fe^{2+} and $Cr_2O_7^{2-}$ will be generated from back reaction. How tiny? An equilibrium constant is needed first.

$$K = \frac{[Cr^{3+}]^2 \cdot [Fe^{3+}]^6}{[Cr_2O_7^{2-}] \cdot [Fe^{2+}]^6 \cdot [H^+]^{14}}$$

The solution is at equilibrium, so:

$$E_{\text{solution}} = E_{Fe} = E_{Cr}$$

$$E_{\text{solution}} = E_{Fe} = E°_{Fe} - 0.059 \log \frac{[Fe^{2+}]}{[Fe^{3+}]}$$

$$E_{\text{solution}} = E_{Cr} = E°_{Cr} - \frac{0.059}{6} \log \frac{[Cr^{3+}]^2}{[Cr_2O_7^{2-}] \cdot [H^+]^{14}}$$

$$E°_{Cr} - \frac{0.059}{6} \log \frac{[Cr^{3+}]^2}{[Cr_2O_7^{2-}] [H^+]^{14}} = E°_{Fe} - 0.059 \log \frac{[Fe^{2+}]}{[Fe^{3+}]}$$

$$6E°_{Cr} - 0.059 \log \frac{[Cr^{3+}]^2}{[Cr_2O_7^{2-}] [H^+]^{14}} = 6E°_{Fe} - (0.059)(6) \log \frac{[Fe^{2+}]}{[Fe^{3+}]}$$

$$6E°_{Cr} - 0.059 \log \frac{[Cr^{3+}]^2}{[Cr_2O_7^{2-}] [H^+]^{14}} = 6E°_{Fe} - 0.059 \log \frac{[Fe^{2+}]^6}{[Fe^{3+}]^6}$$

$$6E°_{Cr} - 6E°_{Fe} = 0.059 \log \frac{[Cr^{3+}]^2}{[Cr_2O_7^{2-}] [H^+]^{14}} - 0.059 \log \frac{[Fe^{2+}]^6}{[Fe^{3+}]^6}$$

$$6(E°_{Cr} - E°_{Fe}) = 0.059 \log \frac{[Cr^{3+}]^2}{[Cr_2O_7^{2-}] [H^+]^{14}} + 0.059 \log \frac{[Fe^{3+}]^6}{[Fe^{2+}]^6}$$

$$6(E°_{Cr} - E°_{Fe}) = 0.059 \left(\log \frac{[Cr^{3+}]^2}{[Cr_2O_7^{2-}] [H^+]^{14}} + \log \frac{[Fe^{3+}]^6}{[Fe^{2+}]^6} \right)$$

$$6(E°_{Cr} - E°_{Fe}) = 0.059 \log \frac{[Cr^{3+}]^2 \cdot [Fe^{3+}]^6}{[Cr_2O_7^{2-}] \cdot [Fe^{2+}]^6 \cdot [H^+]^{14}}$$

$$6(E°_{Cr} - E°_{Fe}) = 0.059 \log K$$

$$\log K = \frac{6(E°_{Cr} - E°_{Fe})}{0.059}$$

$$\log K = \frac{6(1.33 - 0.771)}{0.059}$$

$$\log K = 56.95$$

$$K = 8.9 \times 10^{+56} \quad \text{(That } is \text{ large.)}$$

Now, calculate concentrations.

$$C_{Fe} = [Fe^{3+}] + [Fe^{2+}] = \frac{(\text{total millimoles of Fe})}{(\text{total milliliters of solution})}$$

$$= \frac{(50.00)(0.1000)}{(100.00)} = 5.000 \times 10^{-2}M$$

$$5.000 \times 10^{-2} = [Fe^{3+}] + [Fe^{2+}]$$

$$[Fe^{3+}] = 5.000 \times 10^{-2} - [Fe^{2+}]$$

$$[Fe^{2+}] = 6[Cr_2O_7^{2-}]$$

$$C_{Cr} = [Cr^{3+}] + 2[Cr_2O_7^{2-}] = \frac{(\text{total number of millimoles of Cr})}{(\text{total milliliters of solution})}$$

$$= \frac{(50.00)(0.01667)(2)}{(100.00)} = 1.667 \times 10^{-2}M$$

$$1.667 \times 10^{-2} = [Cr^{3+}] + 2[Cr_2O_7^{2-}]$$

Substitute for $[Cr_2O_7^{2-}]$

$$1.667 \times 10^{-2} = [Cr^{3+}] + \frac{2[Fe^{2+}]}{6}$$

$$[Cr^{3+}] = 1.667 \times 10^{-2} - \frac{[Fe^{2+}]}{3}$$

$$[H^+] = 1.00M$$

Substitute into K

$$K = \frac{[Cr^{3+}]^2[Fe^{3+}]^6}{[Cr_2O_7^{2-}][Fe^{2+}]^6 \cdot [H^+]^{14}}$$

$$K = \frac{\left(1.667 \times 10^{-2} - \frac{[Fe^{2+}]^2}{3}\right) \cdot (5.00 \times 10^{-2} - [Fe^{2+}])^6}{\left(\frac{[Fe^{2+}]}{6}\right)[Fe^{2+}]^6(1.00)^{14}}$$

$$\frac{[Fe^{2+}]^7}{6} = \frac{\left(1.667 \times 10^{-2} - \frac{[Fe^{2+}]}{3}\right)^2 \cdot (5.00 \times 10^{-2} - [Fe^{2+}])^6}{K}$$

$$[Fe^{2+}]^7 = \frac{(6)\left(1.667 \times 10^{-2} - \frac{[Fe^{2+}]}{3}\right)^2 \cdot (5.00 \times 10^{-2} - [Fe^{2+}])^6}{8.9 \times 10^{56}}$$

$$[Fe^{2+}] = \left[\frac{6\left(1.667 \times 10^{-2} - \frac{[Fe^{2+}]}{3}\right)^2 \cdot (5.00 \times 10^{-2} - [Fe^{2+}])^6}{8.9 \times 10^{56}}\right]^{1/7}$$

First approximation:

$$[Fe^{2+}]' = \left[\frac{6(1.667 \times 10^{-2})^2 \cdot (5.00 \times 10^{-2})^6}{8.9 \times 10^{56}}\right]^{1/7}$$

Second approximation:

$$[Fe^{2+}]'' = \left[\frac{6\left(1.667 \times 10^{-2} - \frac{[Fe^{2+}]'}{3}\right)^2 \cdot (5.00 \times 10^{-2} - [Fe^{2+}]')^6}{8.9 \times 10^{56}} \right]^{1/7}$$

$$[Fe^{2+}]' = \left[\frac{6(1.667 \times 10^{-2})^2 \cdot (5.00 \times 10^{-2})^6}{8.9 \times 10^{56}} \right]^{1/7} = 2.2 \times 10^{-10} M$$

$$[Fe^{2+}]'' = \left[\frac{6\left(1.667 \times 10^{-2} - \frac{2.2 \times 10^{-10}}{3}\right)^2 \cdot (5.00 \times 10^{-2} - 2.2 \times 10^{-10})^6}{8.9 \times 10^{56}} \right]^{1/7}$$

$$= 2.2 \times 10^{-10} M$$

$$[Fe^{2+}] = [Fe^{2+}]' = [Fe^{2+}]'' = 2.2 \times 10^{-10} M$$

$$[Cr_2O_7^{2-}] = \frac{[Fe^{2+}]}{6} = \frac{2.2 \times 10^{-10}}{6} = 3.7 \times 10^{-11} M$$

$$[Fe^{3+}] = 5.000 \times 10^{-2} - [Fe^{2+}] = 5.000 \times 10^{-2} - 2.2 \times 10^{-10}$$

$$= 5.000 \times 10^{-2} M$$

$$[Cr^{3+}] = 1.667 \times 10^{-2} - \frac{[Fe^{2+}]}{3} = 1.667 \times 10^{-2} - \frac{2.2 \times 10^{-10}}{3}$$

$$= 1.667 \times 10^{-2} M$$

≡ 5.15 ION-SELECTIVE ELECTRODES ≡

By the 1920s it was known that a potential could be developed across a glass membrane if the solutions on either side of it had different hydrogen ion concentrations. Little use was made of the glass electrode as a pH-sensing device in those early years. The high electrical resistance of glass made the use of conventional potentiometers impractical. Only in the 1930s was electronic instrumentation that could easily be used with glass electrodes developed, made portable, and distributed by laboratory suppliers. The glass electrode, sensitive to hydrogen ion concentrations, became the first of the ion-selective electrodes. Ion-selective electrodes, sometimes called specific ion electrodes, are devices that respond to concentration changes of one sort of ion, with little interference from other species in solution.

The exact mechanism by which a glass electrode—or any other sort of ion-selective electrode—works is not known. That the potential of a glass electrode is linear with pH *is* known, however, and much use may be made of this fact.

The arrangement of a pH-measuring cell is shown in Figure 5.16, together with an expanded view of a glass electrode. The cell may be written as follows.

Ag-AgCl|(ref.) 0.1M HCl|glass|solution of unknown pH||SCE

A potential develops across the glass membrane when the hydrogen ion concentration

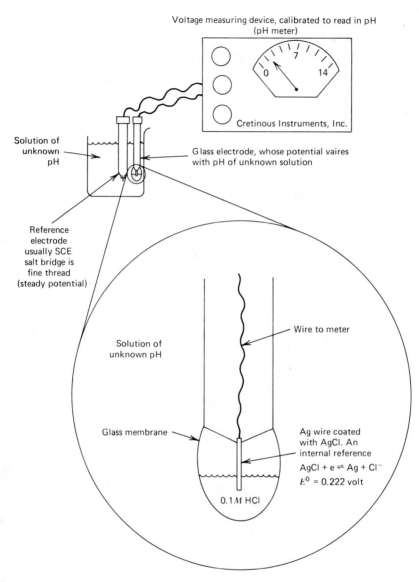

Figure 5.16 *pH Measurement with glass electrode.*

in the unknown solution is different from that inside the glass bulb. The potential, E, is observed on the pH meter, which is simply a voltage measuring device calibrated to read out in pH units. E is given by the equation below:

$$E = E^{o\prime}_{SCE} + E_j - E_{glass}$$

E_j is called a junction potential. It is small and constant.

$$E = \text{const.} - E_{\text{glass}}$$

At 25°C:

$$E_{\text{glass}} = E^{\circ}_{\text{glass}} + 0.059 \log a_H$$

$$a_H = \text{activity of hydrogen ion (a function of } \mu \text{ and } [H^+])$$

$$pH = -\log a_H$$

$$E_{\text{glass}} = E^{\circ}_{\text{glass}} - 0.059 \, pH$$

$$E = \text{const.} - E^{\circ}_{\text{glass}} + 0.059 \, pH$$

or

$$E = \text{const.}' + 0.059 \, pH$$

Thus, the observed potential varies linearly with pH. Because the constant terms are not well known, the pH meter and associated electrodes are calibrated with buffers of known pH.

Hydrogen ion is not the only species to which a glass electrode will respond. At lower hydrogen ion concentrations, and high sodium ion concentrations, such as those encountered after the endpoint of a titration of acid by a strong base, a deviation known as the alkaline error appears. The apparent hydrogen ion concentration is higher than the actual hydrogen ion concentration. The reason for the error is this: at high pH-values, H^+ ions are few in number. Past the endpoint of a titration of strong acid by NaOH, the concentration of sodium ions is high. The glass electrode is sensitive to sodium ions, though not so much as it is to hydrogen ions. When hydrogen ion concentration is low, the sodium ion concentration is high, the electrode mistakes Na^+ for H^+, and gives a falsely high response for hydrogen ion concentration.

The alkaline error, regarded by many as a great pain, can be turned to advantage. In the late 1950's, attention was given to varying the composition of glass membranes. Through extensive experimentation, glasses were discovered that were almost uniquely sensitive to each of the following cations: Na^+, K^+, NH_4^+, Ag^+, and Li^+. These glasses were incorporated into ion-selective electrodes. Interferences can be minimized, and the electrodes can be used, typically, over a concentration range from $1M$ to 10^{-6} or $10^{-8}M$. The actual quantity measured by an electrode is activity, rather than concentration, but this may be compensated for through careful calibration. Many forms of electrodes other than glass membranes are used: immobilized enzymes combined with glasses, liquid ion exchangers, and single crystals.

One of the cleverest ion-selective electrodes is a glass electrode combined with an immobilized enzyme*. Figure 5.17 shows such an arrangement. Urease, an enzyme which hastens the conversion of urea, $NH_2-\overset{\overset{\text{O}}{\|}}{C}-NH_2$, to ammonium ion, NH_4^+

$$\text{urea} \xrightarrow[H_2O]{\text{urease}} NH_4^+ + HCO_3^-$$

* R. A. Durst, American Scientist 59, 353 (1971).

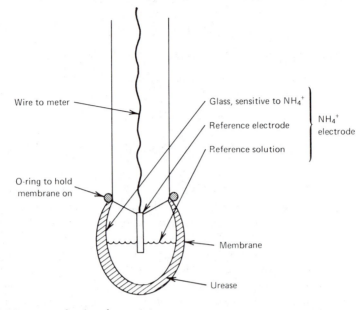

Figure 5.17 *Urea-selective electrode*.*

is secured to the surface of a glass electrode sensitive to ammonium ion with a membrane. When the electrode is dipped into a solution that contains urea, the conversion of urea to NH_4^+ *at the glass surface* is catalyzed by the urease. The ammonium ion electrode responds to the concentration of ammonium ion at its surface. A measurement may be made without sensibly affecting the urea concentration of a solution, for very little urea need be degraded to give a reading.

Conducting crystals may be used for sensors in ion-selective electrodes. One of the earliest electrodes to use a conducting crystal is the fluoride electrode. Instead of a glass membrane, a conducting crystal of LaF_3 doped with europium is the solid across which a potential is developed. A single-crystal fluoride-specific electrode is shown in Figure 5.18*.

The responses of ion-selective electrodes to concentration changes are Nernstian. Suppose that one has a cell like the following.

ref.‖unknown concentration of X∣ind.

(of const.   (indicator electrode,
voltage, like   sensitive to X)
SCE)

$$E_{cell} = E_{ind.} - E_{ref.} + E_j$$

$$E_{cell} = \text{cell potential}$$

$$E_{ind.} = \text{potential of indicator electrode (varies with activity of species } X)$$

$$E_{ref.} = \text{potential of reference electrode (constant)}$$

* *R. A. Durst*, Ibid.

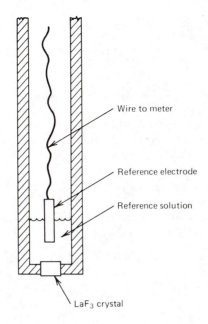

Wire to meter

Reference electrode

Reference solution

LaF$_3$ crystal

Figure 5.18 *Fluoride-selective electrode*.*

$$E_j = \text{junction potential (small and constant)}$$

$$-E_{\text{ref.}} + E_j = \text{const.}'$$

$$E_{\text{cell}} = E_{\text{ind.}} + \text{const.}'$$

The Nernstian behavior of the indicator is shown (25°C) below.

$$E_{\text{ind.}} = E°_{\text{ind.}} - \frac{0.059}{n} \log a_X$$

$$E°_{\text{ind.}} = \text{constant}$$

$$E_{\text{cell}} = E°_{\text{ind.}} - \frac{0.059}{n} \log a_X + \text{const.}'$$

$$E°_{\text{ind.}} + \text{const.}' = \text{const.}$$

$$E_{\text{cell}} = \text{const.} - \frac{0.059}{n} \log a_X$$

By the definition of activity:

$$a_X = f_X [X]$$

$$E_{\text{cell}} = \text{const.} - \frac{0.059}{n} \log (f_X [X])$$

A switch can be thrown on most meters so that an observed potential reading, $E_{\text{obs.}}'$ is equal to $-E_{\text{cell}}$

$$E_{\text{obs.}} = -E_{\text{cell}}$$

* *R. A. Durst*, Ibid.

Let

$$k = -\text{const.}$$

$$-E_{\text{cell}} = -\text{const.} + \frac{0.059}{n} (\log f_X [X])$$

$$E_{\text{obs.}} = k + \frac{0.059}{n} \log (f_X [X]) = k + \frac{0.059}{n} \log a_X$$

Thus, the observed voltage varies linearly with the logarithm of the activity, a_X, of the ion X. There are two ways to use this Nernstian behavior for determining ionic concentrations. The first is a calibration curve technique. The second is a method of standard additions.

In the calibration curve method, a plot of $E_{\text{obs.}}$ versus $[X]$ is made. It is linear if ionic strength is kept constant. Ionic strength can be kept constant by the addition of high and equal concentrations of a noninterfering electrolyte to each sample. With constant ionic strength, the activity coefficient, f_X, of X will remain constant.

$$E_{\text{obs.}} = k + \frac{0.059}{n} \log a_X = k + \frac{0.059}{n} \log (f_X [X])$$

$$= k + \frac{0.059}{n} (\log f_X + \log [X]) = k + \frac{0.059}{n} \log f_X + \frac{0.059}{n} \log [X]$$

Because f_X is made constant,

$$k + \frac{0.059}{n} \log f_X = k' = \text{another constant}$$

$$E_{\text{obs.}} = k' + \frac{0.059}{n} \log [X]$$

Example: A biochemist wanted to determine the rate at which liposomes (little balls of curious fat) lose fluoride. To do this, he had to establish a calibration curve for his fluoride electrode setup. His results are given in Table 5.7.*

Table 5.7
CALIBRATION OF FLUORIDE-SELECTIVE ELECTRODE*

$[F^-]$, M	$E_{\text{obs.}}$, mV
6.00×10^{-4}	+2.5
1.20×10^{-3}	+19.0
2.20×10^{-3}	+34.8
3.69×10^{-3}	+48.0
5.67×10^{-3}	+59.0
8.62×10^{-3}	+69.8
1.25×10^{-2}	+79.2
2.07×10^{-2}	+91.7
2.97×10^{-2}	+102
4.37×10^{-2}	+112

* *Personal communication, S. E. Schullery, Eastern Michigan University, Ypsilanti.*

Table 5.7 continued

[F⁻] , M	$E_{obs.}$, mV
6.16 × 10⁻²	+121
8.73 × 10⁻²	+130
1.19 × 10⁻¹	+138

Courtesy of Prof. S. E. Schullery. All solutions were made 0.1M in NaCl to insure constant ionic strength.

Calculate the molarity of fluoride in a solution which, under the same conditions gave an $E_{obs.}$ value of 96.0 millivolts. To spare taking logarithms, plot the graph on semilog paper, as shown in Figure 5.19. Using the curve like any calibration curve, note that the fluoride ion concentration of the unknown solution is $2.30 \times 10^{-2} M$.

The method of standard additions is not mathematically as straightforward as the calibration curve method, but it may, in practice, be simpler to perform, only two measurements being needed. The potential of an unknown solution, $E_{obs.\ 1}$, is taken. Then, with a standard solution, the concentration of the ion of interest in the unknown solution is increased by a known amount. The potential, $E_{obs.\ 2}$, of the spiked solution (the solution whose concentration has been increased) is taken.

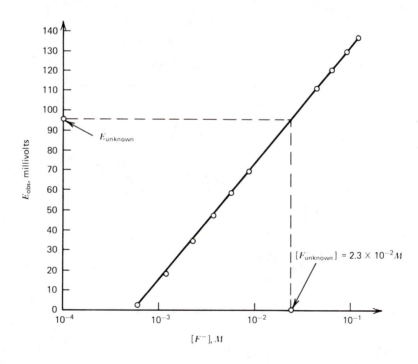

Figure 5.19 *Schullery's calibration for F⁻ electrode. Courtesy of Prof. S. E. Schullery.*

For the unknown: $E_{\text{obs. 1}} = k + \dfrac{0.059}{n} \log a_X$ $\qquad\qquad$ (5.6)

$$a_X = \text{activity of unknown} = f_X C_X$$
$$C_X = \text{concentration of unknown}$$
$$f_X = \text{activity coefficient.}$$

$$E_{\text{obs. 1}} = k + \dfrac{0.059}{n} \log (f_X C_X) \qquad\qquad (5.7)$$

For the calculation of the potential, $E_{\text{obs. 2}}$, of the solution made by adding a known volume, V_S, of a standard of known concentration, C_S, to volume V_X of the unknown of concentration C_X, the concentration C_{XS} of ion X must be known. For this solution, activity, a_{XS}, is given by

$$a_{XS} = f_X C_{XS} = \text{activity of } X \text{ in spiked solution}$$
$$C_{XS} = \text{concentration of } X \text{ in spiked solution}$$
$$f_X = \text{activity coefficient (same for both solutions)}$$

$$E_{\text{obs. 2}} = k + \dfrac{0.059}{n} \log a_{XS}$$

$$E_{\text{obs. 2}} = k + \dfrac{0.059}{n} \log (f_X C_{XS}) \qquad\qquad (5.8)$$

C_{XS} is easily evaluated

$$C_X = \text{concentration of } X \text{ in unknown solution}$$
$$V_X = \text{volume of unknown solution before addition}$$
$$V_S = \text{volume of standard solution added}$$
$$C_S = \text{concentration of } X \text{ in standard solution (known)}$$

$$C_{XS} = \dfrac{C_X V_X + C_S V_S}{V_X + V_S} \qquad\qquad (5.9)$$

Substitute Equation 5.8 into Equation 5.9.

$$E_{\text{obs. 2}} = k + \dfrac{0.059}{n} \log \left[f_X \left(\dfrac{C_X V_X + C_S V_S}{V_X + V_S} \right) \right] \qquad\qquad (5.10)$$

Subtract Equation 5.7 from Equation 5.10 and let

$$\Delta E = E_{\text{obs. 2}} - E_{\text{obs. 1}}$$

$$\Delta E = k + \dfrac{0.059}{n} \log \left[f_X \left(\dfrac{C_X V_X + C_S V_S}{V_X + V_S} \right) \right] - k - \dfrac{0.059}{n} \log (f_X C_X)$$

$$\Delta E = \dfrac{0.059}{n} \log f_X + \dfrac{0.059}{n} \log \left(\dfrac{C_X V_X + C_S V_S}{V_X + V_S} \right) - \dfrac{0.059}{n} \log C_X - \dfrac{0.059}{n} \log f_X$$

$$\Delta E = \dfrac{0.059}{n} \log \left(\dfrac{C_X V_X + C_S V_S}{V_X + V_S} \right) - \dfrac{0.059}{n} \log C_X$$

$$\Delta E = \frac{0.059}{n} \log\left(\frac{\dfrac{C_X V_X + C_S V_S}{V_X + V_S}}{C_X} \right)$$

$$\Delta E = \frac{0.059}{n} \log\left(\frac{\dfrac{C_X V_X}{V_X + V_S} + \dfrac{C_S V_S}{V_X + V_S}}{C_X} \right)$$

$$\Delta E = \frac{0.059}{n} \log\left(\frac{V_X}{V_X + V_S} + \frac{C_S V_S}{V_X + V_S} \cdot \frac{1}{C_X} \right)$$

Now, let

$$\frac{0.059}{n} = S$$

(For the best work, S, is experimentally determined.)

$$\Delta E = S \log \left(\frac{V_X}{V_X + V_S} + \frac{C_S V_S}{V_X + V_S} \cdot \frac{1}{C_X} \right)$$

$$\frac{\Delta E}{S} = \log \left(\frac{V_X}{V_X + V_S} + \frac{C_S V_S}{V_X + V_S} \cdot \frac{1}{C_X} \right)$$

$$10^{\Delta E/S} = \frac{V_X}{V_X + V_S} + \frac{C_S V_S}{V_X + V_S} \cdot \frac{1}{C_X}$$

$$10^{\Delta E/S} - \frac{V_X}{V_X + V_S} = \frac{C_S V_S}{V_X + V_S} \cdot \frac{1}{C_X}$$

$$C_X = \frac{C_S V_S}{V_X + V_S} \left(10^{\Delta E/S} - \frac{V_X}{V_X + V_S} \right)^{-1} \qquad (5.11)$$

If the volume, V_S, of standard solution is small in comparison to the volume, V_X, of the unknown, a simplification is possible.

$$C_X = C_S \cdot \frac{V_S}{V_X} \cdot (10^{\Delta E/S} - 1)^{-1} \qquad (5.12)$$

Example: A fluoride electrode is used with the method of standard additions. The volume of the unknown sample is 100. milliliters. Ten milliliters of 0.1000 NaF are added to the sample. The observed potential before addition is 30. millivolts. The observed potential after addition is 70. millivolts. Calculate the concentration of fluoride in the unknown sample.

It is a good idea to begin problems of this sort by defining terms.

$$C_S = 0.1000M$$
$$V_S = 10.0 \text{ milliliters}$$
$$V_X = 100 \text{ milliliters}$$

Note: units of volts $\quad \Delta E = (70\text{-}30)/1000 = 0.040 \text{ volts}$

$$S = 0.059/n$$

$$n = 1 \text{ (fluoride is monovalent)}$$
$$S = 0.059/1 = 0.059.$$

First, try the full expression, Equation 5.11.

$$c_X = \frac{c_S V_S}{V_X + V_S} \left(10^{\Delta E/S} - \frac{V_X}{V_X + V_S} \right)^{-1}$$

$$= \frac{(0.1000)(10.0)}{(100. + 10.0)} \left(10^{0.040/0.059} - \frac{100.}{100 + 10.0} \right)^{-1}$$

$$= (0.00909)(4.76 - 0.909)^{-1} = 2.4 \times 10^{-3} M$$

Next, see if the condensed expression (Equation 5.12) works.

$$c_X = c_S \frac{V_S}{V_X} (10^{\Delta E/S} - 1)^{-1} = (0.1000) \left(\frac{10.0}{100} \right) (10^{0.040/0.059} - 1)^{-1}$$

$$= (0.01000)(4.76 - 1)^{-1} = 2.6 \times 10^{-3} M$$

The error introduced by using the condensed expression is not large, only about 2 parts in 24 or 26, which translates to about 8 parts in 100, or 8%.

Ion-selective electrodes, like most analytical tools, are subject to interferences. The alkaline error in pH determination has been mentioned. One cannot mindlessly insert an electrode into a sample and expect the reading on the meter to be a key to absolute truth. Two sorts of interferences with fluoride ion determination are shown in Figure 5.20* Figure 5.20 shows fluoride electrode response for solutions of 1 milligram

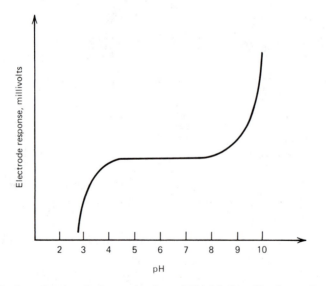

Figure 5.20 *Interferences in determination of F⁻ with fluoride electrode.*

* T. S. Light, "Industrial Analysis and Control with Ion-Selective Electrodes" in Ion-Selective Electrodes, R. A. Durst, Ed., NBS Special Publication 314, November 1969, p. 366.

F⁻/liter as a function of pH of those solutions. At low values of pH, electrode response is quite low. This is because, at lower pH values, much of the total fluoride in any solution is present as undissociated HF. This assertion may be proved by a calculation.

Calculation:

$$\text{For HF}, K_a = 6.7 \times 10^{-4}.$$

What are the values of α_0 and α_1 at pH 3?

Answer:

$$\alpha_0 = \frac{[HF]}{[HF] + [F^-]} = \frac{[H^+]}{[H^+] + K_a} = \frac{10^{-3}}{10^{-3} + 6.7 \times 10^{-4}} = 0.60$$

$$\alpha_1 = \frac{[F^-]}{[HF] + [F^-]} = \frac{K_a}{[H^+] + K_a} = \frac{6.7 \times 10^{-4}}{10^{-3} + 6.7 \times 10^{-4}} = 0.40$$

At pH 3, only 40% of total fluoride is present as fluoride ion. The behavior of the electrode system at low pH values is not really the result of the fluoride electrode's mistaking any other ion for fluoride; much of the fluoride is bound as a molecular species, HF, to which the electrode cannot respond. At higher pH values, where [OH⁻] is large, the electrode *does* respond to OH⁻, mistaking it for F⁻. Thus, hydroxide ion may truly be called an interference.

Interferences are found through careful laboratory study. Often an equation may be written describing the effect of interfering ions. If the activity, a_X, of an ion, X, of interest is being measured, and if ion Y interferes with the measurement, then the observed voltage, $E_{obs.}$, is related to the activities of ions X and Y by the equation:

$$E_{obs.} = k + \frac{0.059}{n} \log (a_X + k_S a_Y^{n/z})$$

$E_{obs.}$ = observed voltage
k = constant
n = number of electrons (ion X)
a_X = activity of ion X
k_S = selectivity coefficient (empirically determined)
a_Y = activity of interferring ion Y
Z = charge on ion Y

Manufacturers supply interference data for electrodes that they sell. Often, interferences may be compensated by the addition of such large amounts of the interfering ion to each standard or sample that its activity is the same for each standard or sample solution.

≡ 5.16 PROBLEMS ≡

1. In a study of the binding of cadmium ion (Cd^{2+}) to liposomes, S. E. Schullery used a cadmium-selective electrode. His calibration data are given below.[†] (Experiment run at constant ionic strength.)

[†] *Personal communication, S. E. Schullery, Eastern Michigan University, Ypsilanti.*

Table 5.8
CALIBRATION OF CADMIUM-SELECTIVE ELECTRODE*

$[Cd^{2+}]$, M	$E_{obs.}$, mV
1.64 × 10^{-4}	364
2.49 × 10^{-4}	369
3.68 × 10^{-4}	374
5.29 × 10^{-4}	379
7.86 × 10^{-4}	384
1.24 × 10^{-3}	390
1.69 × 10^{-3}	393
2.41 × 10^{-3}	398
3.49 × 10^{-3}	402
5.10 × 10^{-3}	406
7.60 × 10^{-3}	411
1.12 × 10^{-2}	415
1.82 × 10^{-2}	421
3.11 × 10^{-2}	427

* Courtesy of Prof. S. E. Schullery.

(a) From these data, calculate the value of cadmium ion concentration in a solution which gives a potential of 400 millivolts.
Ans: *3.0 × $10^{-3}M$*

(b) Does it matter whether n in the Nernst equation for this experiment is 1 or 2?
Ans: *No*

2. A cadmium-selective electrode is used with the method of standard additions. The volume of the unknown sample is 100. milliliters. The observed potential before addition is 379. millivolts. After the addition of 1.00 milliliter of 1.00M Cd^{2+} solution, the potential becomes 414. millivolts. Ionic strength is kept constant. Calculate the concentration of Cd^{2+} in the unknown solution.
Ans: *6.9 × $10^{-4}M$*

≡ 5.17 DETAILED SOLUTIONS TO PROBLEMS ≡

1. (a) Use semilog paper. Plot $E_{obs.}$ on the linear axis and $[Cd^{2+}]$ on the logarithmic axis. From the graph, the concentration of Cd^{2+} in the unknown solution is $3.0 × 10^{-3}M$.

(b) No. Once you have a calibration graph, you need not know the value of n.

2. Use the Equation 5.11.

$$C_X = \frac{C_S V_S}{V_X + V_S} \left(10^{\Delta E/S} - \frac{V_X}{V_X + V_S} \right)^{-1}$$

$$C_S = 1.00M$$

$$V_S = 1.00 \text{ milliliter}$$

$$V_X = 100. \text{ milliliters}$$

$$\Delta E = (414\text{-}379)/1000 \text{ volts}$$

$$\Delta E = 0.035 \text{ volt}$$

$$S = 0.059/n$$

$$n = 2 \quad (\text{Remember, it's } Cd^{2+}!)$$

$$S = 0.059/2$$

$$C_X = \frac{(1.00)(1.00)}{100. + 1.00} \left(10^{0.035/0.0295} - \frac{100}{100 + 1.00} \right)^{-1}$$

$$= 6.9 \times 10^{-4}M$$

Now try this by the abbreviated formula (Equation 5.12).

$$C_X = C_S \left(\frac{V_S}{V_X} \right) (10^{\Delta E/S} - 1)^{-1}$$

$$= \frac{(1.00)(1.00)}{100} (10^{0.035/0.0295} - 1)^{-1}$$

$$= 7.0 \times 10^{-4}M$$

The two answers are quite close.

chapter 6
Separations Not Involving Precipitation

Very often, before a measurement can be taken in analytical laboratory practice, the species to be analyzed must be separated from other species present. Separation by precipitation often comes to mind. There are many other analytical separation techniques, however; these other techniques will be discussed in this chapter. First, the transfer of solutes between two immiscible liquid phases will be treated. Then, multistage solute transfers will be discussed. Continuous transfer of solutes between two immiscible liquid phases (liquid-liquid chromatography) will be described briefly, followed by a longer section on gas-liquid chromatography. In the final section, a technique called sorption chromatography will be described together with some of its applications.

6.2 TRANSFER OF SOLUTES BETWEEN
≡ TWO IMMISCIBLE LIQUID PHASES USING ≡
SEPARATORY FUNNELS

Our concern in this section will be with *extraction* methods that involve the transfer of a solute between two very nearly immiscible liquid phases. We will start with a

414

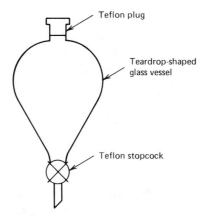

Figure 6.1 *Separatory funnel.*

general treatment of equilibria in a separatory funnel, or in a series of separatory funnels.

First, how does one *do* an extraction? The tool used for simple extractions is called a separatory funnel, and is shown in Figure 6.1. The teardrop shape allows good control over the outflow from the funnel. Teflon plugs and stopcocks are most handy to use with solvents which dissolve the ordinary greases used with glass stopcocks. Let us take an example of an extraction of iodine from water into a denser solvent, carbon tetrachloride, (CCl_4, density = 1.595 grams/milliliter). The procedure is shown in Figure 6.2. After the CCl_4 is put into the funnel, the two phases are equilibrated by agitation. The I_2 dissolves more readily in CCl_4 than in water. The funnel must next be vented, for the heat of solution has vaporized a little of the organic layer. The stopcock is opened with the funnel inverted for venting. Next, the funnel is placed upright, and the iodine-rich layer of CCl_4 drained off. (Note to the student: it is not a smart idea to try this yourself. Carbon tetrachloride can have some nasty physiological effects.) Thus the extraction is completed.

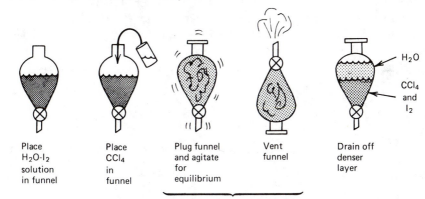

Figure 6.2 *Extraction of I_2 into CCl_4.*

Figure 6.3 *Distribution visualized.*

There are some simple equilibrium laws governing extractions. Suppose that a solute, X, is distributed between two liquid phases, 1 and 2, as shown in Figure 6.3. The concentration of solute X in layer 1 is related to its concentration in layer 2 by a constant called the distribution coefficient, K.

$$K = \frac{[X]_1}{[X]_2}$$ (6.1)

           = distribution coefficient

$[X]_1$ = concentration of solute in solvent 1

$[X]_2$ = concentration of solute in solvent 2

This simple notion of a constant that governs liquid-liquid extractions was developed late in the nineteenth century. It allows us to understand how to control the degree of separation, just as manipulations of chemical concentrations allow us to control the solubility of precipitates.

Common sense suggests that more than one treatment of the layer to be extracted, or *raffinate* with the extracting solvent, the *extractant*, will yield a more complete removal of the solute from the raffinate. Common sense is backed up by common algebra. Suppose that solute X is to be extracted from water (solvent 2) into chloroform (solvent₁), denser than water.*

$$K = \frac{[X]_1}{[X]_2}$$ (6.1)

$[X]_1$ = concentration of X in $CHCl_3$, appropriate units

$[X]_2$ = concentration of X in H_2O, appropriate units

W_0 = grams X in water before extraction

W_1 = grams X left in water after one extraction

---

* *Chloroform, $CHCl_3$ can only be used with great caution.*

W_2 = grams X left in water after two extractions

W_3 = grams X left in water after three extractions

$\cdot$ = $\cdot$

$\cdot$ = $\cdot$

$\cdot$ = $\cdot$

W_n = grams X left in water after n extractions

n = number of extractions

V_2 = volume of water

V_1 = volume of $CHCl_3$

After equilibration of H_2O and $CHCl_3$ layers:

$$W_0 = [X]_1 V_1 + [X]_2 V_2$$

$$[X]_1 = K[X]_2$$

$$W_0 = K[X]_2 V_1 + [X]_2 V_2 = (KV_1 + V_2)[X]_2$$

After the $CHCl_3$ is removed, W_1 grams of X are left in the water layer:

$$W_1 = [X]_2 V_2$$

$$[X]_2 = \frac{W_0}{KV_1 + V_2}$$

$$W_1 = W_0\left(\frac{V_2}{KV_1 + V_2}\right) \tag{6.2}$$

If volume V_1, of fresh $CHCl_3$ is added to the funnel:

$$W_1 = [X]_1' V_1 + [X]_2' V_2$$

$[X]_1'$ = new concentration of X in solvent 1 ($CHCl_3$)

$[X]_2'$ = new concentration of X in solvent 2 (H_2O)

$$K = \frac{[X]_1'}{[X]_2'}$$

$$[X]_1' = K[X]_2'$$

$$W_1 = K[X]_2' V_1 + [X]_2' V_2$$

$$W_1 = (KV_1 + V_2)[X]_2'$$

$$[X]_2' = \frac{W_1}{(KV_1 + V_2)}$$

If the CHCl$_3$ layer is removed, W_2 grams of X are left:

$$W_2 = [X]_2'V_2 = \frac{W_1}{(KV_1 + V_2)} V_2$$

Substitute for W_1 from Equation 6.2

$$W_2 = W_0 \left(\frac{V_2}{KV_1 + V_2}\right)^2 \tag{6.3}$$

Finally, after n extractions, the amount of solute left in the water phase is W_n.

$$W_n = W_0 \left(\frac{V_2}{KV_1 + V_2}\right)^n \tag{6.4}$$

The expression just derived, Equation 6.4, can be put to work.

EXAMPLE:

Suppose that 2.00 grams of solute X are to be extracted from 50.0 milliliters of water (solvent 2) into CHCl$_3$ (solvent 1).

$$K = \frac{[X]_1}{[X]_2} = 3.00$$

If 150.0 milliliters of clean chloroform were available, would it be better, that is, give a more complete extraction, to extract the water layer once with 150.0 milliliters of CHCl$_3$ or three times with 50.0 milliliter portions of CHCl$_3$?

ANSWER:

W_n is the amount of X left in the water layer after n extractions. For the most complete extraction, W_n should be minimized.

First case: one extraction. $V_1 = 150.00$ milliliters

$$V_2 = 50.0 \text{ milliliters}$$

$$W_1 = W_0 \left(\frac{V_2}{KV_1 + V_2}\right)^1 \tag{6.2}$$

$W_0 = 2.00$ grams

$$W_1 = 2.00 \left(\frac{50.0}{(3)(150.0) + 50.0}\right)^1 = 2.00 \left(\frac{50.0}{500.0}\right)$$

$$= 0.200 \text{ grams}$$

Second case: three extractions. V_1 = 50.0 milliliters

V_2 = 50.0 milliliters

$$W_3 = W_0\left(\frac{V_2}{KV_1 + V_2}\right)^3 = 2.00\left(\frac{50.0}{(3)(50.0) + 50.0}\right)^3$$ (6.4)

= 0.0312 grams

Three little extractions give better results than just one big one. This concept is shown graphically in Figure 6.4, which depicts W_n as a function of number of extractions, n, for varying volumes, V_1 of extractant. The data from which the graphs are made are arrayed in Table 6.1. Note that the graphs are linear, plotted on semilog paper. A little algebra will show the reason for the linearity.

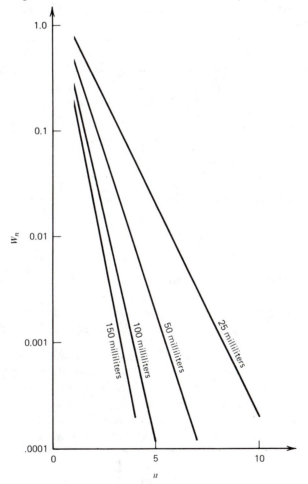

Figure 6.4 W_n as f(n) *for varying* V_1-*values.*

Table 6.1
$V_2 = 50.0$ ml $W = 2.00$ g $K = 3.00$

$V_1 = 25.0$ ml n	W_n	$V_1 = 50.0$ ml n	W_n	$V_1 = 100.0$ ml n	W_n	$V_1 = 150.0$ ml n	W_n
1	0.800	1	0.500	1	0.286	1	0.200
2	0.320	2	0.125	2	0.0408	2	0.0200
3	0.128	3	0.0312	3	0.00583	3	0.00200
4	0.0512	4	0.00781	4	0.000833	4	0.000200
5	0.0205	5	0.00195	5	0.000119	5	0.0000200
6	0.00819	6	0.000488	6		6	
7	0.00328	7	0.000122	7		7	
8	0.00131	8	0.0000305	8		8	
9	0.000524	9					
10	0.000210						

$$W_n = W_0 \left(\frac{V_2}{KV_1 + V_2} \right)^n$$

$$\log W_n = \log W_0 + \log \left(\frac{V_2}{KV_1 + V_2} \right)^n$$

$$= \log W_0 + n \left(\log \frac{V_2}{KV_1 + V_2} \right)$$

$$\log W_n = \left[\log \left(\frac{V_2}{KV_1 + V_2} \right) \right] \cdot n \qquad + \log W_0$$

$$\updownarrow \qquad\qquad \updownarrow \qquad\qquad \updownarrow \qquad\qquad \updownarrow$$

$$y \quad = \qquad (m) \qquad (x) \quad + b$$

$$\updownarrow \qquad\qquad \updownarrow \qquad\qquad \updownarrow \qquad\qquad \updownarrow$$

(dependent = (slope)   (independent + intercept
variable)   variable)

Thus, a plot of $\log W_n$ versus n will yield a straight line.

The picture painted here is a very simple one. Its simplicity depends upon the species X's being present in each phase as X only—not as a dimer, X_2, nor as a protonated species, HX^+, nor as anything but X. We are all familiar enough with chemistry in aqueous systems to realize that the simple picture may not show all the details accurately in certain cases. As an example, take the case of the symbolic weak acid HA. Its distribution between diethyl ether (solvent 1) and water (solvent 2) is given as:

$$K = \frac{[HA]_1}{[HA]_2}$$ (6.1)

The total concentration of all the A-bearing species in the nonpolar ether layer, C_{A_1} is equal to the equilibrium concentration of HA in the ether, $[HA]_1$.

$$\boxed{C_{A_1} = [HA]_1}$$

In the water layer, however, the total concentration of all the A-bearing species, C_{A_2}, is equal to the *sum* of the equilibrium concentrations $[HA]_2$ and $[A^-]_2$.

$$\boxed{C_{A_2} = [HA]_2 + [A^-]_2}$$

It makes sense to define a *conditional* distribution coefficient, K'.

$$\boxed{\begin{aligned} K' &= \frac{C_{A_1}}{C_{A_2}} \\[1em] K' &= \frac{[HA]_1}{[HA]_2 + [A^-]_2} \end{aligned}} \tag{6.5}$$

How is K' related to K mathematically?

$$\boxed{\begin{aligned} &\text{For HA in } H_2O \\[0.5em] &\quad HA_{(2)} \rightleftharpoons H^+_{(2)} + A^-_{(2)} \\[0.5em] &\qquad K_a = \frac{[H^+]_2[A^-]_2}{[HA]_2} \\[0.5em] &\text{From Chapter 2:} \\[0.5em] &\qquad \alpha_0 = \frac{[HA]_2}{[HA]_2 + [A^-]_2} = \frac{[H^+]_2}{[H^+]_2 + K_a} \\[0.5em] &\quad [HA]_2 + [A^-]_2 = \frac{[HA]_2}{\alpha_0} \\[0.5em] &\text{Substitute into Equation 6.5} \\[0.5em] &\qquad K' = \frac{[HA]_1}{[HA]_2/\alpha_0} = \alpha_0 \frac{[HA]_1}{[HA]_2} \\[0.5em] &\qquad K' = \alpha_0 K \end{aligned}} \tag{6.6}$$

The conditional distribution coefficient, K', varies directly with α_0. As α_0 approaches unity, that is, at lower pH values, K' also approaches K. At these low pH values, most of the A$^-$ bearing species are present as HA. At higher and higher pH values, K' shrinks. At higher pH, the equilibrium value of HA concentration in the water decreases, so

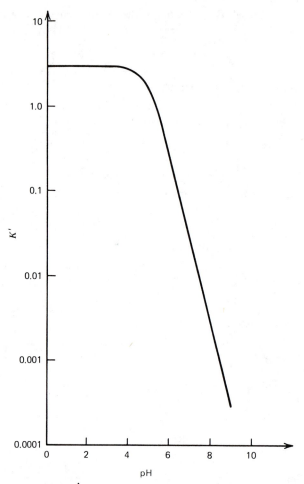

Figure 6.5 *Variation of* K' *as a function of pH.*

Table 6.2
VARIATION OF K' WITH pH ($K = 3.00, K_a = 1.00 \times 10^{-5}$)

pH	α_0	K'	pH	α_0	K'
0	1.00	3.00	6	0.0909	0.273
1	1.00	3.00	7	0.00990	0.0297
2	1.00	3.00	8	0.000999	0.00300
3	0.990	2.97	9	0.000100	0.000300
4	0.909	2.73	10	0.0000100	0.0000300
5	0.500	1.50			

that the equilibrium concentration of HA in the extracting solvent must also decrease. Table 6.2 shows the variation of K' with pH for values of $K = 3.00$ and $K_a = 1.00 \times 10^{-5}$. The data of Table 6.2 are plotted in Figure 6.5.

Sometimes it is desired to separate two species by a process of extraction. Then, the ratios of the distribution constants become important to the experimenter. Suppose that the two species, A and B, are dissolved in aqueous solution. Suppose, moreover, that it is desired to purify A by extracting it into an organic liquid, leaving 99.9% of the B in the aqueous layer. Let the organic layer be designated by subscript (o) and the water layer by subscript (w). The desired value of the distribution constant, K_B, if there are no chemical complications, is shown as follows.

$$K_B = \frac{[B]_o}{[B]_w} = \frac{0.1}{99.9} \cong 10^{-3}$$

It is desired to extract 99.9% of species A into the organic layer. If no chemical complications arise, the value of K_A ought to be as calculated below.

$$K_A = \frac{[A]_o}{[A]_w} = \frac{99.1}{0.1} \cong 10^{+3}$$

The ratio of distribution constants is easily figured.

$$\frac{K_A}{K_B} = \frac{10^{+3}}{10^{-3}} = 10^{+6}$$

Does a ratio of 10^6 guarantee that only 0.1% of species B will be carried over into the organic layer? No. To prove this point, consider a second organic solvent, s.

$$K_{A_s} = \frac{[A]_s}{[A]_w} = 10^4 \cong \frac{99.99}{0.01}$$

$$K_{B_s} = \frac{[B]_s}{[B]_w} = 10^{-2} \cong \frac{1}{99}$$

$$\frac{K_{A_s}}{K_{B_s}} = \frac{10^4}{10^{-2}} = 10^{+6}$$

This ratio is the same as K_A/K_B. Yet, when organic solvent s is used, the A in the organic layer is contaminated with 1 part of B to 99.99 parts of A. In the earlier solvent, the A in the organic layer was only contaminated with 0.1 part of B to 99.9 parts of A. Ratios are useful, but they don't tell the whole story.

Despite all of the seeming complications in theory, extraction is highly important in analytical practice. This is especially true in trace metal analysis. Often, metals are found in natural waters at part-per-million and sub-part-per-million levels. Concentration of trace metals into organic solvents is possible through extraction techniques, as is separation of metals through extraction. How can metal ions—*charged* species floating about in a congenially polar solvent, water—be extracted into highly nonpolar organic solvents? The metals may be combined with organic species to form neutral molecules, which are easily extracted into organic solvents, or the ions may be complexed with neutral ligands, and extracted into organic solvents. Often, the complex formed absorbs light quite strongly in the visible region, allowing spectrophotometric analysis.

Some of the species which form neutral molecules with metal ions are familiar organic precipitating reagents. Most are capable of behaving as weak acids. Their compounds with metal ions are extractable into organic solvents. Some examples are given in Table 6.3.

The metal dithizonates, oxinates, and cupferrates just mentioned are all neutral species. Other metal complexes exist as charged species, and are extracted into organic solvents as ion-association complexes. An example is the ferrous bathophenanthroline complex.

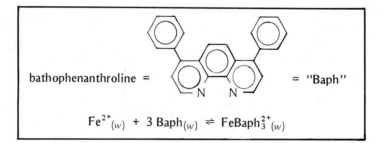

$$Fe^{2+}_{(w)} + 3\,Baph_{(w)} \rightleftharpoons FeBaph_3^{2+}_{(w)}$$

If A^- represents a monovalent anion, it is the ion-association complex, $FeBaph_3A_2$ that is extracted into the organic layer.

$$FeBaph_3^{2+}_{(w)} + 2A^-_{(w)} \rightleftharpoons FeBaph_3A_{2(o)}$$

In the case of the iron-bathophenanthroline complex, the conditional distribution constant, K' is awfully high when the complex is extracted into isoamyl alcohol.

$$K' = \frac{C_{Fe(o)}}{C_{Fe(w)}}$$

Table 6.3
A FEW EXAMPLES OF COMPLEXATION

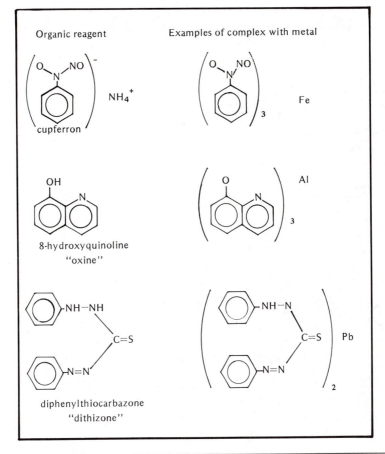

Organic reagent   Examples of complex with metal

cupferron NH_4^+ Fe

8-hydroxyquinoline   Al
"oxine"

diphenylthiocarbazone   Pb
"dithizone"

Said the manufacturer of bathophenanthroline: "The distribution coefficient of the red, bathophenanthroline [complex] between [this] alcohol and water could not readily be determined because of the great solubility of the [complex] in [this] alcohol."[1] Other examples of ion-association complexes are given in Table 6.4.

The list could go on and on. The third edition of Sandell's work* is a massive compendium of information on extraction. Let us now turn our attention to problem solving.

* E. B. Sandell, The Colorimetric Determination of Traces of Metals, 3rd ed., New York Interscience, 1959.

Table 6.4
SOME EXAMPLES OF ION-ASSOCIATION COMPLEXES

Ligand	Complex	Organic Phase	K'	Footnote
2,2'-Biquinoline "Cuproine"	CuL_2A	Isoamyl alcohol $CH_3\!\!>\!\!CH-CH_2-CH_2-OH$ CH_3	1680	3
2,9-Dimethyl 1,10-phenanthroline "Neocuproine"	CuL_2A	Chloroform $CHCl_3$	1200	3
1,10-phenanthroline or ortho-phenanthroline	ML_nA_m for $Zn^{2+}, Cd^{2+},$ $Pb^{2+}, Fe^{2+},$ @ pH 7	Nitrobenzene NO_2	$\cong 10^5$	1, 2

[1] H. Diehl, G. F. Smith, L. McBride, and R. Cryberg, The Iron Reagents, 2nd ed., The G. F. Smith Chemical Company, Columbus, Ohio, 1965.

[2] A. A. Schilt, R. L. Abraham, and J. E. Martin, Anal. Chem. 45, 1808 (1973).

[3] H. Diehl and G. F. Smith, The Copper Reagents, The G. F. Smith Chemical Company, Columbus, Ohio, 1958.

≡ 6.3 PROBLEMS ≡

1. An organic compound, called marmaladone, or Mar for short, is isolated from orange peels by an imaginary chemist. The distribution coefficient of Mar between diethyl ether (solvent 1) and water (solvent 2) is 2.00.

$$K = \frac{[Mar]_1}{[Mar]_2} = 2.00$$

Nothing complicated appears to trouble the extraction—no pH dependence, no dimerization. If 50.0 milliliters of a water layer contain 0.200 gram of marmaladone, how much marmaladone will be left in the aqueous layer after (a) one extraction with 50.0 milliliters of ether, (b) two extractions with 50.0 milliliters of ether, or (c) three extractions with 50.0 milliliters ether? (d) Construct a plot that will enable you to predict how much Mar will be left after four extractions with 50.0 milliliters of ether each.

Ans: *(a) 0.0667 gram; (b) 0.0222 gram; (c) 0.00741 gram; (d) See detailed solutions.*

2. Suppose you had a huge separatory funnel, containing 300. milliliters of water in which is dissolved 1.200 grams of marmaladone. How much Mar would be left in the aqueous layer after (a) one extraction with 300. milliliters of ether, (b) three extractions with 100. milliliters of ether each, and (c) four extractions with 75.0 milliliters of ether each? Note that a total of 300. milliliters of ether is used in each case $(K = 2.00)$.

Ans: *See detailed solutions.*

3. What is the minimum value of K that will allow the extraction of: (a) 99.0% of a solute from 75.0 milliliters of water with four successive extractions by 75.0 milliliters quantities of ether, (b) 99.9% of another solute from 75.0 milliliters of water with four successive extractions by 75.0 milliliter quantities of ether? Assume that no competing equilibria exist.

Ans: *(a) K = 2.16; (b) K = 4.62*

4. Suppose that you have three solutes, A, B, and C. Their distribution coefficients between water and ether are:

(a) for A, $K = \dfrac{[A]_o}{[A]_w} = 1.00$

(b) for B, $K = \dfrac{[B]_o}{[B]_w} = 10.0$

(c) for C, $K = \dfrac{[C]_o}{[C]_w} = 100.$

What is the minimum volume of ether required to remove, in *one* extraction:

(a) 9.0 grams of A from 100.0 milliliters of a water solution originally containing 10.0 grams of A?

(b) 9.0 grams of B from 100.0 milliliters of a water solution originally containing 10.0 grams of B?

(c) 9.0 grams of C from 100.0 milliliters of a water solution originally containing 10.0 grams of C?

Assume that no competing equilibria exist.

Ans: *(a) 9.0 X 10² milliliters; (b) 90. milliliters; (c) 9.0 milliliters*

5. After five extractions with a 100. milliliter quantity of $CHCl_3$ per extraction, 90% of the marmaladone is extracted from a water solution. What percentage of the marmaladone will be extracted by ten extractions with 100. milliliters of $CHCl_3$ per extraction? Marmaladone, like so many ideal solutes, neither dimerizes in chloroform nor dissociates in water.

Ans: *99%*

6. Pyridine, or (structure), is a weak base. In water:

(structure) + H_2O ⇌ (structure) + OH^- or

$$Py + H_2O \rightleftharpoons HPy^+ + OH^-$$

$$K_b = \frac{[HPy^+]_w\,[OH^-]}{[Py]_w} = 3.16 \times 10^{-6}$$

The distribution coefficient for pyridine between water and chloroform is

$$K = \frac{[Py]_o}{[Py]_w} = 2.75 \times 10^4$$

At pH 4.00, the conditional distribution coefficient for pyridine between the aqueous and chloroform layers is 0.87 (data of Golumbic and Orchin*).

$$K' = \frac{[Py]_o}{C_{Py_w}} = 0.87 \qquad \alpha_o = \frac{[Py]}{[HPy^+] + [Py]}$$

$$C_{Py_w} = [Py]_w + [HPy^+]_w$$

The difference between K and K' is very large. Account for it by deriving a relationship among K', K, K_b, and α_o. Assume no dimerization of the pyridine in the chloroform layer.

Ans: $K' = K\alpha_o$

7. Substituted pyridines, like pyridine, smell sickening and distribute themselves between aqueous layers and chloroform. One such substituted pyridine is 2,6-lutidine, CH_3–(structure)–CH_3. It acts as a weak base in water: CH_3–(structure)–CH_3 +

$H_2O \rightleftharpoons CH_3$–(structure)–$CH_3 + OH^-$ or

$$Lut + H_2O \rightleftharpoons HLut^+ + OH^-.$$

$$K_b = \frac{[HLut^+]_w\,[OH^-]}{[Lut]_w} = 1.26 \times 10^{-7}$$

Golumbic and Orchin* determined a series of values of the conditional distribution coefficient, K' for the distribution of 2,6-lutidine between water and chloroform as a function of pH of the aqueous layer.

* Data reprinted with permission from H. Golumbic and M. Orchin, J. Am. Chem. Soc. 72, 4145 (1950). Copyright by the American Chemical Society.

$$K' = \frac{[\text{Lut}]_o}{C_{\text{Lut}_w}}$$

$$C_{\text{Lut}_w} = [\text{Lut}]_w + [\text{HLut}^+]_w$$

2,6-Lutidine does not dimerize in chloroform. Here are their data. From these data, determine K, where $K = \dfrac{[\text{Lut}]_o}{[\text{Lut}]_w}$

pH	log K'
3.20	−1.30
3.60	−0.92
4.00	−0.56

Ans: **3.5 × 10²**

8. Stary and Smizanska[*] studied solvent extraction of metal cupferrates. Figure 6.6 shows % extraction from the aqueous layer as a function of pH. Given this information how would you separate the Hg^{2+}, Pb^{2+}, and Zn^{2+} from each other if you started with 50 milliliters of an aqueous solution 0.01M in each? Use devastatingly simple assumptions.

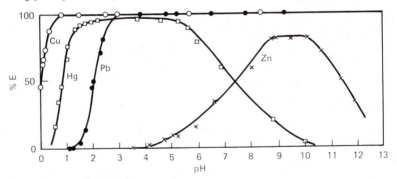

Figure 6.6 Data on cupferrate extraction.

Ans: *4 extractions @ pH 1 with 50 milliliters organic separate Hg^{2+}. Raise pH to 3, and extract once to get the lead out. Zn^{2+} will be left in the aqueous layer.*

9. Cu^+ can be extracted into either isoamyl alcohol *or* chloroform as an ion-association complex, if neo-cuproine is used as the ligand.[†]

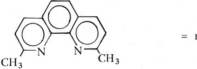

= neocuproine = L

* *Reprinted by permission from J. Stary and J. Smizanska,* Anal. Chim. Acta. 29, 545 *(1963).*

† H. Diehl and G. F. Smith, **The Copper Reagents**, *The G. F. Smith Chemical Company, Columbus, Ohio,* 1958.

$$Cu^+_{(w)} + 2L_{(w)} \rightleftharpoons CuL_2^+{}_{(w)}$$

$$A^- = \text{anion}$$

$$CuL_2^+{}_{(w)} + A^-_{(w)} \rightleftharpoons (CuL_2^+ A^-)_{(o)}$$

The value of K' is about 1500 for extraction into the alcohol, and 1200 for extraction into the chloroform. Why, then, do many workers prefer to use chloroform, despite its toxicity?

Ans: *CHCl$_3$ is denser than H$_2$O.*

10. Suppose that you read an experimental writeup for the extraction of copper as the neocuproine complex into chloroform. It instructs you to extract 20 milliliters of the aqueous solution with one 10 milliliter batch of chloroform and two 5 milliliter batches after that. If K' is 1200 or so, why bother with a second and third extraction?

Ans: *See detailed solutions.*

≡ 6.4 DETAILED SOLUTIONS TO PROBLEMS ≡

1. Remember $W_n = W_0 \left(\dfrac{V_2}{KV_1 + V_2} \right)^n$ $\qquad$ (6.4)

$$W_0 = 0.200 \text{ gram}$$
$$V_2 = \text{volume of } H_2O = 50.0 \text{ milliliters}$$
$$V_1 = \text{volume of ether} = 50.0 \text{ milliliters}$$
$$K = 2.00$$

(a) After one extraction:

$$W_1 = W_0 \left(\frac{V_2}{KV_1 + V_2} \right)^1 = (0.200) \left(\frac{50.0}{(2)(50.0) + 50.0} \right)^1 = 0.0667 \text{ gram}$$

(b) After two extractions:

$$W_2 = W_0 \left(\frac{V_2}{KV_1 + V_2} \right)^2 = (0.200) \left(\frac{50.0}{(2)(50.0) + 50.0} \right)^2 = 0.0222 \text{ gram}$$

(c) After three extractions:

$$W_3 = W_0 \left(\frac{V_2}{KV_1 + V_2} \right)^3 = (0.200) \left(\frac{50.0}{(2)(50.0) + 50.0} \right)^3 = 0.00741 \text{ gram}$$

(d) Remember the relationship derived between $\log W_n$ and n.

$$\log W_n = \left[\log \left(\frac{V_2}{KV_1 + V_2} \right) \right] n + \log W_0$$

Plot $\log W_n$ versus n, as shown in Figure 6.7. From the graph, $W_4 = 0.00247$ gram. Note the usefulness of semilog paper.

2. Remember

$$W_n = W_0 \left(\frac{V_2}{KV_1 + V_2} \right)^n \qquad (6.4)$$

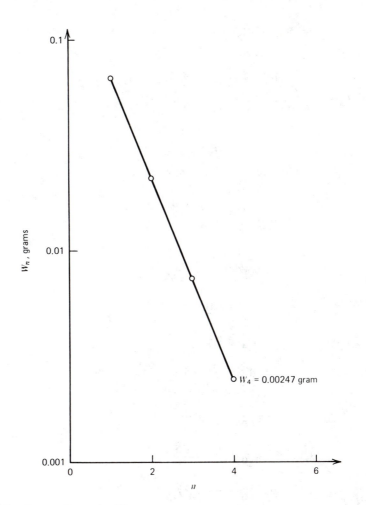

Figure 6.7 *Plot to determine* W_4.

In this problem, V_1, the volume of ether, changes from part (a) to part (b) to part (c).

(a)

$$W_0 = 1.200 \text{ gram}$$

$$V_2 = \text{volume of } H_2O = 300. \text{ milliliters}$$

$$V_1 = \text{volume of ether} = 300. \text{ milliliters}$$

$$K = 2.00$$

$$n = 1$$

$$W_1 = W_0 \left(\frac{V_2}{KV_1 + V_2}\right)^1 = 1.200 \left(\frac{300.}{(2)(300.) + 300.}\right) = 0.400 \text{ gram}$$

(b)

$$W_0 = 1.200 \text{ gram}$$

V_2 = volume of H_2O = 300. milliliters

V_1 = volume of ether = 100. milliliters

$n = 3$

$$W_3 = W_0 \left(\frac{V_2}{KV_1 + V_2} \right)^3 = 1.200 \left(\frac{300}{(2)(100) + 300} \right)^3 = 0.259 \text{ gram}$$

(c) W_0 = 1.200 gram

V_2 = volume of H_2O = 300. milliliters

V_1 = volume of ether = 75.0 milliliters

$n = 4$

$$W_4 = 1.200 \left(\frac{300.}{(2)(75.0) + 300} \right)^4 = 0.237 \text{ gram}$$

3. Remember

$$W_n = W_0 \left(\frac{V_2}{KV_1 + V_2} \right)^n \tag{6.4}$$

(a) $\dfrac{W_4}{W_0} = \dfrac{1.00}{100.0}$ For extraction of 99.0%.

$$W_4 = W_0 \left(\frac{V_2}{KV_1 + V_2} \right)^4$$

$$\frac{W_4}{W_0} = \left(\frac{V_2}{KV_1 + V_2} \right)^4$$

$$\frac{1.00}{100.0} = \left(\frac{75.0}{75.0 + 75.0} \right)^4$$

$$\log \frac{1.00}{100.0} = \log \left(\frac{75}{75K + 75} \right)^4$$

$$\log \frac{1}{100} = 4 \log \left(\frac{75}{75K + 75} \right)$$

$$\frac{1}{4} \log \frac{1.00}{100.0} = -\log 75.0 - \log (75.0K + 75.0)$$

$$+2.375 = +\log (75K + 75)$$

$$237.2 = 75K + 75$$

$$K = 2.16$$

(b)

$$\frac{W_4}{W_0} = \frac{0.100}{100.0}$$

$$\frac{W_4}{W_0} = \left(\frac{75.0}{75.0K + 75.0} \right)^4$$

$$\frac{0.100}{100.0} = \left(\frac{75.0}{75.0K + 75.0} \right)^4$$

$$\log \frac{0.100}{100.0} = \log\left(\frac{75.0}{75.0K + 75.0}\right)^4$$

$$\log \frac{0.100}{100.0} = 4 \log \left(\frac{75.0}{75.0K + 75.0}\right)$$

$$\frac{1}{4} \log \frac{0.100}{100.0} = \log\left(\frac{75.0}{75.0K + 75.0}\right)$$

$$-2.625 = -\log (75.0K + 75.0)$$

$$2.625 = \log (75.0K + 75.0)$$

$$421.7 = 75.0K + 75.0$$

$$K = 4.62$$

4.

$$W_n = W_0 \left(\frac{V_2}{KV_1 + V_2}\right)^n \qquad (6.4)$$

$$W_1 = W_0 \left(\frac{V_2}{KV_1 + V_2}\right)$$

$$\frac{W_1}{W_0} = \left(\frac{V_2}{KV_1 + V_2}\right)$$

$$\frac{W_1}{W_0} (KV_1 + V_2) = V_2$$

$$\frac{W_1}{W_0} KV_1 + \frac{W_1}{W_0} V_2 = V_2$$

$$\frac{W_1}{W_0} KV_1 = V_2 - \frac{W_1}{W_0} V_2$$

$$\frac{W_1}{W_0} KV_1 = V_2 \left(1 - \frac{W_1}{W_0}\right)$$

$$V_1 = \frac{V_2 W_0}{K W_1} \left(1 - \frac{W_1}{W_0}\right)$$

(a)

$W_0 = 10.0$ grams

$W_1 = 1.0$ gram

$V_2 = $ volume of H_2O $= 100.0$ milliliters

$V_1 = $ volume of ether $= ?$

$K = 1.00$

$$V_1 = \left(\frac{100.0}{1.00}\right) \left(\frac{10.0}{1.0}\right) \left(1 - \frac{1.0}{10.0}\right) = 900 \text{ milliliters}$$

(b)

$W_0 = 10.0$ grams

$W_1 = 1.0$ gram

$V_2 = $ volume of H_2O $= 100$ milliliters

$V_1 = $ volume of ether $= ?$

$$K = 10.0$$

$$V_1 = \left(\frac{100.0}{10.00}\right)\left(\frac{10.0}{1.0}\right)\left(1 - \frac{1.0}{10.0}\right) = 90. \text{ milliliters}$$

(c) $W_0 = 10.0$ grams

$W_0 = 1.0$ gram

V_2 = volume of H_2O = 100.0 milliliters

V_1 = volume of ether = ?

$K = 100.$

$$V_1 = \left(\frac{100.0}{100.}\right)\left(\frac{10.0}{1.0}\right)\left(1 - \frac{1.0}{10.0}\right) = 9.0 \text{ milliliters}$$

5.
$$W_n = W_0 \left(\frac{V_2}{KV_1 + V_2}\right)^n \tag{6.4}$$

$$W_5 = W_0 \left(\frac{V_2}{100K + V_2}\right)^5$$

$$\frac{W_5}{W_0} = \frac{10}{100}$$

$$\frac{W_5}{W_0} = \left(\frac{V_2}{100.K + V_2}\right)^5$$

$$\frac{10}{100} = \left(\frac{V_2}{100.K + V_2}\right)^5$$

$$W_{10} = W_0 \left(\frac{V_2}{100.K + V_2}\right)^{10}$$

$$\frac{W_{10}}{W_0} = \left(\frac{V_2}{100K + V_2}\right)^{10}$$

$$\frac{W_{10}}{W_0} = \left[\left(\frac{V_2}{100K + V_2}\right)^5\right]^2$$

Substitute

$$\frac{W_{10}}{W_0} = \left(\frac{10}{100}\right)^2$$

$$\frac{W_{10}}{W_0} = 0.010$$

$$W_{10} = 0.010 \, W_0$$

Thus, 0.99 or 99% of the marmaladone is extracted into the chloroform layer with 10 extractions.

6.
$$K = \frac{[Py]_o}{[Py]_w} = 2.75 \times 10^4$$

$$K' = \frac{[Py]_o}{C_{Py_w}} = 0.87 \text{ at pH } 4.00$$

For Py in H_2O

$$Py_{(w)} + H_2O \rightleftharpoons HPy^+_{(w)} + OH^-$$

$$K_b = \frac{[HPy^+]_w [OH^-]}{[Py]_w} = 3.16 \times 10^{-6}$$

Define α_0 for pyridine:

$$\alpha_0 = \frac{[Py]_w}{[Py]_w + [HPy^+]_w}$$

$$[HPy^+]_w = \frac{K_b}{[OH^-]} [Py]_w$$

$$\alpha_0 = \frac{[Py]_w}{[Py]_w + \frac{K_b}{[OH^-]} [Py]_w}$$

$$\alpha_0 = \frac{[OH^-]}{[OH^-] + K_b}$$

α_0 can be calculated from $[OH^-]$ and K_b.

Remember:

$$K' = \frac{[Py]_o}{[Py]_w + [HPy^+]_w}$$

$$\alpha_0 = \frac{[Py]_w}{[Py]_w + [HPy^+]_w}$$

$$[Py]_w + [HPy^+]_w = \frac{[Py]_w}{\alpha_0}$$

$$K' = \frac{[Py]_o}{\frac{[Py]_w}{\alpha_0}} = \frac{[Py]_o}{[Py]_w} \alpha_0$$

$$K = \frac{[Py]_o}{[Py]_w}$$

$$K' = K\alpha_0$$

at pH 4.00:

$$[H^+] = 1.0 \times 10^{-4} M$$

$$[OH^-] = \frac{K_w}{[H^+]} = \frac{(1.0 \times 10^{-14})}{(1.0 \times 10^{-4})} = 1.0 \times 10^{-10} M$$

$$\alpha_0 = \frac{[OH^-]}{[OH^-] + K_b} = \frac{1.0 \times 10^{-10}}{1.0 \times 10^{-10} + 3.16 \times 10^{-6}} = 3.2 \times 10^{-5}$$

$$K' = K\alpha_0 = (2.75 \times 10^4)(3.2 \times 10^{-5}) = 0.87$$

The derived value of K' is consistent with the value stated in the problem.

7. Just as one can write a relationship for pyridine, one can write a relationship among K, K', and α_0 for 2,6-lutidine.

$$K' = K\alpha_0$$

$$\log K' = \log (K\alpha_0)$$

$$\log K' = \log \alpha_0 + \log K$$

$$\alpha_0 = \frac{[OH^-]}{[OH^-] + K_b}$$

pH	pOH	$[OH^-]$, M	α_0	Log α_0	Log K'
3.20	10.80	1.58×10^{-11}	1.25×10^{-4}	−3.90	−1.30
3.60	10.40	3.98×10^{-11}	3.16×10^{-4}	−3.50	−0.92
4.00	10.00	1.00×10^{-10}	7.94×10^{-4}	−3.10	−0.56

$$\log K' = (\log \alpha_0) + \log K$$

$$\downarrow \qquad \downarrow \qquad \downarrow$$

$$y = m(x) + b$$

where $\log \alpha_0 = 0$ (intercept)

$$\log K' = \log K$$

A plot of $\log K'$ versus $\log \alpha_0$ is made. (Figure 6.8)

where $\log \alpha_0 = 0$

$$\log K' = 2.55$$

$$\log K = 2.55$$

$$K = 3.5 \times 10^2$$

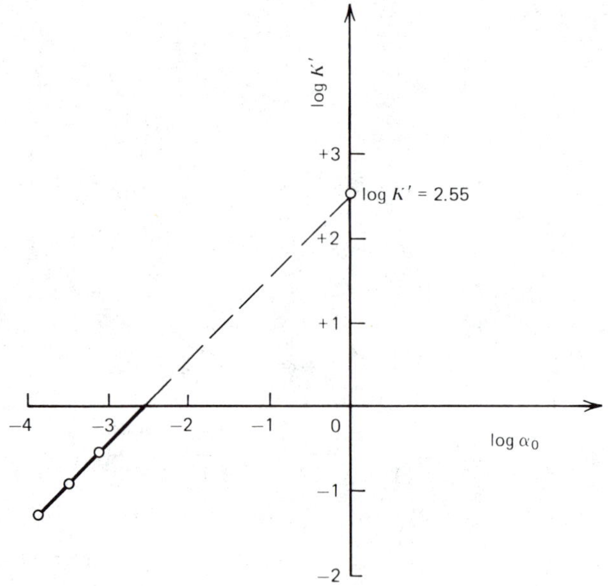

Figure 6.8 *Plot to determine* K' *for*

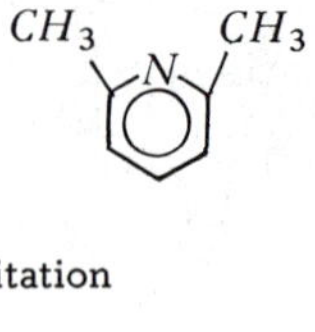

8. The best way to do the separation is to extract the Hg at low pH, leaving Pb and Zn behind in the aqueous layer. The pH may then be raised to pH 3 or so, and the Pb extracted into chloroform, leaving Zn behind in the aqueous layer.

The extraction of Hg is not complete at pH 1, but extraction at a slightly higher pH would contaminate the chloroform layer with some Pb. At pH 1, a value of K'_{Hg} may be estimated from the graph. At pH 1, % extracted = 77.

$$K'_{Hg} = \frac{C_{Hg_O}}{C_{Hg_w}} = \frac{77}{100 - 77}$$

$$K'_{Hg} = 3.3$$

How many extractions are needed to get 99% of the Hg into the chloroform layer? With nothing else to go on, one can hope that the relationship

$$W_n = W_0 \left(\frac{V_2}{KV_1 + V_2} \right)^n \tag{6.4}$$

has some applicability.

$$\frac{W_n}{W_0} = \left(\frac{V_2}{KV_1 + V_2} \right)^n$$

for 99% of Hg to be extracted:

$$W_n = 1$$
$$W_0 = 100$$
$$V_2 = 50 \text{ milliliters}$$
$$V_1 = 50 \text{ milliliters}$$
$$n = ?$$

$$\log \frac{W_n}{W_0} = \log \left(\frac{V_2}{KV_1 + V_2} \right)^n$$

$$\log \frac{W_n}{W_0} = n \log \left(\frac{V_2}{KV_1 + V_2} \right)$$

$$\log \frac{1}{100} = n \log \left(\frac{50}{(3.3)(50) + 50} \right)$$

$$-2 = n \, (-0.63)$$

$$n = 3.15$$

Because $n = 3.15$ is not a possibility in the real world of separatory funnels (just as a family with 2.2 children is unlikely), make $n = 4$. This will—if the relationship is valid—effectively extract the Hg into the chloroform layer. The pH is then raised to 3, and the Pb extracted into $CHCl_3$, leaving Zn behind.

9. Isoamyl alcohol is $\begin{smallmatrix} CH_3 \\ \\ CH_3 \end{smallmatrix}\!\!\!>\!CH-CH_2-CH_2OH$, or 3-methyl-1-butanol. Its density is 0.812 gram/milliliter, and that of chloroform, $CHCl_3$, is 1.498 grams/milliliter. Chloroform, then, sinks to the bottom of a separatory funnel, beneath the

aqueous layer, and may be drained out easily. On the other hand, an isoamyl alcohol layer will be found on *top* of the aqueous layer, and is messy to get out of a separatory funnel. The differences in K'_{Cu} are not big enough to make much difference in the amount of Cu left behind. Why? Suppose $V_1 = V_2$ in the case of extraction by either alcohol or chloroform.

$$W_1 = W_0 \left(\frac{V_2}{KV_1 + V_2} \right)^1$$

$$V_1 = V_2$$

$$\frac{W_1}{W_0} = \frac{1}{K + 1}$$

In $CHCl_3$

$$\frac{W_1}{W_0} = \frac{1}{1200 + 1} = 8.3 \times 10^{-4}$$

In alcohol

$$\frac{W_1}{W_0} = \frac{1}{1500 + 1} = 6.7 \times 10^{-4}$$

The very slight advantage gained by using alcohol is far outweighed by the convenience the chloroform affords. The toxicity of chloroform, however, argues against its use.

10. The second and third extractions are a sort of insurance for (a) complete equilibrium and (b) gathering stray chloroform droplets that might be left behind in the separatory funnel.

≡ 6.5 MULTISTAGE LIQUID-LIQUID EXTRACTIONS ≡

For the extraction of metal complexes into organic solvents, little is needed besides a separatory funnel or two, the chemicals to yield a high value of K' ($\cong 1000$ or so), a steady hand, and a moderately alert brain. Few extractions, usually no more than three or four, are performed on one aqueous layer. Separations, viewed from the trace-metal analyst's point of view, can be very simply done. In biochemistry and in the pharmaceutical industry, matters are very different. Many species that need to be separated from one another have distribution coefficients that are small and close together in almost all available solvent systems. Many of these biologically important species are delicate, and easily decomposed by harsh treatment, for example, heating. Repeated extractions are necessary for separation. It is plain that separatory funnels are awkward vessels for many more than four or five extractions. An ingenious system for multistage liquid-liquid extractions was devised by L. C. Craig, and first described in 1944.[*] Craig, in the original and Williamson and Craig in a subsequent[†] paper, presented the mathematical principles governing separations with the system. The system is often called the Craig countercurrent distribution (abbreviated CCD) or

---

[*] *L. C. Craig*, J. Biol. Chem. 155, *519 (1944)*.

[†] *B. Williamson and L. C. Craig*, J. Biol. Chem. 168, *687 (1947)*.

pseudo-countercurrent distribution, the latter name arising from the batch nature of the process. In this discussion we develop some of the mathematics of the system, give examples of solute distribution and separation, describe the construction of a modern Craig apparatus, and turn, in the problems, to practical applications of the method.

To begin the discussion, assume that we have a number of vessels, each containing the same volume, V_w, of a denser, usually aqueous solvent. No solute has yet been introduced into this system, shown in Figure 6.9. Next, a volume V_o, of a less dense solvent, containing the solute S, is placed in vessel number 0. To make the mathematics easier, suppose that the relationships stated below hold for this extraction system.

$$K = \frac{[S]_o}{[S]_w} = 1$$

$$V_o = V_w$$

After equilibration, which requires much vigorous shaking of vessel 0, the two phases are allowed to separate. Half of the solute S introduced with the less dense phase will remain in the less dense phase, and half will be in the denser phase of vessel 0. This condition is shown in Figure 6.10, on the very top row, where $n = 0$ and $r = 0$. The number of transfers from tube to tube, n, is defined as zero here.

Next, suppose that the lighter layer in vessel 0 is transferred to vessel 1. Then a fresh volume, V_o, of the lighter solvent is placed into vessel 0. Equilibration takes place. The distribution of solute S is shown on the second row of Figure 6.10, where $n = 1$. One-fourth of the solute is found in the top layer of vessel 0, one-fourth is found in the bottom layer of vessel 0, one-fourth is found in the top layer of vessel 1, and one-fourth in the bottom layer of vessel 1.

Next the lighter layer in vessel 1 is transferred to vessel 2. The lighter layer in vessel 0 is transferred to vessel 1. A fresh volume of the lighter layer is placed in vessel 0. Equilibration takes place. The distribution of solute S is shown on the third row of Figure 6.10, where $n = 2$. Distributions of the solute after three, four, five, and six transfers are shown in successive rows.

There seems to be some kind of mathematical order in the distribution of solute S between layers and among tubes. Table 6.5 suggests this order ever more strongly than does Figure 6.10. The coefficients of the powers of 2 in Table 6.5 show some regular-

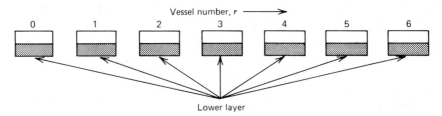

Figure 6.9 *The system* before *either upper layer or solute is added.*

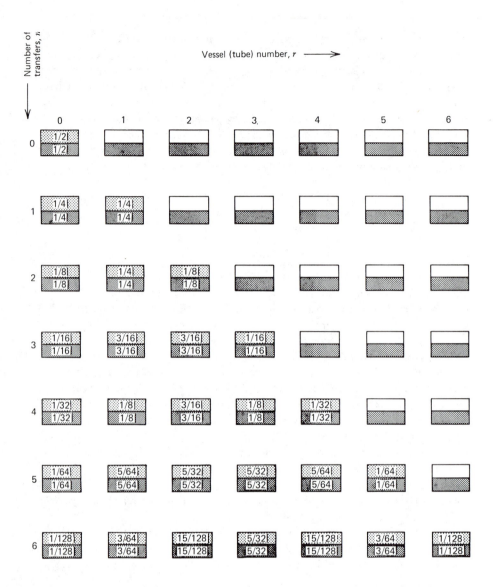

Figure 6.10 *Stages in idealized countercurrent distribution, where* K = 1, *and* V_W = V_0. *Numbers in boxes represent fractions of original amount of solute present in phase.*

ity for values of n from 2 on up. They start with low values, rise, and decline again. In fact, the coefficients in Table 6.5 are just the coefficients of the x and y terms in the binomial expansion of $(x + y)^n$.

Table 6.5
FRACTION OF SOLUTE IN EACH VESSEL (UPPER *AND* LOWER LAYERS) AS A FUNCTION OF VESSEL NUMBER, r AND NUMBER OF TRANSFERS, n.

Number of Transfers, n	Vessel Number, r							
	0	1	2	3	4	5	6	
0	1							$(\times 2^{-0})$
1	1	1						$(\times 2^{-1})$
2	1	2	1					$(\times 2^{-2})$
3	1	3	3	1				$(\times 2^{-3})$
4	1	4	6	4	1			$(\times 2^{-4})$
5	1	5	10	10	5	1		$(\times 2^{-5})$
6	1	6	15	20	15	6	1	$(\times 2^{-6})$

Orderly mathematics can be applied to the Craig process. It is, of course, desirable to generalize the expression for the fraction of solute found in each tube, because the value of K is very seldom exactly unity. Williamson and Craig* showed that the expansion of the binomial

$$\left(\frac{1}{K+1} + \frac{K}{K+1} \right)^n$$

would yield the fraction of solute present in a given tube after a given number of extractions, when volumes of upper and lower layers are equal.

In their general expression, the fraction of solute, $f(n,r)$ present in a given tube, r, after a given number of transfers, n, is:

$$f(n,r) = \left(\frac{n!}{r!\,(n-r)!} \right) \cdot \left(\frac{1}{K+1} \right)^{n-r} \cdot \left(\frac{K}{K+1} \right)^r \qquad † \qquad (6.7)$$

Doing calculations with this expression may be a trifle tiresome, but it is worth working out one or two examples for practice.

EXAMPLE:

A solute, S, has a distribution coefficient of 2.0 between an upper, moving, organic phase and a lower, stationary, aqueous phase.

$$K = \frac{[S]_o}{[S]_w} = 2.0$$

---

* B. Williamson and L. C. Craig, J. Biol. Chem. 168, 167 (1947).
† n! is read as "n factorial" n! = n(n−1)(n−2) . . . 3 · 2 · 1. For example: 5! = 5 · 4 · 3 · 2 · 1. By definition, 1! = 1, 0! = 1.

Calculate the fraction of S in tubes 8, 9, and 10 after 15 transfers.

SOLUTION:

There are three calculations to make:

First: $f(n,r) = f(15,8)$ for $n = 15, r = 8$

Second: $f(n,r) = f(15,9)$ for $n = 15, r = 9$

Third: $f(n,r) = f(15,10)$ for $n = 15, r = 10$

General case:

$$f(n,r) = \frac{n!}{r!(n-r)!} \cdot \left(\frac{1}{K+1}\right)^{n-r} \cdot \left(\frac{K}{K+1}\right)^{r}$$

First: $n = 15, r = 8$

$$f(15,8) = \frac{15!}{8!(15-8)!} \cdot \left(\frac{1}{2+1}\right)^{(15-8)} \cdot \left(\frac{2}{2+1}\right)^{8}$$

$$= 0.1148 \cong 0.11$$

Second: $n = 15, r = 9$

$$f(15,9) = \frac{15!}{9!(15-9)!} \left(\frac{1}{2+1}\right)^{(15-9)} \cdot \left(\frac{1}{2+1}\right)^{9}$$

$$= 0.1786 \cong 0.18$$

Third: $n = 15, r = 10$

$$f(15,10) = \frac{15!}{(10!)(15-10)!} \cdot \left(\frac{1}{2+1}\right)^{(15-10)} \cdot \left(\frac{2}{2+1}\right)^{10}$$

$$= 0.2143 \cong 0.21$$

Computer programs can make these computations easier.

The formidable computational task of evaluating the fraction of solute in each tube from the binomial distribution without benefit of a computer calls for simplification. Williamson and Craig noted that the distribution curve for any substance, for example, a plot of $f(n,r)$ versus r for given values of n and K, looks very Gaussian. Williamson and Craig rewrote the expression for $f(n,r)$ in this way:

Define: $X = \dfrac{K}{K+1}$

$Y = \dfrac{1}{K+1}$

$$f(n,r) = \frac{1}{(2\pi nXY)^{1/2}} \cdot \exp \frac{-(nX-r)^2}{2nXY}$$

           (6.7)

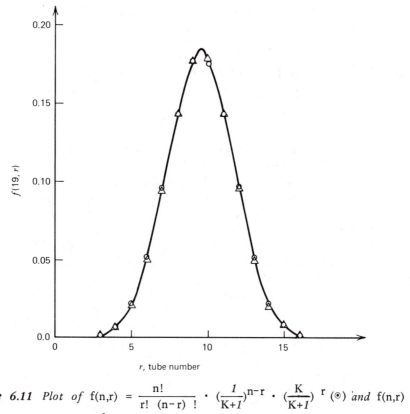

Figure 6.11 Plot of $f(n,r) = \dfrac{n!}{r! \, (n-r)!} \cdot (\dfrac{1}{K+1})^{n-r} \cdot (\dfrac{K}{K+1})^{r}$ (⊙) and $f(n,r) = \dfrac{1}{(2\pi nXY)^{1/2}} \cdot exp \dfrac{-(nX-r)^2}{2nXy}$ (△) versus r for K = 1, n = 19.

Does the Gaussian approximation come close to the binomial reality? In Figure 6.11 below the two sets of data, calculated by binomial expansion and Gaussian approximation, are plotted on the same graph. The Gaussian approximation works even better when $n \geq 25$.

So far, only the distribution of one species throughout the tubes of a Craig apparatus has been considered. How might the method be applied to the separation of two species with K values rather close together? An example will be taken, using those versatile chemical species A and B.

EXAMPLE:

Species A and B are to be separated with a Craig apparatus. The lower phase is aqueous and stationary, and the upper phase is organic and mobile.

$$K_A = 0.5$$
$$K_B = 2.5$$

Draw distribution curves for:

A and B after 5 transfers ($n = 5$), $r = 0$ to 19 (one graph)

A and B after 10 transfers ($n = 10$), $r = 0$ to 19 (another graph)

A and B after 15 transfers ($n = 15$), $r = 0$ to 19 (still another graph)

A and B after 19 transfers ($n = 19$), $r = 0$ to 19 (one more graph)

The distribution curves are shown in Figures 6.12, 6.13, 6.14, and 6.15. From little separation, after 5 transfers (Figure 6.12), the A and B distribution curves move steadily apart as the number of transfers increases. After 19 transfers, good separation is observed (Figure 6.15). Only in tubes 7, 8, 9, 10, 11, 12, and 13 is there significant mixing of A and B. In tubes 1, 2, 3, 4, 5, and 6, nearly pure A is found. In tubes 14, 15, 16, 17, 18, and 19, nearly pure B is found. This separation is possible even when the value of K_B/K_A is 5, and the difference between K_A and K_B is only 2. Countercurrent distribution is indeed a powerful tool for the separation of delicate species, with low, similar values of distribution coefficients.

The early Craig apparatus has been improved upon considerably. Today, hundreds of vessels may be mounted on one rack. Vessels may contain great quantities of liquid; in one procedure. 1-liter vessels were used in the isolation of protogen from *four tons* of mixed beef and pork liver.* Most extractions and distributions are observed on a more modest scale; typically 10 milliliters of lower layer and 10 milliliters of upper layer are the quantities used per vessel.

Figure 6.16 shows one tube of a typical modern Craig apparatus. The tube is shown in two positions. Look first at the left-hand drawing. Chamber 2 is filled roughly half full of the lower layer through opening 1. Chamber 2 of each vessel is filled in like manner, before any of either the lighter, mobile phase, or any of the solute is introduced. Then, a portion of the lighter, mobile phase, containing the solute under study is put into chamber 2 of the vessel through opening 1. The tube is shaken gently about pivot P, sometimes by undergraduate serfs, and sometimes by a motor, to achieve equilibration. Then the tube is set back into its original position (on the left side of Figure 6.16), and the phases are allowed to separate. Next the tube is rocked $90°$ clockwise. The lighter, mobile solvent moves through neck 3 into chamber 4. When the tube is turned back $90°$ counterclockwise, the lighter, mobile solvent flows out drain 5 into inlet $1'$ and chamber $2'$ of the next tube on the rack. Then new lighter solvent is introduced into the first tube through inlet 1, and the process begins anew.

* *E. L. Patterson et al.*, J. Am. Chem. Soc. 76, *1823 (1954).*

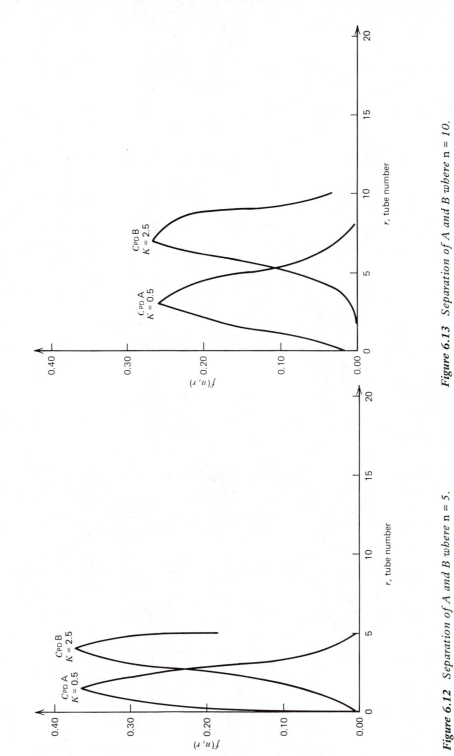

Figure 6.12 Separation of A and B where n = 5.

Figure 6.13 Separation of A and B where n = 10.

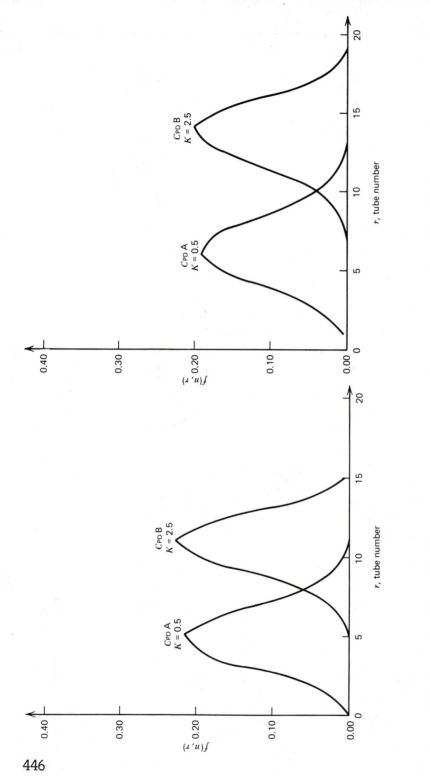

Figure 6.14 Separation of A and B where n = 15.

Figure 6.15 Separation of A and B where n = 19.

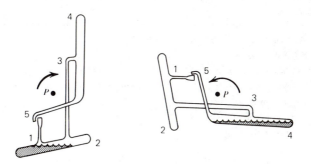

Figure 6.16 *One tube of a modern Craig apparatus.*

≡ 6.6 PROBLEMS ≡

1. Plot $f(n,r)$ versus r for countercurrent distribution when $K = 2.0$, and $n = 19$. Use Table 6.6; don't calculate $f(n,r)$ each time.
   Ans: *See Figure 6.17.*

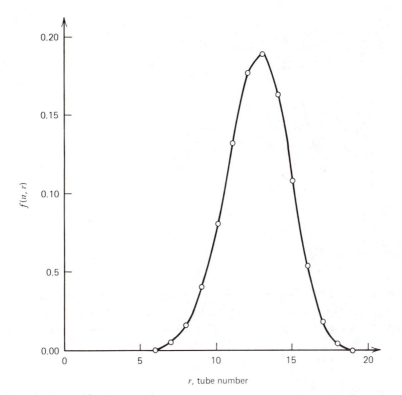

Figure 6.17 *Problem 1.*

2. Four students in a quant lab at Eastern Michigan University, V. Godfroy, K. Helkaa, C. Norrow, and A. Ohno, used a familiar procedure* to determine the distribution constant for bromcresol purple distributed between an aqueous layer and an organic layer, finding that $K = 0.78$, a fact duly recorded in Ms. Helkaa's notebook. (Datum courtesy of K. J. Helkaa.) Plot $f(n,r)$ versus r for the counter-current distribution of bromcresol purple where $n = 5$, $n = 10$, $n = 15$, and $n = 19$, all on the same graph. Use Table 6.6.

Ans: *See Figure 6.18.*

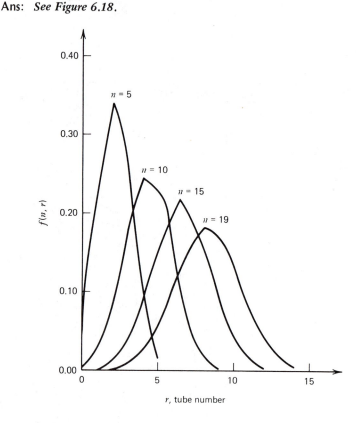

Figure 6.18 *Problem 2.*

3. To see how close the Gaussian approximation is to the exact binomial formulation, plot $f(n,r)$ calculated exactly, and $f(n,r)$ calculated with the Gaussian approximation versus r, on the same graph, where $K = 4.0$, $n = 19$. Use Table 6.6 and Table 6.7.

Ans: *See Figure 6.19.*

---

* *R. M. Scott,* J. Chromatogr. *60, 313 (1971).*

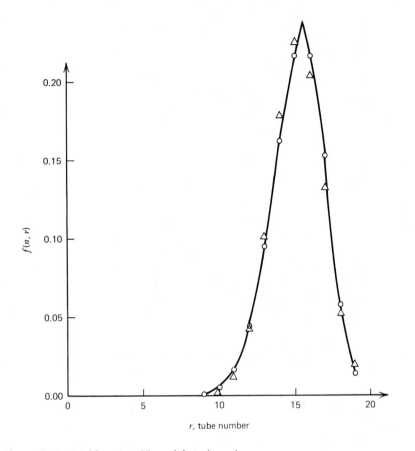

Figure 6.19 *Problem 3:⊙ binomial; △ Gaussian.*

4. Craig et al.* applied countercurrent distribution to penicillins. For benzylpeni-
   cillin, $K = 0.78$, For Δ^2-pentenylpenicillin, $K = 1.18$. Plot $f(n,r)$ versus r for each
   penicillin, on the same graph for $n = 19$. Use Table 6.6. Is separation good after 19
   transfers?
   Ans: *No. See Figure 6.20.*

≡ 6.7 DETAILED SOLUTIONS TO PROBLEMS ≡

Most solutions are graphical.

1. See Figure 6.17.

2. See Figure 6.18. Note how the distribution curve gets fatter and lower with increasing n.

* L. C. *Craig et al.*, J. Biol. Chem. 168, *665 (1947).*

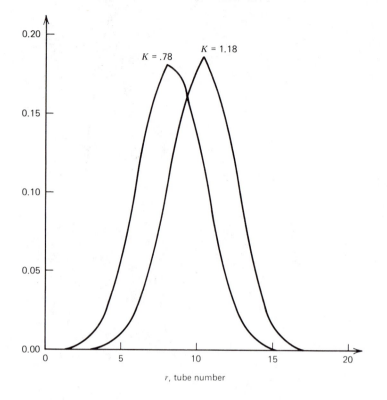

Figure 6.20 Problem 4.

3. There is some difference here. If n were higher, triangles and circles would nearly converge.

4. Separation is very poor after 19 transfers. Fifty or more might be needed.

Table 6.6
SELECTED VALUES OF $f(n,r)$, AS A FUNCTION OF K, n AND r

$K = .78$	$n = 5$	$r = 0$	$f = .05596287$
$K = .78$	$n = 5$	$r = 1$	$f = .21825518$
$K = .78$	$n = 5$	$r = 2$	$f = .34047808$
$K = .78$	$n = 5$	$r = 3$	$f = .26557291$
$K = .78$	$n = 5$	$r = 4$	$f = .10357344$
$K = .78$	$n = 5$	$r = 5$	$f = .01615746$
$K = .78$	$n = 10$	$r = 0$	$f = .00313184$
$K = .78$	$n = 10$	$r = 1$	$f = .02442837$
$K = .78$	$n = 10$	$r = 2$	$f = .08574358$
$K = .78$	$n = 10$	$r = 3$	$f = .17834665$

Table 6.6 Continued

$K = .78$	$n = 10$	$r = 4$	$f = .24344319$
$K = .78$	$n = 10$	$r = 5$	$f = .22786282$
$K = .78$	$n = 10$	$r = 6$	$f = .14811084$
$K = .78$	$n = 10$	$r = 7$	$f = .06601512$
$K = .78$	$n = 10$	$r = 8$	$f = .01930942$
$K = .78$	$n = 10$	$r = 9$	$f = .00334697$
$K = .78$	$n = 10$	$r = 10$	$f = .00026106$
$K = .78$	$n = 15$	$r = 0$	$f = .00017527$
$K = .78$	$n = 15$	$r = 1$	$f = .00205062$
$K = .78$	$n = 15$	$r = 2$	$f = .0111964$
$K = .78$	$n = 15$	$r = 3$	$f = .03784383$
$K = .78$	$n = 15$	$r = 4$	$f = .08855456$
$K = .78$	$n = 15$	$r = 5$	$f = .15195962$
$K = .78$	$n = 15$	$r = 6$	$f = .19754751$
$K = .78$	$n = 15$	$r = 7$	$f = .19811194$
$K = .78$	$n = 15$	$r = 8$	$f = .15452731$
$K = .78$	$n = 15$	$r = 9$	$f = .09374657$
$K = .78$	$n = 15$	$r = 10$	$f = .0438734$
$K = .78$	$n = 15$	$r = 11$	$f = .01555511$
$K = .78$	$n = 15$	$r = 12$	$f = .00404433$
$K = .78$	$n = 15$	$r = 13$	$f = .00072798$
$K = .78$	$n = 15$	$r = 14$	$f = 8.1117694E - 05$
$K = .78$	$n = 15$	$r = 15$	$f = 4.2181203E - 06$
$K = .78$	$n = 19$	$r = 0$	$f = 1.7459018E - 05$
$K = .78$	$n = 19$	$r = 1$	$f = .00025874$
$K = .78$	$n = 19$	$r = 2$	$f = .00181637$
$K = .78$	$n = 19$	$r = 3$	$f = .00802837$
$K = .78$	$n = 19$	$r = 4$	$f = .02504852$
$K = .78$	$n = 19$	$r = 5$	$f = .05861353$
$K = .78$	$n = 19$	$r = 6$	$f = .10667662$
$K = .78$	$n = 19$	$r = 7$	$f = .15452871$
$K = .78$	$n = 19$	$r = 8$	$f = .18079859$
$K = .78$	$n = 19$	$r = 9$	$f = .17236132$
$K = .78$	$n = 19$	$r = 10$	$f = .13444183$
$K = .78$	$n = 19$	$r = 11$	$f = .08579833$
$K = .78$	$n = 19$	$r = 12$	$f = .04461513$
$K = .78$	$n = 19$	$r = 13$	$f = .01873836$
$K = .78$	$n = 19$	$r = 14$	$f = .00626396$
$K = .78$	$n = 19$	$r = 15$	$f = .00162863$
$K = .78$	$n = 19$	$r = 16$	$f = .00031758$
$K = .78$	$n = 19$	$r = 17$	$f = 4.371437E - 05$
$K = .78$	$n = 19$	$r = 18$	$f = 3.7885788E - 06$
$K = .78$	$n = 19$	$r = 19$	$f = 1.5553114E - 07$
$K = 1.18$	$n = 19$	$r = 0$	$f = 3.709596E - 07$
$K = 1.18$	$n = 19$	$r = 1$	$f = 8.316914E - 06$
$K = 1.18$	$n = 19$	$r = 2$	$f = 8.8325626E - 05$

Table 6.6 Continued

$K = 1.18$	$n = 19$	$r = 3$	$f = .0005906$
$K = 1.18$	$n = 19$	$r = 4$	$f = .00278765$
$K = 1.18$	$n = 19$	$r = 5$	$f = .00986828$
$K = 1.18$	$n = 19$	$r = 6$	$f = .02717068$
$K = 1.18$	$n = 19$	$r = 7$	$f = .0595426$
$K = 1.18$	$n = 19$	$r = 8$	$f = .10539039$
$K = 1.18$	$n = 19$	$r = 9$	$f = .15199636$
$K = 1.18$	$n = 19$	$r = 10$	$f = .17935571$
$K = 1.18$	$n = 19$	$r = 11$	$f = .17315978$
$K = 1.18$	$n = 19$	$r = 12$	$f = .13621903$
$K = 1.18$	$n = 19$	$r = 13$	$f = .08655148$
$K = 1.18$	$n = 19$	$r = 14$	$f = .04377032$
$K = 1.18$	$n = 19$	$r = 15$	$f = .01721632$
$K = 1.18$	$n = 19$	$r = 16$	$f = .00507882$
$K = 1.18$	$n = 19$	$r = 17$	$f = .00105759$
$K = 1.18$	$n = 19$	$r = 18$	$f = .00013866$
$K = 1.18$	$n = 19$	$r = 19$	$f = 8.6116167E - 06$
$K = 2$	$n = 19$	$r = 0$	$f = 8.6039149E - 10$
$K = 2$	$n = 19$	$r = 1$	$f = 3.2694876E - 08$
$K = 2$	$n = 19$	$r = 2$	$f = 5.8850777E - 07$
$K = 2$	$n = 19$	$r = 3$	$f = 6.6697547E - 06$
$K = 2$	$n = 19$	$r = 4$	$f = 5.3358038E - 05$
$K = 2$	$n = 19$	$r = 5$	$f = .00032015$
$K = 2$	$n = 19$	$r = 6$	$f = .00149403$
$K = 2$	$n = 19$	$r = 7$	$f = .00554924$
$K = 2$	$n = 19$	$r = 8$	$f = .01664771$
$K = 2$	$n = 19$	$r = 9$	$f = .0406944$
$K = 2$	$n = 19$	$r = 10$	$f = .08138879$
$K = 2$	$n = 19$	$r = 11$	$f = .13318166$
$K = 2$	$n = 19$	$r = 12$	$f = .17757555$
$K = 2$	$n = 19$	$r = 13$	$f = .19123521$
$K = 2$	$n = 19$	$r = 14$	$f = .16391589$
$K = 2$	$n = 19$	$r = 15$	$f = .10927726$
$K = 2$	$n = 19$	$r = 16$	$f = .05463863$
$K = 2$	$n = 19$	$r = 17$	$f = .01928422$
$K = 2$	$n = 19$	$r = 18$	$f = .00428538$
$K = 2$	$n = 19$	$r = 19$	$f = .00045109$
$K = 4$	$n = 19$	$r = 0$	$f = 5.2428792E - 14$
$K = 4$	$n = 19$	$r = 1$	$f = 3.9845882E - 12$
$K = 4$	$n = 19$	$r = 2$	$f = 1.4344517E - 10$
$K = 4$	$n = 19$	$r = 3$	$f = 3.2514239E - 09$
$K = 4$	$n = 19$	$r = 4$	$f = 5.2022781E - 08$
$K = 4$	$n = 19$	$r = 5$	$f = 6.2427338E - 07$
$K = 4$	$n = 19$	$r = 6$	$f = 5.8265516E - 06$
$K = 4$	$n = 19$	$r = 7$	$f = 4.3282954E - 05$
$K = 4$	$n = 19$	$r = 8$	$f = .0002597$
$K = 4$	$n = 19$	$r = 9$	$f = .00126963$

Table 6.6 Continued

K = 4	n = 19	r = 10	f = .00507853
K = 4	n = 19	r = 11	f = .01662065
K = 4	n = 19	r = 12	f = .04432175
K = 4	n = 19	r = 13	f = .09546222
K = 4	n = 19	r = 14	f = .16364952
K = 4	n = 19	r = 15	f = .21819936
K = 4	n = 19	r = 16	f = .21819936
K = 4	n = 19	r = 17	f = .15402308
K = 4	n = 19	r = 18	f = .0684547
K = 4	n = 19	r = 19	f = .01441152

Table 6.7
GAUSSIAN APPROXIMATION

$f(n,r)$	$n = 19$ $K = 4.0$	r
$f(n,r) = 7.1826127E - 18$		0
$f(n,r) = 9.0432578E - 16$		1
$f(n,r) = 8.1942134E - 14$		2
$f(n,r) = 5.3435457E - 12$		3
$f(n,r) = 2.5077941E - 10$		4
$f(n,r) = 8.4702092E - 09$		5
$f(n,r) = 2.0589054E - 07$		6
$f(n,r) = 3.6017918E - 06$		7
$f(n,r) = 4.5346196E - 05$		8
$f(n,r) = .00041087$		9
$f(n,r) = .0026792$		10
$f(n,r) = .01257324$		11
$f(n,r) = .04246483$		12
$f(n,r) = .10321706$		13
$f(n,r) = .18055667$		14
$f(n,r) = .22730844$		15
$f(n,r) = .205948$		16
$f(n,r) = .13428872$		17
$f(n,r) = .06301755$		18
$f(n,r) = .02128254$		19

NOTE: *Tables 6.6 and 6.7 are computer-generated. The student must do the necessary rounding off.*

≡ 6.8 LIQUID-LIQUID CHROMATOGRAPHY ≡ (LLC)

No one will dispute the contention that the Craig apparatus is elegant in both concept and construction. Craig countercurrent distribution, applied intelligently, can separate species that a drawer full of separatory funnels couldn't without an enormous waste of man-hours. Many delicate species would be destroyed by the temperatures involved in gas chromatography, a popular modern method, and must thus be separated by gentler

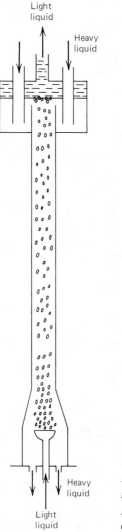

Figure 6.21 *Elgin spray tower from Mass-Transfer Operations by R. E. Treyval. Copyright © (1955 McGraw-Hill Book Co.). Used with permission of McGraw-Hill Book Company.*

means. Craig countercurrent distribution is one of those gentler means, yet it suffers from certain disadvantages. A Craig apparatus is bulky. Each stage is rather complicated, and up to a few hundred stages may be required for separations where distribution constants are very close together. The expense can be enormous, for a simple 20-stage apparatus may cost $1000 or more. Cleaning the apparatus is not a pleasant task.

6.9   Continuous Countercurrent Extraction Methods

It is not always easy to build a truly continuous countercurrent extractor for laboratory use. The denser liquid must flow downward, contacting the less dense liquid as it flows upward. Martin and Synge, in a classic paper, noted the "difficulty of preventing it [a drop of the less dense liquid] from moving in the wrong direction."*

In industry, large scale countercurrent extractors, exemplified by the Elgin tower, have been of considerable use. Figure 6.21 is a diagram of an Elgin tower. The heavier liquid flows in through the two top ports. The lighter liquid is sprayed upward out of a nozzle located at the bottom of the tower. Equilibration of the two phases takes place within the body of the tower. The lighter liquid finally reaches the top of the tower, and may be drawn off. The heavier liquid, after equilibration, is drawn off the bottom of the tower. It is interesting to note that solute may be extracted from the lighter phase into the heavier phase, or from the heavier phase into the lighter phase.

6.10   Real Liquid-Liquid Chromatography (LLC)

6.11   Classical Column Chromatography
≡

Martin and Synge, noting the difficulties associated with trying to use countercurrent extraction in the separation of certain amino acids, devised a new technique, called liquid-liquid chromatography (LLC for short) in 1941.* The original paper describing their work is a classic of the chemical literature. Figure 6.22 illustrates their method. They filled a glass column with purified silica gel. To a reasonable first approximation, silica gel is solid silica, SiO_2, with lots of water attached. Thus, an aqueous phase, H_2O, is attached to a solid support, SiO_2, in silica gel. The silica gel was impregnated with methyl orange, which turns pink in the presence of any of the amino acids separated. Figure 6.22a shows the column ready for action. The mixture of amino acids to be spearated was dissolved in a solution 99 parts chloroform and one part n-butyl alcohol by volume. This solution was placed on top of the column, as

* Reprinted with permission from A. J. P. Martin and R. L. M. Synge, Biochem. Journal 35, 1358 (1941).

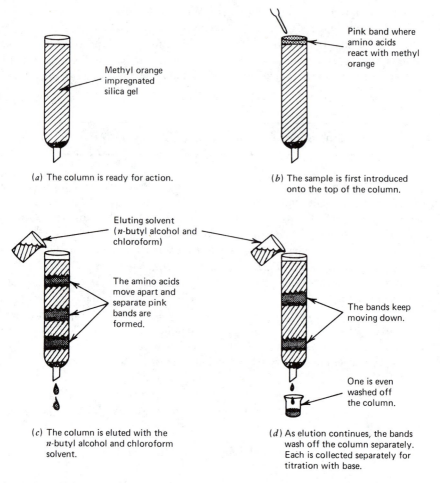

Figure 6.22 Diagram of Martin and Synge's experiment.

shown in Figure 6.22b. A pink band was formed at the top of the column as a result of the amino acids' reacting with the methyl orange, then, as shown in Figure 6.22c, the organic eluting solvent, with *no* amino acids dissolved in it, was washed down the column. Because each amino acid has a different distribution coefficient between water and the organic phase, the amino acids moved down the column at different rates, and a pink band appeared for each amino acid. Figure 6.22d shows what happened as a result of even more washing with the eluting solvent. One band was washed off the column and collected so that it could be titrated with standard base. Other bands were eluted from the column in succession, and also titrated. This became the model liquid-liquid chromatography experiment, one that has been repeated in form, if not in exact substance, many times. From it have grown the variations of paper chromatography, thin-layer chromatography, and high-speed liquid-liquid chromatography, which have proven so useful to workers in many disciplines.

Martin and Synge also investigated the mathematics of liquid-liquid chromatography. They found that quantity of solute, plotted against some function of distance along the column, yielded a Gaussian curve. A Craig countercurrent distribution involving many tubes also yields a Gaussian curve when fraction of solute in a tube is plotted versus tube number. Instead of clearly defined tubes, Martin and Synge had to work with terms called *theoretical plate* and *height equivalent to one theoretical plate.* (HETP). A theoretical plate is analogous to one tube of a Craig apparatus. In the words of Martin and Synge, within a theoretical plate: "perfect equilibrium between the two phases occurs . . . [a] column could be divided up into a number of layers, each of which was equivalent to one theoretical plate, and the height of such a layer was called H.E.T.P. or 'height equivalent to one theoretical plate.' For the present purpose the H.E.T.P. is defined as the thickness of the layer such that the solution issuing from it is in equilibrium with the mean concentration of solute in the non-mobile phase throughout the layer."* The lower the value of the HETP for a given column and separation, the more theoretical plates there are on the column. Thus, chromatographers seek to minimize HETP, especially in high speed liquid-liquid chromatography.

Sometimes, one runs into a term that sounds complicated: "reversed phase chromatography." There is nothing mysterious about this term. It simply is liquid-liquid chromatography in which an organic phase is held stationary on a solid support, while the mobile phase is aqueous. There is nothing at all unnatural about this form of chromatography; it's just different from the earliest varieties of LLC. An example of the use of reversed-phase chromatography may be found in Lee's work in the separation of lithium from other alkali metal ions[†]. It has never been easy to separate alkali metal ions from one another by nonchromatographic techniques. Classical procedures involve very unsatisfactory precipitations whose end products are hard to form and often contaminated. An organic species, troctylphosphine oxide (TOPO or

$$CH_3-(CH_2)_7-\overset{\displaystyle \overset{O}{\|}}{\underset{\displaystyle (CH_2)_7CH_3}{P}}-(CH_2)_7-CH_3\, ,$$

dissolved in another organic species, dibenzoyl methane (DBM or

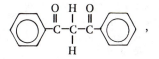

will complex many metal ions, including the alkalis. Lee coated beads of a highly inert fluorocarbon, polytetrafluoroethylene, or Haloport F, with a mixture of DBM

---

* *Reprinted with permission from A. J. P. Martin and R. L. M. Synge,* Biochem. Journal 35, 358 *(1941).*

† D. A. Lee, J. Chromatog. 26, 342 (1967).

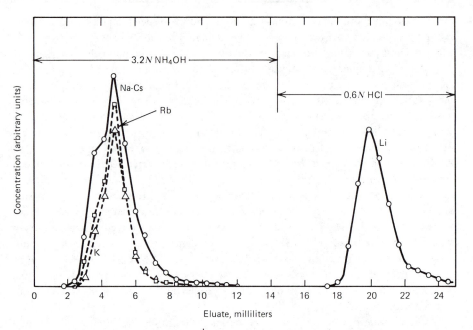

Figure 6.23 *Lee's Separation of Li⁺ from the other alkalis*. Reprinted by permission from D. A. Lee,* J. Chromatog., **26**, 342 (1967).

and TOPO. He introduced a mixture of the alkali metal salts onto the column, and by elution with 3.2 N NH₄OH (3.2 M NH₃ in H₂O) followed by elution with 0.6 N HCl (0.6 M HCl in H₂O), separated Li⁺ quite satisfactorily from the other alkali metal ions. Lee's results are shown in Figure 6.23.

6.12 Paper Chromatography
≡

The solid support for the stationary liquid phase need not consist only of silica or of beads coated with that phase and packed into a glass column. In 1944, Consden, Gordon, and Martin published a paper in which they described a liquid-liquid chromatographic technique applicable to micro samples.[†] The technique is called *paper chromatography*, and has proven its worth wherever micro amounts of delicate— usually biochemical—species are to be separated and identified. In their initial work, Consden, Gordon, and Martin were able to separate and characterize all the amino acids in 200 micrograms of wool. This was and still is, a most impressive accomplishment.

How does paper chromatography work? Briefly, filter paper is saturated with water. Typically, a paper like Whatman No. 41 can contain 22% water by weight. The cellu-

---

[*] D. A. Lee, J. Chromatog. 26, 342 (1967).
[†] R. Consden, A. H. Gordon, and A. J. P. Martin, Biochem. Journal 38, 224 (1944).

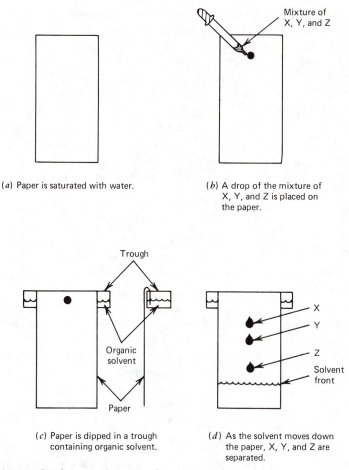

(*a*) Paper is saturated with water.

(*b*) A drop of the mixture of X, Y, and Z is placed on the paper.

Trough

Organic solvent

Paper

X
Y
Z
Solvent front

(*c*) Paper is dipped in a trough containing organic solvent.

(*d*) As the solvent moves down the paper, X, Y, and Z are separated.

Figure 6.24 *What happens in paper chromatography.*

lose of the paper serves as the solid support material, analogous to the silica in liquid-liquid column chromatography. The water saturating the cellulose is the aqueous, stationary, liquid phase. Figure 6.24 shows the paper chromatographic process in simplified form. In Figure 6.24*a*, the paper, saturated with water, but not treated in any other way, is shown. Figure 6.24*a* is rather uninteresting, so a look at Figure 6.24*b* is worthwhile. In Figure 6.24*b*, a drop of a mixture of X, Y, and Z is placed near the top of the paper. The drop stays there, until the top of the paper is placed in a trough of organic solvent, as shown in Figure 6.24*c*. The organic solvent is the mobile phase in liquid-liquid column chromatography. True to its mobile nature, the organic solvent moves down the paper. Each component of the mixture, X, Y, and Z, has a different distribution coefficient between aqueous and organic phases. Hence, X, Y, and Z are separated as the solvent moves down the paper (Figure 6.24*d*). When a proper developing agent (one that reacts with X, Y, and Z to give colored compounds) is applied to the paper, X, Y, and Z appear as spots. Samples of pure X, pure Y, and pure Z can be run on the same paper at the same time as the mixture, and qualitative analysis of the mixture is made easy. This is the case because pure X moves as far down the paper as X in the mixture, pure Y moves as far down the paper as Y in the mixture, and pure Z moves as far down the paper as Z in the mixture. A quantity known as R_F is even more generally useful than distance alone in qualitative identification.

$$\text{For X:} \quad R_{Fx} = \frac{\text{(distance from origin to spot of X)}}{\text{(distance from origin to solvent front)}}$$

$$\text{For Y:} \quad R_{Fy} = \frac{\text{(distance from origin to spot of Y)}}{\text{(distance from origin to solvent front)}}$$

$$\text{For Z:} \quad R_{Fz} = \frac{\text{(distance from origin to spot of Z)}}{\text{(distance from origin to solvent front)}}$$

As might be expected, R_F values are very sensitive to temperature, type of paper, and many other variables. Thus knowns and unknowns are, wherever possible, run on the same chromatogram. Published R_F values can be helpful in estimating what a complete unknown might be.

For the identification of separated amino acids, Consden, Gordon, and Martin dried the paper, sprayed it with a solution of 0.1% ninhydrin

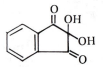

in n-butanol, and dried the paper again. Ninhydrin and amino acids combine to give compounds that are often blue. Thus, location of amino acid spots could be easily established.

Paper chromatography is of immense utility to biochemists and biologists, who must often work with tiny samples of easily destroyed material. It is not limited in its application to purely biochemical species, however, and may even be used to separate metal ions. The technique is elegant, simple, inexpensive, and versatile. It has even led to the development of another separation technique, called *thin-layer chromatography,* or TLC for short.

6.13 Thin-Layer Chromatography (TLC)

Thin-layer chromatography began to be used widely in the late 1950s. The process is illustrated in Figure 6.25. A glass plate is coated with silica gel, as shown in Figure 6.25a. (Most people prefer to buy their plates ready-coated.) Second, in Figure 6.25b,

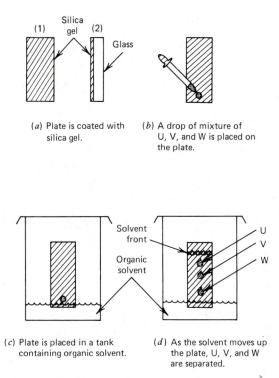

(a) Plate is coated with silica gel.

(b) A drop of mixture of U, V, and W is placed on the plate.

(c) Plate is placed in a tank containing organic solvent.

(d) As the solvent moves up the plate, U, V, and W are separated.

Figure 6.25 *What happens in thin-layer chromatography.*

a drop of a mixture of U, V, and W is placed on the plate. The plate is then placed in a closed jar containing an organic solvent, shown in Figure 6.25c. The solvent moves up the plate, and U, V, and W are separated as shown in Figure 6.25d. Each component of the mixture has a characteristic partition coefficient, which accounts for the separation of the components. The plate may be dried, and a developing reagent, to make the spots of U, V, and W visible, sprayed on. Often, sulfuric acid is used as the developing reagent. It chars organic species, leaving the silica gel unstained. (Sulfuric acid cannot be used on filter paper very easily, for it will char the cellulose the filter paper is made of.) When the plate is dried, U, V, and W appear as spots. The separation and developing process in thin-layer chromatography is more rapid than the process in paper chromatography; thin-layer separations can often be completed in a couple of hours, but one can plan to spend a day working with many of the paper chromatographic separations.

Distribution of a solute between two liquid phases does not always occur in thin-layer chromatography. Indeed, water can be used as a mobile phase in some cases, with an immobile phase that is also aqueous. Separation in such a case is probably due to *sorption* on a solid phase, with water eluting species sorbed onto that solid phase. *Sorption* means either *ad*sorption onto or *ab*sorption within a solid phase. Sorption chromatography, or liquid-solid chromatography, will be discussed later.

6.14 High-Performance Liquid-Liquid Chromatography (HPLC)

≡

Traditional liquid-liquid chromatography had, and still has, one drawback: it is slow. The reason is that one depends on gravity to move the eluting solvent through a tightly packed column. The column requires watching, tying up laboratory personnel. In the 1960s development of a faster method began. The new method preserves liquid-liquid chromatography's advantage of being gentle to delicate species. The new method involves pumping the eluting solvent through a packed column at high speed. Hence, it is often called high-performance liquid-liquid chromatography. In Figure 6.26, a diagram of a high-performance liquid chromatograph is shown. In separatory principle, this device does not vary much from the column of Martin and Synge shown in Figure 6.22. The column still contains a stationary liquid phase on a solid support, and a

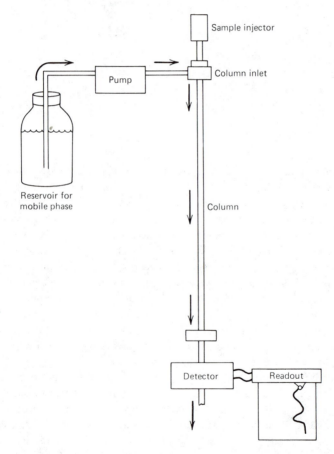

Figure 6.26 Diagram of a high-performance liquid-liquid chromatograph.

mobile liquid phase, together with the sample, moves down the column. What is the difference between the two types of apparatus?

In Figure 6.26, a reservoir for the mobile phase and a pump to force the mobile phase down the column are shown. Inlet pressures of from 500 to 5000 psig (pounds per square inch on gauge; atmospheric pressure is about 14.7 psi) are common. There is a wide variety of pumps available, both mechanical and pneumatic. Many pump arrangements require reservoirs that can withstand high pressures. The sample may be put onto the column by means of a syringe, or by a multiport sample valve.

The column itself must withstand high pressures. A typical column might be made from seamless stainless steel tubing, 500 mm long by 3 mm i.d. by 6.35 mm o.d. The stationary liquid phase within the column is supported on a solid phase. Often, this support and liquid phase arrangement consists of spherical beads coated with the stationary liquid layer. A cross-section of such a bead is shown in Figure 6.27. The combination of bead and liquid is called a *porous layer bead* (PLB), the porous layer being the stationary liquid phase, about 1 μ thick. The bead proper is about 40 μ in diameter. These porous layer beads are analogous to the silica gel particles of Martin and Synge.

Once the separated species leave the column, they must be identified and quantitatively determined. The titration method is slow, and because of the small amounts of material involved, not very practical. Besides, without colored bands moving down a transparent column, one doesn't know just which fraction of the output to titrate. A spectrophotometric detector, sometimes called a UV detector (for ultraviolet), is one of the more popular sorts associated with high speed liquid-liquid chromatography. The device works well for many types of samples, because aromatic compounds (those with benzene rings in them) have enormous molar absorptivities at 254 nm, a UV

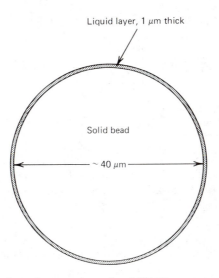

Figure 6.27 *Cross section of porous layer bead (PLB).*

wavelength easily generated and detected. Other organic compounds, and some metal chelates also absorb at 254 nm. The tubing is too narrow to give a good path length. A device called a Z-cell is used to increase path length to a respectable 1 cm or so.[*] An absorption cell is shown in Figure 6.28. The readout device may include a recorder, or a digital display and printout, or some combination of the two.

A typical readout from a high-performance liquid-liquid chromatograph is shown in Figure 6.29.[†] The species separated and detected are substituted urea herbicides

(urea is $H_2N-\overset{\overset{\text{O}}{\|}}{C}-NH_2$). The figure ought to be examined closely. The abscissa is time and the ordinate absorbance. The ordinate is barely indicated by 0.005 As, which just gives the height on the graph corresponding to 0.005 absorbance units. The column is filled with spheres of a CSP (controlled surface porosity) solid support, coated with a solution of a polar species, β, β' oxydipropionitrile. The mobile phase is the nonpolar solvent dibutyl ether. The sample is *tiny*: 1 μl of 67 ng/milliliter of each insecticide in dibutyl ether (1 ng = 10^{-9} gram).

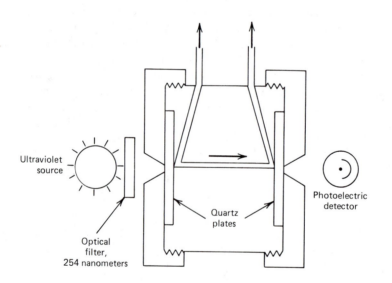

Ultraviolet source

Photoelectric detector

Quartz plates

Optical filter, 254 nanometers

Figure 6.28 UV cell for LLC detection[‡]. Reproduced from the Journal of Chromatographic Science *by permission of Preston Publications, Inc.*

[*] *J. F. K. Huber*, J. Chromatog. Sci. *7, 172 (1969).*

[†] *J. J. Kirkland, ed.*, Modern Practice of Liquid Chromatography, *Wiley-Interscience, New York 1971, p. 105.*

[‡] *J. J. Kirkland*, J. Chromatog. Sci. *7, 7 (1969).*

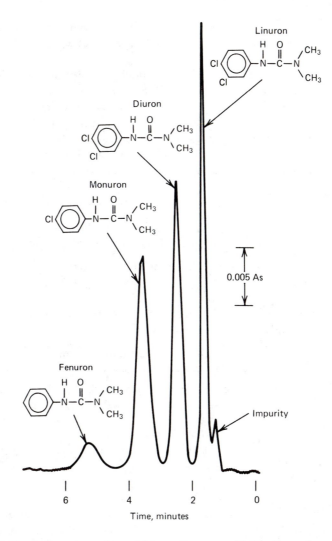

Figure 6.29 *Typical readout from high-performance LLC*. Reproduced from the* Journal of Chromatographic Science *by permission of Preston Publications, Inc.*

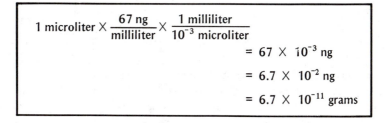

$$1 \text{ microliter} \times \frac{67 \text{ ng}}{\text{milliliter}} \times \frac{1 \text{ milliliter}}{10^{-3} \text{ microliter}}$$

$$= 67 \times 10^{-3} \text{ ng}$$

$$= 6.7 \times 10^{-2} \text{ ng}$$

$$= 6.7 \times 10^{-11} \text{ grams}$$

* *Ibid.*

This sort of sensitivity is awesome. Qualitative identification is made by injection of pure samples of each herbicide onto the column. Each, of course, will have its own characteristic retention time.

Optimizing the conditions under which liquid-liquid chromatography is done occupies much of the chemist's attention. With high-performance liquid-liquid chromatography, the most easily controlled variable is flow rate. The height equivalent to a theoretical plate is the variable which it is desirable to minimize: the lower the value of HETP, the more theoretical plates there are on a column, and the better the separation that can be expected. An equation, applicable to both liquid-liquid and gas-liquid chromatography exists. It is called the van Deemter equation, and is useful in optimizing flow rate.

$$(HETP) \cong A + B/v + Cv$$

(HETP) = height equivalent to a theoretical plate

Constants for a given set of conditions
- v = velocity of mobile phase (carrier)
- A = multiple path term or eddy diffusion term
- B = molecular diffusion term for carrier (very small in LLC)
- C = resistance to mass transfer term

Because B, the molecular diffusion term, is small when the mobile phase is a liquid, it is possible to write, for high values of mobile phase velocities, v, an even simpler form of the equation.

$$(HETP) \cong A + Cv$$
for
LLC

The super simple form of the van Deemter equation is not quite correct at low mobile phase velocities, as Figure 6.30* shows, but it does predict the linear shape observed at higher velocities. Interestingly, HETP rises slowly with carrier velocity. It is thus possible to speed up the carrier velocity—and the analysis—without raising HETP drastically. Practical LLC depends more on solvent composition for optimization of conditions.

6.15 A Reflection

The methods of liquid-liquid chromatography are splendidly suited to the separation of small amounts of easily destroyed chemical species. They range from the very simple, involving the cheapest materials, as in paper chromatography, to methods dependent on recent—and expensive—technological advances in column packings, pumps,

* *J. J. Kirkland, op. cit.*

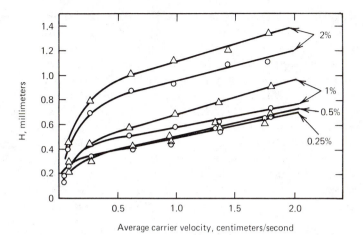

Figure 6.30 *HETP versus* V *curves for various LLC conditions. Percent figures are percentage of β,β'-dioxypropionitrile in liquid layer on bead*. Reproduced from the* Journal of Chromatographic Science *by permission of Preston Publications, Inc.*

and detectors. What began in 1941 in one laboratory, has found applications in many thousands of other laboratories.

≡ 6.16 GAS-LIQUID CHROMATOGRAPHY ≡
(GLC)

The student who enters an analytical chemistry laboratory will very probably be exposed to a device called a gas chromatograph, which has become nearly as common as the pH meter and the spectrophotometer in the teaching and industrial laboratories of the United States. On graduation, the student may take a job with a chemical industry, a clinical laboratory, or a bacteriological laboratory. Everywhere the student will see gas chromatographs and may well use them. What accounts for the wide popularity enjoyed by gas chromatography?

First, a look at what gas chromatography has largely replaced will be useful. In 1968, Williams published in *Analytical Chemistry* that journal's last review of papers in a discipline known as analytical distillation[†]. In William's words:

> *Two years ago, in the biennial review of analytical distillation, we predicted the demise of the art; at best it could retain no more than vestiges of its former usefulness under the continued buffeting of gas-liquid chromatography with the latter's vast virtues of speed, sensitivity, potential accuracy, and very nearly infinitesimal sample size. Analytical distillation was never born; it just grew. If it is to be made respectable by being given a date of birth, it first looked upon the world from the top of a brass tube 0.75 in. in*

---

*   *Ibid.*

†   *Reprinted with permission from F. E. Williams, Anal. Chem.* **40** *[5], 62R (1968). Copyright by the American Chemical Society.*

diameter and 52 feet high in an airshaft at Pond Laboratory at Pennsylvania State University in 1930. With this unit Fenske . . . demonstrated that it was possible to separate two isomeric hydrocarbons boiling only 2° C. apart. Under the impetus of this 'impossible' event, analytical distillation took on stature. In less than half a decade, column packings were developed which made it possible to have an equally powerful separating unit in the confines of a laboratory. Unfortunately, the cult of the instrument maker had not yet been born, or scarcely so, and the clever research and glittering packaging of the gas chromatography era were not available to aid the development of analytical distillation. But it grew and prospered through World War II; it performed notably for that conflict, yet a decade and a half later its day was finished. As an art, analytical distillation joins the alembic among the methods and devices men have used to probe the nature of things.

Williams quite properly noted the large sample size that analytical distillation requires. By contrast, samples of a microliter or two are the rule in gas chromatography. The cost of gas chromatography also makes it an attractive technique. For $500, plus a surplus strip-chart recorder, one can easily separate benzene (b.p. 80. 1°C) and cyclohexane (b.p. 81.4°C) by gas chromatography, a feat not capable of being matched by expensive and bulky distillation apparatus. Azeotrope formation is no problem in gas chromatography, but severely restricts the utility of analytical distillation. (Azeotropes are mixtures whose liquid and vapor compositions are the same, that is, mixtures whose components *cannot* be separated by further distillation.) Gas chromatography on a preparative scale is also possible.

The idea of gas-liquid chromatography is not new. In 1941, in their classic paper on liquid-liquid chromatography, Martin and Synge put forth the basic notion of gas-liquid chromatography[*]. In their words: "The mobile phase need not be a liquid, but may be a vapor. We show below that the efficiency of contact between the phases (theoretical plates per unit length of column) is far greater in the chromatogram than in ordinary distillation or extraction columns. Very refined separations of volatile substances should therefore be possible in a column in which permanent gas is made to flow over gel impregnated with a non-volatile solvent in which the substances to be separated approximately obey Raoult's Law."

For about 10 years, no one followed the suggestion of Martin and Synge, although gas-solid chromatography, depending on adsorption of components from a gas stream onto charcoal, was developed. It remained for Martin to suggest the notion of gas-liquid chromatography to a co-worker. In 1952, James and Martin published a paper entitled "Gas-Liquid Partition Chromatography: The Separation and Micro-Estimation of Volatile Fatty Acids from Formic Acid to Dodecanoic Acid."[†] (In that same year of 1952, Martin and Synge were awarded the Nobel Prize in chemistry.) This paper marked the very beginning of a massive literature of gas chromatography.

James and Martin prepared mixtures of acids to be separated. Typically, a mixture

---

[*] *Reprinted with permission from A. J. P. Martin and R. L. M. Synge,* Biochem. Journal *35, 1358 (1941)*

[†] *A. T. James and A. J. P. Martin,* Biochem Journal *50, 679 (1952).*

might consist of acetic acid (CH_3COOH), propionic acid (CH_3CH_2COOH), *iso*-butyric acid ($CH_3CH\ COOH$), and *n*-butyric acid ($CH_3CH_2CH_2COOH$). The column was a
$$|$$
$$CH_3$$
four-foot length of glass tubing of 4 mm internal diameter. The inert support for the stationary liquid phase was kieselguhr (diatomaceous earth). The stationary liquid phase was made of two nonvolatile substances: stearic acid ($CH_3(CH_2)_{16}COOH$), 10% by weight; and a silicone, 90% by weight. The kieselguhr was coated with this stationary phase, and packed into the glass column. The column was placed in a heated jacket. Nitrogen served as the mobile gas phase. The fatty acid mixture was placed on the column with a micropipette. Because each component of the mixture has a different distribution coefficient between nitrogen and the stationary liquid phase, the components moved along the column at different rates, emerging at different times. Titration with standard base served to identify each acid and to allow estimation of of the quantity of each eluted from the column. The results that James and Martin got for the mixture of acetic, propionic, *iso*-butyric, and *n*-butyric acids are shown in Figure 6.31.* Part B of the figure, called "differential of experimental curve," was obtained very simply by plotting Δ (μg equiv.)/Δ (time) versus time, in much the same way that $\Delta E/\Delta$ (volume) is plotted versus volume in potentiometric titrations. Part B of the figure more nearly resembles a conventional gas chromatogram.

The detection, injection, and heating systems first built by James and Martin were not convenient to use, but in the hands of those scientists, they served to establish gas chromatography. Improvements to the apparatus and technique soon began and are continued to this day.

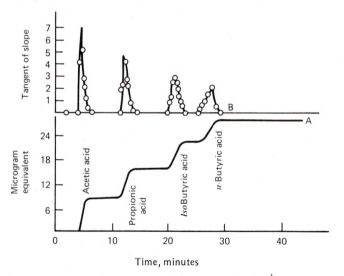

Figure 6.31 *Separation of fatty acids by James and Martin[†].*

* *Reprinted with permission from A. T. James and A. J. P. Martin,* Biochem. Journal 50, 679 (1952).
† *Ibid.*

6.17 Modern Apparatus for Gas-Liquid Chromatography

Figure 6.32 is a rough diagram of a modern set-up for gas-liquid chromatography (GLC). Gas-liquid chromatography is sometimes called simply gas chromatography (GC) or, by many organic chemists, vapor-phase chromatography (VPC). A carrier gas, the mobile phase, is essential to the technique. In the United States, helium, readily obtainable, is frequently used. Outside the United States, hydrogen is often used in place of scarce helium. Careful regulation of pressure at the column inlet is provided by a two-stage tank regulator, and the flow rate may be monitored by other other devices. Without a constant inlet pressure and constant flow rate, reproducibility of results can fall to nil. The gas flows from the regulator to an inlet on the chromatograph into a heated enclosure known as the oven. The temperature of the oven is

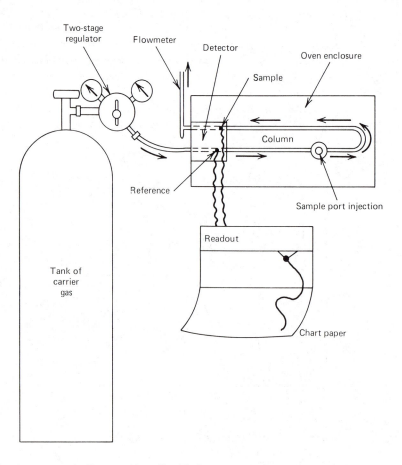

Figure 6.32 *Rough diagram of gas-liquid chromatograph.*

closely regulated, and may be varied. Higher temperatures generally mean lower retention times on the column for all components of an analyte mixture. The carrier gas is still free of sample when it passes through the reference side of a detector, "ref" in Figure 6.32, so that an electronic comparison of carrier without sample to carrier with sample can be made.

Next, the carrier gas flows past the injection port, illustrated sketchily in Figure 6.32. The sample is introduced to the column with a microliter syringe. (A microliter syringe is expensive—they start at about $20—and is very easy to damage. Learn how to use one before you break it out of ignorance.) The syringe, filled with the sample mixture, is pushed through a silicone rubber septum, the syringe plunger is depressed quickly, and the sample vaporizes on the heated block, to be swept up in the stream of carrier gas. Carrier gas and vaporized sample enter the packed column.

The column is a long tube, usually of stainless steel, filled with particles that strongly resemble the porous layer beads of liquid-liquid chromatography. The particles consist of an inert solid, such as kieselguhr, coated with a liquid of very low vapor pressure. This liquid serves as a stationary liquid phase. The components of the sample are separated on the column, much as components are separated on a liquid-liquid chromatographic column. The difference, of course, is that the carrier gas is the mobile phase in gas-liquid chromatography. Also, samples to be separated must be able to withstand the temperatures found in the injection port and the oven.

There is a wide variety of solid supports available for gas chromatography. Many stationary liquid phases are available to the chromatographer. Generally, one uses polar liquid phases to separate polar compounds such as alcohols, and nonpolar liquid phases for the separation of nonpolar compounds such as alkanes. It isn't always necessary to go to all the trouble of buying a solid support and coating it with the appropriate liquid. Little jars of presized solid support, already coated with liquid, are available from a number of supply houses. For those who don't really want to pack and bend their own columns, prepacked columns (sold, to be sure, at a premium) may be bought. For the very best separation, glass capillary columns may be used. These consist of coiled glass capillaries, whose inside walls are coated with appropriate liquid phases. Most of the columns encountered by the student, however, will be made of stainless steel tubing packed with liquid-coated solid support material.

Detection of sample constituents in effluents from chromatograph may be done in many ways. James and Martin* chose an uncomplicated approach, and simply titrated the weak acids emerging from their column. Quicker and more nearly general methods of detection and quantitation were soon developed and are in use today. Almost all of these involve electronic readouts.

6.18 Some Mathematics Applicable to Gas-Liquid Chromatography

Ever since James' and Martin's first paper, gas-liquid chromatographers have been busy deriving mathematical relationships that describe the separations of their craft.

* *A. T. James and A. J. P. Martin,* Biochem. Journal *50, 679 (1952).*

Many of the derivations are singularly involved and difficult, but the results are almost always straightforward and simple. It is these results, and their applications, which will concern us here. We will deal with (1) calculation of the number of theoretical plates on a column and the HETP for a column, (2) calculation of the number of theoretical plates required for the resolution of two sample component peaks, (3) determination of an optimum flow rate by application of the Van Deemter equation, (4) qualitative identification of the peaks in the chromatogram, and (5) quantitative analysis of components separated by gas chromatography. Ability to do these calculations should equip the student very well for the more common chromatographic tasks.

6.19 Calculating the Number of Theoretical Plates on a Column and the HETP for a Column
≡

First, one must understand that calculation of the number of theoretical plates on a column and the value of HETP for the column is valid for one sample species under one specific set of conditions: carrier gas flow rate, column temperature, column packing, column length, and so on. The calculations themselves are easy; but small variations in the variables noted above make the results invalid. The relationship describing the number of theoretical plates on a column is given below. Figure 6.33 shows graphical results.

$$n = 16\left(\frac{t_R}{\Delta t}\right)^2 \tag{6.9}$$

n = number of theoretical plates

t_R = time from injection to center of component peak, usually measured from air peak

Δt = width of component peak

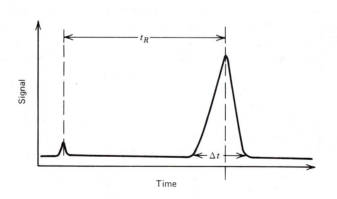

Figure 6.33 Definition of terms.

472 / Separations Not Involving Precipitation

EXAMPLE:

The trace in Figure 6.34 is from a gas chromatogram run with a 4-ft. carbowax column with a n-butanol sample. Calculate n, the number of theoretical plates on the column, and HETP for the column, the latter in cm/plate

$$t_R = \frac{4.5 \text{ in.}}{2.0 \text{ in./min}} = 2.25 \text{ min}$$

$$\Delta t = \frac{0.50 \text{ in.}}{2.0 \text{ in./min}} = 0.25 \text{ min}$$

From Eq. 6.9:

$$n = 16\left(\frac{2.25}{0.25}\right)^2 \cong 1300 \text{ theoretical plates}$$

Actually, n may be calculated more simply than this, because chart distance is directly proportional to time.

$$\frac{t_R}{\Delta t} = \frac{\left(\frac{4.5 \text{ in.}}{2.0 \text{ in./min.}}\right)}{\left(\frac{0.50 \text{ in.}}{2.0 \text{ in./min.}}\right)} = \frac{4.50}{0.50}$$

$$n = 16\left(\frac{4.50}{0.50}\right)^2 \cong 1300 \text{ theoretical plates}$$

To figure HETP is easy:

(column length) = 4 ft

(column length) = (4 ft)(12 in/ft)(2.54 cm/in)

(column length) = 122 cm

$$\text{HETP} = \frac{(\text{column length})}{n} = \frac{122 \text{ cm}}{1300 \text{ plates}} = 0.094 \text{ cm/plate}$$

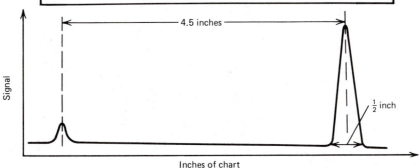

Figure 6.34 *GC trace for estimating HETP.*

6.20   Calculation of the Number of Theoretical Plates
Required for Resolution of Two Sample Component Peaks

≡

The relationship needed for this calculation takes just a little longer to define than that for obtaining the number of theoretical plates on a column, but it is really no more difficult. This relationship is stated below, with Figure 6.35 to aid in definition.

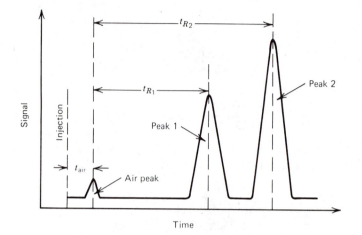

Figure 6.35   *Terms used in resolution calculations.*

$$N_{Req} = 16R^2 \left(\frac{\alpha}{\alpha - 1}\right)^2 \left(\frac{k_2 + 1}{k_2}\right)^2 \tag{6.10}$$

N_{Req} = number of theoretical plates required

R = resolution required. You choose the number.

NOTE:
{
  If R = 1, resolution is roughly 98%
  If R = 1.5, resolution is about 99.7% complete (i.e., the two peaks are separated at the baseline)
}

$\alpha = \dfrac{t_{R_2}}{t_{R_1}}$

t_{R_2} = retention time from air peak to peak 2

t_{R_1} = retention time from air peak to peak 1

$k_2 = \dfrac{t_{R_2}}{t_{air}}$

t_{air} = time from injection to appearance of air peak

EXAMPLE:

Components 1 and 2 were separated on a 10.0-foot column, and the following data were taken.

$$t_{R_2} = 18.0 \text{ min}$$

$$t_{R_1} = 15.0 \text{ min}$$

$$t_{air} = 1.0 \text{ min}$$

$$\Delta t_1 = \text{width of peak 1} = 1.0 \text{ min}$$

What minimum number of theoretical plates is needed for a resolution of 1.5? What minimum column length is needed to get this minimum number of theoretical plates? For this calculation, figure HETP using total retention time $(t_R + t_{air})$ for compound 1.

SOLUTION:

$$N_{Req} = 16R^2\left(\frac{\alpha}{\alpha - 1}\right)^2 \cdot \left(\frac{k_2 + 1}{k_2}\right)^2 \qquad (6.10)$$

$$R = 1.5$$

$$\alpha = \frac{t_{R_2}}{t_{R_1}} = \frac{18.0}{15.0} = 6.0/5.0$$

$$k_2 = \frac{t_{R_2}}{t_{air}} = \frac{18.0}{1.0} = 18.$$

$$N_{Req} = 16(1.5)^2 \cdot \left(\frac{6/5}{6/5 - 1}\right)^2 \cdot \left(\frac{18. + 1}{18.}\right)^2$$

$$N_{Req} \cong 1450 \text{ theoretical plates (tp)}$$

$$n = \text{number of theoretical plates for compound 1}$$

total retention time for cpd 1

$$= t_{R_1} + t_{air} = 15.0 + 1.0$$

$$\Delta t_1 = 1.0 \text{ min}$$

$$n = 16\left(\frac{16.0}{1.0}\right)^2$$

$$n \cong 4100 \text{ tp}$$

$$\text{HETP} = \frac{10.0 \text{ ft}}{4096 \text{ tp}} = 0.0024 \text{ ft/tp}$$

length required = $(\text{HETP})(N_{Req})$ = $(0.0024)(1444)$ = 3.5 ft

(Just to be certain, you could use 4 ft of column.)

6.21 Determination of the Optimum Flow Rate

The Van Deemter equation applies both to liquid-liquid chromatography and to gas-liquid chromatography. It is given again below, and all terms are defined.

$$\text{HETP} = A + B/v + Cv \tag{6.8}$$

HETP = height equivalent to a theoretical plate

v = velocity of mobile phase (carrier gas)

A = multiple path term, or eddy diffusion term

B = molecular diffusion term for carrier (very small in LLC, but about 10^5 times larger than this in GLC)

C = resistance to mass transfer term

Because the molecular diffusion term, B, in liquid-liquid chromatography is rather small, a plot of HETP versus v for liquid chromatography is shaped like that shown in Figure 6.36.

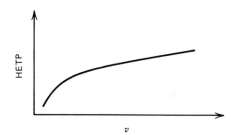

Figure 6.36 *Plot of HETP versus V for liquid-liquid chromatography.*

For gas-liquid chromatography, however, B is rather larger, by a factor of 10^5 or so. The result is that the plot of HETP versus v for gas-liquid chromatography is shaped like that shown in Figure 6.37.

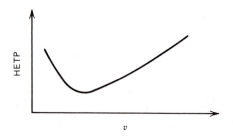

Figure 6.37 *Plot of HETP versus V for gas-liquid chromatography.*

476 / Separations Not Involving Precipitation

In gas-liquid chromatography, then, carrier gas velocity, v, may be manipulated to given an optimum (minimum) value of HETP. This value is found to be just where $v = \sqrt{B/C}$. A derivation that involves calculus can be done to prove this, but a numerical and graphical example will bring the point home.

EXAMPLE:

On a certain column, with helium as the carrier gas, the following values of the Van Deemter terms were found:

A = 0.10 cm

B = 0.35 cm^2/sec (a higher molecular weight gas would give a better value of B)

C = 0.06 sec

Plot HETP versus v for values of v from 1 cm/sec through 10 cm/sec in increments of 1 cm/sec. Find the value of HETP that yields the minimum value of HETP. Is it true that, at minimum HETP, $v = \sqrt{B/C}$?

SOLUTION:

HETP = $A + B/v + Cv$

For v = 1.0 cm/sec

HETP = 0.10 cm + $\dfrac{0.35 \text{ cm}^2/\text{sec}}{1.0 \text{ cm/sec}}$ + (0.06 sec)(1.0 cm/sec)

= 0.51 cm/plate

With nine more calculations, Table 6.8 is constructed.

Table 6.8
VALUES OF HETP AS A
FUNCTION OF v

v, cm/sec	HETP, cm	v, cm/sec	HETP, cm
1.0	0.51	6.0	0.52
2.0	0.40	7.0	0.57
3.0	0.40	8.0	0.62
4.0	0.43	9.0	0.68
5.0	0.47	10.0	0.74

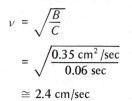

From Table 6.8, Figure 6.38 may be plotted. The plot shows a minimum in the value of HETP near 2.5 cm/sec. HETP is supposed to be at a minimum where $v = \sqrt{B/C}$.

$$v = \sqrt{\frac{B}{C}}$$

$$= \sqrt{\frac{0.35 \text{ cm}^2/\text{sec}}{0.06 \text{ sec}}}$$

$$\cong 2.4 \text{ cm/sec}$$

This calculated value of v is close to the graphically established value of about 2.5 cm/sec. Remember that a value is established graphically from French curve extrapolations. An uncertainty in it of $\pm$ 0.1 cm/sec would not be surprising.

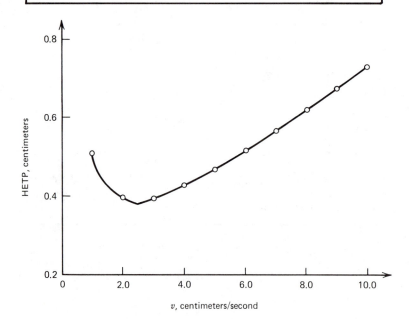

Figure 6.38 *Plot of HETP versus V.*

It is not too hard to determine the values of A, B, and C for a given column, temperature, and species. One has to get three measurements of HETP at three different flow rates. Then, three equations in the three unknowns, A, B, and C are set up and solved for A, B, and C. The little bit of algebra involved might just be worth investigation.

EXAMPLE:

N_2 is used as a carrier gas on a column. The following data are taken for HETP as a function of carrier gas velocity, v.

v, cm/sec	HETP, cm
4.0	0.159
6.0	0.172
8.0	0.189

(a) What are the values of A, B, and C in the Van Deemter equation?

(b) Plot HETP versus v for values of v from 1.0 cm/sec through 10.0 cm/sec in intervals of 1.0 cm/sec. Determine an optimum value of v, one that gives a minimum value of HETP.

SOLUTION:

(a) The Van Deemter equation is:

$$HETP = A + B/v + Cv$$

OR

$$A + B/v + Cv = HETP$$

First, set up three equations in three unknowns.

$$\left\{ \begin{array}{l} A + 1/4B + 4.0C = 0.159 \\ A + 1/6B + 6.0C = 0.172 \\ A + 1/8B + 8.0C = 0.189 \end{array} \right\}$$

Solution for A, B, and C is easy, if one uses the *matrix method*. First, a matrix of coefficients is set up.

$$\begin{vmatrix} 1 & 1/4 & 4.0 & 0.159 \\ 1 & 1/6 & 6.0 & 0.172 \\ 1 & 1/8 & 8.0 & 0.189 \end{vmatrix}$$

The trick is to add, subtract, multiply and divide each line so that the matrix ends up looking like this:

$$\begin{vmatrix} 1 & 0 & 0 & \text{(a number)} \\ 0 & 1 & 0 & \text{(another number)} \\ 0 & 0 & 1 & \text{(still another number)} \end{vmatrix}$$

Once the matrix looks like the one above, the three equations have been solved for A, B, and C.

$$A = \text{a number}$$
$$B = \text{another number}$$
$$C = \text{still another number}$$

The steps in the *inversion of the matrix*, the process of getting it into its final form, are simple.

1. Multiply the second and third rows through by (-1).

$$\begin{vmatrix} 1 & 1/4 & 6.0 & 0.159 \\ -1 & -1/6 & -6.0 & -0.172 \\ -1 & -1/8 & -8.0 & -0.189 \end{vmatrix}$$

2. Add the new second row and the old first row to get a new second row. Get a new third row the same way.

$$\begin{vmatrix} 1 & 1/4 & 4.0 & 0.159 \\ 0 & 2/24 & -2.0 & -0.013 \\ 0 & 1/8 & -4.0 & -0.030 \end{vmatrix}$$

3. Multiply the second row through by (-2).

$$\begin{vmatrix} 1 & 1/4 & 4.0 & 0.159 \\ 0 & -4/24 & 4.0 & 0.026 \\ 0 & 1/8 & -4.0 & -0.030 \end{vmatrix}$$

4. Add the second row to the third row to get a new second row.

$$\begin{vmatrix} 1 & 1/4 & 4.0 & 0.159 \\ 0 & -1/24 & 0 & -0.004 \\ 0 & 1/8 & -4.0 & -0.030 \end{vmatrix}$$

5. Multiply the second row through by (-24).

$$\begin{vmatrix} 1 & 1/4 & 4.0 & 0.159 \\ 0 & 1 & 0 & 0.096 \\ 0 & 1/8 & -4.0 & -0.030 \end{vmatrix}$$

6. Multiply the third row through by (-8).

$$\begin{vmatrix} 1 & 1/4 & 4.0 & 0.159 \\ 0 & 1 & 0 & 0.096 \\ 0 & -1 & 32.0 & 0.240 \end{vmatrix}$$

7. Add the second row to the third row for a new third row.

$$\begin{vmatrix} 1 & 1/4 & 4.0 & 0.159 \\ 0 & 1 & 0 & 0.096 \\ 0 & 0 & 32.0 & 0.336 \end{vmatrix}$$

8. Multiply the third row through by 1/32.

$$\begin{vmatrix} 1 & 1/4 & 4.0 & 0.159 \\ 0 & 1 & 0 & 0.096 \\ 0 & 0 & 1.0 & 0.105 \end{vmatrix}$$

9. Multiply the second row through by $(-1/4)$, and add it to the first row, to make a new first row.

$$\begin{vmatrix} 1 & 0 & 4.0 & 0.135 \\ 0 & 1 & 0 & 0.096 \\ 0 & 0 & 1 & 0.105 \end{vmatrix}$$

10. Multiply the third row through by (-4) and add it to the first row, to give a new first row.

$$\begin{vmatrix} 1 & 0 & 0 & 0.095 \\ 0 & 1 & 0 & 0.096 \\ 0 & 0 & 1 & 0.010 \end{vmatrix}$$

$A = 0.095$ cm

$B = 0.096$ cm^2/sec ⟵ { Note: B for N_2 (MWt = 28) is lower than B for He (MWt = 4)

$C = 0.010$ sec

Thus by simple application of matrix algebra, it is possible to solve equations in three unknowns. There *is* some round-off error. Such error could be reduced by carrying more significant figures, but this is scarcely necessary, because HETP is not often obtained to more than two significant figures.

(b) Now, let us calculate HETP as a function of v.

$$A + B/v + Cv = \text{HETP}$$

$$0.095 + 0.096/v + 0.010v = \text{HETP}$$

For $v = 1.0$ cm/sec

$$0.095 + 0.096/1.0 + (0.010)(1) = \text{HETP}$$

$$0.201 \text{ cm} = \text{HETP}$$

Other values of HETP can be calculated and plotted to yield the graph of Figure 6.39. The optimum value of v, that is, the v-value that gives the minimum value of HETP appears, from Figure 6.39, to be somewhere around 3 cm/sec. From the relationship for the optimum value of v, a calculation may be made. The value obtained, 3.1 cm/sec, is close to the graphically observed value

$$v = \sqrt{\frac{B}{C}} = \sqrt{\frac{0.096}{0.01}} = 3.1 \text{ cm/sec}$$

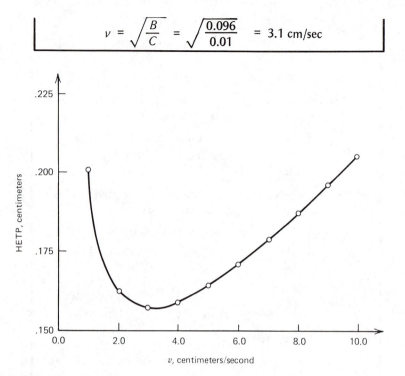

Figure 6.39 *Plot of HETP versus* V.

6.22 Qualitative Analysis

Qualitative analysis with gas-liquid chromatography is not difficult. It is true, of course, that only certain detectors can give nearly complete, positive, and unequivocal indication of what species is emerging from the column at a given time. Most detectors simply let the experimenter know that something or other is emerging from the column. If the chemist has some idea of what unknowns to look for, he or she will be able to run pure standard samples through the chromatograph under the same conditions that the unknown sample is run through. By a process of comparison of chromatograms of unknowns and pure standards, satisfactory identification of an unknown can be made. The retention volume, V_R, and the retention time, t_R, are the quantities that are characteristic of a given compound under a given set of experimental conditions.

t_R = retention time   = time from maximum of air peak to maximum of component peak

V_R = retention volume = volume of carrier gas passed through column from time of in-

jection to time of appearance of
component peak

$$V_R = (t_R)(\text{carrier gas flow rate})$$

Either V_R or t_R can be used for qualitative evaluation, but t_R is usually more convenient to use. An example of qualitative analysis is given below.

EXAMPLE:

A mixture of simple, straight-chain hydrocarbons was made up by a chemist. The mixture consisted of n-hexane $(CH_3(CH_2)_4CH_3)$, n-nonane $(CH_3(CH_2)_7CH_3)$, n-undecane $(CH_3(CH_2)_9CH_3)$, and n-tetradecane $(CH_3(CH_2)_{12}CH_3)$. The flow rate of helium through the nonpolar column was 22 milliliter/min, and the chart tracing for the known mixture was preserved as Figure 6.40. An unknown sample, containing only straight-chain hydrocarbons, was run under conditions identical to those under which the standard was run. The chromatogram of the unknown is shown in Figure 6.41. What are the straight-chain hydrocarbons that make up the unknown?

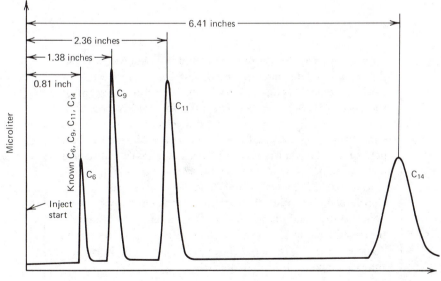

Figure 6.40 *Gas chromatogram of standard containing* n-*hexane,* n-*nonane,* n-*undecane, and* n-*tetradecane.*

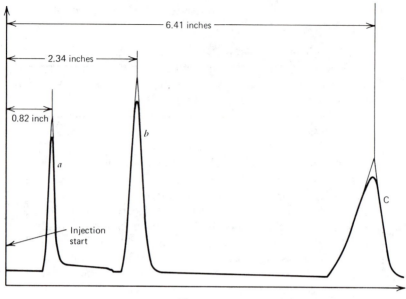

Figure 6.41 *Gas chromatogram of sample containing straight-chain hydrocarbons only. Unknown was run under same conditions as standard of Figure 6.40.*

SOLUTION:

First, observe the chromatogram of the known mixture, Figure 6.40. Note that the abscissa of the chromatogram is marked "time." Actually, on a strip-chart chromatogram of this sort, *time is proportional to distance along the abscissa.* Chart speeds on the recorders used as readout devices in gas chromatography are usually given in units of inches of chart per minute. Thus, there is nothing peculiar about the distances marked along the top edge of the chromatogram: 0.81 in., 1.38 in., 2.36 in., and 6.41 in. These distances are just proportional to the retention times of the components whose peaks appear on the chromatogram. Note that a retention distance is measured from injection of the sample to the apex of a given component peak. Table 6.9 shows relative retention distances for each component of the standard.

Table 6.9
RETENTION DISTANCES FOR COMPOUNDS

Component	Retention Distance, inches
n-hexane (C_6)	0.81
n-nonane (C_9)	1.58
n-undecane (C_{11})	2.36
n-tetradecane (C_{14})	6.41

How does one know that C_6 is the first off the column, C_9 the next, and so forth? One could, for the most rigorous work, run pure n-hexane, note its retention distance or time, and compare this to the chromatogram of the mixture. Then one would run a pure sample of each component, noting retention time, and comparing it to the chromatogram of the mixture. However, it is quite reasonable to assume that, in a sample composed of members of an homologous series, the shortest-chained or lightest member will emerge first, followed by the next-lightest, and so forth. This assumption has been borne out in thousands of laboratories.

Next, observe the chromatogram of Figure 6.41. The retention distances of unknown components a, b, and c are given in Table 6.10 below, together with their probable identities.

Table 6.10
RETENTION DISTANCES FOR COMPONENTS OF UNKNOWN SAMPLE

Component	Retention Distance (in.)	Probable Identity
a	0.82	n-hexane (0.81 in. for known)
b	2.34	n-undecane (2.36 in. for known)
c	6.41	n-tetradecane (6.41 in. for known)

There is a mathematical relationship in gas chromatography that can prove useful in qualitative analysis. If one injects a mixture containing an homologous series of compounds onto a column, the logarithm of retention time will often increase in proportion to the number of carbon atoms in a component. A sort of calibration plot can be made for identification of an unknown that has already been partially identified as a member of a certain homologous series.

EXAMPLE:

A series of straight-chain alcohols was run on a carbowax column. The retention times found are listed below.

Species	Retention Time (sec.)
n-Butyl alcohol	
$CH_3(CH_2)_3OH$	42
n-Hexyl alcohol	
$CH_3(CH_2)_5OH$	73
n-Octyl alcohol	
$CH_3(CH_2)_7OH$	135
n-Decyl alcohol	
$CH_3(CH_2)_9OH$	250

An unknown species, known only to be a straight-chain alcohol, is injected onto the same column, under the same conditions as the standards. Its retention time is 100 sec. What is the unknown?

SOLUTION:

A plot of log (retention time) versus number of carbon atoms is made. Two-cycle semilog paper makes the task easier. Figure 6.42 is the desired plot. From the plot, the unknown is estimated to contain seven carbon atoms. The unknown is most likely n-heptyl alcohol, $CH_3(CH_2)_6OH$. Caution must be used along with such logarithmic plots; much must be known about an unknown before the plot can be relied upon.

6.23 Quantitative Analysis

Quantitative analysis in gas chromatography is certainly important. Often, for example, one will not only want to know what pesticides are present in a milk sample, but the concentrations of these pesticides also. The list of such applications could go

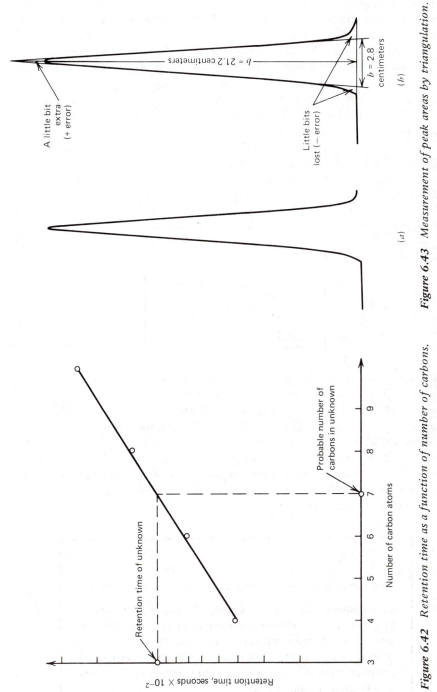

Figure 6.43 Measurement of peak areas by triangulation.

Figure 6.42 Retention time as a function of number of carbons.

on and on. How, then, does one quantitatively interpret a chromatogram? The key to quantitative interpretation is the fact that peak area is proportional to the amount of the compound represented by that peak. Because triangulation is the method most often used by beginners and in laboratories where only the occasional chromatogram is run, it is worth looking at here. Recall that peak shapes in countercurrent distribution and liquid-liquid chromatography are roughly Gaussian. The same sort of theory governs gas chromatography, so that peaks in GC have Gaussian shapes, too. If a Gaussian is a reasonably narrow one, its shape approximates that of a triangle. Figure 6.43a shows a gas chromatographic peak recorded by the author, and Figure 6.43b shows the triangulation of that peak. Note that a small positive error is introduced at the top of the peak, and two small negative errors at the bottom. Sometimes, these errors can come close to canceling each other out. From the measurements made, the area of the triangle is easy to figure.

$$A = \frac{bh}{2}$$

b = length of base = 2.8 cm

h = height = 21.2 cm

A = 30 cm^2

Manipulating the figures for peak areas to give weight percentages, or weights, of components in a mixture is easy. It may be done in a number of ways: (1) absolute calibration, (2) use of previously established multiplicative factors for each mixture component, or (3) a rough method. The first and third methods are illustrated here.

Absolute calibration is the surest means of quantitative analysis. Briefly, a known quantity of a pure substance is injected into a column. The resultant peak area is measured. The peak area resulting from the substance in an unknown mixture is measured. By a simple application of ratio or proportion, the amount of the substance in the unknown mixture may be found.

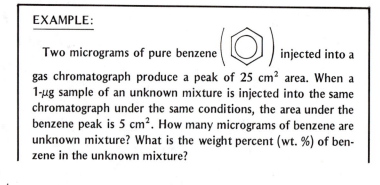

EXAMPLE:

Two micrograms of pure benzene injected into a gas chromatograph produce a peak of 25 cm^2 area. When a 1-μg sample of an unknown mixture is injected into the same chromatograph under the same conditions, the area under the benzene peak is 5 cm^2. How many micrograms of benzene are unknown mixture? What is the weight percent (wt. %) of benzene in the unknown mixture?

SOLUTION:

Let x = no. μg benzene in unknown

5 cm^2 = benzene peak area in unknown

2 μg = no. μg benzene in standard

25 cm^2 = benzene peak area in standard

$$\frac{x}{2} = \frac{5}{25}$$

$$x = \left(\frac{5}{25}\right) \cdot (2)$$

$$= 0.4 \ \mu\text{g benzene}$$

$$\text{wt. \% benzene} = \left(\frac{0.4 \ \mu\text{g in unk}}{1.0 \ \mu\text{g total unk}}\right) \cdot (100\%) = 40\%$$

Another example can be taken to illustrate the use of the absolute calibration method. It involves making up a standard of known amounts of pure components, and using this, rather than singly-injected pure components.

EXAMPLE:

A standard solution is made up of equal weights of n-hexane, n-nonane, n-undecane, and n-tetradecane. A small amount of the solution, determined by density calculation to weigh 0.80 μg, is injected onto a column. The peak areas for each component are shown in Table 6.11.

Table 6.11
PEAK AREAS FOR 0.80 μg MIXTURE

μg of Species	Species	Peak Area (scale units)^2
(1/4)(0.80) = 0.20	n-Hexane (C$_6$)	106.9
(1/4)(0.80) = 0.20	n-Heptane (C$_7$)	102.6
(1/4)(0.80) = 0.20	n-Undecane (C$_{11}$)	80.0
(1/4)(0.80) = 0.20	n-Tetradecane (C$_{14}$)	50.0

Peak areas for the components of an unknown sample run under the same conditions are given in Table 6.12.

Table 6.12
PEAK AREAS FOR UNKNOWN MIXTURE

μg of Species	Species	Peak Area (scale units)^2
?	n-Hexane (C_6)	32.6
?	n-Heptane (C_7)	46.8
?	n-Undecane (C_{11})	73.8
?	n-Tetradecane (C_{14})	84.9

Calculate the amounts of each component in the unknown mixture, and the weight percent of each component in the mixture.

SOLUTION:

Let w = no. μg of C_6 in unknown

$$\frac{w}{32.6} = \frac{0.20}{106.9}$$

$$w = \frac{0.20}{106.9} \cdot (32.6) = 0.061 \ \mu g \ C_6$$

Let x = no. μg of C_7 in unknown

$$\frac{x}{46.8} = \frac{0.20}{102.6}$$

$$x = \left(\frac{0.20}{102.6}\right) \cdot (46.8) = 0.091 \ \mu g \ C_7$$

Let y = no. μg of C_{11} in unknown

$$\frac{y}{80.0} = \frac{0.20}{73.8}$$

$$y = \left(\frac{0.20}{73.8}\right) \cdot (80.0) = 0.22 \ \mu g \ C_{11}$$

Let z = no. μg of C_{14} in unknown

$$\frac{z}{84.9} = \frac{0.20}{50.0}$$

$$z = \left(\frac{0.20}{50.0}\right) \cdot (84.9) = 0.34 \, \mu g \, C_{14}$$

Now, the weight percent of each component in the mixture can be found.

$$\text{total weight} = 0.061 + 0.091 + 0.22 + 0.34$$

$$= 0.71 \, \mu g$$

$$\%C_6 = \left(\frac{0.061}{0.71}\right) \cdot (100\%) = 8.6\%$$

$$\%C_7 = \left(\frac{0.091}{0.71}\right) \cdot (100\%) = 13\%$$

$$\%C_{11} = \frac{0.22}{0.71} \cdot (100\%) = 31\%$$

$$\%C_{14} = \left(\frac{0.34}{0.71}\right) \cdot (100\%) = 48\%$$

The total of weight percents is $8.6 + 13 + 31 + 48 \cong 101\%$. The error is quite small compared to what could arise from area measurement errors.

Next, figures from the example above will be used to illustrate the rough method mentioned at the beginning of the discussion of quantitative analysis. The use of the rough method is only recommended to those who can find no other, or who need only crude approximations. The users assume that the detector is equally sensitive to all species in a mixture. To give a specific example, the users of the rough method postulate that one microgram of n-hexane would give a peak of the same area as that resulting from one microgram of n-tetradecane. Such an assumption *may* be valid for very near neighbors in an homologous series, but there is no compelling reason to suppose that its validity extends beyond that narrow limit.

EXAMPLE:

Using the data in Table 6.12, calculate the weight percentages of C_6, C_7, C_{11}, and C_{14} in the mixture by a rough method.

SOLUTION:

First, sum the areas of all the peaks.

$$\Sigma \text{ areas} = 32.6 + 46.8 + 73.8 + 84.9$$

$$\Sigma \text{ areas} = 238.1$$

Second, assume that the detector is equally sensitive to all mixture components. Then, the following relationship would hold:

weight percent component

$$= \frac{(\text{peak area due to component})}{(\Sigma \text{ areas})} (100\%)$$

Third, do the arithmetic.

$$\%C_6 = \left(\frac{32.6}{238.1}\right) \cdot (100\%)$$

$$= 13.7\%$$

$$\%C_7 = \left(\frac{46.8}{238.1}\right) \cdot (100\%)$$

$$= 19.6\%$$

$$\%C_{11} = \left(\frac{73.8}{238.1}\right) \cdot (100\%)$$

$$= 31.0\%$$

$$\%C_{14} = \left(\frac{84.9}{238.1}\right) \cdot (100\%)$$

$$= 35.7\%$$

In Table 6.13, the results from absolute calibration and the rough method are shown.

Table 6.13
COMPARISON OF PERCENTS OF COMPONENTS OF UNKNOWN CALCULATED BY ABSOLUTE AND ROUGH METHODS

Component	Wt. % Absolute	Wt. % Rough
C_6	8.6	13.7
C_7	13.0	19.6
C_{11}	31.0	31.0
C_{14}	48.0	35.7

As noted above, the rough method is best used only when there is no alternative. It may yield more significant figures but erroneous answers with many significant figures are of little use.

≡ 6.24 PROBLEMS ≡

1. A certain chlorinated hydrocarbon, 1,1,1-trichloroethane, is run through a gas chromatograph. The apex of its peak appears 137 seconds after the apex of the air peak. The trichloroethane peak is 10 seconds wide. What is the value of HETP, in cm, for this species on this column under these conditions, if the column is five feet long?
   Ans: *0.0507 cm*

2. On another column in another chromatograph, 1,1,1-trichloroethane (tric) and carbon tetrachloride (ctet) are separated. The column is ten feet long. The following data are collected.

$$t_{air} = 10. \text{ sec}$$

$$\text{for tric} \quad t_{R_1} = 250. \text{ sec}$$

$$\text{for ctet} \quad t_{R_2} = 300. \text{ sec}$$

$$\text{width of } \Delta t_1 = 8. \text{ sec}$$
$$\text{peak 1}$$
$$\text{(tric)}$$

   If a resolution of 1.5 between tric and ctet is desired, what minimum number of theoretical plates is needed? What minimum column length is needed? [Figure HETP using total retention time $(t_{R_1} + t_{air})$ for tric.]
   Ans: *16,900 t.p.; about a foot long*

3. On a certain chromatographic column, with N_2 used as the carrier gas, the following Van Deemter terms were found:

$$A = 0.10 \text{ cm}$$

$$B = 0.080 \text{ cm}^2/\text{sec}$$

$$C = 0.01 \text{ sec}$$

   Graph HETP versus v for v-values from 1.0 cm/sec to $v = 10.0$ cm/sec in intervals of 1.0 cm/sec. Determine the optimum carrier gas velocity from the graph and from the relationship

$$v = \sqrt{\frac{B}{C}}$$

   Ans: *Make graph. v ≅ 3 cm/sec*

4. On another chromatographic column, used to separate hydrocarbons, with hydrogen as the carrier gas, the following values of HETP as a function of v are found:

v, cm/sec	HETP, cm
1.0	0.625
5.0	0.525
8.0	0.7125

Calculate A, B, and C for this column, and graph HETP versus v for carrier gas velocity values from 1.0 to 10.0 cm/sec in 1.0 cm/sec intervals.

Ans: $A = 0.050$ cm, $B = 0.500$ cm^2/sec, $C = 0.075$ sec; make graph

5. A series of straight-chain alkanes is run through a column in which squalane is the liquid phase. Their retention times are given below.

Species	Retention Time (sec)
n-Hexane $CH_3(CH_2)_4CH_3$	30
n-Octane $CH_3(CH_2)_6CH_3$	62
n-Decane $CH_3(CH_2)_8CH_3$	118
n-Dodecane $CH_3(CH_2)_{10}CH_3$	241

A species known to be a straight-chain alkane has a retention time of 85 seconds on the column. What is it likely to be?

Ans: *n-nonane*

6. A sample of mixed chlorinated hydrocarbons is run through a chromatograph with a thermal conductivity detector. A standard of known composition is run through the same chromatograph.

Standard Species	wt, μg	Peak Area (scale units)^2	Unknown Peak Area (scale units)^2
Chloroethane	0.40	110.0	82.3
1-Chloropropane	0.40	112.2	No peak
1-Chloropentane	0.40	87.3	125.2
1-Chloroheptane	0.40	78.4	180.0

Calculate the weight percents of the components of the unknown sample by the absolute calibration method, and by the rough method described.

	Component	Wt. %, Abs. Calib. Method	Wt. %, Rough Method
Ans:	Chloroethane	17	21
	1-Chloropentane	32	32
	1-Chloroheptane	51	47

≡ 6.25 DETAILED SOLUTIONS TO PROBLEMS ≡

1.

$$n = 16\left(\frac{t_R}{\Delta t}\right)^2 = \text{number of theoretical plates} \qquad (6.9)$$

$t_R = 137 \text{ sec}$

$\Delta t = 10 \text{ sec}$

$$n = 16\left(\frac{137}{10}\right)^2 \cong 3000 \text{ theoretical plates}$$

$$\text{HETP} = \frac{\text{column length}}{n} = \frac{(5\text{ft})(12 \text{ in./ft})(2.54 \text{ cm/in.})}{3000} = 0.05 \text{ cm}$$

2.

$$N_{\text{Req}} = 16R^2\left(\frac{\alpha}{\alpha - 1}\right)^2 \cdot \left(\frac{k_2 + 1}{k_2}\right)^2 \qquad (6.10)$$

$$\alpha = \frac{t_{R_2}}{t_{R_1}} = \frac{300.}{250.} = \frac{6.00}{5.00}$$

$$R = 1.5$$

$$k_2 = \frac{t_{R_2}}{t_{\text{air}}} = \frac{300.}{10.} = 30.$$

$$N_{\text{Req}} = 16\,(1.5)^2 \cdot \left(\frac{6/5}{6/5 - 1}\right)^2 \cdot \left(\frac{30 + 1}{30}\right)^2$$

$$\cong 1400 \text{ theoretical plates}$$

$n = $ number theoretical plates on column

$$= 16\left(\frac{t_{R_1} + t_{\text{air}}}{\Delta t_1}\right)^2 = 16\left(\frac{250 + 10}{8}\right)^2$$

$$\cong 16,900 \text{ theoretical plates}$$

$$\text{HETP} = \frac{\text{column length}}{n} = \frac{10}{16,900} = 5.9 \times 10^{-4} \text{ ft/tp}$$

length needed $= (N_{Req})(HETP)$

$$= (1400 \text{ tp})(5.9 \times 10^{-4} \text{ ft/tp}) = 0.83 \text{ ft}$$

One might use at least a foot-long column.

3.
$$\text{HETP} = A + B/v + Cv \tag{6.8}$$

for $v = 1.0$ cm/sec:

$$\text{HETP} = 0.1 + 0.080/1 + (0.01)(1) = 0.19 \text{ cm}$$

In like manner, HETP is determined for other values of v.

v, cm sec	HETP, cm
1.0	0.19
2.0	0.16
3.0	0.16
4.0	0.16
5.0	0.17
6.0	0.17
7.0	0.18
8.0	0.19
9.0	0.20
10.0	0.21

A plot of HETP versus v shows a minimum of HETP at about $v = 3$ cm/sec. By calculation:

$$v = \sqrt{\frac{B}{C}} = \sqrt{\frac{0.080}{0.01}} \cong 2.8 \text{ cm/sec}$$

4. This is most easily solved by the matrix method. (One could be more certain of the results by determining an experimental curve of HETP versus v for *many* v-values.)

$$A + B/v + Cv = \text{HETP} \tag{6.8}$$

$$\left\{ \begin{array}{l} A + B/1 + 1C = 0.625 \\ A + B/5 + 5C = 0.525 \\ A + B/8 + 8C = 0.7125 \end{array} \right\} \begin{array}{l} \text{three equations} \\ \text{three unknowns} \end{array}$$

the matrix:
$$\begin{vmatrix} 1 & 1 & 1 & 0.625 \\ 1 & 0.2 & 5 & 0.525 \\ 1 & 0.125 & 8 & 0.7125 \end{vmatrix}$$

1. Multiply second row by (-1) and add to first row to get new second row.

$$\begin{vmatrix} 1 & 1 & 1 & 0.625 \\ 0 & 0.8 & -4 & 0.100 \\ 1 & 0.125 & 8 & 0.7125 \end{vmatrix}$$

2. Multiply third row by (-1) and add to first row to get new third row.

$$
\begin{vmatrix}
1 & 1 & 1 & 0.625 \\
0 & 0.8 & -4 & 0.100 \\
0 & 0.875 & -7 & -0.0875
\end{vmatrix}
$$

3. Multiply second row by ($-0.875/8$) and add to third row. Result will be new third row.

$$
\begin{vmatrix}
1 & 1 & 1 & 0.625 \\
0 & 0.8 & -0.4 & 0.100 \\
0 & 0 & -2.625 & -0.196875
\end{vmatrix}
$$

4. Multiply third row by ($-4/2.625$) and add to second row. Result will be new second row.

$$
\begin{vmatrix}
1 & 1 & 1 & 0.625 \\
0 & 0.8 & 0 & 0.400 \\
0 & 0 & -2.625 & -0.196875
\end{vmatrix}
$$

5. Multiply second row by ($-1/0.8$) and add to first row. Result will be new first row.

$$
\begin{vmatrix}
1 & 0 & 1 & 0.125 \\
0 & 0.8 & 0 & 0.400 \\
0 & 0 & -2.625 & -0.196875
\end{vmatrix}
$$

6. Multiply third row by ($1/2.625$) and add to first row. Result will be new first row.

$$
\begin{vmatrix}
1 & 0 & 0 & 0.050 \\
0 & 0.8 & 0 & 0.400 \\
0 & 0 & -2.625 & -0.196875
\end{vmatrix}
$$

7. Multiply second row through by ($1/0.8$). Result will be new second row.

$$
\begin{vmatrix}
1 & 0 & 0 & 0.050 \\
0 & 1 & 0 & 0.500 \\
0 & 0 & -2.625 & -0.196875
\end{vmatrix}
$$

8. Multiply third row through by ($-1/2.625$).

$$
\begin{vmatrix}
1 & 0 & 0 & 0.050 \\
0 & 1 & 0 & 0.500 \\
0 & 0 & 1 & 0.075
\end{vmatrix}
$$

Thus: $A = 0.050$ cm

$B = 0.500$ cm^2/sec

$C = 0.075$ sec

Now, calculate HETP as a function of v.

$$A + B/v + Cv = \text{HETP}$$

$$0.050 + 0.500/v + (0.075)v = \text{HETP}$$

$$\text{For } v = 1.0 \text{ cm/sec}$$

$$0.050 + 0.500/1 + (0.075)1 = \text{HETP}$$

$$\text{HETP} = 0.625 \text{ cm}$$

v, cm/sec	HETP, cm
1.0	0.62
2.0	0.45
3.0	0.44
4.0	0.48
5.0	0.52
6.0	0.58
7.0	0.65
8.0	0.71
9.0	0.78
10.0	0.85

A plot of HETP versus v shows a minimum value of HETP around 2.7 cm/sec carrier gas velocity. The equation for optimum v is:

$$v = \sqrt{\frac{B}{C}} = \sqrt{\frac{0.500}{0.0750}} = 2.6 \text{ cm/sec}$$

Agreement is good between the two values.

5. A plot of log(retention time) versus number of carbon atoms is made. From this plot, one can conclude that the unknown is probably n-nonane (nine carbons).

6. Let $w = \mu g$ of chloroethane in sample

$$\frac{w}{82.3} = \frac{0.40}{110.0}$$

$$w = \left(\frac{0.40}{110.0}\right)(82.3) = 0.30 \ \mu g$$

Let $x = \mu g$ of 1-chloropropane in sample

$$x = 0$$

Let $y = \mu g$ 1-chloropentane in sample

$$\frac{y}{125.2} = \frac{0.40}{87.3}$$

$$y = \left(\frac{0.40}{87.3}\right) \cdot (125.2) = 0.57 \ \mu g$$

Let z = μg 1-chloroheptane in sample

$$\frac{z}{180.0} = \frac{0.40}{78.4}$$

$$z = \left(\frac{0.40}{78.4}\right) \cdot (180.0) = 0.92 \ \mu g$$

$$\text{total sample weight} = 0.30 + 0.57 + 0.92 = 1.79 \ \mu g$$

$$\% \text{ chloroethane} = \left(\frac{0.30}{1.79}\right) \cdot (100\%) = 17\%$$

$$\% \text{ 1-chloropentane} = \left(\frac{0.57}{1.79}\right) \cdot (100\%) = 32\%$$

$$\% \text{ 1-chloroheptane} = \left(\frac{0.92}{1.79}\right) \cdot (100\%) = 51\%$$

By the rough method:

$$\Sigma \text{ areas} = 82.3 + 125.2 + 180.0 = 387.5$$

$$\% \text{ component} = \frac{\text{(component peak area)}}{(\Sigma \text{ areas})} (100\%)$$

$$\% \text{ chloroethane} = \frac{82.3}{387.5} \cdot (100\%) = 21\%$$

$$\% \text{ 1-chloropentane} = \frac{125.2}{387.5} \cdot (100\%) = 32\%$$

$$\% \text{ 1-chloroheptane} = \left(\frac{180.0}{387.5}\right) \cdot (100\%) = 47\%$$

≡ 6.26 SEPARATIONS BY SORPTION ≡
CHROMATOGRAPHY

Many means of chemical separation have been discussed: the precipitation of ionic species and liquid-liquid and gas-liquid separation of metal complexes and organic species with separatory funnels and more complex apparatus. There are many other techniques—some old, some quite new, which may be used in the separation of both organic and inorganic species. The technique we are concerned with here has many names, the simplest of which is *sorption chromatography*.

What is the process of sorption chromatography, or, more specifically, what are the effects of the process properly applied? Figure 6.44 shows the essentials. The column

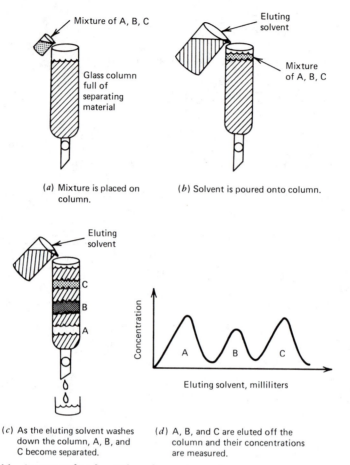

(a) Mixture is placed on column.

(b) Solvent is poured onto column.

(c) As the eluting solvent washes down the column, A, B, and C become separated.

(d) A, B, and C are eluted off the column and their concentrations are measured.

Figure 6.44 *An example of sorption chromatography.*

contains some granular or powdery substance called the separating material. In part (a) of Figure 6.44, it is loaded with a mixture of A, B, and C. An eluting solvent is then poured onto the column, as shown in (b). As the eluting solvent washes down the column, mixture components A, B, and C move apart on the separating material, as shown in (c). If A, B, and C are measured in the outflow, a plot of concentration versus volume of solvent is obtained and is shown in part (d). Thus, A, B, and C are separated on a column of some special material. The term *chromatography* comes from the original work in the field done by Tswett, a Russian botanist, in the early years of this century.* He separated chlorophyll constituents, which are of varying colors. He called the result a chromatogram, a term whose root words mean "color writing."

---

* *M. Tswett, Ber. d. deut. botan. Ges. 24, 384 (1906).*

Tswett's work on chlorophyll involved a very special sort of sorption chromatography—adsorption column chromatography. Adsorption is simply binding to a surface, while absorption means a taking in, or engulfing, far below any surfaces. Tswett placed an extract of plant pigments on a column of $CaCO_3$. The components of the mixture were firmly adsorbed on the $CaCO_3$ granules. Then Tswett eluted the mixture by pouring a solvent, either CS_2 or petroleum ether, onto the column. As the solvent moved down the column, so did the brightly colored constituents of the chlorophyll, each at a different rate. As soon as the constituents were separated from each other, Tswett stopped the flow of solvent, and pushed the column contents carefully out of the tube. He separated the components in a simple way, by slicing out the colored regions of the column with a knife. The reason that a separation could be done is simple. Components of the mixture were adsorbed on the $CaCO_3$ of the column with varying degrees of strength. The solvent, too, was adsorbed onto $CaCO_3$. When solvent was poured onto the column, it slowly replaced the chlorophyll components. The least strongly adsorbed component, like the A of Figure 6.44, was most easily replaced by solvent, and moved farthest down the column. The next-least-strongly adsorbed component, like B, was not quite so easily replaced by solvent, and moved down the column, but quite so far as the first species. The process went on and on, until as many as five components were separated. Chromatography, even in this early version, is simple, elegant, and inexpensive; the cost of a Craig apparatus to perform the same tasks would be very high indeed, as many tubes would be needed. The deed would be prohibitively hard to perform with separatory funnels.

Adsorption column chromatography is widely used in organic chemistry and biochemistry. The familiar alumina (Al_2O_3)-filled column is one example. A contaminated product or solvent is often passed through such a column, and certain impurities are strongly adsorbed on the alumina. Another prosaic example of adsorption column chromatography may be found in the use of a charcoal filter, or column, to remove organic impurities from water. Many organic compounds, found in trace, but objectionable, amounts in drinking water, are strongly adsorbed onto charcoal.

Another form of sorption chromatography of great usefulness is ion-exchange chromatography, a technique practiced with an ion-exchange resin being used as the separating material, and a number of mixtures, from alkali metal salts to transition metal salts to amino acids being the objects of separation. Many natural ion exchangers can be found, but ion-exchange resins can be tailor-made today for the purposes of of the analyst.

Ion exchange resins look to the eye to be composed of many little (50 mesh and smaller) amber beads or spheres. Sizes and colors vary, depending on which resin is used. An ion exchange resin bead is polymeric with functional groups attached here and there on the polymer. The polymeric part of the resin may be formed from styrene $CH=CH_2$ and divinyl benzene $CH=CH_2$. The styrene, on polymerization,

CH=CH₂

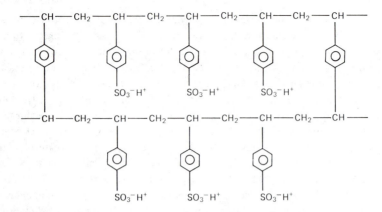

Figure 6.45 Idealized picture of a strong acid cation exchanger.

forms long chains, and the divinyl benzene provides cross-linking. Ion-containing functional groups may be placed on the benzene rings. In the example shown in Figure 6.45, the functional groups are sulfonic acid ($-SO_3^-H^+$) groups. Sulfonic acids are known as strong acids, that is, completely ionized. Resins having sulfonic acid groups are called *strong acid resins*. The hydrogen ions are the only species that can be removed from the resin beads into aqueous solution. All the rest of the resin, including the $-SO_3^-$ groups, is part of the polymeric bead, and not at all soluble in water. Only the H^+ ions may venture into aqueous solution. They may be replaced by other positive ions in a process of *ion exchange.* Ion exchange is the term most commonly used by workers in the field. In the earlier literature, reference was often made to "adsorption" of ions from solution onto the resin. "Adsorption" is not a term much used in today's literature of ion exchange. "Sorption" is the word more commonly used to denote the binding of ion to resin. The process of ion exchange can be represented as an equilibrium.

$$R^-H^+$$

= resin in protonated form, R^- is the organic, nondissolving part of the resin

M^+ = a metal ion in solution

$R^-H^+ + M^+ \rightleftharpoons R^-M^+ + H^+$

$$K = \frac{[R^-M^+][H^+]}{[R^-H^+][M^+]}$$

Terms such as $[R^-M^+]$ and $[R^-H^+]$ are not easy to define. Very often, therefore, a sort of conditional constant—a familiar notion—, or *distribution coefficient*, D, is defined for the sake of convenience.

$$
\begin{array}{l}
C_{MR} = \text{analytical concentration of metal on resin} \\[8pt]
C_M = \text{analytical concentration of metal in aqueous phase} \\[8pt]
D = \dfrac{C_{MR}}{C_M}
\end{array}
$$

D-values are highly dependent on solution conditions. Typically, when values of D are between 0.3 and 1.0, the sorption of an ion onto a resin is said to be slight. When D-values are much greater than unity, sorption is said to be strong.[*]

Other forms of resins besides the strong acid variety exist. Some of these are (a) strong base resins, with quaternary amine groups such as $-N(CH_3)_3^+Cl^-$; (b) weak base resins, with amine groups such as $-NR_2$, where R is a small organic group; and (c) weak acid resins, with carboxylic acid, or $-COOH$ groups. A great variety of resins is available from domestic and foreign manufacturers. Some of these resins, with their properties, are listed in Table 6.14.

Table 6.14
COMMON ION-EXCHANGE RESINS

Type	Trade Names	Capacity, meq/gram
Cation exchange	Dowex 50 ®	4.8
Strong acid	Amberlite IR-120 (H) AR ®	4.2
Cation exchange	Amberlite IRC-150 (H) ®	10.0
Weak acid		
Anion exchange	Dowex 1 ®	3.5
Strong base	Amberlite IRA-400 (Cl) ®	2.3
Anion exchange	Amberlite 4B ®	10.0
Weak base		

The far-right column of Table 6.14 is labeled "capacity, meq/gram." What does this mean? The capacity is defined by the manufacturers as the number of milliequivalents of ions that can be exchanged by one gram of resin.

1 equivalent
        = 1 mole of positive or negative charges

1 milliequivalent
        = 1 millimole of positive or negative charges

(no. milliequivalents)
        = (no. millimoles of anion)(no. charges on that ion)

[*] *J. H. Yoe and H. J. Koch, eds.,* Trace Analysis, *John Wiley and Sons, Inc., New York, 1957, p.89.*

EXAMPLE:

How many milliequivalents of positive charge are there in 50 milliliters of 0.10 M NaCl?

$$(\text{no. moles Na}^+) = \frac{(50\ \text{milliliter})}{(1000\ \text{milliliter/liter})} \cdot (0.1\ \text{mole/liter})$$

$$= 5.0 \times 10^{-3}\ \text{moles}$$

$(\text{no. millimoles Na}^+)$

$$= (5.0 \times 10^{-3}\ \text{moles}) \cdot 10^3\ \frac{\text{millimoles}}{\text{mole}} = 5.0\ \text{millimoles}$$

$(\text{no. milliequivalents})$

$$= (5.0\ \text{millimoles})\left(\frac{1\ \text{meq}}{\text{millimole}}\right) \longleftarrow \begin{array}{l}\text{Na}^+ \text{ has } \textit{one}\\ \text{charge.}\end{array}$$

$$= 5.0\ \text{meq}$$

EXAMPLE:

How many milliequivalents of positive charge are there in 75 milliliter of 0.10 M BaCl$_2$ solution?

$(\text{no. moles Ba}^{2+})$

$$= \frac{75\ \text{milliliter}}{1000\ \text{milliliter/liter}} \cdot (0.1\ \text{mole/liter})$$

$$= 7.5 \times 10^{-3}\ \text{moles}$$

$(\text{no. millimoles Ba}^{2+})$

$$= (7.5 \times 10^{-3}\ \text{moles}) \cdot \left(10^3\ \frac{\text{millimoles}}{\text{mole}}\right)$$

$$= 7.5\ \text{millimoles}$$

$(\text{no. milliequivalents})$

$$= (7.5\ \text{millimoles}) \cdot \left(\frac{2\ \text{meq}}{\text{millimole}}\right) \longleftarrow \begin{array}{l}\text{Ba}^{2+} \text{ has } \textit{two}\\ \text{charges.}\end{array}$$

$$= 15.0\ \text{meq}$$

The capacity of a resin serves chiefly to determine how much resin is needed for a particular task. In practice, much more than the minimum amount of resin needed is taken for a given exchange, to insure that there are plenty of chances for every solution ion to be exchanged to encounter many resin exchange sites.

This much resin will not do the job. There are 500 meq of Na^+ and only 130 meq of exchange sites on the resin.

In studies of water quality, total salt concentrations, in milliequivalents per liter, are useful pieces of information. Total salt concentrations yield no information about what kinds of ions are in the water, but such total salt concentrations can suggest to the water chemist whether further investigation of a water supply is warranted. Low total salt concentrations do not guarantee purity—for precious little Cd^{2+} or Hg^{2+} can do grave damage to people or their livestock—but high salt concentrations automati- cause water samples to be suspect.

after it passes through the column. What was the total salt concentration in the 50.0-milliliter sample placed on the column?

Each milliequivalent of positive ions in the sample attaches to the column, freeing one millimole of H^+, which is flushed out of the column, and then titrated with OH^-. At the endpoint:

$$(\text{no. millimoles NaOH}) = (\text{no. millimoles } H^+)$$

$$(45.00 \text{ milliliter}) \left(1.00 \times 10^{-2} \frac{\text{moles}}{\text{liter}}\right) = (\text{no. millimoles } H^+)$$

$$4.50 \times 10^{-1} = (\text{no. millimoles } H^+)$$

$$(\text{no. millimoles } H^+) = (\text{no. meq salt})$$

$$(\text{no. meq salt}) = 4.50 \times 10^{-1}$$

$$(\text{total salt concentration}) = \frac{(\text{no. meq salt})}{(\text{vol. of sample})}$$

$$= \frac{4.50 \times 10^{-1} \text{ meq}}{5.00 \times 10^{-2} \text{ liter}}$$

$$= 9.00 \text{ meq/liter}$$

Standard solutions of strong acids and bases may be prepared, with little trouble, by the use of ion exchange resins. It really isn't easy to prepare a standard solution of a strong acid from the stock bottle; the molarity of the stock solutions is not often well known. Strong bases are not easily made without later standardization, either. Solid NaOH is not very pure, having about 15% Na_2CO_3 by weight. The NaOH pellets pick up moisture from the air if someone tries to weigh them out, and at the same time pick up CO_2 to form more Na_2CO_3. The following example will show how standard HCl may be made. (A strong base resin, in hydroxide form, could be used to make standard NaOH from NaCl.)

EXAMPLE:

How would you make 500.0 milliliter of 0.01000 M HCl from pure, dry NaCl, a long column of Amberlite IR-120 (H)AR (hydrogen form), and a 500-milliliter volumetric flask?

Measure out the right amount of NaCl. Dissolve in a little water, pour on the column, wash through with water, and collect in the 500-milliliter volumetric flask. Then dilute the contents of the flask to the mark with distilled water.

Calculation of right amount of NaCl:

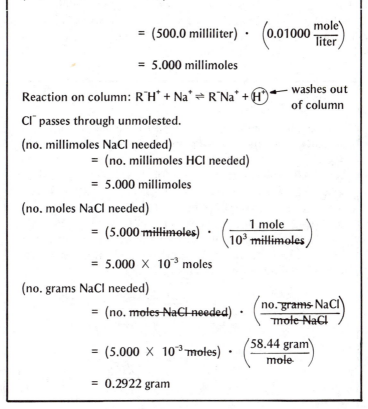

(no. millimoles HCl needed)

$$= (500.0 \text{ milliliter}) \cdot \left(0.01000 \frac{\text{mole}}{\text{liter}}\right)$$

$$= 5.000 \text{ millimoles}$$

Reaction on column: $R^-H^+ + Na^+ \rightleftharpoons R^-Na^+ + (H^+)$ ← washes out of column

Cl^- passes through unmolested.

(no. millimoles NaCl needed)
$$= (\text{no. millimoles HCl needed})$$

$$= 5.000 \text{ millimoles}$$

(no. moles NaCl needed)
$$= (5.000 \text{ millimoles}) \cdot \left(\frac{1 \text{ mole}}{10^3 \text{ millimoles}}\right)$$

$$= 5.000 \times 10^{-3} \text{ moles}$$

(no. grams NaCl needed)
$$= (\text{no. moles NaCl needed}) \cdot \left(\frac{\text{no. grams NaCl}}{\text{mole NaCl}}\right)$$

$$= (5.000 \times 10^{-3} \text{ moles}) \cdot \left(\frac{58.44 \text{ gram}}{\text{mole}}\right)$$

$$= 0.2922 \text{ gram}$$

Thus, you would measure out 0.2922 gram of NaCl, dissolve it, transfer quantitatively to the column, and run the column washings into a 500-milliliter volumetric flask.

The concentration of trace constituents is another, and very important, application of ion exchange. One early example of concentration through ion exchange is the determination of copper in milk. Normally speaking, copper should not be present at a level more than 1 microgram of copper per gram of milk. Cranston and Thompson described a method in 1946.[*] Copper in dried milk solids, and probably in fresh milk, too, is tied up in the protein. Besides protein, whole milk contains quite a lot of fat. Both the copper-binding protein and the greasy butterfat can interfere with an analysis. In the Cranston and Thompson procedure, fat and protein are precipitated from the milk by the addition of $HClO_4$, and copper, as Cu^{2+}, is left in the resultant solution. The solution is filtered, and an aliquot of 100.0 milliliter, adjusted to pH 5.0 and out of an original volume of 250.0 milliliter, is taken. The aliquot is passed

[*] H. A. Cranston and J. B. Thompson, Ind. and Eng. Chem., Anal. Ed. 18, 323 (1946).

through a column of Amberlite IR-100, a strong acid exchanger. The Cu^{2+} sticks to the resin. Then the Cu^{2+} is eluted from the resin with 40.00 milliliter of very low pH solution, 6% HCl. The copper in the washing is analyzed with an instrument called a polarograph, whose response, an electrical current, is linear with increasing copper concentration.

EXAMPLE:

A sample of powdered milk weighs 25. grams. It is dissolved in water. The protein and fat are precipitated. The pH of the filtrate is adjusted to 5.0, and the filtrate volume is made up to 250.0 milliliter in a volumetric flask. A 100.0-milliliter aliquot is taken and run through a column of IR-400 resin. The copper stays on the resin. Then 40.00 milliliter of 6% HCl are passed through the column and collected in a beaker. The beaker contents are analyzed with a polarograph. A standard curve is made up for the polarographic determinations. The data for the curve are shown below.

Cu concentration milligram/liter	Instrument Response, millimeters on scale
1.00	83
1.25	98
1.50	121
1.75	137
2.00	162
2.25	179
2.50	201

The instrument response to the 40.0-milliliter of solution from the milk sample is 170 scale units. What is the concentration of Cu, in μg Cu/gram milk powder, or ppm, in the original milk sample? This problem is mostly one of straightforward stoichiometry and dilution. First, because instrument response is proportional to concentration, a calibration plot is made, as seen in Figure 6.46 below. From this plot, the concentration of copper in the 40.0 milliliters that wash off the column is 2.10 milligram/liter.

Next, find C_{Cu} in the 100-milliliters aliquot.

$$C_{Cu} = \text{concentration Cu in 100 milliliter aliquot}$$

$$(C_{Cu})(100.0 \text{ milliliter})$$
$$= (40.00 \text{ milliliter})(2.10 \text{ milligram/liter})$$
$$C_{Cu} = \frac{40.00 \text{ milliliter}}{100.0 \text{ milliliter}} (2.10 \text{ milligram/liter})$$

This is the → C_{Cu} = 0.840 milligram/liter
concentration
of Cu in *both*
the 100
milliliter
aliquot and
the 250
milliliters

(milligram Cu in 250 milliliter)
$$= (0.840 \text{ milligram/liter})(250 \text{ milliliter})$$
$$(1 \text{ liter/1000 milliliter})$$

$$= 0.210 \text{ milligram}$$

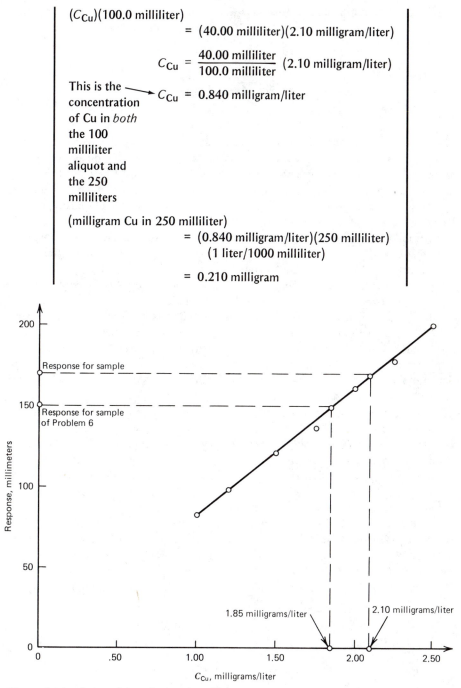

Figure 6.46 *Plot to determine Cu in milk.*

$$\text{(μg Cu in 250 milliliter)}$$
$$= (0.210 \text{ milligram})(1000 \text{ } \mu\text{g/milligram})$$
$$= 210 \text{ } \mu\text{g}$$
$$\frac{\mu\text{g Cu}}{\text{gram powder}} = \frac{210 \text{ } \mu\text{g}}{25 \text{ gram}} = \frac{8.4 \text{ } \mu\text{g}}{\text{gram}}$$

This is quite a high value of Cu concentration in nonfat milk solids. The utility of increasing the sensitivity of the method by concentration of the metal on a resin is plain.

Ion exchange resins can separate one sort of ion from another. The salts of the alkali metals provide an example of such a separation. Sodium and potassium chlorides are not easily separated by conventional means, e.g., precipitation. Traditionally, precipitation of potassium as the perchlorate, with an alcohol involved as one of the solvents, was the method of choice. The method was slow, tedious, and could, with a careless operator, result in an explosion. In 1950, Beukenkamp and Riesmann, using finely divided Dowex-50 resin, separated Na^+ and K^+ (and, hence NaCl and KCl).[*] They used a very low flow rate, and obtained the data plotted in Figure 6.47, showing very good separation of two species that are chemically very similar.

Indeed, ion exchange chromatography has allowed separation of rare earths as early as 1947[†], and the separation of hafnium and zirconium, whose chemistries are very, very similar[‡] in 1949. Today, amino acids are routinely separated by ion exchange.

In an application of competing equilibria, anion exchange resins may be used to separate some metal ions from each other. Kraus and Moore, in a much quoted paper,[§] reported the separation of Mn^{2+}, Fe^{3+}, Co^{2+}, Ni^{2+}, Cu^{2+}, and Zn^{2+} on Dowex 1, a strong base *anion* exchange resin. How was this possible? The salts of the ions were dissolved in 12M HCl. Under this condition, negatively charged chloro complexes of many transition metal ions are easily formed, for example, $CoCl_3^-$, $CoCl_4^{2-}$, $FeCl_4^-$. Formation constants in this highly acidic medium are not well defined. The numbers are not easy to get, but the results are encouraging. Of the metals named, it is safe to say that, in 12M HCl, Mn^{2+}, Fe^{3+}, Co^{2+}, Cu^{2+}, and Zn^{2+} are chiefly present as anionic chloro complexes, while nickel is present as either Ni^{2+}, $NiCl^+$, or $NiCl_2$. If a mixture of these species in 12 M HCl is placed on a column of Dowex 1, an anion exchanger, the Ni-bearing species are easily eluted with 12 M HCl, while all others remain on the column. When 6 M HCl is run through the column, the manganese-bearing species mostly revert to neutral or cationic form, and are eluted. Cobalt is eluted with 4 M HCl, copper with 2.5 M HCl, ferric iron with 0.5 M HCl, and zinc with 0.005 M HCl.

[*] J. Beukenkamp and W. Riesmann, Anal. Chem. 22, 582 (1950).
[†] B. H. Ketelle and G. E. Boyd, J. Am. Chem. Soc. 69, 2800 (1947).
[‡] K. A. Kraus and G. E. Moore, J. Am. Chem. Soc. 71, 3263 (1947).
[§] K. A. Kraus and G. E. Moore, J. Am. Chem. Soc. 75, 1460 (1953).

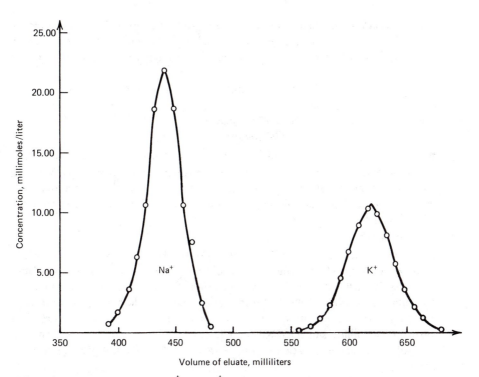

Figure 6.47 *Separation of Na⁺ and K⁺. Reprinted with permission from J. Beuken-kamp and W. Riesman,* Anal. Chem., *22, 582 (1950). Copyright by the American Chemical Society.*

≡ 6.27 PROBLEMS ≡

1. A solution is 0.1011M in NaOH. Can 1.000 liter of this solution be neutralized by passing it slowly through a column containing 50.0 grams of Amberlite® IRC-150(H) weak acid cation exchanger? The capacity of IRC-150(H) is 10. meq/gram. **Ans:** *Yes*

2. Calcium can be a problem in water supplies, because deposits of $CaCO_3$ can successfully clog hot water pipes. The author has on his desk a segment of hot water pipe from Madison, Wisconsin. In the course of six years. $CaCO_3$ deposits reduced the inside diameter through which water could pass from five inches to one-half inch. Deposits of $CaCO_3$ can also "lime up" boilers and hot water tanks, making heating very inefficient. A certain local water supply is heavily loaded with calcium salts. Salts of other metals are almost absent. Calcium concentration is reported to be "4.7 grains per gallon as $CaCO_3$." (One grain per gallon of anything is equiva-

lent to 17.11854 milligrams/liter of that same anything.) From how many liters of this water could all the calcium be removed by a column containing 50. lb of Amberlite® IRC-120 C.P. resin in sodium form? The information below may help.

Reaction: $2R\text{-}SO_3^-Na^+ + Ca^{2+} \rightleftharpoons (R\text{-}SO_3^-)_2Ca^{2+} + 2Na^+$

Capacity: 4.2 meq/gram

For anything: 1 pound weighs 453.6 grams

Ans: *5.9 × 10⁴ liters, or 16,000 gallons*

3. What would be the molarity of an HCl solution prepared by dissolving 3.5000 grams of pure, dry NaCl in a little water, passing the solution through a long column of Dowex® 50 in hydrogen form, collecting the washings in a 500-milliliter volumetric flask, and diluting the contents of the flask to the mark with distilled water?

Ans: *0.1198M*

4. A solution is made up by dissolving $Zn(NO_3)_2$, NaCl, KNO_3, and $Mg(NO_3)_2$ in 50. milliliters of H_2O. The 50. milliliters of solution are passed through a long column of Amberlite® IR-120(H)A.R. in hydrogen form. The column is washed with three 50. milliliter portions of distilled water, and these are collected in the same flask. The flask contents are titrated with 0.01066M NaOH. The endpoint is reached when 23.21 milliliters of the NaOH have been added. What is the total salt concentration of the original solution of $Zn(NO_3)_2$, NaCl, KNO_3, and $Mg(NO_3)_2$ in meq/liter?

Ans: *4.9 meq/liter*

5. A certain ion exchange experiment is done in the quantitative analysis laboratories of a great many schools. Here is the way it works: (1) a student is given exactly 25.00 milliliters of a solution containing NaCl; (2) the student pours this NaCl solution through a column of Amberlite® IR-120(H) resin in hydrogen form; (3) the column is washed with four 25-milliliter portions of deionized water, collecting the washings in the same flask that received the first outflow; and (4) the student titrates the contents of the flask with standard NaOH solution.

One particular student finds that his sample requires 41.68 milliliters of 0.1215M NaOH for titration to the phenolphthalein endpoint. How many grams of NaCl were in the 25.00-milliliter sample he was given?

Ans: *0.2959 gram*

6. A 25.0-gram sample of dried milk is taken, and analyzed for Cu by the procedure of Cranston and Thompson.* The polarograph, still calibrated as shown in Figure 6.46, gives a response of 150 millimeters on the readout scale. What is the concentration of Cu in the milk sample, in μg Cu/gram milk?

Ans: *7.4 μg/gram*

7. Water from a river located down the hill from a plating shop is suspected of containing rather more cadmium than is desirable from either a piscene or a human point of view. A 1-liter sample of the water is taken, and poured through a column of Dowex® 50 in hydrogen form, trapping the Cd^{2+} on the column. The Cd^{2+} is

* *H. A. Cranston and J. B. Thompson, Ind. and Eng. Chem., Anal. Ed. 18, 323 (1946).*

eluted with 100 milliliters of a 5% HCl solution. The 100-milliliters of outflow is analyzed polarographically for Cd^{2+} (no other ions interfere). The response of the polarograph is linear with cadmium concentration. The following results are obtained for the unknown and a number of standard solutions.

C'_{Cd}, M	Instrument Response, mm
2.0×10^{-4}	48
4.0×10^{-4}	101
6.0×10^{-4}	151
8.0×10^{-4}	197
10.0×10^{-4}	249
River sample	125

What is the concentration of Cd^{2+} in the river sample, in milligrams Cd/liter?
Ans: *5.6 milligrams/liter*

8. A chemist decides to separate sodium and potassium ions by the method of Beukenkamp and Riesmann.* He notes that these workers used a low flow rate and a very long column. To make the separation go more quickly, he uses a shorter column, and speeds up the flow rate. His results are shown in the following table. Plot the results. Did his speedup improve separation at all?
Ans: *No*

Vol. of eluate, milliliters	$[Na^+]$, millimoles/liter	$[K^+]$, millimoles/liter
124	0.7	0.0
132	1.6	0.0
140	3.7	0.0
148	6.2	0.0
156	10.6	0.0
164	18.7	0.0
172	21.9	0.0
180	18.9	0.0
188	10.6	0.2
196	5.5	0.5
204	2.5	1.2
212	0.5	4.6
220	0.0	6.6
228	0.0	9.0
236	0.0	10.4
244	0.0	9.9
252	0.0	8.1
260	0.0	5.8
268	0.0	3.7
276	0.0	2.1
284	0.0	1.4
292	0.0	0.5
300	0.0	0.2

* *J. Beukenkamp and W. Riesmann*, Anal. Chem. *22, 582 (1950).*

9. A quant student undertakes the separation of Fe^{3+} and Co^{2+} by the method of Kraus and Moore.* He places a 4.00-milliliter sample, consisting of Fe(III) and Co(II) in 12M HCl on a column of Amberlite® IRA-401S, a strong base anion exchanger in chloride form. Then the chemistry begins.

First he separates the iron and cobalt.

(a) He runs 150 milliliters of 9M HCl through the column and collects it. The outflow contains neither Fe nor Co, according to qualitative tests. He calls this fraction 1, and pours it down the sink.

(b) He then runs 65 milliliters of 4M HCl through the resin and collects it in a beaker. Qualitative tests indicate that it contains cobalt, but no iron. He calls this fraction 2.

(c) Next, he runs 150 milliliters of 0.5M HCl through the column. He collects it in a beaker, and calls it fraction 3. Qualitative tests show that it contains iron, but no cobalt.

(d) Distilled water run through the column dislodges neither iron nor cobalt. All the cobalt is thus in fraction 2, and all the iron in fraction 3.

Second he analyzes his fractions for Fe and Co according to the instructions in his laboratory manual. These are listed as follows.

(a) Transfer fraction 2 to a 250-milliliter volumetric flask and dilute to the mark with deionized water. Keep on calling this fraction 2.

(b) Transfer fraction 3 to a 250-milliliter volumetric flask and dilute to the mark with deionized water. Keep on calling this fraction 3.

(c) Analyze fraction 2 for Co. Add to four 100-milliliter volumetric flasks these amounts of 60 milligrams/liter Co^{2+} standard:

Flask No.	Amt. 60 milligrams/liter Co std.	Co concentration on dilution, milligrams/liter
1	5.00 milliliters	3 milligrams/liter
2	10.00 milliliters	6 milligrams/liter
3	15.00 milliliters	9 milligrams/liter
4	20.00 milliliters	12 milligrams/liter

These will be the standard solutions. Next, take 10 milliliters of fraction 2, and put it in a 100-milliliter volumetric flask. Now, get a Spectronic 20, and set the wavelength dial to 620 nm. Prepare a solution that is three parts acetone (a crude grade will do here) to one part of 50% NH₄SCN solution. Dilute flask 1 to the mark with this solution, and quickly read the absorbance of the resultant blue solution. (The Co^{2+} complex with SCN⁻ is unstable). Repeat this step for flasks 2, 3, 4, and the flask containing 10 milliliters of fraction 2. Plot a Beer's law graph of absorbance versus concentration. Find the concentration of Co in the flask containing 10 milliliters of fraction 2.

* *K. A. Kraus and G. E. Moore*, J. Am. Chem. Soc. 75, *1460 (1953)*.

(d) Analyze fraction 3 for Fe. Add the following reagents to four 100-milliliter volumetric flasks. These are the standard solutions.

Flask No.	Milliliters of 100 milli-grams/liter Fe std.	Milliliters of NH$_4$OH· HCl	Milliliters of o-Phenan-throline (2.5 g/l)	Congo Red Paper	NaAc, 2M	Water	c_{Fe}, milligrams/liter
1	1.00	5	5	Sm. square	Enough to	To mark	1.0
2	2.00	5	5	Sm. square	turn Congo	To mark	2.0
3	3.00	5	5	Sm. square	red paper	To mark	3.0
4	4.00	5	5	Sm. square	red again	To mark	4.0

To another 100 milliliter volumetric flask, add 5 milliliters of fraction 3, and the appropriate amounts of other reagents. Let all the solutions sit for at least 10 min. Get a Spectronic 20, and set the wavelength dial to 510 nm. Read the absorbances of the standard and unknown solutions. Plot a Beer's law graph of absorbance versus concentration of Fe. Find the concentration of Fe. Find the concentration of Fe in the flask made up from fraction 3.

Third: his results are shown below.

(a) Beer's law data for cobalt at 620 nm.

c_{Co}, Milligrams/liter	A
3.0	0.090
6.0	0.190
9.0	0.295
12.0	0.395
Fract. 2 sample	0.260

(b) Beer's law data for iron at 510 nm.

c_{Fe}, Milligrams/liter	A
1.0	0.276
2.0	0.569
3.0	0.777
4.0	1.097
Fract. 3 sample	0.675

From these data, determine the number of milligrams of Co and the number of milligrams of Fe present in the 4.00-milliliter sample that was placed on the column.

Ans: *20.0 milligrams Co, and 12.5 milligrams Fe*

1. The reaction is like this:

$$R\text{-}COOH + Na^+ + OH^- \rightleftharpoons R\text{-}COO^-Na^+ + H_2O$$

All the NaOH can be neutralized *if*

(number meq resin) $>$ (number meq OH^-)

(number millimoles OH^-) = (1000 milliliters)(0.1011 moles/liter)

= 101.1 millimoles

(number meq OH^-) = (101.1 millimoles) $\cdot$ $\left(\dfrac{1\ meq}{millimole} \right)$ ◁ There is *one* charge on OH^-

= 101.0 meq

(number meq resin) = (50.0 grams)(10. meq/gram)

$\cong$ 500 meq

(number meq resin) = (number meq OH^-)

On a stoichiometric basis, there is enough resin to neutralize the solution.

2. The reaction is like this:

$$2R\text{-}SO_3^-Na^+ + Ca^{2+} \rightleftharpoons (R\text{-}SO_3^-)_2Ca^{2+} + 2Na^+$$

First, find out how many meq of resin there are. (We shall carry an extra figure and then round off.)

(number grams resin) = (50. lb.)(453.6 grams/lb.)

= 2.27 $\times$ 10^4 grams

(number meq resin) = (2.27 $\times$ 10^4 grams)(4.2 meq/gram)

= 9.53 $\times$ 10^4 meq

(number meq resin) = (number meq Ca^{2+} that can be replaced)

(number meq Ca^{2+} that can be replaced)
= 9.53 $\times$ 10^4 meq

Now, find the concentration of Ca^{2+} in meq/liter.

($CaCO_3$ concentration, milligrams/liter)

$$= \left(\frac{4.7\ grains}{gallon} \right) \cdot \left(\frac{17.11854\ milligrams/liter}{grains/gallon} \right)$$

= 80.4 milligrams/liter

(CaCO$_3$ concentration, grams/liter)

$$= (80.4 \text{ milligrams/liter}) \cdot \left(\frac{1 \text{ gram}}{1000 \text{ milligrams}} \right)$$

$$= 8.04 \times 10^{-2} \text{ gram/liter}$$

(CaCO$_3$ concentration, moles/liter)

$$= (8.04 \times 10^{-2} \text{ gram/liter}) \cdot \left(\frac{1 \text{ mole}}{100.09 \text{ grams}} \right)$$

$$= 8.03 \times 10^{-4} \text{ moles/liter}$$

Then, remember

$$1\text{Ca}^{2+} \longrightarrow 1\text{CaCO}_3$$

(Ca^{2+} concentration, moles/liter)

$$= (\text{CaCO}_3 \text{ concentration, moles/liter})$$

$$= 8.03 \times 10^{-4} \text{ moles/liter}$$

(Ca^{2+} concentration, millimoles/liter)

$$= (8.03 \times 10^{-4} \text{ moles/liter}) \cdot \left(\frac{1000 \text{ millimoles}}{1 \text{ mole}} \right)$$

$$= 8.03 \times 10^{-1} \text{ millimoles/liter}$$

(Ca^{2+} concentration, meq/liter)

Ca^{2+} has 2 charges

$$= \left(\frac{8.03 \times 10^{-1} \text{ millimoles}}{\text{liter}} \right) \cdot \left(\frac{2 \text{ meq}}{\text{millimole}} \right)$$

$$= 1.61 \text{ meq/liter}$$

Remember

(number meq Ca^{2+} that can be replaced) $= 9.53 \times 10^4$ meq

Then

(Ca^{2+} concentration, meq/liter)(number liters)

$$= (\text{number meq Ca}^{2+} \text{ that can be replaced})$$

$$(\text{number liters}) = \frac{(\text{number meq Ca}^{2+} \text{ that can be replaced})}{|(\text{Ca}^{2+} \text{ concentration, meq/liter})}$$

$$= \frac{9.53 \times 10^4 \text{ meq}}{1.61 \text{ meq/liter}}$$

$$= 5.92 \times 10^4 \text{ liters}$$

$$\cong 5.9 \times 10^4 \text{ liters}$$

$$(\text{number gallons}) = (5.9 \times 10^4 \text{ liters}) \cdot \left(\frac{1 \text{ gallon}}{3.7853 \text{ liters}} \right)$$

$$= 1.6 \times 10^4 \text{ gallons}$$

or

$$(\text{number gallons}) = 16,000 \text{ gallons}$$

3. The reaction is:

$$R\text{-}SO_3^- H^+ + Na^+ + Cl^- \rightleftharpoons R\text{-}SO_3^- Na^+ + H^+ + Cl^-$$

$$(\text{number moles HCl}) = (\text{number moles H}^+) = (\text{number moles NaCl})$$

$$(\text{number moles NaCl}) = (3.500 \text{ grams NaCl}) \left(\frac{1 \text{ mole NaCl}}{58.44 \text{ grams NaCl}} \right)$$

$$= 0.05989 \text{ moles NaCl}$$

$$(\text{number moles HCl}) = (\text{number moles NaCl})$$

$$= 0.05989 \text{ moles HCl}$$

$$(\text{concentration HCl, moles/liter}) = \frac{(0.05989 \text{ moles HCl})}{(0.5000 \text{ liter})} \quad \leftarrow \quad \text{the vol. flask}$$

$$= 0.1198M$$

4. The reactions are:

$$\text{column} \nearrow \begin{array}{l} 2R\text{-}SO_3^- H^+ + Zn^{2+} \rightleftharpoons (R\text{-}SO_3^-)_2 Zn^{2+} + 2H^+ \\ \rightarrow 2R\text{-}SO_3^- H^+ + Na^+ \rightleftharpoons R\text{-}SO_3^- Na^+ + H^+ \end{array}$$

$$\text{column} \searrow \begin{array}{l} R\text{-}SO_3^- H^+ + K^+ \rightleftharpoons R\text{-}SO_3^- K^+ + H^+ \\ 2R\text{-}SO_3^- H^+ + Mg^{2+} \rightleftharpoons (R\text{-}SO_3^-)_2 Mg^{2+} + 2H^+ \end{array}$$

titration flask $\quad H^+ + OH^- \rightleftharpoons H_2O$

(number millimoles NaOH used)

$$= (23.21 \text{ milliliters})(0.01066 \text{ moles/liter})$$

$$= 0.2474 \text{ millimoles}$$

(number meq NaOH used)

OH⁻ has
one charge

$$= (\text{number millimoles NaOH used}) \cdot \left(\frac{1 \text{ meq}}{\text{millimole}} \right)$$

$$= 0.2474 \text{ meq}$$

(number meq NaOH used)

$$= (\text{number meq salts})$$

(number meq salts) = 0.2474 meq

(concentration of salts, meq/liter)

$$= \frac{0.2474 \text{ meq}}{0.050 \text{ liter}} \;\leftarrow\; \text{vol. of solution} \;\cong\; 4.9 \text{ meq/liter}$$

5. The reactions are:

    column $R\text{-}SO_3^- H^+ + Na^+ \rightleftharpoons R\text{-}SO_3^- Na^+ + H^+$

    titration flask $H^+ + OH^- \rightleftharpoons H_2O$

(number millimoles NaOH used) = (41.68 millimoles) · (0.1215 mole/liter)

$$= 5.064 \text{ millimoles}$$

$$(\text{number moles NaOH used}) = (5.064 \text{ millimoles}) \cdot \left(\frac{1 \text{ mole}}{1000 \text{ millimoles}} \right)$$

$$= 5.064 \times 10^{-3} \text{ moles}$$

(number moles NaOH used) = (number moles H^+ in outflow)

(number moles H^+ in outflow) = (number moles NaCl in sample)

(number moles NaCl in sample) = (number moles NaOH used)

$$= 5.064 \times 10^{-3} \text{ moles}$$

(number grams NaCl in sample) = $(5.064 \times 10^{-3}$ moles)(58.44 grams/mole)

$$= 0.2959 \text{ gram}$$

6. From Figure 6.46 the concentration of Cu in the outflow from the column is given as C'_{Cu}.

$$C'_{Cu} = 1.85 \text{ milligrams/liter}$$

This value, C'_{Cu}, is the concentration of copper in 40 milliliters of outflow. Concentration of Cu in the powdered milk sample is easy to figure.

    Let C_{Cu} = concentration Cu in 100 milliliters aliquot from 250 milliliter flask

$(C_{Cu})(100 \text{ milliliters}) = (C'_{Cu})(40 \text{ milliliters})$

$(C_{Cu})(100 \text{ milliliters}) = (1.85 \text{ milligrams/liter})(40 \text{ milliliters})$

$$C_{Cu} = (1.85 \text{ milligrams/liter}) \cdot \left(\frac{40 \text{ milliliters}}{100 \text{ milliliters}} \right)$$

This is the concentration of Cu in *both* the 100-milliliter aliquot *and* the 250-milliliter flask
   $\rightarrow C_{Cu} = 0.740$ milligram/liter

$$\text{(milligrams Cu in 250 milliliters)} = (0.740 \text{ milligrams/liter})(250. \text{ milliliters}) \cdot$$

$$\left(\frac{1 \text{ liter}}{1000. \text{ milliliters}} \right)$$

$$= 0.185 \text{ milligram}$$

$$\text{(}\mu\text{g Cu in 250 milliliters)} = (0.185 \text{ milligram}) \cdot \left(\frac{1000 \, \mu\text{g}}{\text{milligram}} \right)$$

$$= 185. \, \mu\text{g}$$

$$\frac{\mu\text{g Cu}}{\text{gram powder}} = \frac{185 \, \mu\text{g}}{25. \text{ grams}}$$

$$= 7.4 \, \mu\text{g/gram}$$

This is another way to work this problem. It's easier, but it is a sort of substitute for thought. Because both this analysis and the one in the body of the chapter were done in the same way, a simple ratio approach will work.

$$\frac{\text{(Response)}_2}{\text{(Response)}_1} = \frac{\text{(concentration in powder)}_2}{\text{(concentration in powder)}_1}$$

$$\text{(concentration in powder)}_2 = \frac{\text{(Response)}_2}{\text{(Response)}_1} \text{(concentration in powder)}_1$$

$$= \frac{(150 \text{ millimeters})}{(170 \text{ millimeters})} (8.4)(\mu\text{g/gram})$$

$$= 7.4 \, \mu\text{g/gram}$$

7. Because instrument response is linear with concentration, a plot of instrument response versus cadmium concentration is called for. Cd concentration in the outflow is given as C'_{Cd}. From the calibration plot (Figure 6.48):

$$C'_{Cd} = 5.0 \times 10^{-4} M$$

Let $[Cd^{2+}]$ = (Cd concentration in river water, M)

$$[Cd^{2+}](1000 \text{ milliliters}) = (C'_{Cd})(100 \text{ milliliters})$$

$$[Cd^{2+}] = C'_{Cd} \left(\frac{100}{1000} \right) = (5.0 \times 10^{-4}) \cdot \left(\frac{100}{1000} \right)$$

$$= 5.0 \times 10^{-5} M$$

Let C_{Cd} = Cd concentration in river water, milligram/liter

$$C_{Cd} = \left(\frac{5.0 \times 10^{-5} \text{ moles}}{\text{liter}} \right) \left(\frac{112.41 \text{ gram}}{\text{mole}} \right) \left(\frac{1000 \text{ milligram}}{\text{gram}} \right)$$

$$= 5.6 \text{ milligram/liter}$$

This is indeed a high value.

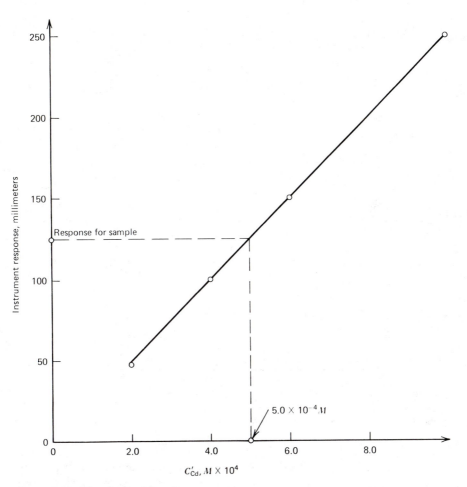

Figure 6.48 *Plot for Problem 7.*

8. The separation was not improved at all; indeed Na^+ and K^+ peaks now overlap slightly. Separation is generally improved by longer columns and slower flow rates.

9. First, plot the Beer's law data for cobalt and iron. These plots are shown in Figures 6.49 and 6.50. From the plots, the concentrations of cobalt and iron in the 100 milliliter volumetric flasks are easily read.

Let C_{Co} = concentration of Co in 100 milliliter vol. flask

C_{Co} = 8.0 milligram/liter

Let C_{Fe} = concentration of Fe in 100 milliliter vol. flask

C_{Fe} = 2.5 milligram/liter

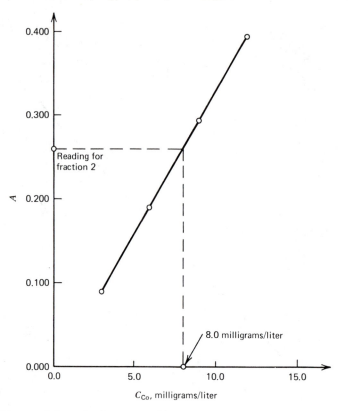

Figure 6.49 *Beer's law plot for Co, Problem 9.*

Now, it's all a matter of dilution algebra to figure the number of milligrams of cobalt and iron in the original 4.00-milliliter sample. One has to pay close attention to the instructions that the student followed.

Let C'_{Co} = conc'n of Co in 250 milliliter flask, which contains *all* the Co of fraction 2, hence *all* the Co

$$(C'_{Co})(10 \text{ milliliter}) = (C_{Co})(100 \text{ milliliter})$$

$$C'_{Co} = (C_{Co}) \frac{100}{10}$$

$$C'_{Co} = (8.0) \frac{100}{10}$$

$$C'_{Co} = 80. \text{ milligram/liter}$$

(no. milligram Co in sample) = (no. milligram Co in 250 milliliter flask)

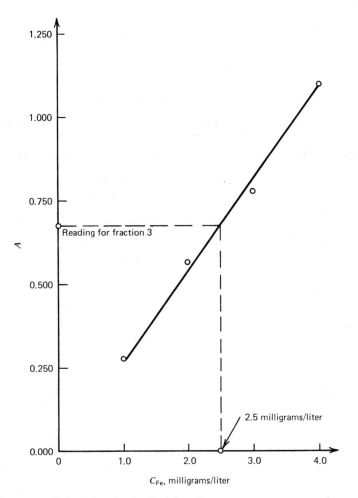

Figure 6.50 *Beer's law plot for Fe, Problem 9.*

(no. milligram Co in 250 milliliter flask) = $\left(\dfrac{80 \text{ milligram}}{\text{liter}}\right) \cdot \left(\dfrac{1 \text{ liter}}{1000 \text{ milliliter}}\right) \cdot$
(250 milliliter)

= 20. milligram

(no. milligram Co in sample) = 20. milligram

Let C'_{Fe} = conc'n of Fe in 250 milliliter flask, which contains *all* the Fe of fraction 3, hence *all* the Fe

$(C'_{Fe})(5 \text{ milliliter})$ = $(C_{Fe})(100 \text{ milliliter})$

$$C'_{Fe} = C_{Fe}\frac{100}{5.0}$$

$$= (2.5)\frac{100}{5.0}$$

$$= 50. \text{ milligram/liter}$$

$$(\text{no. milligram Fe in sample}) = (\text{no. milligram Fe in 250 milliliter flask})$$

$$(\text{no. milligram Fe in 250 milliliter flask}) = \left(\frac{50. \text{ milligram}}{\text{liter}}\right)(250 \text{ milliliter}) \cdot$$
$$\left(\frac{1 \text{ liter}}{1000 \text{ milliliter}}\right)$$

$$= 12.5 \text{ milligram}$$

$$(\text{no. milligram Fe in sample}) = 12.5 \text{ milligram (could be rounded to 12 milligram)}$$

Appendix

FOUR-PLACE LOGARITHMS OF NUMBERS

n	0	1	2	3	4	5	6	7	8	9
10	0000	0043	0086	0128	0170	0212	0253	0294	0334	0374
11	0414	0453	0492	0531	0569	0607	0645	0682	0719	0755
12	0792	0828	0864	0899	0934	0969	1004	1038	1072	1106
13	1139	1173	1206	1239	1271	1303	1335	1367	1399	1430
14	1461	1492	1523	1553	1584	1614	1644	1673	1703	1732
15	1761	1790	1818	1847	1875	1903	1931	1959	1987	2014
16	2041	2068	2095	2122	2148	2175	2201	2227	2253	2279
17	2304	2330	2355	2380	2405	2430	2455	2480	2504	2529
18	2553	2577	2601	2625	2648	2672	2695	2718	2742	2765
19	2788	2810	2833	2856	2878	2900	2923	2945	2967	2989
20	3010	3032	3054	3075	3096	3118	3139	3160	3181	3201
21	3222	3243	3263	3284	3304	3324	3345	3365	3385	3404
22	3424	3444	3464	3483	3502	3522	3541	3560	3579	3598
23	3617	3636	3655	3674	3692	3711	3729	3747	3766	3784
24	3802	3820	3838	3856	3874	3892	3909	3927	3945	3962
25	3979	3997	4014	4031	4048	4065	4082	4099	4116	4133
26	4150	4166	4183	4200	4216	4232	4249	4265	4281	4298
27	4314	4330	4346	4362	4378	4393	4409	4425	4440	4456
28	4472	4487	4502	4518	4533	4548	4564	4579	4594	4609
29	4624	4639	4654	4669	4683	4698	4713	4728	4742	4757
30	4771	4786	4800	4814	4829	4843	4857	4871	4886	4900
31	4914	4928	4942	4955	4969	4983	4997	5011	5024	5038
32	5051	5065	5079	5092	5105	5119	5132	5145	5159	5172
33	5185	5198	5211	5224	5237	5250	5263	5276	5289	5302
34	5315	5328	5340	5353	5366	5378	5391	5403	5416	5428
35	5441	5453	5465	5478	5490	5502	5514	5527	5539	5551
36	5563	5575	5587	5599	5611	5623	5635	5647	5658	5670
37	5682	5694	5705	5717	5729	5740	5752	5763	5775	5786
38	5798	5809	5821	5832	5843	5855	5866	5877	5888	5899
39	5911	5922	5933	5944	5955	5966	5977	5988	5999	6010
40	6021	6031	6042	6053	6064	6075	6085	6096	6107	6117
41	6128	6138	6149	6160	6170	6180	6191	6201	6212	6222
42	6232	6243	6253	6263	6274	6284	6294	6304	6314	6325
43	6335	6345	6355	6365	6375	6385	6395	6405	6415	6425
44	6435	6444	6454	6464	6474	6484	6493	6503	6513	6522
45	6532	6542	6551	6561	6571	6580	6590	6599	6609	6618
46	6628	6637	6646	6656	6665	6675	6684	6693	6702	6712
47	6721	6730	6739	6749	6758	6767	6776	6785	6794	6803

FOUR-PLACE LOGARITHMS OF NUMBERS

n	0	1	2	3	4	5	6	7	8	9
48	6812	6821	6830	6839	6848	6857	6866	6875	6884	6893
49	6902	6911	6920	6928	6937	6946	6955	6964	6972	6981
50	6990	6998	7007	7016	7024	7033	7042	7050	7059	7067
51	7076	7084	7093	7101	7110	7118	7126	7135	7143	7152
52	7160	7168	7177	7185	7193	7202	7210	7218	7226	7235
53	7243	7251	7259	7267	7275	7284	7292	7300	7308	7316
54	7324	7332	7340	7348	7356	7364	7372	7380	7388	7396
55	7404	7412	7419	7427	7435	7443	7451	7459	7466	7474
56	7482	7490	7497	7505	7513	7520	7528	7536	7543	7551
57	7559	7566	7574	7582	7589	7597	7604	7612	7619	7627
58	7634	7642	7649	7657	7664	7672	7679	7686	7694	7701
59	7709	7716	7723	7731	7738	7745	7752	7760	7767	7774
60	7782	7789	7796	7803	7810	7818	7825	7832	7839	7846
61	7853	7860	7868	7875	7882	7889	7896	7903	7910	7917
62	7924	7931	7938	7945	7952	7959	7966	7973	7980	7987
63	7993	8000	8007	8014	8021	8028	8035	8041	8048	8055
64	8062	8069	8075	8082	8089	8096	8102	8109	8116	8122
65	8129	8136	8142	8149	8156	8162	8169	8176	8182	8189
66	8195	8202	8209	8215	8222	8228	8235	8241	8248	8254
67	8261	8267	8274	8280	8287	8293	8299	8306	8312	8319
68	8325	8331	8338	8344	8351	8357	8363	8370	8376	8382
69	8388	8395	8401	8407	8414	8420	8426	8432	8439	8445
70	8451	8457	8463	8470	8476	8482	8488	8494	8500	8506
71	8513	8519	8525	8531	8537	8543	8549	8555	8561	8567
72	8573	8579	8585	8591	8597	8603	8609	8615	8621	8627
73	8633	8639	8645	8651	8657	8663	8669	8675	8681	8686
74	8692	8698	8704	8710	8716	8722	8727	8733	8739	8745
75	8751	8756	8762	8768	8774	8779	8785	8791	8797	8802
76	8808	8814	8820	8825	8831	8837	8842	8848	8854	8859
77	8865	8871	8876	8882	8887	8893	8899	8904	8910	8915
78	8921	8927	8932	8938	8943	8949	8954	8960	8965	8971
79	8976	8982	8987	8993	8998	9004	9009	9015	9020	9025
80	9031	9036	9042	9047	9053	9058	9063	9069	9074	9079
81	9085	9090	9096	9101	9106	9112	9117	9122	9128	9133
82	9138	9143	9149	9154	9159	9165	9170	9175	9180	9186
83	9191	9196	9201	9206	9212	9217	9222	9227	9232	9238
84	9243	9248	9253	9258	9263	9269	9274	9279	9284	9289
85	9294	9299	9304	9309	9315	9320	9325	9330	9335	9340
86	9345	9350	9355	9360	9365	9370	9375	9380	9385	9390
87	9395	9400	9405	9410	9415	9420	9425	9430	9435	9440
88	9445	9450	9455	9460	9465	9469	9474	9479	9484	9489
89	9494	9499	9504	9509	9513	9518	9523	9528	9533	9538

FOUR-PLACE LOGARITHMS OF NUMBERS

n	0	1	2	3	4	5	6	7	8	9
90	9542	9547	9552	9557	9562	9566	9571	9576	9581	9586
91	9590	9595	9600	9605	9609	9614	9619	9624	9628	9633
92	9638	9643	9647	9652	9657	9661	9666	9671	9675	9680
93	9685	9689	9694	9699	9703	9708	9713	9717	9722	9727
94	9731	9736	9741	9745	9750	9754	9759	9763	9768	9773
95	9777	9782	9786	9791	9795	9800	9805	9809	9814	9818
96	9823	9827	9832	9836	9841	9845	9850	9854	9859	9863
97	9868	9872	9877	9881	9886	9890	9894	9899	9903	9908
98	9912	9917	9921	9926	9930	9934	9939	9943	9948	9952
99	9956	9961	9965	9969	9974	9978	9983	9987	9991	9996

Index

Index